VOLUME ONE HUNDRED AND SIXTY EIGHT

INTERNATIONAL REVIEW OF NEUROBIOLOGY

Metabotropic Glutamate Receptors in Psychiatric and Neurological Disorders

INTERNATIONAL REVIEW OF NEUROBIOLOGY

VOLUME 168

SERIES EDITOR

PATRICIA JANAK
Janak Lab, Department of Neuroscience
Dunning Hall, John Hopkins University
Baltimore, MD, USA

PETER JENNER
Division of Pharmacology and Therapeutics
GKT School of Biomedical Sciences
King's College, London, UK

EDITORIAL BOARD

RICHARD L. BELL	HARI SHANKER SHARMA
SHAFIQUR RAHMAN	ARUNA SHARMA
MARIA STAMELOU	GRAŻYNA SÖDERBOM
K RAY CHAUDHURI	NATALIE WITEK
NATALIYA TITOVA	NINA SMYTH
BAI-YUN ZENG	ERIN CALIPARI
TODD E. THIELE	LIISA GALEA

VOLUME ONE HUNDRED AND SIXTY EIGHT

INTERNATIONAL REVIEW OF NEUROBIOLOGY

Metabotropic Glutamate Receptors in Psychiatric and Neurological Disorders

Edited by

LORI A. KNACKSTEDT

*Department of Psychology;
Center for Addiction Research and Education;
Center for OCD, Anxiety, and Related Disorders,
University of Florida, Gainesville, FL, United States*

MAREK SCHWENDT

*Department of Psychology;
Center for Addiction Research and Education;
Center for OCD, Anxiety, and Related Disorders,
University of Florida, Gainesville, FL, United States*

Academic Press is an imprint of Elsevier
50 Hampshire Street, 5th Floor, Cambridge, MA 02139, United States
525 B Street, Suite 1650, San Diego, CA 92101, United States
The Boulevard, Langford Lane, Kidlington, Oxford OX5 1GB, United Kingdom
125 London Wall, London, EC2Y 5AS, United Kingdom

First edition 2023

Copyright © 2023 Elsevier Inc. All rights reserved.

No part of this publication may be reproduced or transmitted in any form or by any means, electronic or mechanical, including photocopying, recording, or any information storage and retrieval system, without permission in writing from the publisher. Details on how to seek permission, further information about the Publisher's permissions policies and our arrangements with organizations such as the Copyright Clearance Center and the Copyright Licensing Agency, can be found at our website: www.elsevier.com/permissions.

This book and the individual contributions contained in it are protected under copyright by the Publisher (other than as may be noted herein).

Notices
Knowledge and best practice in this field are constantly changing. As new research and experience broaden our understanding, changes in research methods, professional practices, or medical treatment may become necessary.

Practitioners and researchers must always rely on their own experience and knowledge in evaluating and using any information, methods, compounds, or experiments described herein. In using such information or methods they should be mindful of their own safety and the safety of others, including parties for whom they have a professional responsibility.

To the fullest extent of the law, neither the Publisher nor the authors, contributors, or editors, assume any liability for any injury and/or damage to persons or property as a matter of products liability, negligence or otherwise, or from any use or operation of any methods, products, instructions, or ideas contained in the material herein.

ISBN: 978-0-323-99394-4
ISSN: 0074-7742

For information on all Academic Press publications
visit our website at https://www.elsevier.com/books-and-journals

Publisher: Zoe Kruze
Acquisitions Editor: Mariana Kuhl
Developmental Editor: Federico Paulo Mendoza
Production Project Manager: Abdulla Sait
Cover Designer: Matthew Limbert

Typeset by STRAIVE, India

Contents

Contributors	*ix*
Preface	*xiii*

1. Metabotropic glutamate receptors in Parkinson's disease   1

Cynthia Kwan, Woojin Kang, Esther Kim, Sébastien Belliveau, Imane Frouni, and Philippe Huot

1. Introduction	2
2. mGlu$_5$ receptors	3
3. mGlu$_4$ receptors	10
4. mGlu$_2$ receptors	18
5. Summary and perspectives	23
References	24

2. The interaction of membrane estradiol receptors and metabotropic glutamate receptors in adaptive and maladaptive estradiol-mediated motivated behaviors in females   33

Caroline S. Johnson and Paul G. Mermelstein

1. Introduction	34
2. Receptor signaling	36
3. Estrogen in motivated behaviors: Reproduction	48
4. Estrogen in motivated behaviors: Dysregulation and drug addiction	59
5. Conclusion	68
References	72

3. Metabotropic glutamate receptor function and regulation of sleep-wake cycles   93

Kimberly M. Holter, Bethany E. Pierce, and Robert W. Gould

1. Introduction	94
2. Overview of metabotropic glutamate receptors (mGlu receptors)	95
3. Sleep	99
4. Discussion	137
References	144

4. The role of metabotropic glutamate receptors in neurobehavioral effects associated with methamphetamine use

177

Peter U. Hámor, Lori A. Knackstedt, and Marek Schwendt

1. Methamphetamine use disorder (MUD)—Characteristics	178
2. Metabotropic glutamate receptors (mGlu)—An overview	182
3. mGlu receptors and methamphetamine-associated neural changes	184
4. mGlu receptors and meth-induced behavioral effects	188
5. mGlu receptor interaction with other neurotransmitter receptors in the context of methamphetamine-associated neural and behavioral effects	198
6. Conclusions	203
Support	204
Conflict of interest	204
References	204

5. The role of mGlu receptors in susceptibility to stress-induced anhedonia, fear, and anxiety-like behavior

221

Cassandra G. Modrak, Courtney S. Wilkinson, Harrison L. Blount,

Marek Schwendt, and Lori A. Knackstedt

1. Introduction	222
2. Animal models of stress induction	224
3. Behavioral tests for anxiety-like behavior, anhedonia, and fear	228
4. Brain-wide mGlu5 and stress-induced anhedonia, fear, and anxiety-like behavior	233
5. The mGlu5-containing circuitry underlying stress-induced anhedonia, fear, and anxiety-like behavior	237
6. Summary of mGlu5 role in anhedonia, fear, and anxiety-like behavior	244
7. mGlu2 and mGlu3 receptors and stress-induced anhedonia, fear, and anxiety-like behavior	247
8. The mGlu2/3-containing circuitry underlying stress-induced anhedonia, fear, and anxiety-like behavior	250
9. mGlu2/3 conclusions	254
10. General conclusions	255
References	256

6. The metabotropic glutamate receptor 5 as a biomarker for psychiatric disorders

265

Ruth H. Asch, Ansel T. Hillmer, Stephen R. Baldassarri, and Irina Esterlis

1. Introduction	266
2. Quantifying mGlu5 with PET	267
3. Mood disorders	268

Contents

vii

4. Anxiety and trauma disorders	277
5. Substance use: Nicotine, cannabis, and alcohol	284
6. Summary and commentary	291
References	292

7. Sex differences and hormonal regulation of metabotropic glutamate receptor synaptic plasticity — **311**

Carly B. Fabian, Marianne L. Seney, and Max E. Joffe

1. Introduction	312
2. Sex differences in mGlu receptor function	314
3. Insight gained from behavioral pharmacology, transgenic mice, and disease models	321
4. Conclusions	334
Acknowledgments	336
Conflict of Interest	336
References	336

8. Roles of metabotropic glutamate receptor 8 in neuropsychiatric and neurological disorders — **349**

Li-Min Mao, Nirav Mathur, Karina Shah, and John Q. Wang

1. Introduction	350
2. mGlu8 receptors	351
3. Anxiety	352
4. Epilepsy	354
5. Parkinson's disease	355
6. Drug addiction	356
7. Chronic pain	357
8. Alzheimer's disease	358
9. Conclusions	359
Author contributions	359
Funding	359
Conflict of interest	359
References	359

9. Metabotropic glutamate receptors and cognition: From underlying plasticity and neuroprotection to cognitive disorders and therapeutic targets — **367**

Brandon K. Hoglund, Vincent Carfagno, M. Foster Olive, and
Jonna M. Leyrer-Jackson

1. Introduction	368
2. Metabotropic glutamate receptor physiology	368

3. Cognition: An executive function	371
4. Mediating cognition: The prefrontal cortex	372
5. Synaptic plasticity and mGlu receptors	373
6. mGlu receptors and cognition: Diseases and associations	375
7. Targeting mGlu receptors as a therapeutic for cognitive therapy	393
8. Conclusion	398
Funding	398
References	398

Contributors

Ruth H. Asch
Department of Psychiatry, Yale University, New Haven, CT, United States

Stephen R. Baldassarri
Yale Program in Addiction Medicine; Department of Internal Medicine, Yale University, New Haven, CT, United States

Sébastien Belliveau
Neurodegenerative Disease Group, Montreal Neurological Institute-Hospital (The Neuro), Montreal, QC, Canada

Harrison L. Blount
Department of Psychology; Center for Addiction Research and Education; Center for OCD, Anxiety, and Related Disorders, University of Florida, Gainesville, FL, United States

Vincent Carfagno
School of Medicine, Midwestern University, Glendale, AZ, United States

Irina Esterlis
Department of Psychiatry; Department of Psychology, Yale University, New Haven; Clinical Neurosciences Division, U.S. Department of Veterans Affairs National Center for Posttraumatic Stress Disorder, Veterans Affairs Connecticut Healthcare System, West Haven, CT, United States

Carly B. Fabian
Center for Neuroscience; Department of Psychiatry; Translational Neuroscience Program, University of Pittsburgh, Pittsburgh, PA, United States

Imane Frouni
Neurodegenerative Disease Group, Montreal Neurological Institute-Hospital (The Neuro); Département de Pharmacologie et Physiologie, Université de Montréal, Montreal, QC, Canada

Robert W. Gould
Department of Physiology and Pharmacology, Wake Forest University School of Medicine, Winston-Salem, NC, United States

Peter U. Hámor
Department of Psychology; Center for Addiction Research and Education, University of Florida, Gainesville, FL; Department of Pharmacology, Weill Cornell Medicine, Cornell University, New York, NY, United States

Ansel T. Hillmer
Department of Psychiatry, Yale University; Department of Radiology and Biomedical Imaging, New Haven, CT, United States

Brandon K. Hoglund
Department of Medical Education, School of Medicine, Creighton University, Phoenix, AZ, United States

Kimberly M. Holter
Department of Physiology and Pharmacology, Wake Forest University School of Medicine, Winston-Salem, NC, United States

Philippe Huot
Neurodegenerative Disease Group, Montreal Neurological Institute-Hospital (The Neuro); Département de Pharmacologie et Physiologie, Université de Montréal; Department of Neurology and Neurosurgery, McGill University; Movement Disorder Clinic, Division of Neurology, Department of Neurosciences, McGill University Health Centre, Montreal, QC, Canada

Max E. Joffe
Center for Neuroscience; Department of Psychiatry; Translational Neuroscience Program, University of Pittsburgh, Pittsburgh, PA, United States

Caroline S. Johnson
Department of Neuroscience, University of Minnesota, Minneapolis, MN, United States

Woojin Kang
Neurodegenerative Disease Group, Montreal Neurological Institute-Hospital (The Neuro), Montreal, QC, Canada

Esther Kim
Neurodegenerative Disease Group, Montreal Neurological Institute-Hospital (The Neuro), Montreal, QC, Canada

Lori A. Knackstedt
Department of Psychology; Center for Addiction Research and Education; Center for OCD, Anxiety, and Related Disorders, University of Florida, Gainesville, FL, United States

Cynthia Kwan
Neurodegenerative Disease Group, Montreal Neurological Institute-Hospital (The Neuro), Montreal, QC, Canada

Jonna M. Leyrer-Jackson
Department of Medical Education, School of Medicine, Creighton University, Phoenix, AZ, United States

Li-Min Mao
Department of Biomedical Sciences, University of Missouri-Kansas City, School of Medicine, Kansas City, MO, United States

Nirav Mathur
Department of Anesthesiology, University of Missouri-Kansas City, School of Medicine, Kansas City, MO, United States

Paul G. Mermelstein
Department of Neuroscience, University of Minnesota, Minneapolis, MN, United States

Cassandra G. Modrak
Department of Psychology; Center for Addiction Research and Education; Center for OCD, Anxiety, and Related Disorders, University of Florida, Gainesville, FL, United States

M. Foster Olive
Department of Psychology, Arizona State University, Tempe, AZ, United States

Contributors

Bethany E. Pierce
Department of Physiology and Pharmacology, Wake Forest University School of Medicine, Winston-Salem, NC, United States

Marek Schwendt
Department of Psychology; Center for Addiction Research and Education; Center for Addiction Research and Education; Center for OCD, Anxiety, and Related Disorders, University of Florida, Gainesville, FL, United States

Marianne L. Seney
Center for Neuroscience; Department of Psychiatry; Translational Neuroscience Program, University of Pittsburgh, Pittsburgh, PA, United States

Karina Shah
Department of Biomedical Sciences, University of Missouri-Kansas City, School of Medicine, Kansas City, MO, United States

John Q. Wang
Department of Biomedical Sciences; Department of Anesthesiology, University of Missouri-Kansas City, School of Medicine, Kansas City, MO, United States

Courtney S. Wilkinson
Department of Psychology; Center for Addiction Research and Education; Center for OCD, Anxiety, and Related Disorders, University of Florida, Gainesville, FL, United States

Preface

Metabotropic glutamate (or mGlu) receptors comprise a diverse group of G-protein–coupled receptors mediating glutamatergic neuromodulation and neuroplasticity. To this date, eight distinct receptors (mGlu1–8) have been identified and characterized in mammalian species. mGlu receptors are subdivided into three groups based on sequence homology and signal transduction pathways. Group I (mGlu1 and 5) activate phospholipase C and are typically expressed postsynaptically, while group II (mGlu2 and 3) and group III (mGlu4 and 6–8) inhibit adenylate cyclase and are mainly localized on presynaptic terminals. Individual mGlu receptors display brain-wide but nonoverlapping brain distribution (with mGlu6 exclusively expressed in retina). It has been well established that dysregulation of individual mGlu receptors leads to aberrant neurotransmission, neuroplasticity, and spectrum of neuropathologies linked to a wide range of disorders from anxiety to epilepsy. Overall, while preclinical research made significant advances into our understanding of neural mGlu receptors, translational potential and clinical utilization of these receptors has been by impeded several factors, including an incomplete understanding of the distribution and function of individual receptor subtypes. Due to similarity in sequence homology, antibodies that could distinguish between receptor subtypes, especially mGlu2 and mGlu3, were slow to be developed. Similarly, early experimental ligands lacked subtype specificity and often displayed unfavorable blood-brain penetration or side-effect profile. And finally, preclinical work has been slow to investigate potential sex differences in mGlu expression, signaling, and role in behavior. As result, clinical testing of several mGlu compounds has been abandoned. However, we believe that recent advances in our understanding of mGlu-dependent neural processes and mGlu pharmacology point to exciting unexplored avenues for the development of novel treatments for various neurological and psychiatric disorders. This volume sets out to provide diverse perspective on mGlu function in health and disease from the perspective of preclinical and clinical neuroscience. The authors review up-to-date literature and discuss what is currently known about sex differences in mGlu receptor function.

In Chapter 1, Kwan, Kang, Kim, Belliveau, Frouni, and Huot describe the roles of mGlu receptors in Parkinson's disease, with a focus on mGlu2–5 receptors. The efficacy of targeting each receptor subtype to reduce specific disease symptoms as well as complications induced by first-line treatments is addressed. A substantial body of preclinical work finds that all four receptor subtypes are involved in Parkinson's disease symptomology. In humans, PET ligands for mGlu receptors have permitted the exploration of changes in these receptors in Parkinson's disease. Clinical trials for several mGlu modulators for treatment of Parkinson's disease are in the beginning stages.

Emerging evidence suggests that interaction with other membrane receptors is a significant factor contributing to region- and sex-specific differences in mGlu receptor function. In this context, Chapter 2 by Johnson and Mermelstein discusses the role of mGlu receptors in mediating estradiol-induced neuroplasticity and motivated behaviors. The chapter provides up-to-date information on the cellular mechanisms of mGlu interaction with membrane-bound estrogen receptors and its relevance for adaptive motivated behavior (reproduction) and maladaptive motivated behavior (addiction).

In Chapter 3, Holter, Pierce, and Gould provide an overview of mGlu receptor subtypes in sleep-wake regulation in healthy populations and in those with psychiatric disorders. The authors provide an overview of known fluctuations in mGlu receptor expression and function across diurnal sleep-wake cycles and the ability of pharmacological manipulation of mGlu receptors to alter cellular physiology and sleep-wake behavior. Data from human studies of mGlu receptor changes in psychiatric and neurodegenerative diseases are reviewed. The chapter provides a rationale for the development of selective mGlu receptor ligands to reduce primary symptoms of psychiatric disorders as well as the sleep disturbances that accompany them. Not only does the chapter summarize the relationship between mGlu receptors, sleep, and brain function, it also highlights the need to consider circadian regulation of mGlu expression and function in all experiments that quantify or manipulate mGlu receptors.

Hámor, Knackstedt, and Schwendt review the large body of research into interactions between mGlu receptors and methamphetamine in Chapter 4. This chapter provides a comprehensive overview of the role of mGlu receptor subtypes in methamphetamine-induced psychomotor activation, reward, reinforcement, and seeking. Both humans and rodent models exhibit cognitive deficits following methamphetamine intake, and the role of mGlu receptors in these deficits is evaluated. A potential role for receptor-receptor interactions involving mGlu receptors and other neurotransmitter receptors in such meth-induced neurobiological and behavioral changes is also considered.

In Chapter 5, Modrak, Wilkinson, Blount, Schwendt, and Knackstedt review the role of mGlu receptors in mediating anxiety, fear, and anhedonia that follow exposure to a stressor, with a focus on the role of these receptors in mediating susceptibility/resilience to the long-term effects of such stress. The authors first provide an overview of the varied preclinical models used to induce stress and to assess anhedonia, fear, and anxiety-like behavior. Next, they summarize the results of studies investigating the role of mGlu2, mGlu3, and mGlu5 (brain-wide and region-specific) in such behaviors. In particular, the medial prefrontal cortex, basolateral amygdala, nucleus accumbens, and ventral hippocampus are key regions where mGlu5, mGlu2, and mGlu3 regulate these behaviors.

In Chapter 6, Asch, Hillmer, Baldassarri, and Esterlis preview evidence gained from positron emission tomography (PET) studies using mGlu5 radioligands as well as postmortem samples, to present a comprehensive commentary of mGlu5 brain changes in the context of symptom severity and duration. Importantly, the utility of mGlu5 PET signal both as a disease biomarker and as a neurobiological "readout" of treatment progress is discussed.

In Chapter 7, Fabian, Seney, and Joffe detail preclinical and clinical studies on the topic of sex differences in the regulation of synaptic plasticity by mGlu receptors. The authors first discuss what is currently known about sex differences in mGlu receptor expression and function and next describe how gonadal hormones regulate mGlu receptor signaling. The main goal of the chapter is to describe data revealing sex-specific mechanisms by which mGlu receptors differentially modulate synaptic plasticity and behavior, with a focus on cognition, social behavior, alcohol and drug use, and anxiety. In doing so, they find that while the existing literature indicates that mGlu receptors are regulators of these behaviors in both male and female rodents, there is evidence that mGlu receptors regulate synaptic plasticity and differently in males and females. However, at present, this body of research is relatively small, and further examination of sex differences in the mGlu signaling underlying physiology and behavior is called for.

Group III metabotropic receptors are the least researched and understood group of mGlu receptors. In Chapter 8, Mao, Mathur, Shah, and Wang provide a comprehensive overview of mGlu8 brain distribution and function under physiological and pathological conditions. The chapter also includes the discussion of an emerging evidence regarding the role of mGlu8 in pathophysiology of several neuropsychiatric and neurological disorders, indicating the urgency to develop specific pharmacological tools to manipulate this receptor *in vivo*.

In Chapter 9, Hoglund, Carfagno, Olive, and Leyrer-Jackson describe the role of mGlu receptors in cognition and in the cognitive dysfunction that accompanies many neurological and psychiatric disorders, including Parkinson's disease, Alzheimer's disease, Fragile X syndrome, posttraumatic stress disorder, and schizophrenia. Preclinical work investigating the efficacy of targeting mGlu receptors with positive and negative allosteric modulators and specific agonists and antagonist to restore cognitive function across these disorders is also discussed. The ability of mGlu receptor signaling to produce neuroprotective effects in particular disease states is also considered.

In summary, we hope that the information and discussions included in this volume will be useful to a broad range of scientists inside and outside of the mGlu receptor research field, who may be looking for comprehensive answers regarding glutamatergic mechanisms underlying neurological and psychiatric disorders. We strove to provide diverse perspectives on mGlu receptors ranging from their role in cellular signaling to their utility as clinical biomarkers and/or diagnostic tools. In several chapters, the expert contributors highlight the necessity to consider interactions between mGlu and other neurotransmitters or sex hormones when studying mGlu receptor functions or considering mGlu-based interventions. Based on the information presented here, we believe that mGlu receptors hold promise for the development of novel effective therapies of a spectrum of neuropsychiatric conditions.

LORI A. KNACKSTEDT
MAREK SCHWENDT
Behavioral and Cognitive Neuroscience Program, Psychology Department
University of Florida, Gainesville, FL

CHAPTER ONE

Metabotropic glutamate receptors in Parkinson's disease

Cynthia Kwan[a], Woojin Kang[a], Esther Kim[a], Sébastien Belliveau[a], Imane Frouni[a,b], and Philippe Huot[a,b,c,d],* [iD]

[a]Neurodegenerative Disease Group, Montreal Neurological Institute-Hospital (The Neuro), Montreal, QC, Canada
[b]Département de Pharmacologie et Physiologie, Université de Montréal, Montreal, QC, Canada
[c]Department of Neurology and Neurosurgery, McGill University, Montreal, QC, Canada
[d]Movement Disorder Clinic, Division of Neurology, Department of Neurosciences, McGill University Health Centre, Montreal, QC, Canada
*Corresponding author: e-mail address: philippe.huot@mcgill.ca

Contents

1. Introduction	2
2. mGlu5 receptors	3
2.1 Distribution and function	3
2.2 Distribution of mGlu5 receptors in PD	4
2.3 Preclinical studies with mGlu5 modulators in PD	4
2.4 Clinical studies	8
3. mGlu4 receptors	10
3.1 Distribution and function	10
3.2 Preclinical studies	12
3.3 Clinical studies	16
4. mGlu2 receptors	18
4.1 Distribution and function	18
4.2 Preclinical studies	19
4.3 Clinical studies	23
5. Summary and perspectives	23
References	24

Abstract

Parkinson's disease (PD) is a complex disorder that leads to alterations in multiple neurotransmitter systems, notably glutamate. As such, several drugs acting at glutamatergic receptors have been assessed to alleviate the manifestation of PD and treatment-related complications, culminating with the approval of the N-methyl-D-aspartate (NMDA) antagonist amantadine for L-3,4-dihydroxyphenylalanine (L-DOPA)-induced dyskinesia. Glutamate elicits its actions through several ionotropic and metabotropic (mGlu) receptors. There are 8 sub-types of mGlu receptors, with sub-types 4 (mGlu4) and 5 (mGlu5) modulators having been tested in the clinic for endpoints pertaining to PD, while

International Review of Neurobiology, Volume 168
ISSN 0074-7742
https://doi.org/10.1016/bs.irn.2022.10.001

Copyright © 2023 Elsevier Inc.
All rights reserved.

sub-types 2 (mGlu$_2$) and 3 (mGlu$_3$) have been investigated in pre-clinical settings. In this book chapter, we provide an overview of mGlu receptors in PD, with a focus on mGlu$_5$, mGlu$_4$, mGlu$_2$ and mGlu$_3$ receptors. For each sub-type, we review, when applicable, their anatomical localization and possible mechanisms underlying their efficacy for specific disease manifestation or treatment-induced complications. We then summarize the findings of pre-clinical studies and clinical trials with pharmacological agents and discuss the potential strengths and limitations of each target. We conclude by offering some perspectives on the potential use of mGlu modulators in the treatment of PD.

1. Introduction

Parkinson's disease (PD) is characterized by degeneration of dopaminergic neurons of the substantia nigra compacta (SNc), which results in a loss of dopamine within the striatum (Hornykiewicz, 1986). This dopamine reduction significantly impacts the transmission along the cortico—basal ganglia—thalamo—cortical loop, which utilizes glutamate and gamma-aminobutyric acid (GABA) as its main neurotransmitters (DeLong & Wichmann, 2007; Wichmann, Bergman, & DeLong, 2018). For instance, synaptic changes occur along striatal projection neurons of the direct and indirect pathways (Shen, Zhai, & Surmeier, 2022), leading to disrupted signaling across the structures of both pathways. Over-active glutamatergic activity within the striatum is regarded as a tenet of L-3,4-dihydroxyphenylalanine (L-DOPA)-induced dyskinesia (Chase & Oh, 2000).

In agreement with abnormal glutamatergic transmission in the parkinsonian and dyskinetic states, several modulators of ionotropic glutamate receptors have been tested in the clinic. The N-methyl-D-aspartate (NMDA) antagonist amantadine is used both to alleviate parkinsonism as monotherapy in early PD and as an anti-dyskinetic agent in advanced PD (Fox et al., 2018). However, the use of amantadine may be limited by its propensity to cause hallucinations (Riederer, Lange, Kornhuber, & Danielczyk, 1991) and there may be tolerance to its anti-dyskinetic action (Thomas et al., 2004). Other NMDA antagonists such as dextromethorphan/quinidine and ketamine have shown promising results as anti-dyskinetic drugs in clinical trials or case reports (Fox et al., 2017; Sherman, Estevez, Magill, & Falk, 2016), but more data are required before these agents can be recommended.

In contrast, the potential utility of antagonists at α-amino-3-hydroxy-5-methyl-4-isoxazolepropionic acid (AMPA) or kainate receptors appears unclear, as most clinical trials conducted so far with molecules harboring this mechanism of action have yielded disappointing results. For instance, topiramate may worsen dyskinesia (Kobylecki et al., 2014), while perampanel was inefficacious at reducing motor fluctuations (Rascol et al., 2012).

There thus appears to be limitations to the benefits conferred by modulators of ionotropic glutamate receptors in PD. Modulation of metabotropic glutamate (mGlu) receptors may potentially be an attractive way to modulate glutamatergic transmission in ways that may benefit PD ant treatment-related complications. There are 8 types of mGlu receptors, divided between 3 groups (Niswender & Conn, 2010). Group I mGlu receptors comprises $mGlu_1$ and $mGlu_5$ receptors, Group II mGlu receptors encompasses $mGlu_2$ and $mGlu_3$ receptors, while Group III mGlu receptors includes $mGlu_4$, $mGlu_6$, $mGlu_7$ and $mGlu_8$ receptors (Reiner & Levitz, 2018). In this book chapter, we aim to provide a summary of the mGlu receptor literature in PD, with a specific focus on $mGlu_5$, $mGlu_4$, $mGlu_2$ and $mGlu_3$ receptors. Indeed, in PD, the most studied mGlu receptors are $mGlu_5$ receptors, for dyskinesia-related endpoints, while $mGlu_4$ receptors were assessed for their potential anti-parkinsonian and anti-dyskinetic effects. More recently, $mGlu_2$ receptors were studied for their possible effects of parkinsonism, dyskinesia and psychosis, while $mGlu_3$ receptors were investigated for possible neuro-protective effects.

2. mGlu$_5$ receptors

2.1 Distribution and function

Group I mGlu receptors are post-synaptic transmembrane G-protein coupled receptors, consisting of $mGlu_1$ and $mGlu_5$ receptors (Lüscher & Huber, 2010). Group I mGlu receptors are coupled to a G_q/G_{11} protein and activate phospholipase C signaling via inositol 1,4,5-triphosphate hydrolysis and secondary messenger production (Berridge & Irvine, 1984; Lüscher & Huber, 2010). mGlu$_5$ receptors have been extensively studied in the context of various developmental and psychiatric disorders (Dölen & Bear, 2008; Lee, Wurtman, Cox, & Nitsch, 1995; Matosin & Newell, 2013; Silverman, Tolu, Barkan, & Crawley, 2010). mGlu$_5$ receptors are subdivided into two isoforms, mGlu$_{5a}$, which is predominantly encountered during the early post-natal period, and mGlu$_{5b}$, which is expressed in adulthood (Minakami, Katsuki, & Sugiyama, 1993; Romano, Van den Pol, & O'Malley, 1996). mGlu$_5$ receptors are mainly found at the post-synaptic membrane of glutamatergic neurons, where they modulate neuronal excitability by regulating currents of ionotropic glutamate receptors

(Lujan, Nusser, Roberts, Shigemoto, & Somogyi, 1996; Shigemoto et al., 1993). It has also been shown that mGlu$_5$ receptors are physically connected to NMDA receptors through the GluN2 subunit by scaffolding proteins such as Homer and Shank (Naisbitt et al., 1999; Tu et al., 1999).

2.2 Distribution of mGlu$_5$ receptors in PD

In a reserpine-treated rat model of PD, mGlu$_5$ mRNA expression was shown to be downregulated in both rostral and caudal areas of the striatum, with no changes in the globus pallidus (GP), subthalamic nucleus (STN), thalamus, substantia nigra (SN), or motor cortex, when compared to normal rats (Ismayilova, Verkhratsky, & Dascombe, 2006). Treatment with reserpine increased mGlu$_5$ receptor maximal binding (B$_{max}$) in the rostral and caudal parts of the striatum compared to vehicle treatment, however no affinity changes were observed (Ismayilova et al., 2006).

In the 1-methyl-4-phenyl-1,2,3,6-tetrahydropyridine (MPTP)-lesioned macaque, no changes in mGlu$_5$ levels were observed when parkinsonian animals were compared to control animals across different brain regions (posterior putamen, GP pars interna [GPi], GP pars externa [GPe] and caudate nucleus) (Morin, Morissette, et al., 2013; Ouattara et al., 2011; Samadi et al., 2008). However, mGlu$_5$ levels increased in the putamen, GPi, GPe and caudate nucleus following treatment with L-DOPA (Morin, Grégoire, et al., 2013; Morin, Morissette, et al., 2013). These increases were no longer present in an additional group of macaques that were administered L-DOPA and the mGlu$_5$ negative allosteric modulator (NAM) 2-nethyl-6-(phenylethynyl)pyridine (MPEP) (Lea IV & Faden, 2006; Morin, Morissette, et al., 2013). Lastly, mGlu$_5$ levels in the caudate nucleus, putamen, GPi and GPe were found to correlate with mean dyskinesia scores (Morin, Grégoire, et al., 2013; Morin, Morissette, et al., 2013).

2.3 Preclinical studies with mGlu$_5$ modulators in PD

2.3.1 Potential neuroprotective effects of mGlu$_5$ antagonists in PD

In MPTP-lesioned mice, the extent of MPTP-induced nigro-striatal degeneration was attenuated by the mGlu$_5$ NAMs MPEP and SIB1893 (Battaglia et al., 2004). In that same study, mGlu$_5$ knock-out mice administered with MPTP showed milder nigro-striatal damage compared to wild-type mice, suggesting a lower sensitivity to MPTP toxicity in these animals (Battaglia et al., 2004). Similarly, in female MPTP-lesioned macaques, chronic administration of 3-[(2-methyl-1,3-thiazol-4-yl)ethynyl]pyridine

(MTEP), a mGlu$_5$ NAM, reduced the extent of MPTP-induced dopaminergic and noradrenergic degeneration (Masilamoni et al., 2011). On the other hand, primary microglia infection with lentiviral vector-α-synuclein and exposition to MTEP (100 μM) resulted in an exacerbation of microglial inflammation and an increased susceptibility of neurons to degeneration, while mGlu$_5$ activation with CHPG (150 μM) protected neurons from neurotoxicity (Zhang et al., 2021). Overall, these results suggest that mGlu$_5$ antagonism may represent a potential strategy to attenuate existing degeneration of monoaminergic neurons in PD, while mGlu$_5$ activation may also act as a preventive treatment for PD neurodegeneration.

2.3.2 Parkinsonism and dyskinesia

In the 6-hydroxydopamine (6-OHDA)-lesioned rat, chronic treatment with MPEP significantly improved akinesia (Breysse, Amalric, & Salin, 2003; Breysse, Baunez, Spooren, Gasparini, & Amalric, 2002) and induced preferentially ipsilateral rotations compared to preferentially contralateral rotations in 6-OHDA-lesioned rats administered with vehicle (Breysse et al., 2002) (Table 1). In another study, the combination of MPEP with the NMDA antagonist MK-801 elicited a pro-kinetic effect that was not observed when MPEP was administered alone (Turle-Lorenzo et al., 2005). In contrast, in a different experiment, when MTEP was administered alone, no effect was observed compared to control in rotarod and cylinder tests (Mela et al., 2007). A possible explanation for these differences between MPEP and MTEP may relate to the affinity of MPEP at mGlu$_4$ receptors, at which it acts as a positive allosteric modulator (PAM) (Mathiesen, Svendsen, Bräuner-Osborne, Thomsen, & Ramirez, 2003), which may underlie its anti-parkinsonian action when administered as monotherapy, while MTEP is devoid of such affinity for mGlu$_4$ receptors (Cosford et al., 2003).

However, when co-administered with L-DOPA, MTEP led to a progressive motor improvement over a 21-day period, with a 10% increase in rotarod performance compared to the L-DOPA alone group (Mela et al., 2007). In the cylinder test, the combination of MTEP and L-DOPA increased contralateral forepaw use by 25% compared to the vehicle, which was slightly lower than that of L-DOPA alone, which reached 40% of contralateral forepaw use (Mela et al., 2007).

In the MPTP-lesioned macaque, MTEP administered alone as monotherapy did not affect severity of parkinsonism (Johnston et al., 2010). Mavoglurant is another mGlu$_5$ NAM (Levenga et al., 2011) that did not elicit any effect on locomotor activity and parkinsonism when administered

Table 1 Pre-clinical studies assessing the effect of mGlu$_5$ ligands on motor symptoms and dyskinesia in experimental parkinsonism.

Drug	Mechanism of action	Animal model	Effect	References
MPEP	NAM	6-OHDA-lesioned rat	- Reversal of akinesia when administered chronically, ipsilateral rotations - Acute administration improved AIMs severity - Chronic administration improved AIMs severity without development of tolerance - Reversal of akinesia when combined with the NMDA antagonist MK-801	Breysse et al. (2002, 2003), Levandis, Bazzini, Armentero, Nappi, and Blandini (2008) and Turle-Lorenzo, Breysse, Baunez, and Amalric (2005)
		MPTP-lesioned macaque	- Reduction of peak dose and global dyskinesia	Morin, Grégoire, Gomez-Mancilla, Gasparini, and Di Paolo (2010)
MTEP	NAM	6-OHDA-lesioned rat	- No effect on rotarod or cylinder test when administered alone - Improvement of rotarod and cylinder test performance in combination with L-DOPA - Acute administration improved AIMs severity - Chronic administration improved AIMs severity without tolerance	Gravius et al. (2008) and Mela et al. (2007)
		MPTP-lesioned macaque	- Reduction of peak dose and global dyskinesia - Reduction of peak L-DOPA anti-parkinsonian action	Johnston, Fox, McIldowie, Piggott, and Brotchie (2010) and Morin et al. (2010)

Dipraglurant NAM	MPTP-lesioned macaque	– Reduction of peak dose dyskinesia and global dyskinesia – No alteration of L-DOPA anti-parkinsonian action	Bezard et al. (2014)
Mavoglurant NAM	MPTP-lesioned macaque	– No significant effect on global dyskinesia – Reduction of peak dose dyskinesia – No effect on locomotor activity and parkinsonism when administered alone – No effect on L-DOPA anti-parkinsonian action	Grégoire et al. (2011)

6-OHDA, 6-hydroxydopamine; AIMs, abnormal involuntary movements; L-DOPA, L-3,4-dihydroxyphenylalanine; MPEP, 2-methyl-6-(phenylethynyl)pyridine; MPTP, 1-methyl-4-phenyl-1,2,3,6-tetrahydropyridine; MTEP, 3-[(2-methyl-1,3-thiazol-4-yl)ethynyl]pyridine; NAM, negative allosteric modulator; NMDA, N-methyl-D-aspartate.

alone, whereas when added to L-DOPA, it further improved locomotion and parkinsonian scores, when compared to L-DOPA alone, in the MPTP-lesioned macaque (Grégoire et al., 2011). However, high doses of MTEP diminished the anti-parkinsonian action of L-DOPA, suggesting that, at higher doses, associated with a greater anti-dyskinetic benefit (see below), mGlu$_5$ antagonists may hinder the efficacy of L-DOPA (Johnston et al., 2010).

Regarding dyskinesia, both MPEP and MTEP alleviated abnormal involuntary movements (AIMs) in the 6-OHDA-lesioned rat (Dekundy, Pietraszek, Schaefer, Cenci, & Danysz, 2006; Gravius et al., 2008; Levandis et al., 2008; Mela et al., 2007). Thus, chronic co-administration of MPEP or MTEP with L-DOPA over 21 and 7 days, respectively, reduced AIMs severity, without indication of tolerance (Gravius et al., 2008; Levandis et al., 2008). Moreover, the addition of MTEP to L-DOPA at the beginning of L-DOPA therapy interfered with the dyskinesia induction process, attenuating the development of AIMs by over 60% compared to L-DOPA alone, over a 21-day period (Mela et al., 2007).

Similarly, MTEP administration reduced peak dose dyskinesia as well as the duration of on-time with disabling dyskinesia when compared to L-DOPA alone, in the MPTP-lesioned macaque (Johnston et al., 2010). In another study in the MPTP-lesioned macaque, both MPEP and MTEP reduced peak dose dyskinesia and global dyskinesia severity (Morin et al., 2010). Dipraglurant (ADX48621), another mGlu$_5$ NAM, also significantly reduced global dyskinesia severity and peak dose dyskinesia in the MPTP-lesioned macaque, without hindering L-DOPA anti-parkinsonian action (Bezard et al., 2014). Conversely, acute administration of mavoglurant failed to decrease global dyskinesia, but did reduce peak dose scores, without impairing L-DOPA anti-parkinsonian action, in the MPTP-lesioned primate (Grégoire et al., 2011).

2.4 Clinical studies

Building on these promising results from pre-clinical studies, a few mGlu$_5$ NAMs have entered clinical testing. As presented in Table 2, in a Phase IIa study, dipraglurant administration significantly decreased peak and global dyskinesia, with a concomitant reduction in daily off-time when compared to placebo (Tison et al., 2016). Recently, dipraglurant was undergoing a randomized, placebo-controlled, quadruple-blind Phase IIb/III trial (NCT04857359) that aimed to recruit 140 patients, with the effect of

Table 2 Clinical studies assessing the effect of mGlu$_5$ ligands in PD.

Drug	Mechanism of action	Study population	Endpoint(s)	Status	Effect	References
Dipraglurant	NAM	- Phase IIa trial - PD subjects with motor fluctuations and dyskinesia	Safety, tolerability and efficacy in the treatment of dyskinesia	Completed	Reduction of peak dose and improvement of global dyskinesia	Tison et al. (2013) and Tison et al. (2016)
Mavoglurant	NAM	- Phase IIa and IIb trials - PD subjects with dyskinesia	Anti-dyskinetic efficacy as measured by the modified AIMS total score	Completed	- Reduction of dyskinesia - Increase of on-time without dyskinesia	Berg et al. (2011), Kumar et al. (2016) and Stocchi et al. (2013a)
Remeglurant	NAM	- Phase I - PD patients		Unknown	Unknown	Wang et al. (2018)

AIMS, Abnormal Involuntary Movement Scale; NAM, negative allosteric modulator; PD, Parkinson's disease.

dipraglurant on the Unified Dyskinesia Rating Scale (UDysRS) as the primary endpoint. This study was terminated because of slow recruitment. An open label study aiming to evaluate the long-term safety and tolerability profile of dipraglurant in PD patients was also recruiting participants (NCT05116813), as possible tolerability issues were raised in a previous study (Tison et al., 2016). Whether the development of dipraglurant for the treatment of dyskinesia will be pursued is unclear at the moment.

Two Phase II trials with mavoglurant (AFQ-056) reported a reduction of dyskinesia without aggravating parkinsonian symptoms in PD patients with moderate to severe dyskinesia (Berg et al., 2011). Indeed, mavoglurant was well tolerated by PD patients, with minimal decrease in motor function (Berg et al., 2011; Stocchi et al., 2013). A larger Phase IIb study examined the effects of different doses of mavoglurant, and found that daily administration of 200 mg mavoglurant reduced dyskinesia (Stocchi et al., 2013). In addition, cessation of treatment was accompanied by worsening of dyskinesia severity, with some patients developing delusions or psychosis (Stocchi et al., 2013). In another study, mavoglurant extended the duration of on-time without troublesome dyskinesia when compared to placebo, although the study was terminated early owing to enrolment issues (Kumar et al., 2016). Despite these promising results, a meta-analysis concluded that adding mavoglurant to L-DOPA is not efficacious in reducing dyskinesia (Negida et al., 2021)

Lastly, a Phase I study evaluated the effect of remeglurant (MRZ-8456) for the treatment of dyskinesia in PD patients (Wang et al., 2018), however no results have been disclosed as of yet.

3. mGlu₄ receptors

3.1 Distribution and function

Along with mGlu$_6$, mGlu$_7$, and mGlu$_8$ receptors, mGlu$_4$ receptors form Group III of mGlu receptors, all of which perform modulatory roles in the brain except for mGlu$_6$ receptors (Conn & Pin, 1997). Group III mGlu receptors are predominantly expressed on pre-synaptic GABAergic and glutamatergic synapses (Bradley, Standaert, Levey, & Conn, 1999; Corti, Aldegheri, Somogyi, & Ferraguti, 2002; Matsui & Kita, 2003; Valenti et al., 2003) and are coupled to G_i/G_o proteins to regulate synaptic transmission (Duty, 2010). In the rodent brain, mGlu$_4$ levels vary from low expression in the neocortex and hippocampus, to moderate expression in the SN pars reticulata (SNr) and entopeduncular nucleus, with high expression in the GP and cerebellum (Bradley et al., 1999; Corti et al., 2002). In addition to their neuronal localization, mGlu$_4$ receptors are also expressed on glial cells (Taylor, Diemel, & Pocock, 2003; Yao et al., 2005), where they are hypothesized to mediate anti-inflammatory responses (Fallarino et al., 2010).

In PD, increased GABAergic inhibition at striato-pallidal synapses is thought to play an important role in motor dysfunction (Marino et al., 2003). Based on the high expression of mGlu$_4$ receptors in the GP, it may be inferred that selective receptor activation at the pre-synaptic terminal may inhibit neurotransmission at striato-pallidal synapses (Doller, Bespalov, Miller, Pietraszek, & Kalinichev, 2020) by reducing GABA release in the GPe (Charvin et al., 2018), thereby providing relief of parkinsonian deficits. Indeed, the non-selective Group III mGlu agonist L-(+)-2-amino-4-phosphonobutyric acid (L-AP4) dampened GABAergic transmission at the striato-pallidal synapse of wild-type mice, but this effect was absent in mGlu$_4$ knockout mice (Valenti et al., 2003). Similar results were obtained in rodent brain slices with the Group III mGlu receptor orthosteric agonist LSP1-2111 (Beurrier et al., 2009) and the selective mGlu$_4$ PAM N-phenyl-7-(hydroxyimino)cyclopropa[b]chromen1a-carboxamide (PHCCC) (Maj et al., 2003; Marino et al., 2003). These results collectively reinforce the involvement of mGlu$_4$ receptors at the striato-pallidal synapse (Duty, 2010).

In addition to being found on pre-synaptic GABAergic terminals in the GP, there is evidence that mGlu$_4$ receptors are expressed on post-synaptic glutamatergic terminals in the primate GP (Bogenpohl, Galvan, Hu, Wichmann, & Smith, 2013). mGlu$_4$ receptors may be expressed at the STN-GPe synapse, further modulating glutamatergic transmission in the GP (Bogenpohl et al., 2013). Accordingly, in the 6-OHDA-lesioned rat, activation of Group III mGlu receptors decreased both GABA and glutamate levels in the GP (Deltheil, Turle-Lorenzo, & Amalric, 2011). Given the reduced inhibition of the STN and ensuing enhanced subthalamofugal transmission in PD (Blandini, Greenamyre, & Nappi, 1996; Greenamyre, 2001; Smith, Wichmann, Factor, & DeLong, 2012), mGlu$_4$ activation may inhibit glutamate release at the subthalamo-pallidal synapse and restore the imbalance between the direct and indirect basal ganglia pathways to improve motor deficits (Bogenpohl et al., 2013).

In PD, enhanced activity at the cortico-striatal synapse is thought to contribute to increased firing of the indirect pathway (Obeso et al., 2008). Some mGlu$_4$ receptors are localized at the pre-synaptic terminal of cortico-striatal glutamatergic synapses (Bradley et al., 1999; Duty, 2010) and predominantly connect with neurons of the indirect pathway (Iskhakova & Smith, 2016), where they dampen excitatory cortical input (Bennouar et al., 2013; Cuomo et al., 2009; Gubellini, Melon, Dale, Doller, & Kerkerian-Le Goff, 2014). Therefore, mGlu$_4$ activation may preferentially inhibit the hyperactive indirect pathway while maintaining excitation of the direct pathway, ultimately normalizing output from the basal ganglia (Charvin et al., 2018).

3.2 Preclinical studies

3.2.1 Potential neuroprotective effects of mGluR$_4$ agonists in PD

The neuro-protective effect of the broad-spectrum Group III mGlu agonist L-AP4 was first demonstrated following supra-nigral injection in the 6-OHDA-lesioned rat, and is presumably due to inhibition of glutamate release in the SN (Austin et al., 2010). In the same animal model, supra-nigral injection of the mGlu$_4$ PAM VU0155041 provided protection against 6-OHDA-induced nigro-striatal denervation, with 40% of tyrosine hydroxylase (TH)-positive cells remaining in the intact hemisphere after intra-SNc injection of 6-OHDA (Betts, O'Neill, & Duty, 2012). There was evidence of reduced inflammatory markers, such as ionized calcium binding adaptor molecule 1 (IBA-1) and glial fibrillary acidic protein (GFAP), in the brains of animals treated with VU0155041, which may represent an additional mechanism underlying the neuro-protective effect of mGlu$_4$ activation (Betts et al., 2012). Consistent with these findings, systemic administration of the mGlu$_4$ PAM PHCCC attenuated MPTP-induced nigro-striatal denervation in wild-type mice (Battaglia et al., 2006); these findings were not replicated in mGlu$_4$ knockout mice, suggesting the neuroprotection conferred by PHCCC was mediated by mGlu$_4$ receptors (Battaglia et al., 2006).

3.2.2 Parkinsonism and dyskinesia

Activation of mGlu$_4$ receptors with orthosteric agonists or structurally diverse PAMs has consistently demonstrated efficacy for treatment of motor deficits in PD (Charvin et al., 2018; Doller et al., 2020) (Table 3). In reserpine-treated rats, PHCCC significantly reversed akinesia (Broadstock et al., 2012; Marino et al., 2003). Similarly, the mGlu$_4$ PAM VU015504 dose-dependently decreased haloperidol-induced catalepsy and reserpine-induced akinesia in rats (Niswender et al., 2008). Several other mGlu$_4$ PAMs, e.g., VU0364770, ADX88178 and LuAF21934, also effectively improved motor impairment of haloperidol-induced catalepsy in rats

Table 3 Pre-clinical studies assessing the effect of mGlu$_4$ ligands on motor symptoms and dyskinesia in experimental parkinsonism.

Drug	Mechanism of action	Animal model	Effect	References
PHCCC	PAM	- Reserpine-induced akinesia rat	- Reversal of akinesia	Broadstock, Austin, Betts, and Duty (2012) and Marino et al. (2003)
VU0155041	PAM	- Haloperidol-induced catalepsy rat - Reserpine-induced akinesia rat	- Dose-dependently decreased catalepsy - Reversal of akinesia	Niswender et al. (2008)
VU0364770	PAM	- Haloperidol-induced catalepsy rat - Unilateral 6-OHDA-lesioned rat	- Improved catalepsy - No significant effect on the onset, duration or intensity of AIMs - Reversed forelimb asymmetry both as monotherapy and with sub-therapeutic doses of L-DOPA	Iderberg et al. (2015) and Jones et al. (2012)
ADX88178	PAM	- Haloperidol-induced catalepsy rat - Bilateral 6-OHDA-lesioned rat - Unilateral 6-OHDA-lesioned rat - MitoPark mouse	- Improved catalepsy - Dose-dependent reversal of forelimb akinesia when administered with a low dose of L-DOPA - No effect on the onset, duration or intensity of AIMs - Potentiated L-DOPA-induced locomotion	Le Poul et al. (2012)
LuAF21934	PAM	- Haloperidol-induced catalepsy rat - Unilateral 6-OHDA-lesioned rat	- Improved catalepsy - No effect when administered alone - Improved forelimb akinesia when administered with sub-threshold doses of L-DOPA - Decreased the incidence but not the severity of AIMs	Bennouar et al. (2013)
VU0418506	PAM	- Haloperidol-induced catalepsy rat - Unilateral 6-OHDA-lesioned rat	- Improved catalepsy - Reversed forelimb asymmetry both as monotherapy and with sub-therapeutic doses of L-DOPA	Niswender et al. (2016)

Continued

Table 3 Pre-clinical studies assessing the effect of mGlu$_4$ ligands on motor symptoms and dyskinesia in experimental parkinsonism.—cont'd

Drug	Mechanism of action	Animal model	Effect	References
PXT001687	PAM	- Haloperidol-induced catalepsy rat - Bilateral 6-OHDA-lesioned rat	- Improved catalepsy - Improved L-DOPA-induced locomotion	Charvin et al. (2017)
LSP1-2111	Orthosteric agonist	MPTP-lesioned marmoset	- No effect on dyskinesia severity - Anti-parkinsonian action as monotherapy - No added anti-parkinsonian effect when administered with L-DOPA	Mann et al. (2020)
Foliglurax	PAM	MPTP-lesioned macaque	- Mild anti-parkinsonian effect as monotherapy - Enhanced the anti-parkinsonian action of a sub-optimal dose of L-DOPA - Diminished dyskinesia severity	Charvin et al. (2018)

6-OHDA, 6-hydroxydopamine; AIMs, abnormal involuntary movements; L-DOPA, L-3,4-dihydroxyphenylalanine; MPTP, 1-methyl-4-phenyl-1,2,3,6-tetrahydropyridine; PAM, positive allosteric modulator.

(Bennouar et al., 2013; Charvin et al., 2017; Jones et al., 2012; Le Poul et al., 2012; Niswender et al., 2016).

In the unilateral 6-OHDA-lesioned rat, VU0364770 and VU0418506 reversed forelimb asymmetry, both alone and with low doses of L-DOPA (Jones et al., 2012; Niswender et al., 2016). In contrast, the mGlu$_4$ PAM LuAF21934 had no effect when administered alone, but effectively improved akinesia in a dose-dependent manner when administered with low doses of L-DOPA (Bennouar et al., 2013). In the bilateral 6-OHDA-lesioned rat, monotherapy with the mGlu$_4$ PAM ADX88178 had no significant effect on forelimb akinesia, but led to a significant reversal in forelimb akinesia when administered with a sub-threshold dose of L-DOPA (Le Poul et al., 2012). In the same model, administration of the mGlu$_4$ PAM PXT001687 with a low dose of L-DOPA markedly improved locomotor activity (Charvin et al., 2017). In the MitoPark mouse, a model of progressive dopaminergic deficits (Ekstrand et al., 2007), ADX88178 treatment in combination with a therapeutic dose of L-DOPA potentiated locomotion (Le Poul et al., 2012).

Contrary to findings on the anti-parkinsonian efficacy of mGlu$_4$ ligands, studies have reported mixed results on their anti-dyskinetic efficacy (Table 3). In the 6-OHDA-lesioned rat, LuAF21934 decreased the incidence but not the severity of dyskinesia (Bennouar et al., 2013). On the other hand, in the same model, treatment with either ADX88178 or VU0364770 had no significant effect on the onset, duration or intensity of dyskinesia (Iderberg et al., 2015; Le Poul et al., 2012). Similarly, when the Group III mGlu orthosteric agonist LSP1-2111 was administered with L-DOPA to MPTP-lesioned marmosets, it failed to significantly improve dyskinesia severity (Mann et al., 2020). Conversely, in MPTP-lesioned macaques, the mGlu$_4$ PAM PXT002331 (foliglurax) dose-dependently alleviated parkinsonism, including bradykinesia, tremor, and posture, when administered with a sub-optimal dose of L-DOPA (Charvin et al., 2018). Foliglurax also significantly improved dyskinesia severity in the same animal model (Charvin et al., 2018). Results from this study were used to support the translation of foliglurax to the clinic (Doller et al., 2020).

3.2.3 Neuro-psychiatric symptoms
The effect of mGlu$_4$ ligands on neuro-psychiatric symptoms has not been studied in the context of PD, but they have demonstrated efficacy in animal models of anxiety. In a rat model of anxiety using the conflict drinking Vogel test, the mGlu$_4$ PAM PHCCC led to dose-dependent anxiolytic effects

(Stachowicz, Kłak, Kłodzińska, Chojnacka-Wojcik, & Pilc, 2004). Moreover, the mGlu$_4$ PAM ADX88178 improved performance in the elevated plus maze and marble burying paradigms in rodent anxiety models (Lindsley & Hopkins, 2012). These results collectively provide support for mGlu$_4$ PAMs as a treatment strategy for anxiety (Stachowicz et al., 2004), possibly mediated by the inhibition of glutamate release (Schoepp, 2001). Given these favorable findings, studies assessing the effects of mGlu$_4$ activation on PD patient anxiety, as well as other neuropsychiatric symptoms, may be of interest.

3.3 Clinical studies

As shown in Table 4, to date, only a few clinical studies have assessed the effects of mGlu$_4$ ligands in PD and the translational potential of these ligands remains uncertain. The novel PET radioligand [^{11}C]-PXT012253 was first administered in an open–label Phase I trial in healthy subjects to assess the binding of foliglurax to mGlu$_4$ receptors in the human brain (NCT03826134). This was followed by an open–label trial that assessed [^{11}C]-PXT012253 binding to mGlu$_4$ receptors in healthy and PD subjects (NCT04175132). However, the results from this study have not been disclosed and it was later withdrawn. On the other hand, a multicenter, double–blind, randomized, placebo–controlled Phase IIa trial assessed the efficacy and safety of foliglurax in reducing off–time and dyskinesia in 157 PD subjects (NCT03162874). Over the 28-day treatment course, foliglurax (20 or 60 mg daily) failed to significantly improve off–time or dyskinesia compared to placebo (Rascol et al., 2022). Another multicenter, double–blind, randomized, placebo–controlled Phase IIa trial similarly evaluated the effect of foliglurax on L-DOPA-induced dyskinesia and wearing–off in PD (NCT03331848). The trial sought to assess the efficacy, safety and tolerability of an 8–week oral treatment of foliglurax (20 mg daily), but was withdrawn for business reasons. Despite the lack of success of foliglurax, clinical development of another mGlu$_4$ PAM, AP-472 has been ongoing. A randomized, double–blind, placebo–controlled Phase I trial is being conducted in healthy adults to evaluate the safety, tolerability, and pharmacokinetics of AP-472 (PRNewswire, 2021). The results from this trial will inform dose selection for Phase Ib and II trials in subjects with PD. The outcome of these trials is presently unknown, last being updated on October 4th 2021 (PRNewswire, 2021).

Table 4 Clinical studies assessing the effect of mGlu$_4$ ligands in PD.

Drug	Mechanism of action	Study population	Endpoint(s)	Status	Effect	Reference
[^{11}C]-PXT012253	PAM	– Open-label trial – Healthy and PD subjects	Assess binding of foliglurax to mGluR$_4$ in the human brain	Results were undisclosed and study was withdrawn	Unknown	NCT04175132
Foliglurax	PAM	– Phase IIa trial – PD subjects with motor fluctuations and dyskinesia	Change from baseline to end of treatment period in the daily awake off-time	Completed	No significant improvement in off-time duration or dyskinesia severity	NCT03162874 (Rascol, Medori, Baayen, Such, & Meulien, 2022)
		– Phase IIa trial – PD subjects with dyskinesia	Change from baseline in UDysRS total score	Withdrawn for business reasons	Unknown	NCT03331848
AP-472	PAM	– Phase I trial – Healthy subjects	Safety, tolerability and pharmacokinetics	Unknown, last updated 4 October 2021	Unknown	PRNewswire (2021)

PAM, positive allosteric modulator; PD, Parkinson's disease; UDysRS, Unified Dyskinesia Rating Scale.

4. mGlu$_2$ receptors
4.1 Distribution and function

Using [^{3}H]-LY-354,740, a Group II mGlu orthosteric agonist, high density mGlu$_{2/3}$ receptor binding was found in the accessory olfactory bulb, striatum, cerebellar granular layer, molecular layer of the dentate gyrus, molecular layer and stratum lucidum of the hippocampus, the antero-ventral thalamic nuclei, as well as the cerebral cortex (higher density at layers I, III and IV) in rat (Richards et al., 2005). In brains from healthy humans, an autoradiographic study using the orthosteric antagonist [^{3}H]-LY-341,495 to image mGlu$_{2/3}$ receptors discovered higher binding in the caudate nucleus compared to the putamen; binding levels in the caudate nucleus and putamen were both higher than in the GP (Samadi et al., 2009). Within the GP, [^{3}H]-LY-341,495 binding was higher in the GPe compared to the GPi (Samadi et al., 2009).

Binding of [^{3}H]-LY-341,495 to mGlu$_{2/3}$ receptors was assessed in the striatum, GPe and GPi of MPTP-lesioned macaques treated with saline, or L-DOPA, with MPTP-naïve macaques as controls (Samadi et al., 2008). Neither administration of MPTP nor MPTP + L-DOPA induced changes in the areas analyzed, when compared to control animals. Autoradiographic mGlu$_{2/3}$ binding with [^{3}H]-LY-341,495 was performed on post-mortem brain tissue of healthy subjects and 4 different groups of L-DOPA-treated PD patients: patients with dyskinesia, patients without dyskinesia, patients with motor fluctuations, and patients without motor fluctuations (Samadi et al., 2009). There was a significant decrease in [^{3}H]-LY-341,495 binding in the caudate nucleus of PD patients, regardless of their motor complications, when compared to control subjects (Samadi et al., 2009). Binding of [^{3}H]-LY-341,495 in the caudate nucleus of PD patients with and without dyskinesia was not different to that of controls (Samadi et al., 2009). However, there was a significant reduction of [^{3}H]-LY-341,495 binding in the caudate nucleus of PD patients without motor fluctuations when compared to controls, and when compared to PD patients with motor fluctuations (Samadi et al., 2009). There were no significant changes in [^{3}H]-LY-341,495 binding in the putamen or the GPe when PD patients, regardless of their group, were compared to controls (Samadi et al., 2009). [^{3}H]-LY-341,495 binding was significantly reduced in the GPi when PD patients (encompassing all four PD patient groups) were pooled together and compared to the control group (Samadi et al., 2009). Although reduction in $[^3H]$-LY-341,495 binding did not reach significance when PD patients with or without dyskinesia were compared to controls, PD patients without motor fluctuations exhibited a significant binding reduction in the GPi when compared to control individuals (Samadi et al., 2009), while PD patients with motor fluctuations did not present significant changes in binding density when compared to control patients (Samadi et al., 2009). In summary, $mGlu_{2/3}$ receptor levels are diminished in the caudate nucleus and GPi of PD patients without motor fluctuations, and are reduced in the GPi in PD. The results of these experiments do not suggest that altered $mGlu_{2/3}$ levels in the regions of the basal ganglia examined underlie dyskinesia or motor fluctuations.

4.2 Preclinical studies

4.2.1 Potential neuroprotective effects of mGlu$_{2/3}$ agonists in PD

A few studies have examined the neuroprotective potential of $mGlu_{2/3}$ activation in experimental models of PD. The $mGlu_{2/3}$ orthosteric agonist LY-379,268 provided some protection against nigral or striatal infusion of 6-OHDA (Murray et al., 2002). Thus, chronic treatment with LY-379,268 reduced the decline of striatal TH optical density observed after 6-OHDA administration (Murray et al., 2002). Other studies have suggested that activation of $mGlu_3$ receptors underlies the neuro-protective effects (Battaglia et al., 2003, 2009; Corti et al., 2007), notably through the production of neurotrophic factors such as glial cell line-derived neurotrophic factor (GDNF) and transforming growth factor beta (TGF-β) (Battaglia et al., 2009; Corti et al., 2007).

4.2.2 Parkinsonism and dyskinesia

There is mounting evidence that $mGlu_2$ activation is beneficial to both parkinsonian disability and dyskinesia, in rodent and non-human primate models of PD (Table 5). For instance, the $mGlu_{2/3}$ orthosteric agonist 2R,4R-APDC improved forelimb use asymmetry and reduced (+)-amphetamine-induced rotation compared to vehicle in the 6-OHDA-lesioned rat (Chan et al., 2010). In the MPTP-lesioned non-human primate, the $mGlu_2$ PAMs LY-487,379 (Sid-Otmane et al., 2020) and its derivative CBiPES (Frouni et al., 2021), which harbors improved pharmacokinetic properties (Johnson et al., 2005), both enhanced the anti-parkinsonian action of a therapeutic dose of L-DOPA.

In 6-OHDA-lesioned rats and MPTP-lesioned marmosets, activation of $mGlu_2$ receptors has consistently alleviated L-DOPA induced dyskinesia.

Table 5 Pre-clinical studies assessing the effect of mGlu$_{2/3}$ ligands on motor symptoms, dyskinesia and psychosis-like behaviors in experimental parkinsonism.

Drug	Mechanism of action	Animal model	Effect	References
LY-379,268	Orthosteric agonist	– Reserpine-induced akinesia rat – 6-OHDA-lesioned rat	– No effect on reserpine-induced akinesia – No induction of rotational behavior as monotherapy – No effect on AIMs severity – Reduces the time spent on rotarod, when given in combination with L-DOPA	Murray et al. (2002) and Rylander et al. (2009)
LY-487,379	PAM	– 6-OHDA-lesioned rat – MPTP-lesioned marmoset	– Reduction of peak dose and global AIMs – Reduction of AIMs development – No effect on anti-parkinsonian action of L-DOPA – Reduction of peak dose dyskinesia – Reduction of peak dose psychosis-like behaviors – Enhancement of the anti-parkinsonian action of L-DOPA	Hamadjida et al. (2020) and Sid-Otmane et al. (2020)
LY-354,740	Orthosteric agonist	– 6-OHDA-lesioned rat – MPTP-lesioned marmoset	– Reduction AIMs severity – Attenuation of AIMs development – No effect on anti-parkinsonian action of L-DOPA – Reduction of peak dose dyskinesia – Reduction of peak dose psychosis-like behaviors- Enhancement of the anti-parkinsonian action of L-DOPA	Frouni et al. (2019)

LY-354,740 + LY-487,379	Orthosteric agonist + PAM	MPTP-lesioned marmoset	Combination of sub-optimal doses provided mild extra anti-dyskinetic effect, but did not provide additive anti-psychotic or anti-parkinsonian effects	Nuara et al. (2020)
2R,4R-APDC	Orthosteric agonist	6-OHDA-lesioned rat	– Reduction of dopaminergic neuronal loss in the SNc – Attenuation of striatal dopamine loss – Improvement of forelimb use asymmetry (intra-SNc injection)	Chan et al. (2010)
CBiPES	PAM	MPTP-lesioned marmoset	– Reduction of peak dose and global dyskinesia – Reduction of peak dose and global psychosis-like behaviors – Improvement of the anti-parkinsonian action of L-DOPA	Frouni et al. (2021)

6-OHDA, 6-hydroxydopamine; AIMs, abnormal involuntary movements; L-DOPA, L-3,4-dihydroxyphenylalanine; MPTP, 1-methyl-4-phenyl-1,2,3,6-tetrahydropyridine; PAM, positive allosteric modulator; SNc, substantia nigra pars compacta.

Administration of the $mGlu_2$ PAM LY-487,379 with L-DOPA significantly attenuated established dyskinesia in 6-OHDA-lesioned rats, and attenuated the development of dyskinesia when started concurrently with L-DOPA (Hamadjida et al., 2020). Similar anti-dyskinetic results were obtained with LY-487,379 in the MPTP-lesioned marmoset (Sid-Otmane et al., 2020). The $mGlu_2$ PAM CBiPES, similarly had significant efficacy at reducing dyskinesia in MPTP-lesioned marmosets (Frouni et al., 2021). Activation of $mGlu_{2/3}$ receptors via orthosteric stimulation with LY-354,740 also produced anti-dyskinetic effects in both the 6-OHDA-lesioned rat and the MPTP-lesioned marmoset (Frouni et al., 2019). In addition, when begun concurrently with the first dose of L-DOPA, LY-354,740 attenuated the development of dyskinesia (Frouni et al., 2019). Further studies combining $mGlu_{2/3}$ orthosteric stimulation and $mGlu_2$ positive allosteric modulation have discovered an additive anti-dyskinetic effect conferred by the combination of an orthosteric agonist and a PAM, in the MPTP-lesioned marmoset (Nuara et al., 2020).

Antagonism of serotonin (5-HT) type 2A ($5-HT_{2A}$) receptors has been shown to be an effective approach to relieve L-DOPA-induced dyskinesia (Hamadjida et al., 2018). However, there seems to be a limit to the reduction of dyskinesia that can be achieved through $5-HT_{2A}$ blockade (Kwan et al., 2019). As $5-HT_{2A}$ and $mGlu_2$ receptors have been shown to form functional hetero-complexes (González-Maeso et al., 2008) at which $5-HT_{2A}$ antagonism and $mGlu_2$ activation produce equivalent downstream signaling (Delille et al., 2012), we hypothesized, and then demonstrated, that simultaneous activation of $mGlu_2$ and antagonism of $5-HT_{2A}$ receptors elicits an additive pharmacological effect, whereby the combination of both approaches diminishes dyskinesia to a greater extent than $mGlu_2$ activation or $5-HT_{2A}$ blockade alone, in the MPTP-lesioned marmoset (Kwan et al., 2021). Lastly, the functional interaction between $5-HT_{2A}$ and $mGlu_2$ receptors was further hinted when it was shown that antagonism of $mGlu_2$ receptors reversed the anti-dyskinetic benefit of $5-HT_{2A}$ antagonism, in the MPTP-lesioned marmoset (Kwan et al., 2021).

4.2.3 Neuro-psychiatric symptoms

Selective $mGlu_2$ activation with the PAMs LY-487,379 (Sid-Otmane et al., 2020) and CBiPES (Frouni et al., 2021) or non-selective $mGlu_{2/3}$ activation with the orthosteric agonist LY-354,740 (Frouni et al., 2019) both improved psychosis-like behaviors in the MPTP-lesioned marmoset, suggesting that stimulation of $mGlu_2$ receptors might be a way to alleviate PD psychosis, in addition to dyskinesia and parkinsonism.

4.3 Clinical studies

To this date, no mGlu$_{2/3}$ orthosteric agonist or mGlu$_2$ PAM has entered clinical testing for endpoints related to PD. That being said, several clinical trials with agents activating mGlu$_{2/3}$ or mGlu$_2$ receptors have been tested in the clinic, and were generally well tolerated (Adams et al., 2013; Liu et al., 2012; Mehta et al., 2018; Patil et al., 2007; Zhang et al., 2015), suggesting that they might also be well tolerated by the PD population.

5. Summary and perspectives

This book chapter aimed to provide an overview of mGlu receptors in PD. It should be noted that only 4 of the 8 mGlu receptors have been tested either pre-clinically or clinically for endpoint related to parkinsonism, dyskinesia, psychosis and neuro-protection. Only mGlu$_5$ antagonism and mGlu$_4$ activation have been tested in the clinic, the former for dyskinesia, the latter for both parkinsonism and dyskinesia. In contrast, mGlu$_2$ and mGlu$_3$ receptors have not been modulated in the context of clinical trials.

Regarding mGlu$_5$ receptors, two clinical trials with the mGlu$_5$ NAM dipraglurant were on-going (NCT04857359, NCT05116813). However, slow recruitment led to their termination, and issues regarding the tolerability of dipraglurant have previously been observed (Tison et al., 2016). Moreover, concerns were raised, in the MPTP-lesioned primate, that drugs acting at this target might interfere with the anti-parkinsonian action of L-DOPA (Johnston et al., 2010). As such, it remains to be seen whether antagonizing mGlu$_5$ receptors will be efficacious and well tolerated in the clinic. Regarding mGlu$_4$ receptors, the potential efficacy in the clinic of ligands acting at this target remains uncertain, as foliglurax failed to demonstrate efficacy in a Phase II trial (Doller et al., 2020). Whether this failure was due to an issue with the drug or the target remains uncertain, but optimism is warranted, given the numerous studies that have demonstrated a therapeutic potential for mGlu$_4$ PAMs and orthosteric agonists in experimental parkinsonism (Charvin, 2018). Concerning mGlu$_2$ and mGlu$_3$ receptors, whereas promising, studies with agents acting at these two receptors remain pre-clinical, and their efficacy, or lack thereof, has not been demonstrated at the clinical level. Nevertheless, they both harbor significant therapeutic potential. Indeed, mGlu$_2$ activation appears to hold promise to address each of parkinsonism, L-DOPA-induced dyskinesia and PD psychosis, while mGlu$_3$ stimulation might confer neuroprotective benefits.

References

Adams, D. H., Kinon, B. J., Baygani, S., Millen, B. A., Velona, I., Kollack-Walker, S., et al. (2013). A long-term, phase 2, multicenter, randomized, open-label, comparative safety study of pomaglumetad methionil (LY2140023 monohydrate) versus atypical antipsychotic standard of care in patients with schizophrenia. *BMC Psychiatry, 13,* 143.

Austin, P. J., Betts, M. J., Broadstock, M., O'Neill, M. J., Mitchell, S. N., & Duty, S. (2010). Symptomatic and neuroprotective effects following activation of nigral group III metabotropic glutamate receptors in rodent models of Parkinson's disease. *British Journal of Pharmacology, 160,* 1741–1753.

Battaglia, G., Busceti, C. L., Molinaro, G., Biagioni, F., Storto, M., Fornai, F., et al. (2004). Endogenous activation of mGlu5 metabotropic glutamate receptors contributes to the development of nigro-striatal damage induced by 1-methyl-4-phenyl-1, 2, 3, 6-tetrahydropyridine in mice. *The Journal of Neuroscience, 24,* 828–835.

Battaglia, G., Busceti, C. L., Molinaro, G., Biagioni, F., Traficante, A., Nicoletti, F., et al. (2006). Pharmacological activation of mglu4 metabotropic glutamate receptors reduces nigrostriatal degeneration in mice treated with 1-methyl-4-phenyl-1,2,3,6-tetrahydropyridine. *The Journal of Neuroscience, 26,* 7222–7229.

Battaglia, G., Busceti, C. L., Pontarelli, F., Biagioni, F., Fornai, F., Paparelli, A., et al. (2003). Protective role of group-II metabotropic glutamate receptors against nigro-striatal degeneration induced by 1-methyl-4-phenyl-1,2,3,6-tetrahydropyridine in mice. *Neuropharmacology, 45,* 155–166.

Battaglia, G., Molinaro, G., Riozzi, B., Storto, M., Busceti, C. L., Spinsanti, P., et al. (2009). Activation of mGlu3 receptors stimulates the production of GDNF in striatal neurons. *PLoS One, 4,* e6591.

Bennouar, K. E., Uberti, M. A., Melon, C., Bacolod, M. D., Jimenez, H. N., Cajina, M., et al. (2013). Synergy between L-DOPA and a novel positive allosteric modulator of metabotropic glutamate receptor 4: Implications for Parkinson's disease treatment and dyskinesia. *Neuropharmacology, 66,* 158–169.

Berg, D., Godau, J., Trenkwalder, C., Eggert, K., Csoti, I., Storch, A., et al. (2011a). AFQ056 treatment of levodopa-induced dyskinesias: Results of 2 randomized controlled trials. *Movement Disorders, 26,* 1243–1250.

Berg, D., Godau, J., Trenkwalder, C., Eggert, K., Csoti, I., Storch, A., et al. (2011b). AFQ056 treatment of levodopa-induced dyskinesias: Results of 2 randomized controlled trials. *Movement Disorders, 26,* 1243–1250.

Berridge, M. J., & Irvine, R. F. (1984). Inositol trisphosphate, a novel second messenger in cellular signal transduction. *Nature, 312,* 315–321.

Betts, M. J., O'Neill, M. J., & Duty, S. (2012). Allosteric modulation of the group III mGlu4 receptor provides functional neuroprotection in the 6-hydroxydopamine rat model of Parkinson's disease. *British Journal of Pharmacology, 166,* 2317–2330.

Beurrier, C., Lopez, S., Révy, D., Selvam, C., Goudet, C., Lhérondel, M., et al. (2009). Electrophysiological and behavioral evidence that modulation of metabotropic glutamate receptor 4 with a new agonist reverses experimental parkinsonism. *The FASEB Journal, 23,* 3619–3628.

Bezard, E., Pioli, E. Y., Li, Q., Girard, F., Mutel, V., Keywood, C., et al. (2014). The mGluR5 negative allosteric modulator dipraglurant reduces dyskinesia in the MPTP macaque model. *Movement Disorders, 29,* 1074–1079.

Blandini, F., Greenamyre, J. T., & Nappi, G. (1996). The role of glutamate in the pathophysiology of Parkinson's disease. *Functional Neurology, 11,* 3–15.

Bogenpohl, J., Galvan, A., Hu, X., Wichmann, T., & Smith, Y. (2013). Metabotropic glutamate receptor 4 in the basal ganglia of parkinsonian monkeys: Ultrastructural localization and electrophysiological effects of activation in the striatopallidal complex. *Neuropharmacology, 66,* 242–252.

Bradley, S. R., Standaert, D. G., Levey, A. I., & Conn, P. J. (1999). Distribution of group III mGluRs in rat basal ganglia with subtype-specific antibodies. *Annals of the New York Academy of Sciences, 868,* 531–534.

Breysse, N., Amalric, M., & Salin, P. (2003). Metabotropic glutamate 5 receptor blockade alleviates akinesia by normalizing activity of selective basal-ganglia structures in parkinsonian rats. *The Journal of Neuroscience, 23,* 8302–8309.

Breysse, N., Baunez, C., Spooren, W., Gasparini, F., & Amalric, M. (2002). Chronic but not acute treatment with a metabotropic glutamate 5 receptor antagonist reverses the akinetic deficits in a rat model of parkinsonism. *The Journal of Neuroscience, 22,* 5669–5678.

Broadstock, M., Austin, P. J., Betts, M. J., & Duty, S. (2012). Antiparkinsonian potential of targeting group III metabotropic glutamate receptor subtypes in the rodent substantia nigra pars reticulata. *British Journal of Pharmacology, 165,* 1034–1045.

Chan, H., Paur, H., Vernon, A. C., Zabarsky, V., Datla, K. P., Croucher, M. J., et al. (2010). Neuroprotection and functional recovery associated with decreased microglial activation following selective activation of mGluR2/3 receptors in a rodent model of Parkinson's disease. *Parkinson's Disease, 2010,* 190450.

Charvin, D. (2018). mGlu(4) allosteric modulation for treating Parkinson's disease. *Neuropharmacology, 135,* 308–315.

Charvin, D., Di Paolo, T., Bezard, E., Gregoire, L., Takano, A., Duvey, G., et al. (2018). An mGlu4-positive allosteric modulator alleviates parkinsonism in primates. *Movement Disorders, 33,* 1619–1631.

Charvin, D., Pomel, V., Ortiz, M., Frauli, M., Scheffler, S., Steinberg, E., et al. (2017). Discovery, structure–activity relationship, and antiparkinsonian effect of a potent and brain-penetrant chemical series of positive allosteric modulators of metabotropic glutamate receptor 4. *Journal of Medicinal Chemistry, 60,* 8515–8537.

Chase, T. N., & Oh, J. D. (2000). Striatal dopamine-and glutamate-mediated dysregulation in experimental parkinsonism. *Trends in Neurosciences, 23,* S86–S91.

Conn, P. J., & Pin, J. P. (1997). Pharmacology and functions of metabotropic glutamate receptors. *Annual Review of Pharmacology and Toxicology, 37,* 205–237.

Corti, C., Aldegheri, L., Somogyi, P., & Ferraguti, F. (2002). Distribution and synaptic localisation of the metabotropic glutamate receptor 4 (mGluR4) in the rodent CNS. *Neuroscience, 110,* 403–420.

Corti, C., Battaglia, G., Molinaro, G., Riozzi, B., Pittaluga, A., Corsi, M., et al. (2007). The use of knock-out mice unravels distinct roles for mGlu2 and mGlu3 metabotropic glutamate receptors in mechanisms of neurodegeneration/neuroprotection. *The Journal of Neuroscience, 27,* 8297.

Cosford, N. D., Tehrani, L., Roppe, J., Schweiger, E., Smith, N. D., Anderson, J., et al. (2003). 3-[(2-Methyl-1, 3-thiazol-4-yl) ethynyl]-pyridine: A potent and highly selective metabotropic glutamate subtype 5 receptor antagonist with anxiolytic activity. *Journal of Medicinal Chemistry, 46,* 204–206.

Cuomo, D., Martella, G., Barabino, E., Platania, P., Vita, D., Madeo, G., et al. (2009). Metabotropic glutamate receptor subtype 4 selectively modulates both glutamate and GABA transmission in the striatum: Implications for Parkinson's disease treatment. *Journal of Neurochemistry, 109,* 1096–1105.

Dekundy, A., Pietraszek, M., Schaefer, D., Cenci, M. A., & Danysz, W. (2006). Effects of group I metabotropic glutamate receptors blockade in experimental models of Parkinson's disease. *Brain Research Bulletin, 69,* 318–326.

Delille, H. K., Becker, J. M., Burkhardt, S., Bleher, B., Terstappen, G. C., Schmidt, M., et al. (2012). Heterocomplex formation of 5-HT2A-mGlu2 and its relevance for cellular signaling cascades. *Neuropharmacology, 62,* 2184–2191.

DeLong, M. R., & Wichmann, T. (2007). Circuits and circuit disorders of the basal ganglia. *Archives of Neurology, 64,* 20–24.

Deltheil, T., Turle-Lorenzo, N., & Amalric, M. (2011). Orthosteric group III mGlu receptor agonists for Parkinson's disease treatment: Symptomatic and neurochemical action on pallidal synaptic transmission. *Current Neuropharmacology, 9*, 14–15.

Dölen, G., & Bear, M. F. (2008). Role for metabotropic glutamate receptor 5 (mGluR5) in the pathogenesis of fragile X syndrome. *The Journal of Physiology, 586*, 1503–1508.

Doller, D., Bespalov, A., Miller, R., Pietraszek, M., & Kalinichev, M. (2020). A case study of foliglurax, the first clinical mGluR4 PAM for symptomatic treatment of Parkinson's disease: Translational gaps or a failing industry innovation model? *Expert Opinion on Investigational Drugs, 29*, 1323–1338.

Duty, S. (2010). Therapeutic potential of targeting group III metabotropic glutamate receptors in the treatment of Parkinson's disease. *British Journal of Pharmacology, 161*, 271–287.

Ekstrand, M. I., Terzioglu, M., Galter, D., Zhu, S., Hofstetter, C., Lindqvist, E., et al. (2007). Progressive parkinsonism in mice with respiratory-chain-deficient dopamine neurons. *Proceedings of the National Academy of Sciences of the United States of America, 104*, 1325–1330.

Fallarino, F., Volpi, C., Fazio, F., Notartomaso, S., Vacca, C., Busceti, C., et al. (2010). Metabotropic glutamate receptor-4 modulates adaptive immunity and restrains neuroinflammation. *Nature Medicine, 16*, 897–902.

Fox, S. H., Katzenschlager, R., Lim, S. Y., Barton, B., De Bie, R. M., Seppi, K., et al. (2018). International Parkinson and movement disorder society evidence-based medicine review: Update on treatments for the motor symptoms of Parkinson's disease. *Movement Disorders, 33*, 1248–1266.

Fox, S. H., Metman, L. V., Nutt, J. G., Brodsky, M., Factor, S. A., Lang, A. E., et al. (2017). Trial of dextromethorphan/quinidine to treat levodopa-induced dyskinesia in Parkinson's disease. *Movement Disorders, 32*, 893–903.

Frouni, I., Hamadjida, A., Kwan, C., Bédard, D., Nafade, V., Gaudette, F., et al. (2019). Activation of mGlu2/3 receptors, a novel therapeutic approach to alleviate dyskinesia and psychosis in experimental parkinsonism. *Neuropharmacology, 158*, 107725.

Frouni, I., Kwan, C., Nuara, S. G., Belliveau, S., Kang, W., Hamadjida, A., et al. (2021). Effect of the mGlu2 positive allosteric modulator CBiPES on dyskinesia, psychosis-like behaviours and parkinsonism in the MPTP-lesioned marmoset. *Journal of Neural Transmission, 128*, 73–81.

González-Maeso, J., Ang, R. L., Yuen, T., Chan, P., Weisstaub, N. V., López-Giménez, J. F., et al. (2008). Identification of a serotonin/glutamate receptor complex implicated in psychosis. *Nature, 452*, 93–97.

Gravius, A., Dekundy, A., Nagel, J., More, L., Pietraszek, M., & Danysz, W. (2008). Investigation on tolerance development to subchronic blockade of mGluR5 in models of learning, anxiety, and levodopa-induced dyskinesia in rats. *Journal of Neural Transmission, 115*, 1609–1619.

Greenamyre, J. T. (2001). Glutamatergic influences on the basal ganglia. *Clinical Neuropharmacology, 24*, 65–70.

Grégoire, L., Morin, N., Ouattara, B., Gasparini, F., Bilbe, G., Johns, D., et al. (2011). The acute antiparkinsonian and antidyskinetic effect of AFQ056, a novel metabotropic glutamate receptor type 5 antagonist, in l-Dopa-treated parkinsonian monkeys. *Parkinsonism & Related Disorders, 17*, 270–276.

Gubellini, P., Melon, C., Dale, E., Doller, D., & Kerkerian-Le Goff, L. (2014). Distinct effects of mGlu4 receptor positive allosteric modulators at corticostriatal vs. striatopallidal synapses may differentially contribute to their antiparkinsonian action. *Neuropharmacology, 85*, 166–177.

Hamadjida, A., Nuara, S. G., Bédard, D., Gaudette, F., Beaudry, F., Gourdon, J. C., et al. (2018). The highly selective 5-HT(2A) antagonist EMD-281,014 reduces dyskinesia and psychosis in the l-DOPA-treated parkinsonian marmoset. *Neuropharmacology, 139*, 61–67.

Hamadjida, A., Sid-Otmane, L., Kwan, C., Frouni, I., Nafade, V., Bédard, D., et al. (2020). The highly selective mGlu2 receptor positive allosteric modulator LY-487,379 alleviates l-DOPA-induced dyskinesia in the 6-OHDA-lesioned rat model of Parkinson's disease. *European Journal of Neuroscience, 51*, 2412–2422.

Hornykiewicz, O. (1986). Biochemical pathophysiology of Parkinson's disease. Parkinson's disease. *Advances in Neurology, 45*, 19–34.

Iderberg, H., Maslava, N., Thompson, A. D., Bubser, M., Niswender, C. M., Hopkins, C. R., et al. (2015). Pharmacological stimulation of metabotropic glutamate receptor type 4 in a rat model of Parkinson's disease and L-DOPA-induced dyskinesia: Comparison between a positive allosteric modulator and an orthosteric agonist. *Neuropharmacology, 95*, 121–129.

Iskhakova, L., & Smith, Y. (2016). mGluR4-containing corticostriatal terminals: Synaptic interactions with direct and indirect pathway neurons in mice. *Brain Structure and Function, 221*, 4589–4599.

Ismayilova, N., Verkhratsky, A., & Dascombe, M. (2006). Changes in mGlu5 receptor expression in the basal ganglia of reserpinised rats. *European Journal of Pharmacology, 545*, 134–141.

Johnson, M. P., Barda, D., Britton, T. C., Emkey, R., Hornback, W. J., Jagdmann, G. E., et al. (2005). Metabotropic glutamate 2 receptor potentiators: Receptor modulation, frequency-dependent synaptic activity, and efficacy in preclinical anxiety and psychosis model(s). *Psychopharmacology, 179*, 271–283.

Johnston, T. H., Fox, S. H., McIldowie, M. J., Piggott, M. J., & Brotchie, J. M. (2010). Reduction of L-DOPA-induced dyskinesia by the selective metabotropic glutamate receptor 5 antagonist 3-[(2-methyl-1, 3-thiazol-4-yl) ethynyl] pyridine in the 1-methyl-4-phenyl-1, 2, 3, 6-tetrahydropyridine-lesioned macaque model of Parkinson's disease. *The Journal of Pharmacology and Experimental Therapeutics, 333*, 865–873.

Jones, C. K., Bubser, M., Thompson, A. D., Dickerson, J. W., Turle-Lorenzo, N., Amalric, M., et al. (2012). The metabotropic glutamate receptor 4-positive allosteric modulator VU0364770 produces efficacy alone and in combination with L-DOPA or an adenosine 2A antagonist in preclinical rodent models of Parkinson's disease. *The Journal of Pharmacology and Experimental Therapeutics, 340*, 404–421.

Kobylecki, C., Burn, D. J., Kass-Iliyya, L., Kellett, M. W., Crossman, A. R., & Silverdale, M. A. (2014). Randomized clinical trial of topiramate for levodopa-induced dyskinesia in Parkinson's disease. *Parkinsonism & Related Disorders, 20*, 452–455.

Kumar, R., Hauser, R. A., Mostillo, J., Dronamraju, N., Graf, A., Merschhemke, M., et al. (2016). Mavoglurant (AFQ056) in combination with increased levodopa dosages in Parkinson's disease patients. *International Journal of Neuroscience, 126*, 20–24.

Kwan, C., Frouni, I., Bédard, D., Nuara, S. G., Gourdon, J. C., Hamadjida, A., et al. (2019). 5-HT2A blockade for dyskinesia and psychosis in Parkinson's disease: Is there a limit to the efficacy of this approach? A study in the MPTP-lesioned marmoset and a literature mini-review. *Experimental Brain Research, 237*, 435–442.

Kwan, C., Frouni, I., Nuara, S. G., Belliveau, S., Kang, W., Hamadjida, A., et al. (2021). Combined 5-HT(2A) and mGlu(2) modulation for the treatment of dyskinesia and psychosis in Parkinson's disease. *Neuropharmacology, 186*, 108465.

Le Poul, E., Boléa, C., Girard, F., Poli, S., Charvin, D., Campo, B., et al. (2012). A potent and selective metabotropic glutamate receptor 4 positive allosteric modulator improves movement in rodent models of Parkinson's disease. *The Journal of Pharmacology and Experimental Therapeutics, 343*, 167–177.

Lea, P. M., IV, & Faden, A. I. (2006). Metabotropic glutamate receptor subtype 5 antagonists MPEP and MTEP. *CNS Drug Reviews, 12*, 149–166.

Lee, R., Wurtman, R. J., Cox, A. J., & Nitsch, R. M. (1995). Amyloid precursor protein processing is stimulated by metabotropic glutamate receptors. *Proceedings of the National Academy of Sciences, 92*, 8083–8087.

Levandis, G., Bazzini, E., Armentero, M.-T., Nappi, G., & Blandini, F. (2008). Systemic administration of an mGluR5 antagonist, but not unilateral subthalamic lesion, counteracts l-DOPA-induced dyskinesias in a rodent model of Parkinson's disease. *Neurobiology of Disease, 29*, 161–168.

Levenga, J., Hayashi, S., de Vrij, F. M., Koekkoek, S. K., van der Linde, H. C., Nieuwenhuizen, I., et al. (2011). AFQ056, a new mGluR5 antagonist for treatment of fragile X syndrome. *Neurobiology of Disease, 42*, 311–317.

Lindsley, C. W., & Hopkins, C. R. (2012). Metabotropic glutamate receptor 4 (mGlu4)-positive allosteric modulators for the treatment of Parkinson's disease: Historical perspective and review of the patent literature. *Expert Opinion on Therapeutic Patents, 22*, 461–481.

Liu, W., Downing, A. C. M., Munsie, L. M., Chen, P., Reed, M. R., Ruble, C. L., et al. (2012). Pharmacogenetic analysis of the mGlu2/3 agonist LY2140023 monohydrate in the treatment of schizophrenia. *The Pharmacogenomics Journal, 12*, 246–254.

Lujan, R., Nusser, Z., Roberts, J. D. B., Shigemoto, R., & Somogyi, P. (1996). Perisynaptic location of metabotropic glutamate receptors mGluR1 and mGluR5 on dendrites and dendritic spines in the rat hippocampus. *European Journal of Neuroscience, 8*, 1488–1500.

Lüscher, C., & Huber, K. M. (2010). Group 1 mGluR-dependent synaptic long-term depression: Mechanisms and implications for circuitry and disease. *Neuron, 65*, 445–459.

Maj, M., Bruno, V., Dragic, Z., Yamamoto, R., Battaglia, G., Inderbitzin, W., et al. (2003). (-)-PHCCC, a positive allosteric modulator of mGluR4: Characterization, mechanism of action, and neuroprotection. *Neuropharmacology, 45*, 895–906.

Mann, E., Jackson, M., Lincoln, L., Fisher, R., Rose, S., & Duty, S. (2020). Antiparkinsonian effects of a metabotropic glutamate receptor 4 agonist in MPTP-treated marmosets. *Journal of Parkinson's Disease, 10*, 959–967.

Marino, M. J., Williams, D. L., Jr., O'Brien, J. A., Valenti, O., McDonald, T. P., Clements, M. K., et al. (2003). Allosteric modulation of group III metabotropic glutamate receptor 4: A potential approach to Parkinson's disease treatment. *Proceedings of the National Academy of Sciences of the United States of America, 100*, 13668–13673.

Masilamoni, G. J., Bogenpohl, J. W., Alagille, D., Delevich, K., Tamagnan, G., Votaw, J. R., et al. (2011). Metabotropic glutamate receptor 5 antagonist protects dopaminergic and noradrenergic neurons from degeneration in MPTP-treated monkeys. *Brain, 134*, 2057–2073.

Mathiesen, J. M., Svendsen, N., Bräuner-Osborne, H., Thomsen, C., & Ramirez, M. T. (2003). Positive allosteric modulation of the human metabotropic glutamate receptor 4 (hmGluR4) by SIB-1893 and MPEP. *British Journal of Pharmacology, 138*, 1026–1030.

Matosin, N., & Newell, K. A. (2013). Metabotropic glutamate receptor 5 in the pathology and treatment of schizophrenia. *Neuroscience and Biobehavioral Reviews, 37*, 256–268.

Matsui, T., & Kita, H. (2003). Activation of group III metabotropic glutamate receptors presynaptically reduces both GABAergic and glutamatergic transmission in the rat globus pallidus. *Neuroscience, 122*, 727–737.

Mehta, M. A., Schmechtig, A., Kotoula, V., McColm, J., Jackson, K., Brittain, C., et al. (2018). Group II metabotropic glutamate receptor agonist prodrugs LY2979165 and LY2140023 attenuate the functional imaging response to ketamine in healthy subjects. *Psychopharmacology, 235*, 1875–1886.

Mela, F., Marti, M., Dekundy, A., Danysz, W., Morari, M., & Cenci, M. A. (2007). Antagonism of metabotropic glutamate receptor type 5 attenuates l-DOPA-induced dyskinesia and its molecular and neurochemical correlates in a rat model of Parkinson's disease. *Journal of Neurochemistry, 101*, 483–497.

Minakami, R., Katsuki, F., & Sugiyama, H. (1993). A variant of metabotropic glutamate receptor subtype 5: An evolutionally conserved insertion with no termination codon. *Biochemical and Biophysical Research Communications, 194*, 622–627.

Morin, N., Grégoire, L., Gomez-Mancilla, B., Gasparini, F., & Di Paolo, T. (2010). Effect of the metabotropic glutamate receptor type 5 antagonists MPEP and MTEP in parkinsonian monkeys. *Neuropharmacology*, *58*, 981–986.

Morin, N., Grégoire, L., Morissette, M., Desrayaud, S., Gomez-Mancilla, B., Gasparini, F., et al. (2013). MPEP, an mGlu5 receptor antagonist, reduces the development of L-DOPA-induced motor complications in de novo parkinsonian monkeys: Biochemical correlates. *Neuropharmacology*, *66*, 355–364.

Morin, N., Morissette, M., Grégoire, L., Gomez-Mancilla, B., Gasparini, F., & Di Paolo, T. (2013). Chronic treatment with MPEP, an mGlu5 receptor antagonist, normalizes basal ganglia glutamate neurotransmission in L-DOPA-treated parkinsonian monkeys. *Neuropharmacology*, *73*, 216–231.

Murray, T. K., Messenger, M. J., Ward, M. A., Woodhouse, S., Osborne, D. J., Duty, S., et al. (2002). Evaluation of the mGluR2/3 agonist LY379268 in rodent models of Parkinson's disease. *Pharmacology Biochemistry and Behavior*, *73*, 455–466.

Naisbitt, S., Kim, E., Tu, J. C., Xiao, B., Sala, C., Valtschanoff, J., et al. (1999). Shank, a novel family of postsynaptic density proteins that binds to the NMDA receptor/PSD-95/GKAP complex and cortactin. *Neuron*, *23*, 569–582.

Negida, A., Ghaith, H. S., Fala, S. Y., Ahmed, H., Bahbah, E. I., Ebada, M. A., et al. (2021). Mavoglurant (AFQ056) for the treatment of levodopa-induced dyskinesia in patients with Parkinson's disease: A meta-analysis. *Neurological Sciences*, *42*, 1–9.

Niswender, C. M., & Conn, P. J. (2010). Metabotropic glutamate receptors: Physiology, pharmacology, and disease. *Annual Review of Pharmacology and Toxicology*, *50*, 295–322.

Niswender, C. M., Johnson, K. A., Weaver, C. D., Jones, C. K., Xiang, Z., Luo, Q., et al. (2008). Discovery, characterization, and antiparkinsonian effect of novel positive allosteric modulators of metabotropic glutamate receptor 4. *Molecular Pharmacology*, *74*, 1345–1358.

Niswender, C. M., Jones, C. K., Lin, X., Bubser, M., Thompson Gray, A., Blobaum, A. L., et al. (2016). Development and antiparkinsonian activity of VU0418506, a selective positive allosteric modulator of metabotropic glutamate receptor 4 homomers without activity at mGlu2/4 heteromers. *ACS Chemical Neuroscience*, *7*, 1201–1211.

Nuara, S. G., Hamadjida, A., Kwan, C., Bédard, D., Frouni, I., Gourdon, J. C., et al. (2020). Combined mGlu2 orthosteric stimulation and positive allosteric modulation alleviates l-DOPA-induced psychosis-like behaviours and dyskinesia in the parkinsonian marmoset. *Journal of Neural Transmission*, *127*, 1023–1029.

Obeso, J. A., Rodriguez-Oroz, M. C., Benitez-Temino, B., Blesa, F. J., Guridi, J., Marin, C., et al. (2008). Functional organization of the basal ganglia: Therapeutic implications for Parkinson's disease. *Movement Disorders*, *23*(Suppl 3), S548–S559.

Ouattara, B., Grégoire, L., Morissette, M., Gasparini, F., Vranesic, I., Bilbe, G., et al. (2011). Metabotropic glutamate receptor type 5 in levodopa-induced motor complications. *Neurobiology of Aging*, *32*, 1286–1295.

Patil, S. T., Zhang, L., Martenyi, F., Lowe, S. L., Jackson, K. A., Andreev, B. V., et al. (2007). Activation of mGlu2/3 receptors as a new approach to treat schizophrenia: A randomized Phase 2 clinical trial. *Nature Medicine*, *13*, 1102–1107.

PRNewswire. (2021). *Appello pharmaceuticals begins phase 1 clinical trial*. https://www.prnewswire.com/news-releases/appello-pharmaceuticals-begins-phase-1-clinical-trial-301391724.html. (Accessed 19 October 2022).

Rascol, O., Barone, P., Behari, M., Emre, M., Giladi, N., Olanow, C. W., et al. (2012). Perampanel in Parkinson disease fluctuations: A double-blind randomized trial with placebo and entacapone. *Clinical Neuropharmacology*, *35*, 15–20.

Rascol, O., Medori, R., Baayen, C., Such, P., & Meulien, D. (2022). A randomized, double-blind, controlled phase II study of foliglurax in Parkinson's disease. *Movement Disorders*, *37*, 1088–1093.

Reiner, A., & Levitz, J. (2018). Glutamatergic signaling in the central nervous system: Ionotropic and metabotropic receptors in concert. *Neuron, 98*, 1080–1098.

Richards, G., Messer, J., Malherbe, P., Pink, R., Brockhaus, M., Stadler, H., et al. (2005). Distribution and abundance of metabotropic glutamate receptor subtype 2 in rat brain revealed by [3H]LY354740 binding in vitro and quantitative radioautography: Correlation with the sites of synthesis, expression, and agonist stimulation of [35S] GTPgammas binding. *The Journal of Comparative Neurology, 487*, 15–27.

Riederer, P., Lange, K. W., Kornhuber, J., & Danielczyk, W. (1991). Pharmacotoxic psychosis after memantine in Parkinson's disease. *Lancet, 338*, 1022–1023.

Romano, C., Van den Pol, A. N., & O'Malley, K. L. (1996). Enhanced early developmental expression of the metabotropic glutamate receptor mGluR5 in rat brain: Protein, mRNA splice variants, and regional distribution. *Journal of Comparative Neurology, 367*, 403–412.

Rylander, D., Recchia, A., Mela, F., Dekundy, A., Danysz, W., & Cenci, M. A. (2009). Pharmacological modulation of glutamate transmission in a rat model of L-DOPA-induced dyskinesia: Effects on motor behavior and striatal nuclear signaling. *The Journal of Pharmacology and Experimental Therapeutics, 330*, 227–235.

Samadi, P., Grégoire, L., Morissette, M., Calon, F., Tahar, A. H., Bélanger, N., et al. (2008a). Basal ganglia group II metabotropic glutamate receptors specific binding in non-human primate model of L-Dopa-induced dyskinesias. *Neuropharmacology, 54*, 258–268.

Samadi, P., Grégoire, L., Morissette, M., Calon, F., Tahar, A. H., Dridi, M., et al. (2008b). mGluR5 metabotropic glutamate receptors and dyskinesias in MPTP monkeys. *Neurobiology of Aging, 29*, 1040–1051.

Samadi, P., Rajput, A., Calon, F., Grégoire, L., Hornykiewicz, O., Rajput, A. H., et al. (2009). Metabotropic glutamate receptor II in the brains of Parkinsonian patients. *Journal of Neuropathology and Experimental Neurology, 68*, 374–382.

Schoepp, D. D. (2001). Unveiling the functions of presynaptic metabotropic glutamate receptors in the central nervous system. *The Journal of Pharmacology and Experimental Therapeutics, 299*, 12–20.

Shen, W., Zhai, S., & Surmeier, D. J. (2022). Striatal synaptic adaptations in Parkinson's disease. *Neurobiology of Disease, 167*, 105686.

Sherman, S. J., Estevez, M., Magill, A. B., & Falk, T. (2016). Case reports showing a long-term effect of subanesthetic ketamine infusion in reducing l-DOPA-induced dyskinesias. *Case Reports in Neurology, 8*, 53–58.

Shigemoto, R., Nomura, S., Ohishi, H., Sugihara, H., Nakanishi, S., & Mizuno, N. (1993). Immunohistochemical localization of a metabotropic glutamate receptor, mGluR5, in the rat brain. *Neuroscience Letters, 163*, 53–57.

Sid-Otmane, L., Hamadjida, A., Nuara, S. G., Bédard, D., Gaudette, F., Gourdon, J. C., et al. (2020). Selective metabotropic glutamate receptor 2 positive allosteric modulation alleviates L-DOPA-induced psychosis-like behaviours and dyskinesia in the MPTP-lesioned marmoset. *European Journal of Pharmacology, 873*, 172957.

Silverman, J. L., Tolu, S. S., Barkan, C. L., & Crawley, J. N. (2010). Repetitive self-grooming behavior in the BTBR mouse model of autism is blocked by the mGluR5 antagonist MPEP. *Neuropsychopharmacology, 35*, 976–989.

Smith, Y., Wichmann, T., Factor, S. A., & DeLong, M. R. (2012). Parkinson's disease therapeutics: New developments and challenges since the introduction of levodopa. *Neuropsychopharmacology, 37*, 213–246.

Stachowicz, K., Kłak, K., Kłodzińska, A., Chojnacka-Wojcik, E., & Pilc, A. (2004). Anxiolytic-like effects of PHCCC, an allosteric modulator of mGlu4 receptors, in rats. *European Journal of Pharmacology, 498*, 153–156.

Stocchi, F., Rascol, O., Destee, A., Hattori, N., Hauser, R. A., Lang, A. E., et al. (2013a). AFQ056 in Parkinson patients with levodopa-induced dyskinesia: 13-Week, randomized, dose-finding study. *Movement Disorders, 28*, 1838–1846.

Stocchi, F., Rascol, O., Destee, A., Hattori, N., Hauser, R. A., Lang, A. E., et al. (2013b). AFQ056 in Parkinson patients with levodopa-induced dyskinesia: 13-Week, randomized, dose-finding study. *Movement Disorders, 28*, 1838–1846.

Taylor, D. L., Diemel, L. T., & Pocock, J. M. (2003). Activation of microglial group III metabotropic glutamate receptors protects neurons against microglial neurotoxicity. *The Journal of Neuroscience, 23*, 2150–2160.

Thomas, A., Iacono, D., Luciano, A., Armellino, K., Di Iorio, A., & Onofrj, M. (2004). Duration of amantadine benefit on dyskinesia of severe Parkinson's disease. *Journal of Neurology, Neurosurgery & Psychiatry, 75*, 141–143.

Tison, F., Durif, F., Corvol, J. C., Eggert, K., Trenkwalder, C., Lew, M., et al. (2013). Safety, tolerability and anti-dyskinetic efficacy of dipraglurant, a novel mGluR5 negative allosteric modulator (NAM). In *Parkinson's disease (PD) patients with levodopa-induced dyskinesia (LID)(S23. 004)* AAN Enterprises.

Tison, F., Keywood, C., Wakefield, M., Durif, F., Corvol, J. C., Eggert, K., et al. (2016). A phase 2a trial of the novel mglur5-negative allosteric modulator dipraglurant for levodopa-induced dyskinesia in Parkinson's disease. *Movement Disorders, 31*, 1373–1380.

Tu, J. C., Xiao, B., Naisbitt, S., Yuan, J. P., Petralia, R. S., Brakeman, P., et al. (1999). Coupling of mGluR/Homer and PSD-95 complexes by the Shank family of postsynaptic density proteins. *Neuron, 23*, 583–592.

Turle-Lorenzo, N., Breysse, N., Baunez, C., & Amalric, M. (2005). Functional interaction between mGlu 5 and NMDA receptors in a rat model of Parkinson's disease. *Psychopharmacology, 179*, 117–127.

Valenti, O., Marino, M. J., Wittmann, M., Lis, E., DiLella, A. G., Kinney, G. G., et al. (2003). Group III metabotropic glutamate receptor-mediated modulation of the striatopallidal synapse. *The Journal of Neuroscience, 23*, 7218–7226.

Wang, W. W., Zhang, X. R., Zhang, Z. R., Wang, X. S., Chen, J., Chen, S. Y., et al. (2018). Effects of mGluR5 antagonists on Parkinson's patients with L-Dopa-induced dyskinesia: A systematic review and meta-analysis of randomized controlled trials. *Frontiers in Aging Neuroscience, 10*, 262.

Wichmann, T., Bergman, H., & DeLong, M. R. (2018). Basal ganglia, movement disorders and deep brain stimulation: Advances made through non-human primate research. *Journal of Neural Transmission, 125*, 419–430.

Yao, H. H., Ding, J. H., Zhou, F., Wang, F., Hu, L. F., Sun, T., et al. (2005). Enhancement of glutamate uptake mediates the neuroprotection exerted by activating group II or III metabotropic glutamate receptors on astrocytes. *Journal of Neurochemistry, 92*, 948–961.

Zhang, Y.-N., Fan, J.-K., Gu, L., Yang, H.-M., Zhan, S.-Q., & Zhang, H. (2021). Metabotropic glutamate receptor 5 inhibits α-synuclein-induced microglia inflammation to protect from neurotoxicity in Parkinson's disease. *Journal of Neuroinflammation, 18*, 1–24.

Zhang, W., Mitchell, M. I., Knadler, M. P., Long, A., Witcher, J., Walling, D., et al. (2015). Effect of pomaglumetad methionil on the QT interval in subjects with schizophrenia. *International Journal of Clinical Pharmacology and Therapeutics, 53*, 462–470.

CHAPTER TWO

The interaction of membrane estradiol receptors and metabotropic glutamate receptors in adaptive and maladaptive estradiol-mediated motivated behaviors in females

Caroline S. Johnson and Paul G. Mermelstein*

Department of Neuroscience, University of Minnesota, Minneapolis, MN, United States
*Corresponding author: e-mail address: pmerm@umn.edu

Contents

1. Introduction	34
2. Receptor signaling	36
2.1 Estrogen receptors	36
2.2 Membrane-bound estrogen receptors	37
2.3 Metabotropic glutamate receptors	41
2.4 mER-mGlu signaling	43
3. Estrogen in motivated behaviors: Reproduction	48
3.1 Luteinizing hormone surge	49
3.2 mER-mGlu signaling in the luteinizing hormone surge	50
3.3 Lordosis	52
3.4 mER-mGlu signaling in lordosis	54
3.5 Limbic circuits in reproduction	56
4. Estrogen in motivated behaviors: Dysregulation and drug addiction	59
4.1 Estradiol in drug addiction	60
4.2 Reward system neurocircuitry	62
4.3 mER-mGlu signaling in the striatum	65
5. Conclusion	68
References	72

International Review of Neurobiology, Volume 168
ISSN 0074-7742
https://doi.org/10.1016/bs.irn.2022.11.001

Copyright © 2023 Elsevier Inc.
All rights reserved.

Abstract

Estrogen receptors were initially identified as intracellular, ligand-regulated transcription factors that result in genomic change upon ligand binding. However, rapid estrogen receptor signaling initiated outside of the nucleus was also known to occur via mechanisms that were less clear. Recent studies indicate that these traditional receptors, estrogen receptor α and estrogen receptor β, can also be trafficked to act at the surface membrane. Signaling cascades from these membrane-bound estrogen receptors (mERs) can rapidly alter cellular excitability and gene expression, particularly through the phosphorylation of CREB. A principal mechanism of neuronal mER action has been shown to occur through glutamate-independent transactivation of metabotropic glutamate receptors (mGlu), which elicits multiple signaling outcomes. The interaction of mERs with mGlu has been shown to be important in many diverse functions in females, including driving motivated behaviors. Experimental evidence suggests that a large part of estradiol-induced neuroplasticity and motivated behaviors, both adaptive and maladaptive, occurs through estradiol-dependent mER activation of mGlu. Herein we will review signaling through estrogen receptors, both "classical" nuclear receptors and membrane-bound receptors, as well as estradiol signaling through mGlu. We will focus on how the interactions of these receptors and their downstream signaling cascades are involved in driving motivated behaviors in females, discussing a representative adaptive motivated behavior (reproduction) and maladaptive motivated behavior (addiction).

1. Introduction

Motivated behaviors are those which occur in pursuit of a goal. At the most fundamental level, motivated behaviors are those in pursuit of basic needs that allow for the survival of the individual and/or species (Simpson & Balsam, 2015; Watts & Swanson, 2002). This includes adaptive behaviors such as ingestive behaviors (seeking food, water) (Watts, Kanoski, Sanchez–Watts, & Langhans, 2022), reproductive behaviors (including copulation and parenting) (Chen et al., 2019; Meisel & Mullins, 2006; Micevych & Meisel, 2017), and social interactions (including defensive behaviors) (Lischinsky & Lin, 2020), among others. However, motivated behaviors can also be maladaptive such as drug seeking (Nestler, 2001; Simpson & Balsam, 2015). Ultimately, for the individual, these behaviors are typically rewarding (Meisel & Mullins, 2006; Simpson & Balsam, 2015).

Manifestation of motivated behaviors, as with all behaviors, requires the complex integration of a wide variety of signals through numerous neural circuits. Not only must there be integration of signals between neural and musculature systems to physically produce the behavior, but before that, within the brain itself, the myriad internal and external cues must be processed and integrated. Sensory information from the environment must be incorporated with the current behavioral state, which is further modulated by hormonal and other interosensory information (Watts & Swanson, 2002). In many motivated behaviors, the integration between hypothalamic and limbic circuits in the forebrain is essential. The hypothalamus is critically important, driving many homeostatic functions and their associated behaviors (Chen et al., 2019; Chen & Hong, 2018; Hashikawa, Hashikawa, Falkner, & Lin, 2017; Micevych & Meisel, 2017; Tanimura & Watts, 1998; Watts et al., 2022). Positive and negative feedback loops provide information about the physiological state of the organism, and neuroendocrine regulation drives homeostasis (Watts & Swanson, 2002). However, the development of motivated behaviors is also critically reliant on learning, memory, and reward processes. Behaviors that are aimed at reaching a particular goal are more likely to occur again in the future if they are rewarded (Watts & Swanson, 2002), and reward circuitry in the limbic system integrates this information with homeostatic drive from the hypothalamus.

Motivated behaviors are also modulated by gonadal hormones. One such modifier is estradiol (Yoest, Cummings, & Becker, 2015). Estradiol belongs to a group of steroid hormones, i.e., estrogens, consisting of estrone (E_1), estradiol (E_2), estriol (E_3), and estretrol (E_4) (Fuentes & Silveyra, 2019). While the gonads primarily release estrogens in females and androgens in males, both classes of steroids are found in both sexes. Furthermore, the brain can synthesize these steroid hormones directly, termed neurosteroids (Baulieu, 1991, 1998). It is important to note that while estrogens occur in low levels in males, they are not inactive. In particular, many effects of androgens in the male brain occur due to the aromatization of testosterone to estradiol (Wu et al., 2009).

Of these four hormones, estradiol (E_2, 17β-estradiol) is the physiologically predominant estrogen in circulation in sexually mature females and is perhaps best understood in its role in female reproductive behavior (Long, Serey, & Sinchak, 2014; McEwen, Jones, & Pfaff, 1987; Micevych, Mermelstein, & Sinchak, 2017; Micevych & Sinchak, 2013; Mills, Sohn, & Micevych, 2004; Sinchak et al., 2003; Sinchak & Micevych, 2001). However, estradiol plays a role in other motivated behaviors in females, as well as in disorders of motivated behaviors, through the modulation of neurotransmission and neuroplasticity. As examples, estradiol and progesterone also modulate motivation for food intake (Cummings & Becker, 2012; Eckel, 2011; Richard, López-Ferreras, Anderberg, Olandersson, & Skibicka, 2017). When circulating levels of estradiol are high, the motivation to seek food is low (Richard et al., 2017; Yoest, Cummings, & Becker, 2019). In contrast, the motivation to seek psychostimulant drugs is elevated. During the follicular phase, as levels of circulating estradiol rise, women with a history of cocaine use self-report greater motivation to seek the drug, along with subjectively greater feelings of euphoria, as compared with self-reports during the luteal phase, wherein estradiol production is lowered and progesterone concentrations rise (Evans, Haney, & Foltin, 2002).

A growing body of evidence suggests that a significant portion of estradiol-induced neuroplasticity, both in adaptive and maladaptive motivated behaviors, occurs through estradiol-dependent estrogen receptor (ER) activation of metabotropic glutamate receptors (mGlu) (Micevych & Mermelstein, 2008). Here we will review signaling through estrogen receptors, both "classical" nuclear acting and membrane-bound, as well as estradiol signaling through the mGlu family of glutamate receptors, and the interactions and signaling cascades initiated by mER-mGlu complexes. We will also review how signaling through these signaling complexes is in involved in a representative adaptive motivated behavior (reproduction), and how these same signaling mechanisms can drive disorders of motivated behaviors (addiction).

2. Receptor signaling
2.1 Estrogen receptors

At present, there have been several types of estrogen receptors identified. The "classical" receptors, estrogen receptor α (ERα) (Koike, Sakai, & Muramatsu, 1987; Spreafico, Bettini, Pollio, & Maggi, 1992) and estrogen receptor β (ERβ) (Shughrue, Komm, & Merchenthaler, 1996; Shughrue, Lane, & Merchenthaler, 1997), were initially determined to be intracellular, ligand-regulated transcription factors (McKenna, Lanz, & O'Malley, 1999), members of the nuclear receptor superfamily (Evans, 1988; Giguère, Yang, Segui, & Evans, 1988). As a steroid receptor, upon ligand binding, changes in gene expression occur via the dimerized receptor binding to DNA estrogen response elements (EREs) (O'Malley & Tsai, 1992). Both ERα and ERβ are found throughout the brain and periphery in both males and females (Cooke, Nanjappa, Ko, Prins, & Hess, 2017; Heldring et al., 2007). Within the brain, initial *in situ* hybridization studies indicated key estrogen-concentrating regions, including multiple hypothalamic regions (Pfaff & Keiner, 1973; Simerly, Swanson, Chang, & Muramatsu, 1990). In adult female rats these regions include those involved in reproduction, notably the medial preoptic area, the anteroventral periventricular nucleus, the ventromedial nucleus, and the arcuate nucleus (Simerly et al., 1990). These ERs are also found in the cortex, hippocampus, and midbrain (Merchenthaler, Lane, Numan, & Dellovade, 2004; Shughrue et al., 1997).

2.2 Membrane-bound estrogen receptors

ERα- and ERβ-mediated gene expression through the aforementioned signaling mechanism is often termed classical or genomic signaling. However, it has since been determined that many of the genes regulated by estrogens lack ERE sequences (Marino, Galluzzo, & Ascenzi, 2006; Vrtačnik, Ostanek, Mencej-Bedrač, & Marc, 2014). This observation led to the discovery of indirect genomic actions elicited by estrogen through protein-protein interactions with other response elements and other transcription factors (Aranda & Pascual, 2001; Göttlicher, Heck, & Herrlich, 1998). However, even with the discovery of multiple pathways involved in estrogen–induced genomic changes, there were additional unidentified mechanisms by which estradiol could affect cellular change seemingly by acting outside the nucleus.

Historically, while activity through intracellular ERα and ERβ was understood to elicit genomic changes on the timescale of hours to days (at least at the level of experimental measurement), much more rapid actions of estrogens were also observed, ranging within the timescale of seconds to minutes (Kelly, Moss, & Dudley, 1976; Szego & Davis, 1967). And though the rapid actions of estradiol have now been noted for decades, studies at first were met with skepticism. Initial reports from Szego and Davis (1967) noted that following ovariectomy (ovx), acute exogenous estradiol treatment in rats increased cAMP in the uterus to concentrations that were indistinguishable from intact animals. Moreover, this increase occurred within seconds of administration. These findings suggested that estradiol might be initiating signaling cascades that originate at the membrane, rather than solely through nuclear–initiated actions. Building on this, findings of the rapid actions of estradiol in the brain were soon reported. In female preoptic-septal neurons in the hypothalamus, bath application of estradiol immediately modulated neuronal firing rates. These firing rates returned to experimental baseline upon removal of the steroid, again implicating a rapid mechanism for estradiol signaling (Kelly et al., 1976).

Additional work testing the hypothesis that estradiol was acting at the surface membrane of cells has followed these initial reports. One impactful tool was the use of estradiol conjugated to bovine serum albumen (BSA), a large, hydrophilic protein. By conjugating 17β-estradiol to BSA, the steroid is only accessible to the surface membrane. Utilizing this compound in conjunction with a ligand blotting assay, the presence of estrogen-specific membrane binding proteins was found in the hypothalamus of adult female rats, as well as the olfactory bulb and cerebellum (Zheng & Ramirez, 1997), strongly suggesting the presence of an estrogen receptor localized at the membrane. However, skepticism remained, as there was suspicion that estradiol might be cleaved from BSA (Stevis, Deecher, Suhadolnik, Mallis, & Frail, 1999). To further study the effects of membrane versus nuclear ERs, large dendrimer macromolecules conjugated to multiple estrogen molecules were produced (Harrington et al., 2006). These estrogen-dendrimer conjugates used robust chemical linkages to avoid the potential for cleaving, and due to both their size and charge were unable to cross the cellular membrane. As such, these conjugates were precluded from activating nuclear ERs, but still capable of resulting in rapid estradiol signaling (Harrington et al., 2006). Additionally, a modified, membrane-targeted ER, which lacked a nuclear localization sequence, was generated in human breast cancer cells, and compared to wild-type nuclear ERs (Rai, Frolova, Frasor, Carpenter, & Katzenellenbogen, 2005). In the absence of estradiol, immunocytochemistry indicated that the putative membrane ERs (mERs) were in fact localized at the membrane. Application of estradiol resulted in differential downstream effects of the two ERs; estradiol activation of the mER did not stimulate genomic effects, but did stimulate rapid, non–genomic effects (Rai et al., 2005). Overexpression of both ERα and ERβ in Chinese hamster ovary (CHO) cells indicated that at least of portion of these receptors are located at the membrane (Razandi, Pedram, Greene, & Levin, 1999). Estradiol binding to either CHO-ERα or CHO-ERβ resulted in the activation of G-protein signaling cascades, including MAPK and ERK activity. Interestingly, estradiol binding to membrane-ERα (mERα) inhibited c-Jun kinase activity, while estradiol stimulated the kinase when bound to mERβ (Razandi et al., 1999). Importantly, it was found that both membrane-bound and nuclear ERs can be derived from the same transcript, suggesting it is posttranslational modifications that target the receptor to the membrane (see *mER-mGlu Signaling* section below) (Razandi et al., 1999). Though it was found that there were significantly more nuclear receptors, both membrane and nuclear receptors appear to be similar in size and have the same affinity for 17β-estradiol (Razandi et al., 1999). In many regions with high nuclear ER expression, it may be that fewer mERs are necessary, given the robust amplification effects of the recruited second messenger systems (Hammes & Levin, 2007; Levin, 2002).

Around the same time, a novel estrogen receptor was discovered. Specifically, the receptor GPR30 (Filardo, Quinn, Bland, & Frackelton, 2000), now referred to as GPER1 (Prossnitz & Barton, 2011). GPER1 is a G protein–coupled receptor (GPCR), located both within and outside of the nervous system (Filardo et al., 2000; Prossnitz & Barton, 2011). GPCRs are the largest family of receptors in the nervous system (Niswender & Conn, 2010). These membrane-bound receptors possess seven transmembrane domains, and couple with G-proteins to transduce intracellular ligand-initiated signals through many diverse intracellular signaling cascades (Niswender & Conn, 2010). The associated G-protein is comprised of heterotrimeric complex consisting of α, β, and γ subunits. When the receptor complex is inactive, the α subunit is bound to the β/γ complex as well as GDP. Upon activation, guanosine diphosphate (GDP) bound to the receptor is exchanged for guanosine triphosphate (GTP), the α subunit splits from the heterotrimeric complex, and various cell signaling cascades are initiated (Niswender & Conn, 2010). Activation of GPER1 by estradiol may occur through the activation of either $G_{\alpha s}$ or $G_{\alpha i}$, and the subsequent transactivation of the epidermal growth factor receptor (Filardo et al., 2000), resulting in downstream activation of ERK1/2 (Epidermal Growth Factor, Methods and Protocols, 2005), as well as the production of cAMP (Filardo, Quinn, Frackelton, & Bland, 2002; Thomas, Pang, Filardo, & Dong, 2005), which can ultimately regulate transcription (Meyer & Barton, 2009). Interestingly, while this signaling cascade is indicative of a membrane-bound receptor, studies have shown that the majority of GPER1 is found in intracellular locations, notably the endoplasmic reticulum, rather than at the cell membrane (Prossnitz, Arterburn, & Sklar, 2007; Revankar, Cimino, Sklar, Arterburn, & Prossnitz, 2005). General expression patterns of the receptor remain unclear. It appears as though age, species, and sex may affect GPER1 expression patterns (Luo & Liu, 2020). Furthermore, GPER1 demonstrates a much lower binding capacity for estradiol than do the classical receptors, and it remains unclear if this receptor plays a significant role in mediating the actions of estrogen *in vivo* (Luo & Liu, 2020). Significant research will help clarify the location and function(s) of GPER1.

The use of transgenic knockout mice lacking functional ERα and/or ERβ allowed researchers to further explore rapid estradiol signaling in the central nervous system *in vivo* (ÁbrahÁm, Todman, Korach, & Herbison, 2004). In ovx wildtype mice, estradiol administration rapidly (within 15 min) increased the number of cells immunoreactive (−ir) for phosphorylated CREB (pCREB) in a brain region-specific manner. These wildtype mice showed increases in rapid estradiol-induced pCREB-ir in the ventrolateral part of the ventromedial nucleus of the hypothalamus (VMHvl), the medial septum, and the medial preoptic nucleus (MPN). Estradiol did not alter pCREB-ir in the granulate cortex or in the caudate-putamen. In contrast to the wildtype mice, knockout of ERα, ERβ, or both exhibited altered pCREB responses 15 min after estradiol treatment in a brain region-specific manner. In the VMHvl, there was no estradiol-induced change in pCREB in ERα knockout mice, though the response was still present in ERβ knockout mice. The opposite held true in the medial septum, where the increase in pCREB-ir was unaffected in the ERα knockout mice but was absent in mice lacking ERβ. Interestingly, the rapid estradiol-induced increase in pCREB-ir in the MPN requires the expression of both receptors, as mice expressing only one of either ERα or ERβ still exhibited this increase. In mice devoid of both ERα and ERβ, this increase was attenuated (ÁbrahÁm et al., 2004). Taken together, these data strongly indicated that the activation of ERα and/or ERβ is directly responsible for rapidly eliciting changes in a brain region-specific manner, although there was still the unlikely possibility that they were somehow independently responsible for the expression of a distinct membrane estrogen receptor in a brain region-specific manner.

Studies utilizing both immunohistochemistry (Dewing et al., 2007) and co-immunoprecipitation (Dewing, Christensen, Bondar, & Micevych, 2008) have further confirmed that a subpopulation of traditional estrogen receptors are found at the membrane of hypothalamic neurons. Notably, being located at the membrane does not preclude mER effects on gene expression. Estradiol signaling through membrane receptors has been shown to elicit changes in gene expression and protein synthesis through various transcription factors. Of particular importance is estradiol activation of the protein kinase C (PKC)-MAPK signaling cascade, ultimately leading to the phosphorylation of CREB (Boulware et al., 2005; Gu, Rojo, Zee, Yu, & Simerly, 1996; Szego et al., 2006; Wade & Dorsa, 2003). While it was initially proposed that mERα and mERβ may bind directly to G-proteins and operate through GPCR signaling pathways (Razandi et al., 1999). neither the structure of the mERs nor the multitude of cell signaling cascades activated by estradiol support this hypothesis.

Rather, it seems as though these mERs have the ability to transactivate other GPCRs, notably metabotropic glutamate receptors (mGlu).

2.3 Metabotropic glutamate receptors

Metabotropic glutamate receptors are Class C GPCRs (Niswender & Conn, 2010). There are eight distinct mGlu, divided into three subgroups (Group I, II, and III) (Table 1), based upon the G proteins to which they couple, their ligand selectivity, and receptor sequence homology (Niswender & Conn, 2010). Like ionotropic glutamate receptors, mGlu bind glutamate to effect changes in the cell. Unlike ionotropic receptors, however, mGlu activation results in a slower, more long-lasting type of modulation, and contributes to neuronal and synaptic plasticity (Dewing et al., 2007; Gross et al., 2016). As with all GPCRs, ligand–dependent activation of the mGlu results in the dissociation of the G_α subunit from the $G_{\beta/\gamma}$ subunits to recruit various downstream signaling pathways. One significant distinction between the three groups of mGlu is the particular G_α subunit to which they couple; Group I mGlu, mGlu1 and mGlu5, associate primarily with the G_q alpha subunit while Group II and III mGlu are primarily associated with the $G_{i/o}$ G_α proteins (Niswender & Conn, 2010).

In Group 1 mGlu (mGlu1 and mGlu5), found post-synaptically along with some glial expression, activation of the receptor results in G_q subunit activation of phospholipase C (PLC). PLC cleaves phosphatidylinositol 4,5-bisphosphate (PIP_2) into diacylglycerol (DAG) and inositol 1,4,5-triphosphate (IP_3), which lead to the activation of protein kinase C (PKC) and the release of intracellular calcium, respectively (Niswender & Conn, 2010). Along with this classical pathway, Group I mGlu can activate other

Table 1 Metabotropic glutamate receptor types by group and their associated G_a protein.

Group	Receptor	G_a protein
I	mGlu1 mGlu5	G_q
II	mGlu2 mGlu3	$G_{i/o}$
III	mGlu4 mGlu6 mGlu7 mGlu8	$G_{i/o}$

downstream signaling cascades, including the MAPK/ERK and mTOR/p70 S6 kinase pathways, important in synaptic plasticity (Niswender & Conn, 2010; Page et al., 2006). Importantly, Group I mGlu activity has been shown to phosphorylate CREB through the same pathway as estradiol (Choe & Wang, 2001; Warwick, Nahorski, & Challiss, 2005). Group I signaling often results in long-lasting modulation of both long-term potentiation (Anwyl, 2009) and long-term depression (Lüscher & Huber, 2010) through neuronal and synaptic modulation. These two Group I mGlu are found throughout the nervous system and are often found within the same region, including within the hippocampus, striatum, and hypothalamus (Boulware et al., 2005; Dewing et al., 2007; Gross et al., 2016; Huang & Woolley, 2012). Interestingly, while these two mGlu are similar in structure, expression location, and downstream signaling cascades, their activation can result in dissimilar, and sometimes opposing, molecular, cellular, and even behavioral effects (Gross et al., 2016).

Group II mGlu (mGlu2, mGlu3) are largely found pre-synaptically, with some post-synaptic and glial expression, throughout the nervous system, including, like the Group I receptors, in the hippocampus (Meitzen & Mermelstein, 2011), striatum (Grove-Strawser, Boulware, & Mermelstein, 2010), and hypothalamus (Wong et al., 2019). Unlike Group I receptors, however, Group II receptors are primarily associated with the $G_{i/o}$ G_α proteins (Niswender & Conn, 2010). When activated, the $G_{i/o}$ proteins inhibit adenylyl cyclase (AC) and subsequently the production of cAMP. This reduces protein kinase A (PKA) signaling, which can ultimately affect amongst other targets, both voltage-gated calcium and potassium channels. These mGlu can also function as auto-receptors at glutamate synapses to reduce glutamate release (Cartmell & Schoepp, 2000), and heteroreceptors at GABA synapses to modulate GABA release (Hirono, Yoshioka, & Konishi, 2001; Hoffpauir & Gleason, 2002). The pre-synaptic activity of Group II receptors is important in long-term depression at excitatory synapses (Kahn, Alonso, Robbe, Bockaert, & Manzoni, 2001; Nicholls et al., 2006).

Finally, Group III mGlu (mGlu4, mGlu6, mGlu7, mGlu8) act in much the same way as Group II mGlu. These receptors associate with $G_{i/o}$ proteins, which leads to downstream inhibition of AC (Niswender & Conn, 2010). However, the role of group III mGlu has received considerably less attention than group I and group II receptors, and have not been implicated in estradiol-mediated activation of second messenger signaling and behavior.

2.4 mER-mGlu signaling

Membrane ER (mER) signaling through mGlu has been found throughout the brain (Boulware et al., 2005; Boulware, Kordasiewicz, & Mermelstein, 2007; Chaban, Li, McDonald, Rapkin, & Micevych, 2011; Christensen & Micevych, 2012; Dewing et al., 2007; Grove-Strawser et al., 2010; Kuo, Hamid, Bondar, Prossnitz, & Micevych, 2010; Mahavongtrakul, Kanjiya, Maciel, Kanjiya, & Sinchak, 2013; Spampinato et al., 2012). Structural interactions are supported by immunohistochemical experiments indicating colocalization of mERα and mGlu1a in addition to co-immunoprecipitation of the two receptors in membrane fractions (Dewing et al., 2007, 2008). As mGlu are functionally coupled to multiple downstream effectors, membrane ER activation of mGlu has been shown to activate diverse cell signaling pathways, often leading to distinct outcomes.

Estradiol activation of group I mGlu through mERα results in CREB phosphorylation, providing the putative mechanism by which estradiol was originally observed activate this transcription factor (Choe & Wang, 2001; Warwick et al., 2005) through activation of PLC, PKC, and the release of intracellular calcium (Boulware et al., 2007; Kuo, Hariri, Bondar, Ogi, & Micevych, 2009; Wong, Abrams, & Micevych, 2015). In the initial characterization, cultured female hippocampal neurons, estradiol activation of mGlu1a through ERα was shown to elicit MAPK-dependent CREB phosphorylation in the nucleus within seconds of application, an effect that was blocked by MAPK/ERK kinase inhibitors (Boulware et al., 2005). This estradiol-induced increase of CREB phosphorylation via MAPK signaling was not seen in age-matched male hippocampal cells, suggesting this particular pathway is sex-specific. Furthermore, this same study found that estradiol is also capable of activating group II mGlu. Estradiol activation of mGlu2/3, through either ERα or ERβ, decreased the concentration of cAMP within the cell through Gi/o signaling. Furthermore, activation of mGlu2/3 was sufficient to decrease protein kinase A (PKA) activity, resulting in the dephosphorylation of L-type calcium channels, ultimately reducing CREB phosphorylation mediated through membrane depolarization (Boulware et al., 2005).

While the specific mechanisms involved in the ER engagement of mGlu remain to be fully elucidated, it is hypothesized that activation occurs through a direct protein-protein interaction (Micevych & Mermelstein, 2008). Regardless of specifics, the interaction between mERs with mGlu requires caveolin proteins (Boulware et al., 2007; Christensen & Micevych, 2012; Razandi, Oh, Pedram, Schnitzer, & Levin, 2002).

Caveolins (CAVs) are a family of scaffolding proteins found throughout the body and are involved in trafficking receptors to the cell membrane (Anderson, 1998). There are three isoforms, CAV1–3 (Krajewska & Masłowska, 2004; Williams & Lisanti, 2004). Outside of the nervous system it was shown that ERα interacts with CAV1 (Razandi et al., 2002; Schlegel, Wang, Katzenellenbogen, Pestell, & Lisanti, 1999; Schlegel, Wang, Pestell, & Lisanti, 2001), and that this pairing was necessary to track ERα to the membrane (Razandi et al., 2002).

In the early 2000s, it was determined that caveolins were expressed in the brain (Cameron, Ruffin, Bollag, Rasmussen, & Cameron, 1997; Galbiati et al., 2001; Ikezu et al., 1998; Mikol, Hong, Cheng, & Feldman, 1999), though its relevance was less clear. Boulware et al. (2007) identified all three CAV isoforms in hippocampal tissue. To understand if CAVs associated with mERs in the hippocampus, cultured neurons were transfected with ERα DNA in which the receptor expressed a single point mutation that disrupts ERα interactions with CAV1 (Boulware et al., 2007). Outside the nervous system this mutation decreases ERα located at the membrane (Razandi et al., 2002). In these transfected hippocampal neurons, application of estradiol failed to increase CREB phosphorylation. Furthermore, transfecting hippocampal neurons with a dominant-negative form of CAV1, or reducing CAV1 expression through siRNA knockdown, was also sufficient to preclude estradiol-induced CREB phosphorylation. However, activation of mGlu1a alone still resulted in phosphorylation of CREB (Boulware et al., 2007), indicating the necessity of CAV1 for ERα interactions with mGlu1a. CAV1 mediates ERα interactions with group I mGlu in the arcuate nucleus of the hypothalamus (ARH) as well (Christensen & Micevych, 2012). Knockdown of CAV1 in the hypothalamus of female rats prevented ERα from being located at the membrane, without affecting the levels of intracellular ERα. In ovx female rats, siRNA knockdown of CAV1 was also sufficient to prevent the estradiol-dependent display of sexual receptivity behavior (see Section 3; Christensen & Micevych, 2012).

Interestingly, none of the ERα-CAV1 manipulations in hippocampal cell culture affected L-type calcium channel-dependent CREB phosphorylation, demonstrating that CAV1 was not involved in ERα-mGlu2/3 signaling (Boulware et al., 2007). In all cases (ERα point mutation, dominant negative CAV1, and siRNA CAV1 knockdown), the estradiol-induced L-type calcium channel-dependent decrease in CREB phosphorylation remained intact (Boulware et al., 2007). Outside of the nervous system,

L-type calcium channels are associated with CAV3 (Balijepalli, Foell, Hall, Hell, & Kamp, 2006; Couchoux, Allard, Legrand, Jacquemond, & Berthier, 2007), and because the estradiol activation of mGlu2/3 in hippocampal neurons affects these channels, the role of CAV3 was next explored. Similar to manipulations with CAV1, a single point mutation in CAV3 was sufficient to result in a dominant-negative variant. In neurons expressing this mutation, estradiol no longer affected L-type calcium channel-dependent CREB phosphorylation. However, estradiol-induced CREB phosphorylation through mGlu1 was unaffected (Boulware et al., 2007). In the absence of functional CAV3, application of an mGlu2/3 agonist was still sufficient to result in attenuated L-type calcium channel CREB phosphorylation, suggesting that CAV3 is associated with ER signaling (both ERα and ERβ) through group II mGlu. Finally, reducing CAV3 expression via siRNA knockdown attenuated the effect of estradiol on L-type calcium channel-mediated phosphorylation of CREB (Boulware et al., 2007). Taken together, results from these two studies indicate that signaling through the particular ER–mGlu pairing is mediated through the CAV isoform associated with the ER; CAV1 mediates ERα interactions with group I mGlu, and CAV3 is involved with mERs pairing with group II mGlu (Fig. 1).

While the experiments regarding CAV function revealed that these proteins are important for clustering the receptor at the membrane, the precise mechanism of action linking ERs to their associated mGlu remained unclear. Palmitoylation, a reversible, post-transcriptional modification involved in the trafficking and functioning of proteins both within and outside of the nervous system (Fukata & Fukata, 2010), was an appealing mechanism. Outside of the nervous system, palmitoylation of ERα via palmitoyl acyltransferase proteins DHHC-7 and DHHC-21 is crucial in trafficking the receptor to the membrane (Acconcia et al., 2005; La Rosa, Pesiri, Leclercq, Marino, & Acconcia, 2012; Pedram, Razandi, Deschenes, & Levin, 2012). In hippocampal cells taken from female rat pups, global blockade of palmitoylation, through a palmitoylation inhibitor, eliminated both the estradiol-induced CREB phosphorylation through the mGlu1 pathway, and the estradiol-induced attenuation of L-type calcium channel-dependent phosphorylation of CREB through mGlu2 (Meitzen et al., 2013). To determine which protein(s) needed to be palmitoylated for estradiol-induced mGlu signaling, neurons in culture were transfected with DNA to overexpress ERα where the cysteine palmitoylation site was eliminated.

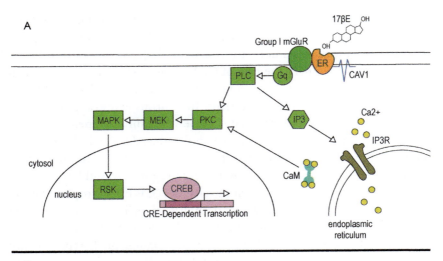

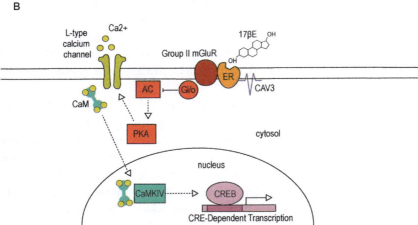

Fig. 1 17β-Estradiol (17βE) binds to membrane-bound estrogen receptors (ER) to activate distinct signaling pathways via Group I (Panel A) or Group II (Panel B) mGlu. (A) ER interactions with group I mGlu, dependent on CAV1, activates Gq-mediated signaling through protein lipase C (PLC) and protein kinase C (PKC), activation of MAPK/ERK, RSK, and ultimately the phosphorylation of CREB. PLC also activates IP3, resulting in the release of intracellular calcium. (B) ER activation of Group II mGlu, dependent on CAV3, results in the Gi/o-mediated inhibition of adenylyl cyclase (AC), decreasing activity (indicated by dashed lines) of protein kinase A (PKA), resulting in reduced L-type calcium channel currents, and decreases in L-type calcium channel-dependent CREB phosphorylation. Ca^{2+}, calcium; CaM, Calmodulin; CaMKIV, calmodulin-dependent protein kinase IV; CAV, caveolin; IP3, inositol triphosphate; IP3R, IP3 receptor; MEK, MAPK/ERK kinase; MAPK, mitogen-activated protein kinase; RSK, ribosomal S6 kinase.

In these experiments, estradiol was no longer able to result in the phosphorylation of CREB. However, signaling through ERβ was unaffected. Overexpressing the ERβ variant was sufficient to attenuate estradiol effects on L-type calcium channels, without affecting ERα-mediated CREB phosphorylation (Meitzen et al., 2013). Using siRNA, the roles of palmitoyl acyltransferase proteins DHHC-7 and DHHC-21 were next investigated. Knockdown of either DHHC-7 or DHHC-21 eliminated CREB phosphorylation by estradiol activation of both ERα and ERβ, without affecting mGlu signaling or CAV expression (Meitzen et al., 2013). However, caveolin proteins have also been shown to be palmitoylated (Dietzen, Hastings, & Lublin, 1995). In hippocampal cells taken from female rat pups, siRNA knockdown of both DHHC-7 and DHHC-21, but not DHHC-7 alone, was sufficient to decrease palmitoylation of CAV1 and to prevent its association with ERα (Eisinger, Larson, Boulware, Thomas, & Mermelstein, 2018; Eisinger et al., 2018). All together, these data indicate that palmitoylation is an important part of mER–mGlu signaling, involved in both the association of mERs with CAV proteins and the coupling of mERs with mGlu.

While both the CAV isoform and the palmitoylation of ER and CAV1 are crucial components of estradiol-induced mER–mGlu signaling, these signaling cascades also depend on the brain region in which they are expressed, and the synaptic location of the signaling complex. In the hippocampus, estradiol activation results in two distinct signaling pathways, activating either group I mGlu (mGlu1 and mGlu5) or group II mGlu (mGlu2 and mGlu3) (Boulware et al., 2007). While the same group I and group II mGlu are present in the female striatum, estradiol here utilizes different mER–mGlu signaling cascades (Grove-Strawser et al., 2010). Though activation of mGlu1a alone in cultured striatal neurons was sufficient to increase CREB phosphorylation, application of an mGlu1a antagonist did not prevent the rapid estradiol-induced phosphorylation of CREB. Rather, inhibiting the activity of mGlu5 was sufficient to attenuate this phosphorylation, indicating that in the striatum, estradiol increases CREB phosphorylation through ERα transactivation of mGlu5. These results held true *in vivo*, when adult ovx female rats were given an injection of an mGlu5 antagonist preceding a bolus of estradiol. Alone, estradiol was sufficient to increase pCREB in the striatum. In the presence of the mGlu5 antagonists, this increase in pCREB was abolished (Grove-Strawser et al., 2010). The estradiol activation of group II mGlu in the striatum is distinct from that in the hippocampus, as well. Like in the hippocampus, estradiol acting

Table 2 Interaction and outcome of mER-mGlu signaling by region.

Region	mGlu group	mGlu receptor	Estrogen receptor	Estradiol effect	CAV
Hippocampus	Group I	mGlu1a	ERα	Increase in CREB phosphorylation	CAV1
Hippocampus	Group II	mGlu2	ERα, ERβ	Reduction in L-type Ca^{2+} channel CREB phosphorylation	CAV3
Striatum	Group I	mGlu5	ERα	Increase in CREB phosphorylation	CAV1
Striatum	Group II	mGlu3	ERα, ERβ	Reduction in L-type Ca^{2+} channel CREB phosphorylation	CAV3

through either ERα or ERβ is sufficient to attenuate the L-type calcium channel-mediated phosphorylation of CREB. However, only siRNA knockdown of mGlu3, and not mGlu2, was sufficient to attenuate this (Grove-Strawser et al., 2010). That is, in the striatum, estradiol activates an mER-mGlu3 signaling cascade, as opposed to the now presumed mER-mGlu2 cascade seen in the hippocampus (Table 2). Though mRNA expression of group II receptors is low in the striatum, physiological targeting in cultured neurons appears to indicate that activity via these receptors is sufficient to affect an outcome, likely driven, in part, by signal amplification processes intrinsic to second messenger signaling cascades.

3. Estrogen in motivated behaviors: Reproduction

In females, the interaction of ERs with mGlu has enormous functional consequences in driving motivated behaviors. Female reproduction and reproductive behavior are examples of motivated behaviors that are mediated by this interaction. The release of estrogen and progesterone from the ovaries is crucial for female reproduction, with many reproductive processes orchestrated by hormone-sensitive circuits within the brain. Both ovulation, the central event in female reproduction, and lordosis, a stereotypical female reproductive behavior seen in many species, occur through the coordination of multiple neuroanatomical circuits, with these processes dependent on mER-mGlu signaling.

3.1 Luteinizing hormone surge

Ovulation is the central event in female reproduction, initiated by the surge release of luteinizing hormone (LH) from the anterior pituitary. Ovulation is controlled by a network of neurons in the hypothalamus that act as a pattern generator, releasing gonadotropin releasing hormone (GnRH) onto LH neurons in the anterior pituitary in small, rhythmic pulses (Herbison, 2018; Kauffman & Smith, 2013). Throughout most of the estrous cycle, low levels of circulating estradiol exert negative feedback on the hypothalamic-pituitary-gonadal (HPG) axis, maintaining a low level of activity in GnRH neurons in the hypothalamus. Simultaneously, the release of gonadotrophs from the anterior pituitary is also low. Rising estradiol concentrations from ovarian release trigger a switch from an estrogen-negative to an estrogen-positive feedback loop. This positive feedback loop, which is unique to females, is crucial in the surge release of LH that ultimately triggers ovulation (Wintermantel et al., 2006). The preovulatory rise in circulating estradiol results in a sharp increase in GnRH neuronal activity and the release of LH from the pituitary to elicit ovulation (Ferin, Tempone, Zimmering, & Wiele, 1969; Labhsetwar, 1970) (Fig. 2).

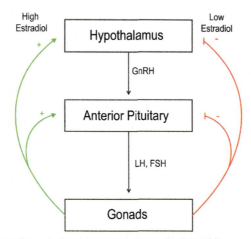

Fig. 2 The hypothalamic-pituitary-gondal (HPG) axis. Gonadotropin releasing hormone (GnRH) is released from the hypothalamus into the hypophyseal portal system where it reaches the anterior pituitary to stimulate the release of both luteinizing hormone (LH) and follicle stimulating hormone (FSH) into the general circulation. LH and FSH reach the gonads, ovaries in females, where they act on receptors. Estradiol released by the ovaries is part of both a negative and positive feedback cycle. When circulating estradiol levels are low, negative feedback is exerted on the HPG axis. As circulating levels rise throughout the estrous/menstrual cycle, estradiol begins to exert positive feedback, increasing the pulsatile release of GnRH and resulting in the characteristic LH surge that ultimately elicits ovulation.

While GnRH neurons do not express ERα or nuclear progesterone receptors (Herbison & Theodosis, 1992; Shivers, Harlan, Morrell, & Pfaff, 1983; Wintermantel et al., 2006), kisspeptin neurons that are upstream regulators of GnRH signaling (Irwig et al., 2005; Messager et al., 2005) express the necessary steroid receptors (Clarkson, d'Anglemont de Tassigny, Moreno, Colledge, & Herbison, 2008; Navarro et al., 2009; Wakabayashi et al., 2010). Two key populations of kisspeptin neurons important in the regulation of the HPG axis are found in the hypothalamus. Those in the arcuate nucleus (ARH), also co-expressing neurokinin B and dynorphin (Goodman et al., 2007) and so-termed KNDy neurons (Cheng, Coolen, Padmanabhan, Goodman, & Lehman, 2010), are important in estradiol-negative feedback, while those in the anteroventral periventricular nucleus (AVPV) are involved in estradiol-positive feedback (Clarkson et al., 2008; Kauffman & Smith, 2013; Mohr et al., 2021).

Classically, it has been understood that both estradiol and progesterone released from the ovaries orchestrate the LH surge. While blocking either progesterone receptors or progesterone synthesis prevents the LH surge and attenuates the estrous cycle, removing both the ovaries and the adrenals (another source of peripheral progesterone) blunts, but does not completely inhibit, the estradiol-induced LH surge (Mann, Korowitz, & Barraclough, 1975; Mann & Barraclough, 1973; Micevych, Rissman, Gustafsson, & Sinchak, 2003; Micevych et al., 2003), suggesting that there must be another source of progesterone. Furthermore, levels of progesterone in circulation prior to the LH surge are quite low (Feder, Brown-Grant, & Corker, 1971; Kalra & Kalra, 1974; Smith, Freeman, & Neill, 1975), suggesting that pre-LH surge progesterone is not synthesized in the periphery. Rather, it has become apparent that progesterone is also synthesized *de novo* in the brain (Kuo et al., 2010; Micevych et al., 2007; Sinchak et al., 2003), and that it is this neuroprogesterone (neuroP) that is vital in the LH surge that ultimately leads to ovulation (Micevych & Sinchak, 2011).

3.2 mER-mGlu signaling in the luteinizing hormone surge

The role of progesterone in ovulation is important. Recently, it has determined that neuroP plays a critical role in regulating the estrogen-positive feedback that leads ovulation. Specifically, by activating progesterone receptors on kisspeptin neurons in the AVPV, neuroP promotes release of kisspeptin onto GnRH neurons, producing the LH surge that drives ovulation (Delhousay, Chuon, Mittleman-Smith, Micevych, & Sinchak, 2019;

Micevych, Rissman, et al., 2003; Micevych, Sinchak, et al., 2003; Sinchak, Mohr, & Micevych, 2020). While female rats without ovaries or adrenals exhibit a blunted estradiol-induced LH surge (Mann et al., 1975; Mann & Barraclough, 1973;Micevych, Rissman, et al., 2003; Micevych, Sinchak, et al., 2003), complete inhibition only occurs following inhibition of neuroP synthesis, via jugular vein administration of the 3β-HSD inhibitor trilostane (Micevych, Rissman, et al., 2003; Micevych, Sinchak, et al., 2003). A wide range of cell types within the central nervous system are capable of steroidogenesis, including astrocytes that can synthesize and release progesterone (Zwain & Yen, 1999). Primary astrocyte cultures treated with 17β-estradiol showed an estradiol-dependent increase in progesterone concentration (Sinchak et al., 2003), suggesting that it is these cells that are involved in initiating the LH surge.

Within the hypothalamus, astrocytes regulate releasing factors as well as synthesize neurosteroids (Jung-Testas et al., 1999; Mahesh, Dhandapani, & Brann, 2006; Zwain & Yen, 1999), and the role of astrocytes in the generation of the LH surge has recently begun to be understood. Hypothalamic astrocytes express ERα and ERβ, both on the membrane and intracellularly (Bondar, Kuo, Hamid, & Micevych, 2009; Chaban, Lakhter, & Micevych, 2004; Garcia-Segura, Naftolin, Hutchison, Azcoitia, & Chowen, 1999; Pawlak, Karolczak, Krust, Chambon, & Beyer, 2005; Quesada, Romeo, & Micevych, 2007). In these cells, estradiol acting through mERs results in the rapid release of intracellular calcium ($[Ca^{2+}]_i$) stores (Chaban et al., 2004), an effect that can blocked by either the ER antagonist ICI 182,780, or by the mGlu1a antagonist LY 367385 (Kuo et al., 2009), implicating estradiol signaling through the recruitment of mGlu1a. Furthermore, activation of mGlu1a on its own is also sufficient to affect $[Ca^{2+}]_i$ release (Kuo et al., 2009). Though estradiol is sufficient to activate mGlu1a without the presence of glutamate in astrocytes, when mGlu1a activity is elicited concurrently with estradiol, as with a selective agonist, the estradiol-induced response is augmented (Kuo et al., 2009, 2010), suggesting that the presence of both estradiol and glutamate allows for the appropriate physiological release of intracellular calcium stores though a cooperative pathway.

Coimmunoprecipitation studies further confirmed the interaction of mERα and mGlu1a at the level of the membrane in these astrocytes (Kuo et al., 2009). The activation of the mERα-mGlu1a signaling cascade in hypothalamic astrocytes resulting in the increase of $[Ca^{2+}]_i$ release occurs via the production of inositol triphosphate (IP$_3$), (Chen, Kuo, Wong, & Micevych, 2014; Kuo et al., 2009). The increase in $[Ca^{2+}]_i$ further activates a Ca^{2+}-sensitive adenylyl cyclase (AC-1), leading to the production of cAMP, and activation of PKA. While estradiol activation in neurons requires PKC signaling, in astrocytes this role is played by PKA. The phosphorylation of PKA further activates steroid acute regulatory protein (StAR) and translocator protein (TSPO) to mediate cholesterol transport and ultimately allow for steroidogenesis (Chen et al., 2014) and the synthesis of neuroP (Kuo et al., 2009; Micevych et al., 2007; Micevych & Sinchak, 2011; Sinchak et al., 2003). In the absence of estradiol, activating mGlu1a alone is sufficient to result in the release of $[Ca^{2+}]_i$ (Kuo et al., 2009) and subsequent neuroP release, while inhibiting any part of the cell signaling cascade initiated by the ERα activation of mGlu1a in hypothalamic astrocytes (e.g., IP_3 receptors, AC-1 activity, PKA activity) is sufficient to inhibit neuroP synthesis (Chen et al., 2014; Kuo et al., 2009, 2010). Additionally, within hypothalamic astrocytes, like neurons, trafficking and insertion of mERα at the membrane is dependent upon both estradiol and mGlu1a; blocking mGlu1a activity prevents insertion of mERα into the membrane, even in the presence of estradiol (Bondar et al., 2009). Ultimately, it is the signaling cascade initiated by the activation of the mER-mGlu complex in these hypothalamic astrocytes that allows for the synthesis of neuroP, which is vital for the LH surge and ultimately ovulation (Micevych & Sinchak, 2011).

3.3 Lordosis

Perhaps the most obvious physical display of reproductive behavior in females is lordosis. Lordosis is a reflexive behavior characterized by the arching of the spine, the raising of the head and hindquarters, and the lifting of the tail to allow for intromission by a male (Beach, 1976; Pfaff et al., 2002). The release of ovarian hormones regulates the display of this receptive behavior through the priming neural circuits and the integration of appropriate external sensory cues. Within the central nervous system, estrogen and progesterone work both in concert to result in a specific and measurable time course of the expression of the behavior (Sinchak & Micevych, 2001) through activation of intracellular ERs as well as mER-mGlu (Christensen, Dewing, & Micevych, 2011; Micevych, Rissman, et al., 2003; Micevych, Sinchak, et al., 2003). These hormonal cues are integrated with external sensory cues to allow the behavior to occur at a time in the estrous cycle that maximizes the potential for reproductive success.

A core circuit in the production of lordosis is found within the hypothalamus. Here, the connections between the arcuate nucleus (ARH), the medial preoptic nucleus (MPN), and the ventromedial nucleus of the hypothalamus (VMH) have been shown to be fundamental in the expression of lordosis (Christensen et al., 2012; Johnson, Hong, & Micevych, 2020; Micevych & Meisel, 2017; Mills et al., 2004; Sinchak et al., 2013; Sinchak & Micevych, 2003) (Fig. 3). Within this circuit, estradiol first acts on ERα-containing neuropeptide Y (NPY) neurons in the ARH (Dewing et al., 2007; Mills et al., 2004; Sar, Sahu, Crowley, & Kalra, 1990), resulting in the release of NPY onto NPY-Y1 receptors in proopiomelanocortin (POMC) neurons, also found within the ARH. It is a subset of these POMC neurons that project to the MPN and release β-endorphin (β-End) onto μ-opioid receptors (MORs) in the MPN. When activated, MORs are internalized within the cell. Though activation and internalization are separate events, internalization is a sufficient marker of opioid receptor activation (Allen et al., 1997; Micevych, Eckersell, Brecha, & Holland, 1997; Trafton, Abbadie, Marek, & Basbaum, 2000). Additionally, the release of β-End from POMC neurons through the use

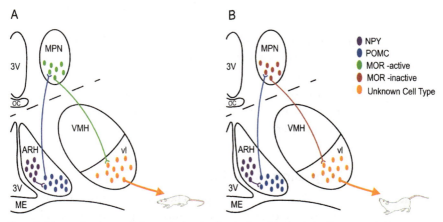

Fig. 3 Neurocircuitry of hypothalamic lordosis circuit. Initially estradiol acts via mER-mGlu1a on NPY neurons in the arcuate (ARH), which further activate POMC/beta-endorphin neurons also in the ARH. From here, projections reach the medial preoptic nucleus (MPN) to activate neurons containing the mu opioid receptor (MOR), which further project to the ventrolateral portion (vl) of the ventromedial nucleus of the hypothalamus (VMH). The VMHvl is the final integrative hub in the hypothalamus. When active (A), MOR neurons ultimately inhibit the expression of lordosis. Following progesterone inactivation of MOR neurons (B), lordosis is able to be expressed. Above the dashed line indicates rostral regions of the hypothalamus, below the dashed line indicates caudal regions. 3V, third ventricle; ME, median eminence; OC, optic chiasm.

of optogenetics has been shown to induce internalization of the MOR and prohibit the display of lordosis in an otherwise receptive mouse (Johnson et al., 2020).

The activation/internalization of these MORs is a key component in the estradiol-induced stimulation of this circuit, and is specifically dependent on ERα (Micevych, Rissman, et al., 2003; Micevych, Sinchak, et al., 2003). Transgenic mice devoid of ERα did not exhibit estradiol-induced MOR internalization in hypothalamic circuits in response to estradiol treatment, while both wild-type mice and mice lacking ERβ displayed estradiol-induced MOR internalization in these regions (Micevych, Rissman, et al., 2003; Micevych, Sinchak, et al., 2003). Throughout the estrous cycle the internalization of the receptor is out of phase with the expression of sexual receptivity (Sinchak & Micevych, 2003). That is, when MOR is internalized, lordosis is prohibited until progesterone levels increase to a level sufficient to inactive the receptors (Sinchak & Micevych, 2001), ultimately allowing the behavior to be expressed. Following ovariectomy (ovx) and hormone replacement, rats and mice remain unreceptive 30–48 h following an initial administration of estradiol (delivered as estradiol benzoate (EB)) if progesterone is delivered around the 26–44-h mark before behavioral testing (Micevych & Sinchak, 2013; Sinchak & Micevych, 2001). Alternatively, in rats, 6 days of EB alone is sufficient to induce lordosis, through a progesterone-independent manner (Micevych & Sinchak, 2013). While many long-term changes occur to result in the production of lordosis, much of the machinery involved is fast-acting, utilizing fast mERα-mGlu signaling interactions.

3.4 mER-mGlu signaling in lordosis

A subset of these ARH NPY neurons express both mERα and mGlu1a, and both immunohistochemical data and coimmunoprecipitation of membrane fractions have indicated an interaction between these two receptors (Dewing et al., 2007, 2008). The internalization of MOR in the MPN has been shown to be dependent upon mERα signaling in the ARH. The administration of membrane-impermeable E6-biotin was sufficient to induce MOR internalization in the MPN, an effect that was blocked by the ER antagonist ICI 182,780. Furthermore, antagonizing mGlu1a activity in the ARH of adult, ovx female before estradiol treatment attenuates both the internalization of MOR in the MPN and the subsequent expression of lordosis. Activation of ARH mGlu1a alone, in the absence of estradiol, was sufficient to internalize MPN MOR to the same degree as estradiol alone, or estradiol along with an mGlu1a agonist (Dewing et al., 2007), suggesting that the estradiol-dependent internalization of MORs in the MPN occurs through mERα-mGlu1a signaling.

This internalization also appears to depend at least in part through PKC signaling. Infusion of BIS, a general PKC inhibitor, directly into the ARH of adult, ovx female rats 30 min before estradiol treatment attenuated both the internalization of MOR in the MPN and the display of lordosis. Inducing PKC signaling in the ARH through direct infusion of a PKC agonist was sufficient to facilitate lordosis in the absence of estradiol (Dewing et al., 2008). To determine if this requisite PKC signaling was downstream of mGlu1a, an mGlu1a agonist was infused directly into the ARH following infusion of BIS. Activation of mGlu1a without PKC signaling was unable to internalize MOR. However, inhibition of mGlu1a signaling via an antagonist with subsequent agonist-induced PKC signaling was capable of internalizing the MORs in the MPN (Dewing et al., 2008). Taken together, estradiol signaling through the mERα-mGlu1a activates PKC signaling, which is necessary for the estradiol-induced activation/internalization of MOR and the subsequent display of lordosis. As in hippocampal culture, ERα in the ARH is clustered at the membrane via CAV1, where the scaffolding protein permits the interaction with mGlu1a (Christensen & Micevych, 2012). Disrupting this process, through siRNA knockdown of CAV1, is sufficient to reduce mERα levels, without affecting intracellular ERs. This loss of mERα is sufficient to attenuate the estradiol-induced internalization of MOR, and the ensuing display of lordosis in adult female ovx, steroid-primed rats (Christensen & Micevych, 2012).

An important consequence of mER-mGlu signaling is morphological adaptation. Throughout the brain, estradiol has been shown to induce neuronal changes, including the generation and/or pruning of dendrites and dendritic spines (Calizo & Flanagan-Cato, 2000; Christensen et al., 2011; Garcia-Segura, Baetens, & Naftolin, 1986; Gould, Woolley, Frankfurt, & McEwen, 1990; Gross & Mermelstein, 2020; Matsumoto & Arai, 1981; Meisel & Luttrell, 1990; Woolley & McEwen, 1992). These changes can be induced rapidly through mERα-mGlu signaling. In the ARH, estradiol activation of mGlu1a via mERα in adult female rats has been shown to substantially increase the number of dendritic spines within 4 h, and these spines persist for at least 48 h. While the newly formed spines presented an immature filopodia morphology initially, 20 h after formation there was in increase in the number of mushroom-shaped spines, indicating functional synapses. The time course of these morphological changes follows that of lordosis. Dendritic spine remodeling relies in part on actin, and the activity of the actin isoform of β-actin has been shown to be involved in the estradiol-dependent change in ARH spine density (Christensen et al., 2011). Inhibiting β-actin polymerization in adult, ovx female rats 1 h before estradiol treatment significantly decreased β-actin density and attenuated the display of lordosis, indicating that estradiol-induced β-actin polymerization is involved in female reproductive behavior. Furthermore, cofilin, an actin-binding protein, must be deactivated via phosphorylation in order for new spines to form (Bamburg, 1999). Estradiol treatment increases the levels of phosphorylated cofilin in the ARH of adult, ovx female rats, an effect that can be blocked by administration of an mGlu1a antagonist (Christensen et al., 2011). These data indicate that inhibition of mGlu1a activity in the ARH is sufficient to prevent rapid estradiol-induced spinogenesis, as well as the display of lordosis (Christensen et al., 2011; Dewing et al., 2007).

3.5 Limbic circuits in reproduction

Projections from hypothalamic nuclei robustly innervate the limbic system, and these connections are involved in both reward and incentive salience. Limbic-hypothalamic circuitry is important in the motivation to engage in reproductive behaviors in both sexes (Meisel & Mullins, 2006; Micevych & Meisel, 2017). Though the long-term consequence of reproduction is the survival of the species through the production of offspring, the short-term motivation of reproduction is typically driven not with the continuation of the species in mind, but rather through the immediate drive, or motivation, for the rewarding aspects of the behavior. While the hypothalamic circuit is a crucial component in the physiological regulation of reproduction and the lordosis reflex, there is also an aspect of "wanting" that drives the production of behavior, in females as much as in males (Meisel & Mullins, 2006).

The reproductive limbic circuit consists of the MPN, the ventral tegmental area (VTA), and the nucleus accumbens (NAc), divided into core (NAcC) and shell (NAcSh) subregions. A key node connecting the hypothalamus to the limbic component is the medial preoptic area (MPOA), including the MPN. Projections from the MPOA reciprocally innervate the mesolimbic dopamine (DA) system, including the VTA (Simerly & Swanson, 1988) (Fig. 4). In reward modulation, the VTA plays an important role, and heavily innervates the NAc (Kokane & Perrotti, 2020). In females,

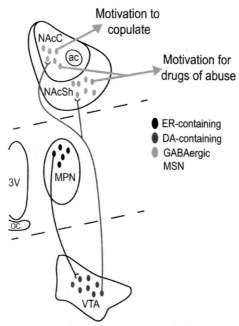

Fig. 4 Inputs to the nucleus accumbens core (NAcC) and shell (NAcSh) subregions mediate different forms of motivation. Estrogen receptor (ER)-containing neurons in the medial preoptic nucleus (MPN) project to the ventral tegmental area (VTA) to synapse onto dopamine (DA)-containing neurons. In females, the cyclic fluctuations in estradiol impact DA signaling. Neurons from the VTA project forward to the NAcC and NAcSh where they are integrated with other signals to impact motivated behaviors, including the drive for reproduction and seeking drugs of abuse.

the majority of the projections from the MPOA to the VTA arise from cells that contain ERs, and estradiol in the MPOA has been shown to mediate the release of DA from VTA neurons that innervate the NAc (Tobiansky et al., 2016). Dopamine has long been known to play a role in the neural control of motor function, but is also important in reward salience, learning, and motivation. Several hypotheses regarding DA function in motivation have been put forward, and it has come to be accepted that this is an important neurotransmitter in pursuing goal-directed motivated behaviors (Berridge, 1996; Wise, 2004). Importantly, sex differences in DA mechanisms mediating reward have been noted, including in the motivation for reproduction (Yoest et al., 2015).

While most work studying female reproductive behavior has focused on the lordosis reflex, pre-copulatory behaviors from the female, such as hopping or darting indicate at least some level of control over the mating process, since copulation cannot occur until the female stops her motion and permits mounting (Erskine, 1989; McClintock & Adler, 1978). This activity, which is called "pacing," is done to pace the rate of male intromissions to a rate which is optimal for the female to become pregnant (Adler, 1978), and ultimately elicits a DA response in the NAc in response to mating (Becker, Rudick, & Jenkins, 2001; Bradley & Meisel, 2001; Guarraci, Megroz, & Clark, 2004; Jenkins & Becker, 2003a; Meisel, Camp, & Robinson, 1993; Mermelstein & Becker, 1995).

The NAc is perhaps one of the most crucial integrative structures in the mediation of reward circuitry, in both males and females. Initially studied in males, in rats it was found that mesolimbic DA release in the NAc was associated with motivational aspects of sexual behavior, as opposed to strictly the consummatory components (Everitt, 1990; Pfaus & Phillips, 1991). Study of the role of mesolimbic DA in the NAc of females soon followed. In Syrian hamsters, extracellular levels of DA increased rapidly upon introduction of a male, which appeared to be dissociable from the expression of lordosis (Meisel et al., 1993). Studies using a modified operant chamber, in which female rats have the ability to both access and escape from the side in which the male resides, but the male is confined to one side, have indicated that female rodents develop a conditioned place preference for the side of the chamber in which the male resides (Jenkins & Becker, 2003b; Meisel & Joppa, 1994). Levels of extracellular DA in both the NAc and the striatum are elevated during paced mating encounters, an effect that appears to be enhanced by the presence of estradiol (Jenkins & Becker, 2003a; Meisel et al., 1993; Mermelstein & Becker, 1995). This increase in extracellular DA is greater in females that have encountered paced mating before, as compared with those that are sexually naïve (Kohlert & Meisel, 1999).

Furthermore, estrogen acts differentially in the striatum and the NAc to modulate specific aspects of pacing behavior. Implanting estradiol bilaterally into the striatum of adult, ovx rats increased the percent of exits from the side of the chamber containing the male, an effect that was diminished upon application of the estrogen receptor antagonist ICI 182,780. In contrast, bilateral estradiol implantation in the NAc was sufficient to increase the latency to return to the male's side of the chamber. Here, application of ICI 182,780 decreased the return latency. Finally, estradiol activity in the striatum did not affect return latency, nor did estradiol in the NAc affect percent exits, indicating that the effect of estradiol in these two regions is separable (Xiao & Becker, 1997). Lesioning these two regions in ovx, hormone primed rats further elucidated the dissociable effects of estrogen on pacing behavior. Lesions to the striatum were sufficient to decrease the exits the female made from the male's side of the chamber, effectively decreasing pacing behavior by leaving her in proximity to the male for a greater amount of time. The loss of the striatum had no effect on seeking reproductive behavior, nor on the production of lordosis. Lesioning the NAc, and particularly the shell subregion, was sufficient to prevent the female from seeking out the male to mate, but, like the loss of the striatum, had no effect on the display of lordosis (Jenkins & Becker, 2001). In female rats, NAc lesions have also been shown to increase the rejection of mounts by a male, again without affecting the physical display of lordosis (Rivas & Mir, 1991). Taken together, these data indicate that estradiol within the female NAc is important in the motivation to mate with a male.

While the complete mechanisms of estradiol modulation on motivational circuitry have yet to be fully elucidated, these circuits appear to lack nuclear estrogen receptors (Pfaff & Keiner, 1973), which may indicate a role for mER-mGlu signaling cascades. While the role of mER-mGlu signaling in the drive to copulate remains unclear, the changes in DA signaling in the limbic structures during paced mating is akin to that seen following repeated exposure to drugs of abuse (Robinson & Berridge, 2003). This suggests that mER-mGlu signaling may be important in the drive to seek reward and reinforcement from these two different motivational sources.

4. Estrogen in motivated behaviors: Dysregulation and drug addiction

Behaviors that activate the reward system motivate animals to continue those behaviors. When these are adaptive behaviors (e.g., reproduction), the ultimate outcome is the continuation of the survival of the individual and/or the species. However, not all behaviors that activate the reward system are adaptive; certain behaviors, including drug-taking, gambling, and the overindulgence of palatable foods, to name a few, can highjack the reward system and reinforce ultimately maladaptive behaviors. As these behaviors elicit a DA response, repeated reinforcement of the behavior can transition a human (or animal) from casual to compulsive activity. One of the most detrimental maladaptive behaviors for both an individual and society is addiction to drugs of abuse (Reid, Lingford-Hughes, Cancela, & Kalivas, 2012). Substance use disorder is a neuropsychiatric illness characterized by chronic compulsive drug seeking and intake at the expense of motivation for other behaviors, with abstinence from the drug often punctuated by periods of relapse. The pharmacological characteristics of the drug, along with environmental, societal, genetic, and hormonal factors all play a role in the development of a substance use disorder (Kokane & Perrotti, 2020; Reid et al., 2012). That said, all drugs of abuse provide acute reward and reinforcement, and long-term intake leads to long-lasting neuroplastic changes (Reid et al., 2012). It should be noted, however, that not all experiences with drugs of abuse end in addiction; in both humans and rodents, only a minority of individuals exposed to these substances will go on to experience addiction (Belin, Balado, Piazza, & Deroche-Gamonet, 2009; Brady & Randall, 1999; Deroche-Gamonet, Belin, & Piazza, 2004). In females, this path to addiction appears to be facilitated by estradiol, with a role for estradiol signaling through mGlu. The following sections review some of the key neurocircuitry affected by drugs of abuse, with a focus on cocaine, amphetamines, and morphine, to illustrate how estradiol plays a role in these structural and functional changes, and particularly how signaling through mER-mGlu complexes helps facilitate the development of substances use disorders.

4.1 Estradiol in drug addiction

In women, the first instance of drug use is often a form of self-medication in response to negative life events or as a means for coping with psychological factors such as anxiety and/or depression. This is in contrast to men, who more frequently report that initial drug use occurred in a social setting (Annis & Graham, 1995; Brady & Randall, 1999). While the overall number of individuals experiencing substance use disorder does not seem to differ between the sexes (McHugh, Votaw, Sugarman, & Greenfield, 2018), patterns of drug use and abuse seem to show some variances (Cotto et al., 2010). In particular, women tend to escalate drug use from initial consumption to compulsive use at a faster rate than do men, a phenomenon called "telescoping" (Becker & Chartoff, 2019; Brady & Randall, 1999; Haas & Peters, 2000; Kosten, Gawin, Kosten, & Rounsaville, 1993). Once compulsive use is reached, neuroplastic changes in both the reward and stress systems appear to show sex differences as well (Becker & Chartoff, 2019; Kokane & Perrotti, 2020).

During abstinence from a drug, women often report greater cravings and more unpleasant withdrawal symptoms than men do (Fox, Sofuoglu, Morgan, Tuit, & Sinha, 2013; Hudson & Stamp, 2011; Kosten et al., 1993). These negative outcomes are modulated at least in part by the hormonal cycle, as estradiol seems to influence motivation for drugs of abuse, and its activity may underlie some of the neuroplasticity seen in addiction (Weinberger et al., 2015). Notably, the subjective responses to cocaine vary across the menstrual cycle. During the follicular phase, when estradiol is rising, women report subjectively greater euphoria, a more intense high, and motivation to seek more of the drug than they do during the luteal phase, in which progesterone dominates (Evans et al., 2002). Additionally, women who have gone through treatment for substance use disorders report more instances of spontaneous relapse than do men, which may be due in part to being more sensitive to environmental cues and in part due to the influence of the hormonal cycle (Janes et al., 2010; Ruda-Kucerova et al., 2015). It should be noted that estradiol plays a role in the compulsive intake of other substances of abuse, as well as gambling (Bobzean, DeNobrega, & Perrotti, 2014; Maria, Flanagan, & Brady, 2014; Tavares et al., 2003).

In female rodents, estradiol appears to increase the motivation for psychostimulants. Repeated exposure to psychostimulants in turn enhances incentive sensitization and behavioral sensitization (Kawa & Robinson, 2019; Strakowski, Sax, Setters, & Keck, 1996) through the modulation of the ascending dopamine system by estradiol (Becker et al., 2001; Hu & Becker, 2003). Behavioral sensitization to both cocaine and amphetamine, including an increase stereotypy and rotational movement, is greatest in ovx females when treated with exogenous estradiol (Becker, 1990a, 1990b; Becker et al., 2001; Peris, Decambre, Coleman-Hardee, & Simpkins, 1991; Schultz et al., 2009; Sell, Thomas, & Cunningham, 2002). In female rats, sensitization to cocaine occurs at lower doses of the drug than in males (Post, Lockfeld, Squillace, & Contel, 1981). Furthermore, females require less cocaine and less time to establish a conditioned place preference for the drug than do males (Zakharova, Wade, & Izenwasser, 2009), and take longer to extinguish cocaine-seeking behaviors (Kerstetter, Aguilar, Parrish, & Kippin, 2008). Intact females also demonstrate greater motivation for cocaine in progressive ratio tasks (Kawa & Robinson, 2019) and exhibit a higher breaking point during these tasks than do males (Cummings et al., 2011; Roberts, Bennett, & Vickers, 1989). Motivation is greatest when levels of circulating estradiol are highest (Becker & Hu, 2008; Becker & Koob, 2016; Roberts et al., 1989). In ovx rats, this motivation for cocaine is lower than those rats with intact ovaries (Hu & Becker, 2008; Perry, Westenbroek, & Becker, 2013), though estradiol treatment is sufficient to enhance cocaine acquisition (Hu, Crombag, Robinson, & Becker, 2004; Hu & Becker, 2008; Lynch, Roth, Mickelberg, & Carroll, 2001).

Estradiol administration in males does not facilitate cocaine acquisition (Jackson, Robinson, & Becker, 2006). Finally, in females, reinstatement of drug-taking behaviors is also modulated by hormones. Estradiol enhances cue-, stress-, and drug-induced reinstatement, while progesterone attenuates reinstatement (Anker & Carroll, 2010a, 2010b; Feltenstein, Henderson, & See, 2011; Feltenstein & See, 2007; Fuchs, Evans, Mehta, Case, & See, 2005; Swalve, Smethells, Zlebnik, & Carroll, 2016).

4.2 Reward system neurocircuitry

Drugs of abuse highjack the motivation and reward circuitry, particularly the mesolimbic DA system. The ascending mesolimbic DA pathways originate in the midbrain, sending projections from the ventral tegmental area (VTA) and the substantia nigra to many forebrain regions, including notably the striatum, prefrontal cortex, hippocampus, and amygdala (Morales & Margolis, 2017; Wise, 2004). DAergic input to these regions ultimately modulates motivated behaviors. In turn, the prefrontal cortex, hippocampus, and amygdala largely contain glutamate neurons that innervate the NAc (Giacometti & Barker, 2020). The VTA contains a heterogenous population of neurons, many of which are also dopaminergic. There exists further heterogeneity within the VTA DA neurons, with some neurons being GABAergic and others glutamatergic (Morales & Margolis, 2017). Importantly, inherent sex differences in the VTA have been found. In rodents, females have a higher proportion of DAergic neurons (Kritzer & Creutz, 2008), and differences in the size, shape, volume, and distribution of these neurons have been found (McArthur, McHale, & Gillies, 2007). In the VTA, DAergic neurons have been shown to express both ERα and ERβ (Kritzer & Creutz, 2008), and the firing rates of VTA neurons also appears to be sensitive to estradiol and fluctuates with the estrous cycle. During proestrus, when circulating levels of estradiol are at their highest, basal firing rates are at their lowest. In estrus, after circulating estradiol levels have dropped, firing rates are at their highest (Zhang, Yang, Yang, Jin, & Zhen, 2008). Local activation of DA neurons in the VTA through the use of optogenetics is sufficient to result in operant reinforcement of behavior, indicating that these neurons provide reward signaling (Kim et al., 2012). Furthermore, optogenetic targeting of DAergic VTA neurons that terminate in the nucleus accumbens (NAc) is sufficient to drive intra-cranial self-stimulation (Steinberg et al., 2014), a model in which animals perform an action to receive exogenous stimulation to a particular brain region.

Dopaminergic inputs from the VTA innervate many forebrain regions involved in motivated behavior and drug addiction, including the medial prefrontal cortex (mPFC), hippocampus, amygdala, and NAc. Glutamatergic prefrontal cortical regions, including the prelimbic and infralimbic areas, receive DAergic input from the VTA, and ultimately send reciprocal projections (Geisler & Wise, 2008). These excitatory inputs modulate DA firing rate in the VTA (Floresco, West, Ash, Moore, & Grace, 2003; Murase, Grenhoff, Chouvet, Gonon, & Svensson, 1993), as well as DAergic activity in the NAc (Koob & Volkow, 2016). Much of the function of the prefrontal regions is in executive control and emotional regulation (Miller, 2000). The prefrontal cortex also may play a role in the internal representation of reward, as well as associative learning (Chudasama & Robbins, 2003). Within the PFC, the infralimbic cortex may play a role in decision making (Roughley & Killcross, 2021; Sweis, Larson, Redish, & Thomas, 2018). As such, dysfunction in mPFC circuits may play a role in the reinstatement of habits after extinction, and particularly in seeking drugs after a period of abstinence (Peters, Kalivas, & Quirk, 2009) that may be exacerbated in females (Doncheck et al., 2018).

Similarly, given the well-documented role of the hippocampus in learning and memory (Opitz, 2014; Voss, Bridge, Cohen, & Walker, 2017), the region has been shown to play an important role in learning in response to reward, as well as in drug-addiction (Kutlu & Gould, 2016). Glutamatergic projections from different subregions of the hippocampus differentially innervate the NAc; projections from the dorsal hippocampus reach the NAc core (NAcC) while projections from the ventral hippocampus innervate the NAc shell (NAcSh) (Groenewegen, der Zee, te Kortschot, & Witter, 1987; Kelley & Domesick, 1982). Estradiol affects both ionotropic glutamate receptor density (Palomero-Gallagher, Bidmon, & Zilles, 2003) and dendritic spine density (Woolley & McEwen, 1992) in the hippocampus.

The NAc is a highly specialized structure, capable of the complex integration of various inputs, and is a key node in reward circuitry. This region is important in reward and incentive salience, and in driving motivated behaviors (Meredith, Baldo, Andrezjewski, & Kelley, 2008). As whole, the NAc is composed of the core (NAcC) and the shell (NAcSh) (Groenewegen, Wright, Beijer, & Voorn, 1999; Zahm & Brog, 1992). The NAcC is largely involved with motor circuitry and the execution of conditioned behaviors, while the NAcSh is chiefly involved in the interaction with other limbic structures and the reinforcement of reward (Eisinger, Larson, et al., 2018; Eisinger, Woolfrey, et al., 2018; Meredith et al., 2008). The predominant output neuron in the NAc, and indeed the striatum as a whole, is the GABAergic medium spiny neuron (MSN)—so-called due to the density of spines it possesses (Yager, Garcia, Wunsch, & Ferguson, 2015). MSNs in the NAc receive inputs from many neural regions. In terms of motivation, DA input from the ascending DA pathways and glutamatergic inputs from the hippocampus, prefrontal cortex, and amygdala are particularly salient. Importantly, estradiol plays an important role in modulating both DAergic (Becker, 1990a, 1990b; Eisinger, Larson, et al., 2018; Eisinger, Woolfrey, et al., 2018; Mermelstein, Becker, & Surmeier, 1996; Thompson & Moss, 1994) and glutamatergic (Forlano & Woolley, 2010) activity in the NAc. The core subregion in particular exhibits estrous cycle–dependent changes, as well as intrinsic sex differences. Female MSNs have a greater density of dendritic spines, and these spines are larger, than those found in male MSNs (Forlano & Woolley, 2010). Accordingly, in females, the MSNs in this region receive more excitatory glutamatergic inputs than do males (Sazdanović et al., 2013). Furthermore, the intrinsic electrophysiological properties of the MSN change with the rodent estrous cycle (Proaño, Morris, Kunz, Dorris, & Meitzen, 2018).

Through the integration of inputs, the NAc is involved in both cognitive and motor behaviors through projections to numerous other brain structures, including sending projections back to the prefrontal cortex and VTA (Kokane & Perrotti, 2020; Smith, Lobo, Spencer, & Kalivas, 2013; Yager et al., 2015). Another important aspect modulated by the NAc is incentive salience and reward learning (Belin & Everitt, 2008), and DA signaling here is important in learning environmental cues that are associated with reward (Berridge, 2007). Importantly, each subdivision of the NAc displays fundamentally distinct properties, including regulating distinct functions in motivated behaviors, as well as exhibiting different efferent and afferent projection patterns. While both subdivisions receive mesolimbic DAergic input from the VTA, the forebrain sources of glutamate input to each subdivision are distinct. The core receives much of its forebrain input from the prelimbic cortex, while the shell receives a majority of its input from the infralimbic cortex (Berendse, Graaf, & Groenewegen, 1992; Wright & Groenewegen, 1996). The NAc sends reciprocal projections back to hindbrain mesolimbic DA regions, though input to the VTA is largely from the shell, while projections reaching the substantia nigra originate from the core (Ikemoto, 2007).

Though the NAc receives substantial attention in terms of reward circuitry and drugs of abuse, the dorsal striatum is also involved. The dorsal striatum can be divided into dorsomedial and dorsolateral subdivisions, each with a particular role in reward processing. The dorsomedial subdivision appears to play a role in goal-directed learning and the integration of action-outcome associations to further influence motor outcomes (Yin, Knowlton, & Balleine, 2005), while the dorsolateral striatum is involved in stimulus-response learning, including the development of habitual behavior (Yin, Knowlton, & Balleine, 2006). While reward stimuli activate the NAc, integration of signals between the ventral and dorsal striatum is a key process in the development of conditioned learning, in habit formation, and indeed in the development of drug addiction. Activity in these intrastriatal pathways allows for reward-associated cues to eventually activate circuits within the dorsal striatum even in the absence of the rewarding stimulus. Addiction to drugs of abuse is associated with greater DA activity in the dorsal striatum as opposed to the ventral striatum (Becker, 2016). Within the female dorsal striatum, estradiol (and progesterone) modulate DA release (Yoest, Quigley, & Becker, 2018). Within the region, estradiol rapidly potentiates amphetamine-stimulated DA release (Becker & Beer, 1986). Furthermore, estradiol in the female dorsolateral striatum has been shown to enhance many components of drug-seeking, including the acquisition of, motivation for, and the reinstatement of the behavior (Becker, 2016). Overall, it is the convergence of DAergic and glutamatergic inputs, modulated by estradiol in females, within the NAc that appear to be crucial in driving motivated behavior, both adaptive and maladaptive (Fig. 4).

4.3 mER-mGlu signaling in the striatum

Within the circuits involved in the motivation to consume drugs of abuse, one of the most critical regions involved in the acquisition of drug addiction is the dorsal striatum, and importantly the NAc in the ventral striatum. The output neuron in striatum is the GABAergic MSN (Yager et al., 2015). Estradiol modulates glutamatergic and DAergic input to the region, through acting both on receptors located on the cell bodies from regions projecting to the NAc, as well as locally, and repeated exposure to psychostimulants is able to enhance both inputs through dendritic spine neuroplasticity (D'Souza, 2015; Nestler, 2001). Within the NAc, estradiol stimulates significant synaptic structural changes akin to that seen by psychostimulants alone (Dumitriu et al., 2012; Staffend, Loftus, & Meisel, 2011). Throughout both the core and shell regions, estradiol signaling through mGlu modulates the structural and functional circuitry of the NAc in response to drugs of abuse, and likely acts through mER–mGlu signaling, as MSNs are largely devoid of nuclear ERs (Almey, Filardo, Milner, & Brake, 2012; Pfaff & Keiner, 1973).

While all four group I and group II mGlu are found in the striatum, estradiol signaling through mGlu here appears to recruit different pathways than those utilized in the hippocampus. While much of the mER–mGlu signaling in the hippocampus occurs through either ER–mGlu1a or ER–mGlu2 pathways, striatal neurons appear to utilize ER–mGlu5 and ER–mGlu3 (Grove-Strawser et al., 2010). That is, while estradiol-induced CREB phosphorylation occurs through mER–mGlu1 signaling in the hippocampus, in the striatum this same outcome depends on ERa-mGlu5 signaling. Like in the hippocampus, this requires CAV1, as well as MAPK activity. Furthermore, estradiol attenuation of L-type calcium channel-mediated CREB phosphorylation can be mimicked through the activation of either ERα or ERβ in striatal neurons, though in the striatum this relies upon mGlu3, rather than mGlu2, activity (Grove-Strawser et al., 2010).

Within the NAc, both group I receptors, mGlu1a and mGlu5, are expressed, and are colocalized in approximately half of the dendritic spines (Mitrano & Smith, 2007). While mGlu1a and mGlu5 recruit many of the same downstream signaling components following estradiol transactivation, it has become apparent that these two mGlu can have opposing effects on neuronal architecture even within the same brain region. In ovx rats, positive allosteric modulation of mGlu5 decreases dendritic spine density both the shell and core regions of the NAc. Furthermore, bilateral injection of the mGlu5 agonist CHPG is sufficient to mimic the decrease in spine density elicited by the positive allosteric modulator. On the other hand, positive modulation of mGlu1a in the NAc increases spine density in both the core and shell. Together, these data indicate that mERs are paired with mGlu5 in the core, and mGlu1a in the shell. Additionally, the effects of mGlu signaling on spine density also appear to be brain region-specific, as the positive modulation of either group I mGlu was insufficient to alter dendritic structure in the dorsal striatum (Gross et al., 2016) (Table 3).

Table 3 Differential effects of estradiol on spine density in the core and shell subregions of the nucleus accumbens are mediated by different Group I mGlu.

Region	Estradiol effect on dendritic density	Estrogen receptor	mGlu	CAV
Core	↓	ERα	mGlu5	CAV1
Shell	↑/−	ERα	mGlu1a	CAV1

Estradiol also influences spine density in the NAc. In the female NAcC, estradiol decreases spine density (Staffend et al., 2011; Peterson, Mermelstein, & Meisel, 2015), an effect that can be blocked by antagonism of mGlu5 (Peterson et al., 2015). Estradiol effects on dendritic morphology in the shell have been more varied, with either no change (Staffend et al., 2011) or an increase (Peterson et al., 2015) having been reported. Unlike in the NAcC, this effect does not appear to occur through mGlu5 signaling, but rather mGlu1a (Peterson et al., 2015). The mER-mGlu5 mediated decrease in NAcC spine density in females is reminiscent of the decrease in NAcC spine density seen in males in response to cocaine administration (Dumitriu et al., 2012; (Waselus et al., 2013). As such, the activity of mER-mGlu5 signaling is of particular interest in understanding the effects of drugs of abuse on the NAc, and indeed in female motivation for these substances. Activation of mGlu5 in the NAc has been shown to be required for several drug-seeking behaviors (Besheer et al., 2009; Bossert, Gray, Lu, & Shaham, 2006; Kumaresan et al., 2009; Wang, Moussawi, Knackstedt, Shen, & Kalivas, 2013), though most of these studies focused on males.

In female rats, estradiol facilitates cocaine-induced locomotor sensitization (Segarra et al., 2010; Sircar & Kim, 1999). This appears to be dependent upon mGlu5 activity, as an intraperitoneal injection of the mGlu5 antagonist MPEP preceding estradiol replacement in ovx rats was sufficient to attenuate this cocaine-induced increase in locomotion (Martinez, Peterson, Meisel, & Mermelstein, 2014). This is reminiscent of the effect of mGlu5 antagonism in males, which was also able to attenuate the cocaine-induced increase in locomotion (Herzig & Schmidt, 2004; McGeehan, Janak, & Olive, 2004). Furthermore, in females, the locomotive response of ER-mGlu5 signaling following cocaine administration requires endocannabinoid (eCB) activity (Peterson, Martinez, Meisel, & Mermelstein, 2016). Activation of group I mGlu results in the release of eCBs, which bind to type 1 cannabinoid receptors (CB1R) located on pre-synaptic terminals (Wilson & Nicoll, 2002). Both eCBs and CB1R have been implicated in drug addiction (Olière, Jolette-Riopel, Potvin, & Jutras-Aswad, 2013), and estradiol has been shown to enhance the activity of eCBs in females (de Fonseca, Cebeira, Ramos, Martín, & Fernández-Ruiz, 1994; Maccarrone, 2004). Administration of a CB1 inverse agonist was sufficient to not only attenuate cocaine-induced locomotor sensitization, but also to prohibit the estradiol-induced decrease in spine density in the NAcC (Peterson et al., 2016). Estradiol facilitates other effects of cocaine as well, including an increase in self-administration of the drug (Hu & Becker, 2008; Larson, Anker, Gliddon, Fons, & Carroll, 2007; Ramôa, Doyle, Naim, & Lynch, 2013). Estradiol treatment in ovx rats increases cocaine intake during a self-administration task, while inhibiting mGlu5 signaling via intraperitoneal injection in these animals before estradiol administration is sufficient to block the increase (Martinez et al., 2016). This suggests that like cocaine-induced locomotor sensitization, estradiol signaling through mGlu5 is involved in another key aspect of reward neurocircuitry.

Activity of mER-mGlu signaling in the dorsal striatum has also been shown to be important. Estradiol in the dorsal striatum facilitates rotational behavior in females following amphetamine administration (Becker, 1990b; Becker & Beer, 1986). This effect is both rapid and mediated by mGlu5; estradiol alters the amphetamine-induced DA response within 30 min (Becker, 1990b) and this response can be blocked by either an ER antagonist or an mGlu5 antagonist (Song, Yang, Peckham, & Becker, 2019).

While lesser studied in terms of drug addiction, estradiol activation of mER-mGlu1a in the NAc also appears to play a role in females. As noted above, estradiol activation of this signaling complex results in an opposite outcome of mGlu5 stimulation; an increase in dendritic spine density in both the core and shell regions (Gross et al., 2016). This increase in spine density is akin to that seen in both the hypothalamus (Micevych & Christensen, 2012) and hippocampus (Huang & Woolley, 2012). In male rats it has been shown that prolonged withdrawal from cocaine results in an increase in calcium permeable AMPA receptors in the NAc, and that activation of mGlu1 and subsequent PKC activity, is sufficient to reduce this accumulation (McCutcheon et al., 2011). Blocking mGlu1 signaling early in the withdrawal phase accelerated the accumulation of these particular AMPA receptors (Loweth et al., 2014). Whether these results hold true in females, and if this mGlu1 activation might occur through estradiol activation remains to be seen.

5. Conclusion

In females, many motivated behaviors are modulated by estradiol. Classical intracellular estrogen receptors, ERα and ERβ, have been shown to also be trafficked to the surface membrane, where signaling cascades activated by these membrane-bound estrogen receptors (mERs) can rapidly affect cellular excitability, gene expression, neuroplasticity, and ultimately behavior. This rapid estradiol signaling has since been found throughout the body and brain. A principal mechanism of neuronal mER action is through glutamate-independent transactivation of metabotropic glutamate receptors (mGlu), which elicits multiple signaling outcomes. The signaling cascades initiated by the mER-mGlu complex has been shown to be involved in many physiological functions in both sexes, but particularly in females.

A growing body of evidence suggests that a significant portion of estradiol-induced neuroplasticity occurs through the estradiol-dependent interaction of mERs and mGlu (Micevych & Mermelstein, 2008), and can drive both adaptive and maladaptive motivated behaviors. Within this signaling cascade, uncovering the roles of CAV and palmitoylation has led to further understanding of the complexities involved in mER-mGlu signaling. Herein we have reviewed membrane-initiated estrogen receptor signaling in females, with a focus on the interactions between these mERs and mGlu, and how this interaction drives both adaptive motivated behaviors, represented by reproduction, and maladaptive behaviors, including addiction.

In female reproduction, ovulation, lordosis, and the motivational drive to copulate have been shown to be dependent upon mER-mGlu signaling. While both estradiol and progesterone released from the ovaries facilitate the LH surge that leads to ovulation, it has become apparent that progesterone synthesized de novo from hypothalamic astrocytes plays a crucial role (Kuo et al., 2010; Micevych et al., 2007; Micevych & Sinchak, 2011; Sinchak et al., 2003). Removing peripheral sources of progesterone blunts, but does not entirely inhibit, the LH surge. However, blocking hypothalamic neuroP as well is sufficient to eliminate the surge entirely. The synthesis and release of neuroP has been found to be reliant upon estradiol activation of mERα-mGlu1a signaling, initiated at the membrane in hypothalamic astrocytes. Blocking any part of the signaling cascade (mGlu1a, IP$_3$ receptors, AC-1 activity, PKA activity) is sufficient to inhibit neuroP synthesis (Chen et al., 2014; Kuo et al., 2010, 2009). Without neuroP the LH surge, and furthermore ovulation, cannot occur (Micevych & Sinchak, 2011).

Following ovulation, the display of lordosis is also reliant upon the activation of both intracellular ERs and mER-mGlu signaling, and further integrated with external sensory cues. The estradiol-induced activation/internalization of MORs in the MPN and the resulting temporary inhibition of lordosis is a key component in the lordosis response, and is dependent on mERα-mGlu1a signaling in the ARH (Dewing et al., 2007, 2008; Micevych, Rissman, et al., 2003; Micevych, Sinchak, et al., 2003). Signaling through the ARH mER-mGlu1a complex results in an increase the number of dendritic spines within 4 h, persisting for at least 48 h, a time course which tracks the expression of lordosis (Christensen et al., 2011). Disrupting the mER-mGlu1a signaling cascade in the ARH (Christensen et al., 2011; Dewing et al., 2007), or CAV1 expression there (Christensen & Micevych, 2012), is sufficient to inhibit the behavior.

While the long-term consequence of reproduction is the continuation of the species, at the level of the individual there is a more immediate rewarding aspect that drives the motivation to copulate, in females as much as in males (Meisel & Mullins, 2006). The engagement of the limbic system and dopamine signaling therein appears also to involve mER-mGlu signaling. Though the complete role of mER-mGlu signaling in the female drive for copulation remains to be elucidated, the limbic circuits involved appear to lack nuclear estrogen receptors (Pfaff & Keiner, 1973). Furthermore, the changes in DA signaling in limbic structures seen in response to paced mating are similar to those seen in response to repeated exposure to drugs of abuse, suggesting a similarity in signaling mechanisms.

Not all motivated behaviors are adaptive; certain behaviors can highjack the reward system and reinforce ultimately maladaptive behaviors. In females, this can be facilitated by estradiol. In women, during the follicular phase, as levels of circulating estradiol rise, those individuals with a history of cocaine use self-report an increased drive to seek the drug, along with subjectively greater feelings of euphoria following consumption. This is in contrast to self-reports during the luteal phase, in which estradiol levels are lower and progesterone dominates (Evans et al., 2002). This estradiol-induced motivation drive for cocaine is seen in female rodents, as well (Becker, 1990b; Becker et al., 2001; Hu & Becker, 2003; Peris et al., 1991; Schultz et al., 2009; Sell et al., 2002). Estradiol plays an important role in modulating DAergic (Becker, 1990a; Mermelstein et al., 1996; Thompson & Moss, 1994) and glutamatergic (Forlano & Woolley, 2010) activity in the limbic system, affecting mesolimbic DA signaling. The ascending mesolimbic DA projections to the forebrain originate in the VTA, a region that displays inherent sex differences. Females have a higher proportion of VTA neurons that express the catecholamine neurons (Kritzer & Creutz, 2008), and these neurons, whose firing rates differ across the estrous cycle (Zhang et al., 2008), are also morphologically different than those found in males (McArthur et al., 2007). Inherent sex differences are also seen in the NAc, a key region in reward circuitry that receives direct projections from the VTA. These differences are particularly noticeable in the core subregion of the NAc, wherein female MSNs display different dendritic spine morphology than males

(Forlano & Woolley, 2010), and exhibit intrinsic electrophysiological changes throughout the estrous cycle (Proaño et al., 2018).

Furthermore, it is the NAc that has received much attention in understanding the pathophysiology underlying drug addiction. In females, estradiol dramatically modulates neuronal structure, as well as the DAergic and glutamatergic inputs to the region. Many of these estradiol-induced changes are reminiscent of structural changes induced by drugs of abuse alone (Dumitriu et al., 2012). Throughout the NAc, estradiol appears to signal through mERs, as the region is largely devoid of nuclear ERs (Pfaff & Keiner, 1973). Indeed, estradiol signaling through mER transactivation of mGlu here has been shown to modulate estradiol-induced structural changes as well as responses to drugs of abuse. In the NAc, striatal neurons express both group I and group II mGlu, and signaling through each of these results in different outcomes on neuronal morphology. In this region, estradiol-induced CREB phosphorylation occurs through mERα-mGlu5 signaling, while estradiol attenuation of L-type calcium channel-mediated CREB phosphorylation can be induced through the activation of either ERα or ERβ activation of mGlu3 (Grove-Strawser et al., 2010). These signaling cascades also appear to be region specific, even within a nucleus. Within the NAc, signaling through each of the group I mGlu, mGlu1a or mGlu5, differently affects spine density. Signaling through mGlu5 decreases dendritic spin density throughout the NAc, while signaling through mGlu1a increases spine density (Gross et al., 2016). This mGlu5-mediated decrease in spine density in the NAcC is akin to that seen following estradiol administration (Staffend et al., 2011; Peterson et al., 2015), and indeed inhibiting mGlu5 signaling is sufficient to block the estradiol-induced morphological changes (Peterson et al., 2015). Furthermore, estradiol facilitates cocaine-induced locomotor sensitization (Segarra et al., 2010; Sircar & Kim, 1999) and an increase in self-administration of the drug (Hu & Becker, 2008; Larson et al., 2007; Ramôa et al., 2013), through mGlu5 signaling (Martinez et al., 2014, 2016). This mER-mGlu5 facilitation of cocaine-induced locomotor sensitization is further dependent on eCB activity (Peterson et al., 2016). As such, the role of mGlu5, and in particular the mER-mGlu5 signaling cascade in females, may prove to be a therapeutic target in the treatment of drug addiction.

Overall, progress has been made in understanding function of rapid estradiol signaling, including the signaling cascades initiated by ER transactivation of mGlu. While this mechanism is found in both sexes, the interaction of these two receptors has been shown to be intricately involved in many physiological functions in females, including mediating reproduction and motivation, both adaptive and maladaptive. Current and future research will continue to add to our understanding of mER–mGlu signaling and its physiological and behavioral outcomes.

References

Ábrahám, I. M., Todman, M. G., Korach, K. S., & Herbison, A. E. (2004). Critical in vivo roles for classical estrogen receptors in rapid estrogen actions on intracellular signaling in mouse brain. *Endocrinology, 145*(7), 3055–3061. https://doi.org/10.1210/en.2003-1676.

Acconcia, F., Ascenzi, P., Bocedi, A., Spisni, E., Tomasi, V., Trentalance, A., et al. (2005). Palmitoylation-dependent estrogen receptor α membrane localization: Regulation by 17β-estradiol. *Molecular Biology of the Cell, 16*(1), 231–237. https://doi.org/10.1091/mbc.e04-07-0547.

Adler, N. (1978). On the mechanisms of sexual behavior and their evolutionary constraints. *Biological Determinants of Sexual Behavior,* 657–695.

Allen, B. J., Rogers, S. D., Ghilardi, J. R., Menning, P. M., Kuskowski, M. A., Basbaum, A. I., et al. (1997). Noxious cutaneous thermal stimuli induce a graded release of endogenous substance P in the spinal cord: Imaging peptide action in vivo. *The Journal of Neuroscience, 17*(15), 5921–5927. https://doi.org/10.1523/jneurosci.17-15-05921.1997.

Almey, A., Filardo, E. J., Milner, T. A., & Brake, W. G. (2012). Estrogen receptors are found in glia and at extranuclear neuronal sites in the dorsal striatum of female rats: Evidence for cholinergic but not dopaminergic colocalization. *Endocrinology, 153*(11), 5373–5383. https://doi.org/10.1210/en.2012-1458.

Anderson, R. G. (1998). The caveolae membrane system. *Annual Review of Biochemistry, 67*(1), 199–225.

Anker, J. J., & Carroll, M. E. (2010a). Reinstatement of cocaine seeking induced by drugs, cues, and stress in adolescent and adult rats. *Psychopharmacology, 208*(2), 211–222. https://doi.org/10.1007/s00213-009-1721-2.

Anker, J. J., & Carroll, M. E. (2010b). Sex differences in the effects of allopregnanolone on yohimbine-induced reinstatement of cocaine seeking in rats. *Drug and Alcohol Dependence, 107*(2–3), 264–267. https://doi.org/10.1016/j.drugalcdep.2009.11.002.

Annis, H. M., & Graham, J. M. (1995). Profile types on the inventory of drinking situations: Implications for relapse prevention counseling. *Psychology of Addictive Behaviors, 9*(3), 176–182. https://doi.org/10.1037/0893-164x.9.3.176.

Anwyl, R. (2009). Metabotropic glutamate receptor-dependent long-term potentiation. *Neuropharmacology, 56*(4), 735–740. https://doi.org/10.1016/j.neuropharm.2009.01.002.

Aranda, A., & Pascual, A. (2001). Nuclear hormone receptors and gene expression. *Physiological Reviews, 81*(3), 1269–1304. https://doi.org/10.1152/physrev.2001.81.3.1269.

Balijepalli, R. C., Foell, J. D., Hall, D. D., Hell, J. W., & Kamp, T. J. (2006). Localization of cardiac L-type Ca2+ channels to a caveolar macromolecular signaling complex is required for β2-adrenergic regulation. *Proceedings of the National Academy of Sciences, 103*(19), 7500–7505. https://doi.org/10.1073/pnas.0503465103.

Bamburg, J. R. (1999). Proteins of the ADF/cofilin family: Essential regulators of actin dynamics. *Annual Review of Cell and Developmental Biology, 15*(1), 185–230. https://doi.org/10.1146/annurev.cellbio.15.1.185.

Baulieu, É.-É. (1991). Neurosteroids: A new function in the brain. *Biology of the Cell, 71*(1–2), 3–10. https://doi.org/10.1016/0248-4900(91)90045-o.

Baulieu, É.-É. (1998). Neurosteroids: A novel function of the brain. *Psychoneuroendocrinology, 23*(8), 963–987. https://doi.org/10.1016/s0306-4530(98)00071-7.

Beach, F. A. (1976). Sexual attractivity, proceptivity, and receptivity in female mammals. *Hormones and Behavior*, 7(1), 105–138.

Becker, J. B. (1990a). Direct effect of 17β-estradiol on striatum: Sex differences in dopamine release. *Synapse*, 5(2), 157–164. https://doi.org/10.1002/syn.890050211.

Becker, J. B. (1990b). Estrogen rapidly potentiates amphetamine-induced striatal dopamine release and rotational behavior during microdialysis. *Neuroscience Letters*, 118(2), 169–171. https://doi.org/10.1016/0304-3940(90)90618-j.

Becker, J. B. (2016). Sex differences in addiction. *Dialogues in Clinical Neuroscience*, 18(4), 395–402. https://doi.org/10.31887/dcns.2016.18.4/jbecker.

Becker, J. B., & Beer, M. E. (1986). The influence of estrogen on nigrostriatal dopamine activity Behavioral and neurochemical evidence for both pre- and postsynaptic components. *Behavioural Brain Research*, 19(1), 27–33. https://doi.org/10.1016/0166-4328(86)90044-6.

Becker, J. B., & Chartoff, E. (2019). Sex differences in neural mechanisms mediating reward and addiction. *Neuropsychopharmacology*, 44(1), 166–183. https://doi.org/10.1038/s41386-018-0125-6.

Becker, J. B., & Hu, M. (2008). Sex differences in drug abuse. *Frontiers in Neuroendocrinology*, 29(1), 36–47. https://doi.org/10.1016/j.yfrne.2007.07.003.

Becker, J. B., & Koob, G. F. (2016). Sex differences in animal models: Focus on addiction. *Pharmacological Reviews*, 68(2), 242–263. https://doi.org/10.1124/pr.115.011163.

Becker, J. B., Rudick, C. N., & Jenkins, W. J. (2001). The role of dopamine in the nucleus accumbens and striatum during sexual behavior in the female rat. *The Journal of Neuroscience*, 21(9), 3236–3241. https://doi.org/10.1523/jneurosci.21-09-03236.2001.

Belin, D., Balado, E., Piazza, P. V., & Deroche-Gamonet, V. (2009). Pattern of intake and drug craving predict the development of cocaine addiction-like behavior in rats. *Biological Psychiatry*, 65(10), 863–868. https://doi.org/10.1016/j.biopsych.2008.05.031.

Belin, D., & Everitt, B. J. (2008). Cocaine seeking habits depend upon dopamine-dependent serial connectivity linking the ventral with the dorsal striatum. *Neuron*, 57(3), 432–441. https://doi.org/10.1016/j.neuron.2007.12.019.

Berendse, H. W., Graaf, Y. G., & Groenewegen, H. J. (1992). Topographical organization and relationship with ventral striatal compartments of prefrontal corticostriatal projections in the rat. *Journal of Comparative Neurology*, 316(3), 314–347. https://doi.org/10.1002/cne.903160305.

Berridge, K. C. (1996). Food reward: Brain substrates of wanting and liking. *Neuroscience & Biobehavioral Reviews*, 20(1), 1–25. https://doi.org/10.1016/0149-7634(95)00033-b.

Berridge, K. C. (2007). The debate over dopamine's role in reward: The case for incentive salience. *Psychopharmacology*, 191(3), 391–431. https://doi.org/10.1007/s00213-006-0578-x.

Besheer, J., Grondin, J. J. M., Salling, M. C., Spanos, M., Stevenson, R. A., & Hodge, C. W. (2009). Interoceptive effects of alcohol require mGlu5 receptor activity in the nucleus accumbens. *Journal of Neuroscience*, 29(30), 9582–9591. https://doi.org/10.1523/jneurosci.2366-09.2009.

Bobzean, S. A. M., DeNobrega, A. K., & Perrotti, L. I. (2014). Sex differences in the neurobiology of drug addiction. *Experimental Neurology*, 259, 64–74. https://doi.org/10.1016/j.expneurol.2014.01.022.

Bondar, G., Kuo, J., Hamid, N., & Micevych, P. (2009). Estradiol-induced estrogen receptor-α trafficking. *Journal of Neuroscience*, 29(48), 15323–15330. https://doi.org/10.1523/jneurosci.2107-09.2009.

Bossert, J. M., Gray, S. M., Lu, L., & Shaham, Y. (2006). Activation of group II metabotropic glutamate receptors in the nucleus accumbens shell attenuates context-induced relapse to heroin seeking. *Neuropsychopharmacology*, 31(10), 2197–2209. https://doi.org/10.1038/sj.npp.1300977.

Boulware, M. I., Kordasiewicz, H., & Mermelstein, P. G. (2007). Caveolin proteins are essential for distinct effects of membrane estrogen receptors in neurons. *Journal of Neuroscience*, *27*(37), 9941–9950.

Boulware, M. I., Weick, J. P., Becklund, B. R., Kuo, S. P., Groth, R. D., & Mermelstein, P. G. (2005). Estradiol activates group I and II metabotropic glutamate receptor signaling, leading to opposing influences on cAMP response element-binding protein. *Journal of Neuroscience*, *25*(20), 5066–5078. https://doi.org/10.1523/jneurosci.1427-05.2005.

Bradley, K. C., & Meisel, R. L. (2001). Sexual behavior induction of c-Fos in the nucleus accumbens and amphetamine-stimulated locomotor activity are sensitized by previous sexual experience in female syrian hamsters. *The Journal of Neuroscience*, *21*(6), 2123–2130. https://doi.org/10.1523/jneurosci.21-06-02123.2001.

Brady, K. T., & Randall, C. L. (1999). Gender differences in substance use disorders. *Psychiatric Clinics of North America*, *22*(2), 241–252. https://doi.org/10.1016/s0193-953x(05)70074-5.

Calizo, L. H., & Flanagan-Cato, L. M. (2000). Estrogen selectively regulates spine density within the dendritic arbor of rat ventromedial hypothalamic neurons. *The Journal of Neuroscience*, *20*(4), 1589–1596. https://doi.org/10.1523/jneurosci.20-04-01589.2000.

Cameron, P. L., Ruffin, J. W., Bollag, R., Rasmussen, H., & Cameron, R. S. (1997). Identification of caveolin and caveolin-related proteins in the brain. *The Journal of Neuroscience*, *17*(24), 9520–9535. https://doi.org/10.1523/jneurosci.17-24-09520.1997.

Cartmell, J., & Schoepp, D. D. (2000). Regulation of neurotransmitter release by metabotropic glutamate receptors. *Journal of Neurochemistry*, *75*(3), 889–907. https://doi.org/10.1046/j.1471-4159.2000.0750889.x.

Chaban, V. V., Lakhter, A. J., & Micevych, P. (2004). A membrane estrogen receptor mediates intracellular calcium release in astrocytes. *Endocrinology*, *145*(8), 3788–3795. https://doi.org/10.1210/en.2004-0149.

Chaban, V., Li, J., McDonald, J. S., Rapkin, A., & Micevych, P. (2011). Estradiol attenuates the adenosine triphosphate-induced increase of intracellular calcium through group II metabotropic glutamate receptors in rat dorsal root ganglion neurons. *Journal of Neuroscience Research*, *89*(11), 1707–1710. https://doi.org/10.1002/jnr.22718.

Chen, P., & Hong, W. (2018). Neural circuit mechanisms of social behavior. *Neuron*, *98*(1), 16–30. https://doi.org/10.1016/j.neuron.2018.02.026.

Chen, P. B., Hu, R. K., Wu, Y. E., Pan, L., Huang, S., Micevych, P. E., et al. (2019). Sexually dimorphic control of parenting behavior by the medial amygdala. *Cell*, *176*(5), 1206–1221. e18 https://doi.org/10.1016/j.cell.2019.01.024.

Chen, C., Kuo, J., Wong, A., & Micevych, P. (2014). Estradiol modulates translocator protein (TSPO) and steroid acute regulatory protein (StAR) via protein kinase A (PKA) signaling in hypothalamic astrocytes. *Endocrinology*, *155*(8), 2976–2985. https://doi.org/10.1210/en.2013-1844.

Cheng, G., Coolen, L. M., Padmanabhan, V., Goodman, R. L., & Lehman, M. N. (2010). The kisspeptin/neurokinin B/dynorphin (KNDy) cell population of the arcuate nucleus: Sex differences and effects of prenatal testosterone in sheep. *Endocrinology*, *151*(1), 301–311. https://doi.org/10.1210/en.2009-0541.

Choe, E. S., & Wang, J. Q. (2001). Group I metabotropic glutamate receptor activation increases phosphorylation of cAMP response element-binding protein, Elk-1, and extracellular signal-regulated kinases in rat dorsal striatum. *Molecular Brain Research*, *94*(1–2), 75–84. https://doi.org/10.1016/s0169-328x(01)00217-0.

Christensen, A., Bentley, G. E., Cabrera, R., Ortega, H. H., Perfito, N., Wu, T. J., et al. (2012). Hormonal regulation of female reproduction. *Hormone and Metabolic Research*, *44*(08), 587–591. https://doi.org/10.1055/s-0032-1306301.

Christensen, A., Dewing, P., & Micevych, P. (2011). Membrane-initiated estradiol signaling induces spinogenesis required for female sexual receptivity. *Journal of Neuroscience*, *31*(48), 17583–17589. https://doi.org/10.1523/jneurosci.3030-11.2011.

Christensen, A., & Micevych, P. (2012). CAV1 siRNA reduces membrane estrogen receptor-α levels and attenuates sexual receptivity. *Endocrinology*, *153*(8), 3872–3877. https://doi.org/10.1210/en.2012-1312.

Chudasama, Y., & Robbins, T. W. (2003). Dissociable contributions of the orbitofrontal and infralimbic cortex to pavlovian autoshaping and discrimination reversal learning: Further evidence for the functional heterogeneity of the rodent frontal cortex. *The Journal of Neuroscience*, *23*(25), 8771–8780. https://doi.org/10.1523/jneurosci.23-25-08771.2003.

Clarkson, J., d'Anglemont de Tassigny, X., Moreno, A. S., Colledge, W. H., & Herbison, A. E. (2008). Kisspeptin-GPR54 signaling is essential for preovulatory gonadotropin-releasing hormone neuron activation and the luteinizing hormone surge. *Journal of Neuroscience*, *28*(35), 8691–8697. https://doi.org/10.1523/jneurosci.1775-08.2008.

Cooke, P. S., Nanjappa, M. K., Ko, C., Prins, G. S., & Hess, R. A. (2017). Estrogens in male physiology. *Physiological Reviews*, *97*(3), 995–1043. https://doi.org/10.1152/physrev.00018.2016.

Cotto, J. H., Davis, E., Dowling, G. J., Elcano, J. C., Staton, A. B., & Weiss, S. R. B. (2010). Gender effects on drug use, abuse, and dependence: A special analysis of results from the national survey on drug use and health. *Gender Medicine*, 7(5), 402–413. https://doi.org/10.1016/j.genm.2010.09.004.

Couchoux, H., Allard, B., Legrand, C., Jacquemond, V., & Berthier, C. (2007). Loss of caveolin-3 induced by the dystrophy-associated P104L mutation impairs L-type calcium channel function in mouse skeletal muscle cells. *The Journal of Physiology*, *580*(3), 745–754. https://doi.org/10.1113/jphysiol.2006.124198.

Cummings, J. A., & Becker, J. B. (2012). Quantitative assessment of female sexual motivation in the rat: Hormonal control of motivation. *Journal of Neuroscience Methods*, *204*(2), 227–233. https://doi.org/10.1016/j.jneumeth.2011.11.017.

Cummings, J. A., Gowl, B. A., Westenbroek, C., Clinton, S. M., Akil, H., & Becker, J. B. (2011). Effects of a selectively bred novelty-seeking phenotype on the motivation to take cocaine in male and female rats. *Biology of Sex Differences*, *2*(1), 3. https://doi.org/10.1186/2042-6410-2-3.

de Fonseca, F. R., Cebeira, M., Ramos, J. A., Martín, M., & Fernández-Ruiz, J. J. (1994). Cannabinoid receptors in rat brain areas: Sexual differences, fluctuations during estrous cycle and changes after gonadectomy and sex steroid replacement. *Life Sciences*, *54*(3), 159–170. https://doi.org/10.1016/0024-3205(94)00585-0.

Delhousay, L. K., Chuon, T., Mittleman-Smith, M., Micevych, P., & Sinchak, K. (2019). RP3V kisspeptin neurons mediate neuroprogesterone induction of the luteinizing hormone surge in female rat. *BioRxiv*, *700435*. https://doi.org/10.1101/700435.

Deroche-Gamonet, V., Belin, D., & Piazza, P. V. (2004). Evidence for addiction-like behavior in the rat. *Science*, *305*(5686), 1014–1017. https://doi.org/10.1126/science.1099020.

Dewing, P., Boulware, M. I., Sinchak, K., Christensen, A., Mermelstein, P. G., & Micevych, P. (2007). Membrane estrogen receptor-α interactions with metabotropic glutamate receptor 1a modulate female sexual receptivity in rats. *Journal of Neuroscience*, *27*(35), 9294–9300. https://doi.org/10.1523/jneurosci.0592-07.2007.

Dewing, P., Christensen, A., Bondar, G., & Micevych, P. (2008). Protein kinase C signaling in the hypothalamic arcuate nucleus regulates sexual receptivity in female rats. *Endocrinology*, *149*(12), 5934–5942. https://doi.org/10.1210/en.2008-0847.

Dietzen, D. J., Hastings, W. R., & Lublin, D. M. (1995). Caveolin is palmitoylated on multiple cysteine residues: Palmitoylation is not necessary for localization of caveolin to caveolae. *Journal of Biological Chemistry*, *270*(12), 6838–6842. https://doi.org/10.1074/jbc.270.12.6838.

Doncheck, E. M., Urbanik, L. A., DeBaker, M. C., Barron, L. M., Liddiard, G. T., Tuscher, J. J., et al. (2018). 17β-Estradiol potentiates the reinstatement of cocaine seeking in female rats: Role of the prelimbic prefrontal cortex and cannabinoid type-1 receptors. *Neuropsychopharmacology*, *43*(4), 781–790. https://doi.org/10.1038/npp.2017.170.

D'Souza, M. S. (2015). Glutamatergic transmission in drug reward: Implications for drug addiction. *Frontiers in Neuroscience*, *9*, 404. https://doi.org/10.3389/fnins.2015.00404.

Dumitriu, D., LaPlant, Q., Grossman, Y. S., Dias, C., Janssen, W. G., Russo, S. J., et al. (2012). Subregional, dendritic compartment, and spine subtype specificity in cocaine regulation of dendritic spines in the nucleus accumbens. *Journal of Neuroscience*, *32*(20), 6957–6966. https://doi.org/10.1523/jneurosci.5718-11.2012.

Eckel, L. A. (2011). The ovarian hormone estradiol plays a crucial role in the control of food intake in females. *Physiology & Behavior*, *104*(4), 517–524. https://doi.org/10.1016/j.physbeh.2011.04.014.

Eisinger, K. R. T., Larson, E. B., Boulware, M. I., Thomas, M. J., & Mermelstein, P. G. (2018). Membrane estrogen receptor signaling impacts the reward circuitry of the female brain to influence motivated behaviors. *Steroids*, *133*, 53–59. https://doi.org/10.1016/j.steroids.2017.11.013.

Eisinger, K. R. T., Woolfrey, K. M., Swanson, S. P., Schnell, S. A., Meitzen, J., Dell'Acqua, M., et al. (2018). Palmitoylation of caveolin-1 is regulated by the same DHHC acyltransferases that modify steroid hormone receptors. *Journal of Biological Chemistry*, *293*(41), 15901–15911. https://doi.org/10.1074/jbc.ra118.004167.

Epidermal Growth Factor Methods and Protocols. (2005). *British Journal of Cancer*. https://doi.org/10.1385/159745012x.

Erskine, M. S. (1989). Solicitation behavior in the estrous female rat: A review. *Hormones and Behavior*, *23*(4), 473–502. https://doi.org/10.1016/0018-506x(89)90037-8.

Evans, R. M. (1988). The steroid and thyroid hormone receptor superfamily. *Science*, *240*(4854), 889–895. https://doi.org/10.1126/science.3283939.

Evans, S. M., Haney, M., & Foltin, R. W. (2002). The effects of smoked cocaine during the follicular and luteal phases of the menstrual cycle in women. *Psychopharmacology*, *159*(4), 397–406. https://doi.org/10.1007/s00213-001-0944-7.

Everitt, B. J. (1990). Sexual motivation: A neural and behavioural analysis of the mechanisms underlying appetitive and copulatory responses of male rats. *Neuroscience & Biobehavioral Reviews*, *14*(2), 217–232. https://doi.org/10.1016/s0149-7634(05)80222-2.

Feder, H. H., Brown-Grant, K., & Corker, C. S. (1971). Pre-ovulatory progesterone, the adrenal cortex and the "critical period" for luteinizing hormone release in rats. *Journal of Endocrinology*, *50*(1), 29–39. https://doi.org/10.1677/joe.0.0500029.

Feltenstein, M. W., Henderson, A. R., & See, R. E. (2011). Enhancement of cue-induced reinstatement of cocaine-seeking in rats by yohimbine: Sex differences and the role of the estrous cycle. *Psychopharmacology*, *216*(1), 53–62. https://doi.org/10.1007/s00213-011-2187-6.

Feltenstein, M. W., & See, R. E. (2007). Plasma progesterone levels and cocaine-seeking in freely cycling female rats across the estrous cycle. *Drug and Alcohol Dependence*, *89*(2–3), 183–189. https://doi.org/10.1016/j.drugalcdep.2006.12.017.

Ferin, M., Tempone, A., Zimmering, P. E., & Wiele, R. L. V. (1969). Effect of antibodies to 17 β -estradiol and progesterone on the estrous cycle of the rat. *Endocrinology*, *85*(6), 1070–1078. https://doi.org/10.1210/endo-85-6-1070.

Filardo, E. J., Quinn, J. A., Bland, K. I., & Frackelton, A. R. (2000). Estrogen-induced activation of Erk-1 and Erk-2 requires the G protein-coupled receptor homolog, GPR30, and occurs via trans-activation of the epidermal growth factor receptor through release of HB-EGF. *Molecular Endocrinology*, *14*(10), 1649–1660. https://doi.org/10.1210/mend.14.10.0532.

Filardo, E. J., Quinn, J. A., Frackelton, A. R., & Bland, K. I. (2002). Estrogen action via the G protein-coupled receptor, GPR30: Stimulation of adenylyl cyclase and cAMP-mediated attenuation of the epidermal growth factor receptor-to-MAPK signaling axis. *Molecular Endocrinology, 16*(1), 70–84. https://doi.org/10.1210/me.16.1.70.

Floresco, S. B., West, A. R., Ash, B., Moore, H., & Grace, A. A. (2003). Afferent modulation of dopamine neuron firing differentially regulates tonic and phasic dopamine transmission. *Nature Neuroscience, 6*(9), 968–973. https://doi.org/10.1038/nn1103.

Forlano, P. M., & Woolley, C. S. (2010). Quantitative analysis of pre- and postsynaptic sex differences in the nucleus accumbens. *Journal of Comparative Neurology, 518*(8), 1330–1348. https://doi.org/10.1002/cne.22279.

Fox, H. C., Sofuoglu, M., Morgan, P. T., Tuit, K. L., & Sinha, R. (2013). The effects of exogenous progesterone on drug craving and stress arousal in cocaine dependence: Impact of gender and cue type. *Psychoneuroendocrinology, 38*(9), 1532–1544. https://doi.org/10.1016/j.psyneuen.2012.12.022.

Fuchs, R. A., Evans, K. A., Mehta, R. H., Case, J. M., & See, R. E. (2005). Influence of sex and estrous cyclicity on conditioned cue-induced reinstatement of cocaine-seeking behavior in rats. *Psychopharmacology, 179*(3), 662–672. https://doi.org/10.1007/s00213-004-2080-7.

Fuentes, N., & Silveyra, P. (2019). Chapter Three: Estrogen receptor signaling mechanisms. *Advances in Protein Chemistry and Structural Biology, 116*, 135–170. https://doi.org/10.1016/bs.apcsb.2019.01.001.

Fukata, Y., & Fukata, M. (2010). Protein palmitoylation in neuronal development and synaptic plasticity. *Nature Reviews Neuroscience, 11*(3), 161–175. https://doi.org/10.1038/nrn2788.

Galbiati, F., Engelman, J. A., Volonte, D., Zhang, X. L., Minetti, C., Li, M., et al. (2001). Caveolin-3 null mice show a loss of caveolae, changes in the microdomain distribution of the dystrophin-glycoprotein complex, and T-tubule abnormalities. *Journal of Biological Chemistry, 276*(24), 21425–21433. https://doi.org/10.1074/jbc.m100828200.

Garcia-Segura, L. M., Baetens, D., & Naftolin, F. (1986). Synaptic remodelling in arcuate nucleus after injection of estradiol valerate in adult female rats. *Brain Research, 366*(1–2), 131–136. https://doi.org/10.1016/0006-8993(86)91287-4.

Garcia-Segura, L. M., Naftolin, F., Hutchison, J. B., Azcoitia, I., & Chowen, J. A. (1999). Role of astroglia in estrogen regulation of synaptic plasticity and brain repair. *Journal of Neurobiology, 40*(4), 574–584.

Geisler, S., & Wise, R. A. (2008). Functional implications of glutamatergic projections to the ventral tegmental area. *Reviews in the Neurosciences, 19*(4–5), 227–244. https://doi.org/10.1515/revneuro.2008.19.4-5.227.

Giacometti, L., & Barker, J. (2020). Sex differences in the glutamate system: Implications for addiction. *Neuroscience & Biobehavioral Reviews, 113*, 157–168. https://doi.org/10.1016/j.neubiorev.2020.03.010.

Giguère, V., Yang, N., Segui, P., & Evans, R. M. (1988). Identification of a new class of steroid hormone receptors. *Nature, 331*(6151), 91–94. https://doi.org/10.1038/331091a0.

Goodman, R. L., Lehman, M. N., Smith, J. T., Coolen, L. M., de Oliveira, C. V. R., Jafarzadehshirazi, M. R., et al. (2007). Kisspeptin neurons in the arcuate nucleus of the ewe express both dynorphin A and neurokinin B. *Endocrinology, 148*(12), 5752–5760. https://doi.org/10.1210/en.2007-0961.

Göttlicher, M., Heck, S., & Herrlich, P. (1998). Transcriptional cross-talk, the second mode of steroid hormone receptor action. *Journal of Molecular Medicine, 76*(7), 480–489. https://doi.org/10.1007/s001090050242.

Gould, E., Woolley, C. S., Frankfurt, M., & McEwen, B. S. (1990). Gonadal steroids regulate dendritic spine density in hippocampal pyramidal cells in adulthood. *The Journal of Neuroscience: The Official Journal of the Society for Neuroscience, 10*(4), 1286–1291.

Groenewegen, H. J., der Zee, E. V.-V., te Kortschot, A., & Witter, M. P. (1987). Organization of the projections from the subiculum to the ventral striatum in the rat. A study using anterograde transport of Phaseolus vulgaris leucoagglutinin. *Neuroscience, 23*(1), 103–120. https://doi.org/10.1016/0306-4522(87)90275-2.

Groenewegen, H. J., Wright, C. I., Beijer, A. V. J., & Voorn, P. (1999). Convergence and segregation of ventral striatal inputs and outputs. *Annals of the New York Academy of Sciences, 877*(1), 49–63. https://doi.org/10.1111/j.1749-6632.1999.tb09260.x.

Gross, K. S., Brandner, D. D., Martinez, L. A., Olive, M. F., Meisel, R. L., & Mermelstein, P. G. (2016). Opposite effects of mGluR1a and mGluR5 activation on nucleus accumbens medium spiny neuron dendritic spine density. *PLoS One, 11*(9), e0162755. https://doi.org/10.1371/journal.pone.0162755.

Gross, K. S., & Mermelstein, P. G. (2020). Estrogen receptor signaling through metabotropic glutamate receptors. *Vitamins and Hormones, 114*, 211–232. https://doi.org/10.1016/bs.vh.2020.06.003.

Grove-Strawser, D., Boulware, M. I., & Mermelstein, P. G. (2010). Membrane estrogen receptors activate the metabotropic glutamate receptors mGluR5 and mGluR3 to bidirectionally regulate CREB phosphorylation in female rat striatal neurons. *Neuroscience, 170*(4), 1045–1055. https://doi.org/10.1016/j.neuroscience.2010.08.012.

Gu, G., Rojo, A. A., Zee, M. C., Yu, J., & Simerly, R. B. (1996). Hormonal regulation of CREB phosphorylation in the anteroventral periventricular nucleus. *The Journal of Neuroscience, 16*(9), 3035–3044. https://doi.org/10.1523/jneurosci.16-09-03035.1996.

Guarraci, F. A., Megroz, A. B., & Clark, A. S. (2004). Paced mating behavior in the female rat following lesions of three regions responsive to vaginocervical stimulation. *Brain Research, 999*(1), 40–52. https://doi.org/10.1016/j.brainres.2003.10.056.

Haas, A. L., & Peters, R. H. (2000). Development of substance abuse problems among drug-involved offenders: Evidence for the telescoping effect. *Journal of Substance Abuse, 12*(3), 241–253. https://doi.org/10.1016/s0899-3289(00)00053-5.

Hammes, S. R., & Levin, E. R. (2007). Extranuclear steroid receptors: Nature and actions. *Endocrine Reviews, 28*(7), 726–741. https://doi.org/10.1210/er.2007-0022.

Harrington, W. R., Kim, S. H., Funk, C. C., Madak-Erdogan, Z., Schiff, R., Katzenellenbogen, J. A., et al. (2006). Estrogen dendrimer conjugates that preferentially activate extranuclear, nongenomic versus genomic pathways of estrogen action. *Molecular Endocrinology, 20*(3), 491–502.

Hashikawa, Y., Hashikawa, K., Falkner, A. L., & Lin, D. (2017). Ventromedial hypothalamus and the generation of aggression. *Frontiers in Systems Neuroscience, 11*, 94. https://doi.org/10.3389/fnsys.2017.00094.

Heldring, N., Pike, A., Andersson, S., Matthews, J., Cheng, G., Hartman, J., et al. (2007). Estrogen receptors: How do they signal and what are their targets. *Physiological Reviews, 87*(3), 905–931. https://doi.org/10.1152/physrev.00026.2006.

Herbison, A. E. (2018). The gonadotropin-releasing hormone pulse generator. *Endocrinology, 159*(11), 3723–3736. https://doi.org/10.1210/en.2018-00653.

Herbison, A. E., & Theodosis, D. T. (1992). Localization of oestrogen receptors in preoptic neurons containing neurotensin but not tyrosine hydroxylase, cholecystokinin or luteinizing hormone-releasing hormone in the male and female rat. *Neuroscience, 50*(2), 283–298. https://doi.org/10.1016/0306-4522(92)90423-y.

Herzig, V., & Schmidt, W. J. (2004). Effects of MPEP on locomotion, sensitization and conditioned reward induced by cocaine or morphine. *Neuropharmacology, 47*(7), 973–984. https://doi.org/10.1016/j.neuropharm.2004.07.037.

Hirono, M., Yoshioka, T., & Konishi, S. (2001). GABAB receptor activation enhances mGluR-mediated responses at cerebellar excitatory synapses. *Nature Neuroscience, 4*(12), 1207–1216. https://doi.org/10.1038/nn764.

Hoffpauir, B. K., & Gleason, E. L. (2002). Activation of mGluR5 modulates GABAA receptor function in retinal amacrine cells. *Journal of Neurophysiology, 88*(4), 1766–1776. https://doi.org/10.1152/jn.2002.88.4.1766.

Hu, M., & Becker, J. B. (2003). Effects of sex and estrogen on behavioral sensitization to cocaine in rats. *The Journal of Neuroscience, 23*(2), 693–699. https://doi.org/10.1523/jneurosci.23-02-00693.2003.

Hu, M., & Becker, J. B. (2008). Acquisition of cocaine self-administration in ovariectomized female rats: Effect of estradiol dose or chronic estradiol administration. *Drug and Alcohol Dependence, 94*(1–3), 56–62. https://doi.org/10.1016/j.drugalcdep.2007.10.005.

Hu, M., Crombag, H. S., Robinson, T. E., & Becker, J. B. (2004). Biological basis of sex differences in the propensity to self-administer cocaine. *Neuropsychopharmacology, 29*(1), 81–85. https://doi.org/10.1038/sj.npp.1300301.

Huang, G. Z., & Woolley, C. S. (2012). Estradiol acutely suppresses inhibition in the hippocampus through a sex-specific endocannabinoid and mGluR-dependent mechanism. *Neuron, 74*(5), 801–808. https://doi.org/10.1016/j.neuron.2012.03.035.

Hudson, A., & Stamp, J. A. (2011). Ovarian hormones and propensity to drug relapse: A review. *Neuroscience & Biobehavioral Reviews, 35*(3), 427–436. https://doi.org/10.1016/j.neubiorev.2010.05.001.

Ikemoto, S. (2007). Dopamine reward circuitry: Two projection systems from the ventral midbrain to the nucleus accumbens–olfactory tubercle complex. *Brain Research Reviews, 56*(1), 27–78. https://doi.org/10.1016/j.brainresrev.2007.05.004.

Ikezu, T., Ueda, H., Trapp, B. D., Nishiyama, K., Sha, J. F., Volonte, D., et al. (1998). Affinity-purification and characterization of caveolins from the brain: Differential expression of caveolin-1, -2, and -3 in brain endothelial and astroglial cell types. *Brain Research, 804*(2), 177–192. https://doi.org/10.1016/s0006-8993(98)00498-3.

Irwig, M. S., Fraley, G. S., Smith, J. T., Acohido, B. V., Popa, S. M., Cunningham, M. J., et al. (2005). Kisspeptin activation of gonadotropin releasing hormone neurons and regulation of Kiss-1 mRNA in the male rat. *Neuroendocrinology, 80*(4), 264–272. https://doi.org/10.1159/000083140.

Jackson, L. R., Robinson, T. E., & Becker, J. B. (2006). Sex differences and hormonal influences on acquisition of cocaine self-administration in rats. *Neuropsychopharmacology, 31*(1), 129–138. https://doi.org/10.1038/sj.npp.1300778.

Janes, A. C., Pizzagalli, D. A., Richardt, S., de Frederick, B., Holmes, A. J., Sousa, J., et al. (2010). Neural substrates of attentional bias for smoking-related cues: An fMRI study. *Neuropsychopharmacology, 35*(12), 2339–2345. https://doi.org/10.1038/npp.2010.103.

Jenkins, W. J., & Becker, J. B. (2001). Role of the striatum and nucleus accumbens in paced copulatory behavior in the female rat. *Behavioural Brain Research, 121*(1–2), 119–128. https://doi.org/10.1016/s0166-4328(00)00394-6.

Jenkins, W. J., & Becker, J. B. (2003a). Dynamic increases in dopamine during paced copulation in the female rat. *European Journal of Neuroscience, 18*(7), 1997–2001. https://doi.org/10.1046/j.1460-9568.2003.02923.x.

Jenkins, W. J., & Becker, J. B. (2003b). Female rats develop conditioned place preferences for sex at their preferred interval. *Hormones and Behavior, 43*(4), 503–507. https://doi.org/10.1016/s0018-506x(03)00031-x.

Johnson, C., Hong, W., & Micevych, P. (2020). Optogenetic activation of β-endorphin terminals in the medial preoptic nucleus regulates sexual receptivity. *ENeuro, 7*(1). https://doi.org/10.1523/eneuro.0315-19.2019. ENEURO.0315-19.2019.

Jung-Testas, I., Thi, A. D., Koenig, H., Désarnaud, F., Shazand, K., Schumacher, M., et al. (1999). Progesterone as a neurosteroid: Synthesis and actions in rat glial cells. Proceedings of 10th International Congress on Hormonal Steroids, Quebec, Canada, 17–21 June 1998. *The Journal of Steroid Biochemistry and Molecular Biology, 69*(1–6), 97–107. https://doi.org/10.1016/s0960-0760(98)00149-6.

Kahn, L., Alonso, G., Robbe, D., Bockaert, J., & Manzoni, O. J. (2001). Group 2 metabotropic glutamate receptors induced long term depression in mouse striatal slices. *Neuroscience Letters*, *316*(3), 178–182. https://doi.org/10.1016/s0304-3940(01)02397-7.

Kalra, S. P., & Kalra, P. S. (1974). Temporal interrelationships among circulating levels of estradiol, progesterone and lh during the rat estrous cycle: Effects of exogenous progesterone 1. *Endocrinology*, *95*(6), 1711–1718. https://doi.org/10.1210/endo-95-6-1711.

Kauffman, A. S., & Smith, J. T. (Eds.). (2013). *Kisspeptin signaling in reproductive biology (Vol. 342)*. New York: Springer.

Kawa, A. B., & Robinson, T. E. (2019). Sex differences in incentive-sensitization produced by intermittent access cocaine self-administration. *Psychopharmacology*, *236*(2), 625–639. https://doi.org/10.1007/s00213-018-5091-5.

Kelley, A. E., & Domesick, V. B. (1982). The distribution of the projection from the hippocampal formation to the nucleus accumbens in the rat: An anterograde and retrograde-horseradish peroxidase study. *Neuroscience*, *7*(10), 2321–2335. https://doi.org/10.1016/0306-4522(82)90198-1.

Kelly, M. J., Moss, R. L., & Dudley, C. A. (1976). Differential sensitivity of preoptic-septal neurons to microelectrophoressed estrogen during the estrous cycle. *Brain Research*, *114*(1), 152–157. https://doi.org/10.1016/0006-8993(76)91017-9.

Kerstetter, K. A., Aguilar, V. R., Parrish, A. B., & Kippin, T. E. (2008). Protracted time-dependent increases in cocaine-seeking behavior during cocaine withdrawal in female relative to male rats. *Psychopharmacology*, *198*(1), 63–75. https://doi.org/10.1007/s00213-008-1089-8.

Kim, K. M., Baratta, M. V., Yang, A., Lee, D., Boyden, E. S., & Fiorillo, C. D. (2012). Optogenetic mimicry of the transient activation of dopamine neurons by natural reward is sufficient for operant reinforcement. *PLoS One*, *7*(4), e33612. https://doi.org/10.1371/journal.pone.0033612.

Kohlert, J. G., & Meisel, R. L. (1999). Sexual experience sensitizes mating-related nucleus accumbens dopamine responses of female Syrian hamsters. *Behavioural Brain Research*, *99*(1), 45–52. https://doi.org/10.1016/s0166-4328(98)00068-0.

Koike, S., Sakai, M., & Muramatsu, M. (1987). Molecular cloning and characterization of rat estrogen receptor cDNA. *Nucleic Acids Research*, *15*(6), 2499–2513. https://doi.org/10.1093/nar/15.6.2499.

Kokane, S. S., & Perrotti, L. I. (2020). Sex differences and the role of estradiol in mesolimbic reward circuits and vulnerability to cocaine and opiate addiction. *Frontiers in Behavioral Neuroscience*, *14*, 74. https://doi.org/10.3389/fnbeh.2020.00074.

Koob, G. F., & Volkow, N. D. (2016). Neurobiology of addiction: A neurocircuitry analysis. *The Lancet Psychiatry*, *3*(8), 760–773. https://doi.org/10.1016/s2215-0366(16)00104-8.

Kosten, T. A., Gawin, F. H., Kosten, T. R., & Rounsaville, B. J. (1993). Gender differences in cocaine use and treatment response. *Journal of Substance Abuse Treatment*, *10*(1), 63–66. https://doi.org/10.1016/0740-5472(93)90100-g.

Krajewska, W. M., & Masłowska, I. (2004). Caveolins: Structure and function in signal transduction. *Cellular & Molecular Biology Letters*, *9*(2), 195–220.

Kritzer, M. F., & Creutz, L. M. (2008). Region and sex differences in constituent dopamine neurons and immunoreactivity for intracellular estrogen and androgen receptors in mesocortical projections in rats. *Journal of Neuroscience*, *28*(38), 9525–9535. https://doi.org/10.1523/jneurosci.2637-08.2008.

Kumaresan, V., Yuan, M., Yee, J., Famous, K. R., Anderson, S. M., Schmidt, H. D., et al. (2009). Metabotropic glutamate receptor 5 (mGluR5) antagonists attenuate cocaine priming- and cue-induced reinstatement of cocaine seeking. *Behavioural Brain Research*, *202*(2), 238–244. https://doi.org/10.1016/j.bbr.2009.03.039.

Kuo, J., Hamid, N., Bondar, G., Prossnitz, E. R., & Micevych, P. (2010). Membrane estrogen receptors stimulate intracellular calcium release and progesterone synthesis in hypothalamic astrocytes. *Journal of Neuroscience, 30*(39), 12950–12957. https://doi.org/10.1523/jneurosci.1158-10.2010.

Kuo, J., Hariri, O. R., Bondar, G., Ogi, J., & Micevych, P. (2009). Membrane estrogen receptor-α interacts with metabotropic glutamate receptor type 1a to mobilize intracellular calcium in hypothalamic astrocytes. *Endocrinology, 150*(3), 1369–1376. https://doi.org/10.1210/en.2008-0994.

Kutlu, M. G., & Gould, T. J. (2016). Effects of drugs of abuse on hippocampal plasticity and hippocampus-dependent learning and memory: Contributions to development and maintenance of addiction. *Learning & Memory, 23*(10), 515–533. https://doi.org/10.1101/lm.042192.116.

La Rosa, P., Pesiri, V., Leclercq, G., Marino, M., & Acconcia, F. (2012). Palmitoylation regulates 17β-estradiol-induced estrogen receptor-α degradation and transcriptional activity. *Molecular Endocrinology, 26*(5), 762–774. https://doi.org/10.1210/me.2011-1208.

Labhsetwar, A. P. (1970). Role of estrogens in ovulation: A study using the estrogenantagonist, I.C.I. 46, 474. *Endocrinology, 87*(3), 542–551. https://doi.org/10.1210/endo-87-3-542.

Larson, E. B., Anker, J. J., Gliddon, L. A., Fons, K. S., & Carroll, M. E. (2007). Effects of estrogen and progesterone on the escalation of cocaine self-administration in female rats during extended access. *Experimental and Clinical Psychopharmacology, 15*(5), 461–471. https://doi.org/10.1037/1064-1297.15.5.461.

Levin, E. R. (2002). Cellular functions of plasma membrane estrogen receptors. *Steroids, 67*(6), 471–475. https://doi.org/10.1016/s0039-128x(01)00179-9.

Lischinsky, J. E., & Lin, D. (2020). Neural mechanisms of aggression across species. *Nature Neuroscience, 23*(11), 1317–1328. https://doi.org/10.1038/s41593-020-00715-2.

Long, N., Serey, C., & Sinchak, K. (2014). 17β-estradiol rapidly facilitates lordosis through G protein-coupled estrogen receptor 1 (GPER) via deactivation of medial preoptic nucleus μ-opioid receptors in estradiol primed female rats. *Hormones and Behavior, 66*(4), 663–666. https://doi.org/10.1016/j.yhbeh.2014.09.008.

Loweth, J. A., Scheyer, A. F., Milovanovic, M., LaCrosse, A. L., Flores-Barrera, E., Werner, C. T., et al. (2014). Synaptic depression via mGluR1 positive allosteric modulation suppresses cue-induced cocaine craving. *Nature Neuroscience, 17*(1), 73–80. https://doi.org/10.1038/nn.3590.

Luo, J., & Liu, D. (2020). Does GPER really function as a G protein-coupled estrogen receptor in vivo? *Frontiers in Endocrinology, 11*, 148.

Lüscher, C., & Huber, K. M. (2010). Group 1 mGluR-dependent synaptic long-term depression: Mechanisms and implications for circuitry and disease. *Neuron, 65*(4), 445–459. https://doi.org/10.1016/j.neuron.2010.01.016.

Lynch, W. J., Roth, M. E., Mickelberg, J. L., & Carroll, M. E. (2001). Role of estrogen in the acquisition of intravenously self-administered cocaine in female rats. *Pharmacology Biochemistry and Behavior, 68*(4), 641–646. https://doi.org/10.1016/s0091-3057(01)00455-5.

Maccarrone, M. (2004). Sex and drug abuse: A role for retrograde endocannabinoids? *Trends in Pharmacological Sciences, 25*(9), 455–456. https://doi.org/10.1016/j.tips.2004.07.001.

Mahavongtrakul, M., Kanjiya, M. P., Maciel, M., Kanjiya, S., & Sinchak, K. (2013). Estradiol dose-dependent regulation of membrane estrogen receptor-α, metabotropic glutamate receptor-1a, and their complexes in the arcuate nucleus of the hypothalamus in female rats. *Endocrinology, 154*(9), 3251–3260. https://doi.org/10.1210/en.2013-1235.

Mahesh, V. B., Dhandapani, K. M., & Brann, D. W. (2006). Role of astrocytes in reproduction and neuroprotection. *Molecular and Cellular Endocrinology, 246*(1–2), 1–9. https://doi.org/10.1016/j.mce.2005.11.017.

Mann, D. R., & Barraclough, C. A. (1973). Role of estrogen and progesterone in facilitating LH release in 4-Day cyclic rats. *Endocrinology*, *93*(3), 694–699. https://doi.org/10.1210/endo-93-3-694.

Mann, D. R., Korowitz, C. D., & Barraclough, C. A. (1975). Adrenal gland involvement in synchronizing the preovulatory release of LH in rats 1. *Proceedings of the Society for Experimental Biology and Medicine*, *150*(1), 115–120. https://doi.org/10.3181/00379727-150-38985.

Maria, M. M. M.-S., Flanagan, J., & Brady, K. (2014). Ovarian hormones and drug abuse. *Current Psychiatry Reports*, *16*(11), 511. https://doi.org/10.1007/s11920-014-0511-7.

Marino, M., Galluzzo, P., & Ascenzi, P. (2006). Estrogen signaling multiple pathways to impact gene transcription. *Current Genomics*, *7*(8), 497–508. https://doi.org/10.2174/138920206779315737.

Martinez, L. A., Gross, K. S., Himmler, B. T., Emmitt, N. L., Peterson, B. M., Zlebnik, N. E., et al. (2016). Estradiol facilitation of cocaine self-administration in female rats requires activation of mGluR5. *ENeuro*, *3*(5). https://doi.org/10.1523/eneuro.0140-16.2016. ENEURO.0140-16.2016.

Martinez, L. A., Peterson, B. M., Meisel, R. L., & Mermelstein, P. G. (2014). Estradiol facilitation of cocaine-induced locomotor sensitization in female rats requires activation of mGluR5. *Behavioural Brain Research*, *271*, 39–42. https://doi.org/10.1016/j.bbr.2014.05.052.

Matsumoto, A., & Arai, Y. (1981). Neuronal plasticity in the deafferented hypothalamic arcuate nucleus of adult female rats and its enhancement by treatment with estrogen. *Journal of Comparative Neurology*, *197*(2), 197–205. https://doi.org/10.1002/cne.901970203.

McArthur, S., McHale, E., & Gillies, G. E. (2007). The size and distribution of midbrain dopaminergic populations are permanently altered by perinatal glucocorticoid exposure in a sex- region- and time-specific manner. *Neuropsychopharmacology*, *32*(7), 1462–1476. https://doi.org/10.1038/sj.npp.1301277.

McClintock, M. K., & Adler, N. T. (1978). The role of the female during copulation in wild and domestic Norway rats (rattus norvegicus). *Behaviour*, *67*(1–2), 67–95. https://doi.org/10.1163/156853978x00260.

McCutcheon, J. E., Loweth, J. A., Ford, K. A., Marinelli, M., Wolf, M. E., & Tseng, K. Y. (2011). Group I mGluR activation reverses cocaine-induced accumulation of calcium-permeable ampa receptors in nucleus accumbens synapses via a protein kinase C-dependent mechanism. *Journal of Neuroscience*, *31*(41), 14536–14541. https://doi.org/10.1523/jneurosci.3625-11.2011.

McEwen, B. S., Jones, K. J., & Pfaff, D. W. (1987). Hormonal control of sexual behavior in the female rat: Molecular, cellular and neurochemical studies. *Biology of Reproduction*, *36*(1), 37–45. https://doi.org/10.1095/biolreprod36.1.37.

McGeehan, A. J., Janak, P. H., & Olive, M. F. (2004). Effect of the mGluR5 antagonist 6-methyl-2-(phenylethynyl)pyridine (MPEP) on the acute locomotor stimulant properties of cocaine, d-amphetamine, and the dopamine reuptake inhibitor GBR12909 in mice. *Psychopharmacology*, *174*(2), 266–273. https://doi.org/10.1007/s00213-003-1733-2.

McHugh, R. K., Votaw, V. R., Sugarman, D. E., & Greenfield, S. F. (2018). Sex and gender differences in substance use disorders. *Clinical Psychology Review*, *66*, 12–23. https://doi.org/10.1016/j.cpr.2017.10.012.

McKenna, N. J., Lanz, R. B., & O'Malley, B. W. (1999). Nuclear receptor coregulators: Cellular and molecular biology. *Endocrine Reviews*, *20*(3), 321–344. https://doi.org/10.1210/edrv.20.3.0366.

Meisel, R. L., Camp, D. M., & Robinson, T. E. (1993). A microdialysis study of ventral striatal dopamine during sexual behavior in female Syrian hamsters. *Behavioural Brain Research*, *55*(2), 151–157. https://doi.org/10.1016/0166-4328(93)90111-3.

Meisel, R. L., & Joppa, M. A. (1994). Conditioned place preference in female hamsters following aggressive or sexual encounters. *Physiology & Behavior, 56*(5), 1115–1118. https://doi.org/10.1016/0031-9384(94)90352-2.

Meisel, R. L., & Luttrell, V. R. (1990). Estradiol increases the dendritic length of ventromedial hypothalamic neurons in female Syrian Hamsters. *Brain Research Bulletin, 25*(1), 165–168. https://doi.org/10.1016/0361-9230(90)90269-6.

Meisel, R. L., & Mullins, A. J. (2006). Sexual experience in female rodents: Cellular mechanisms and functional consequences. *Brain Research, 1126*(1), 56–65. https://doi.org/10.1016/j.brainres.2006.08.050.

Meitzen, J., Luoma, J. I., Boulware, M. I., Hedges, V. L., Peterson, B. M., Tuomela, K., et al. (2013). Palmitoylation of estrogen receptors is essential for neuronal membrane signaling. *Endocrinology, 154*(11), 4293–4304. https://doi.org/10.1210/en.2013-1172.

Meitzen, J., & Mermelstein, P. G. (2011). Estrogen receptors stimulate brain region specific metabotropic glutamate receptors to rapidly initiate signal transduction pathways. *Journal of Chemical Neuroanatomy, 42*(4), 236–241. https://doi.org/10.1016/j.jchemneu.2011.02.002.

Merchenthaler, I., Lane, M. V., Numan, S., & Dellovade, T. L. (2004). Distribution of estrogen receptor α and β in the mouse central nervous system: In vivo autoradiographic and immunocytochemical analyses. *Journal of Comparative Neurology, 473*(2), 270–291. https://doi.org/10.1002/cne.20128.

Meredith, G. E., Baldo, B. A., Andrezjewski, M. E., & Kelley, A. E. (2008). The structural basis for mapping behavior onto the ventral striatum and its subdivisions. *Brain Structure and Function, 213*(1–2), 17–27. https://doi.org/10.1007/s00429-008-0175-3.

Mermelstein, P. G., & Becker, J. B. (1995). Increased extracellular dopamine in the nucleus accumbens and striatum of the female rat during paced copulatory behavior. *Behavioral Neuroscience, 109*(2), 354–365. https://doi.org/10.1037/0735-7044.109.2.354.

Mermelstein, P. G., Becker, J. B., & Surmeier, D. J. (1996). Estradiol reduces calcium currents in rat neostriatal neurons via a membrane receptor. *The Journal of Neuroscience: The Official Journal of the Society for Neuroscience, 16*(2), 595–604.

Messager, S., Chatzidaki, E. E., Ma, D., Hendrick, A. G., Zahn, D., Dixon, J., et al. (2005). Kisspeptin directly stimulates gonadotropin-releasing hormone release via G protein-coupled receptor 54. *Proceedings of the National Academy of Sciences, 102*(5), 1761–1766. https://doi.org/10.1073/pnas.0409330102.

Meyer, M. R., & Barton, M. (2009). ERα, ERβ, and gp ER: Novel aspects of oestrogen receptor signalling in atherosclerosis. *Cardiovascular Research, 83*(4), 605–610. https://doi.org/10.1093/cvr/cvp187.

Micevych, P. E., Chaban, V., Ogi, J., Dewing, P., Lu, J. K. H., & Sinchak, K. (2007). Estradiol stimulates progesterone synthesis in hypothalamic astrocyte cultures. *Endocrinology, 148*(2), 782–789. https://doi.org/10.1210/en.2006-0774.

Micevych, P., & Christensen, A. (2012). Membrane-initiated estradiol actions mediate structural plasticity and reproduction. *Frontiers in Neuroendocrinology, 33*(4), 331–341. https://doi.org/10.1016/j.yfrne.2012.07.003.

Micevych, P. E., Eckersell, C. B., Brecha, N., & Holland, K. L. (1997). Estrogen modulation of opioid and cholecystokinin systems in the limbic-hypothalamic circuit. *Brain Research Bulletin, 44*(4), 335–343. https://doi.org/10.1016/s0361-9230(97)00212-8.

Micevych, P. E., & Meisel, R. L. (2017). Integrating neural circuits controlling female sexual behavior. *Frontiers in Systems Neuroscience, 11*, 42. https://doi.org/10.3389/fnsys.2017.00042.

Micevych, P. E., & Mermelstein, P. G. (2008). Membrane estrogen receptors acting through metabotropic glutamate receptors: An emerging mechanism of estrogen action in brain. *Molecular Neurobiology, 38*(1), 66. https://doi.org/10.1007/s12035-008-8034-z.

Micevych, P. E., Mermelstein, P. G., & Sinchak, K. (2017). Estradiol membrane-initiated signaling in the brain mediates reproduction. *Trends in Neurosciences, 40*(11), 654–666. https://doi.org/10.1016/j.tins.2017.09.001.

Micevych, P. E., Rissman, E. F., Gustafsson, J., & Sinchak, K. (2003). Estrogen receptor-α is required for estrogen-induced µ-opioid receptor internalization. *Journal of Neuroscience Research, 71*(6), 802–810. https://doi.org/10.1002/jnr.10526.

Micevych, P., & Sinchak, K. (2011). The neurosteroid progesterone underlies estrogen positive feedback of the LH surge. *Frontiers in Endocrinology, 2,* 90. https://doi.org/10.3389/fendo.2011.00090.

Micevych, P., & Sinchak, K. (2013). Temporal and concentration-dependent effects of oestradiol on neural pathways mediating sexual receptivity. *Journal of Neuroendocrinology, 25*(11), 1012–1023. https://doi.org/10.1111/jne.12103.

Micevych, P., Sinchak, K., Mills, R. H., Tao, L., LaPolt, P., & Lu, J. K. H. (2003). The luteinizing hormone surge is preceded by an estrogen-induced increase of hypothalamic progesterone in ovariectomized and adrenalectomized rats. *Neuroendocrinology, 78*(1), 29–35. https://doi.org/10.1159/000071703.

Mikol, D. D., Hong, H. L., Cheng, H., & Feldman, E. L. (1999). Caveolin-1 expression in schwann cells. *Glia, 27*(1), 39–52. https://doi.org/10.1002/(sici)1098-1136(199907)27:1.

Miller, E. K. (2000). The prefontral cortex and cognitive control. *Nature Reviews Neuroscience, 1*(1), 59–65. https://doi.org/10.1038/35036228.

Mills, R. H., Sohn, R. K., & Micevych, P. E. (2004). Estrogen-induced µ-opioid receptor internalization in the medial preoptic nucleus is mediated via neuropeptide Y-Y1 receptor activation in the arcuate nucleus of female rats. *The Journal of Neuroscience, 24*(4), 947–955. https://doi.org/10.1523/jneurosci.1366-03.2004.

Mitrano, D. A., & Smith, Y. (2007). Comparative analysis of the subcellular and subsynaptic localization of mGluR1a and mGluR5 metabotropic glutamate receptors in the shell and core of the nucleus accumbens in rat and monkey. *Journal of Comparative Neurology, 500*(4), 788–806. https://doi.org/10.1002/cne.21214.

Mohr, M. A., Wong, A. M., Sukumar, G., Dalgard, C. L., Hong, W., Wu, T. J., et al. (2021). RNA-sequencing of AVPV and ARH reveals vastly different temporal and transcriptomic responses to estradiol in the female rat hypothalamus. *PLoS One, 16*(8), e0256148. https://doi.org/10.1371/journal.pone.0256148.

Morales, M., & Margolis, E. B. (2017). Ventral tegmental area: Cellular heterogeneity, connectivity and behaviour. *Nature Reviews Neuroscience, 18*(2), 73–85. https://doi.org/10.1038/nrn.2016.165.

Murase, S., Grenhoff, J., Chouvet, G., Gonon, F. G., & Svensson, T. H. (1993). Prefrontal cortex regulates burst firing and transmitter release in rat mesolimbic dopamine neurons studied in vivo. *Neuroscience Letters, 157*(1), 53–56. https://doi.org/10.1016/0304-3940(93)90641-w.

Navarro, V. M., Gottsch, M. L., Chavkin, C., Okamura, H., Clifton, D. K., & Steiner, R. A. (2009). Regulation of gonadotropin-releasing hormone secretion by kisspeptin/dynorphin/neurokinin B neurons in the arcuate nucleus of the mouse. *Journal of Neuroscience, 29*(38), 11859–11866. https://doi.org/10.1523/jneurosci.1569-09.2009.

Nestler, E. J. (2001). Molecular basis of long-term plasticity underlying addiction. *Nature Reviews Neuroscience, 2*(2), 119–128. https://doi.org/10.1038/35053570.

Nicholls, R. E., Zhang, X., Bailey, C. P., Conklin, B. R., Kandel, E. R., & Stanton, P. K. (2006). mGluR2 acts through inhibitory Gα subunits to regulate transmission and long-term plasticity at hippocampal mossy fiber-CA3 synapses. *Proceedings of the National Academy of Sciences, 103*(16), 6380–6385. https://doi.org/10.1073/pnas.0601267103.

Niswender, C. M., & Conn, P. J. (2010). Metabotropic glutamate receptors: Physiology, pharmacology, and disease. *Annual Review of Pharmacology and Toxicology, 50*(1), 295–322. https://doi.org/10.1146/annurev.pharmtox.011008.145533.

Olière, S., Jolette-Riopel, A., Potvin, S., & Jutras-Aswad, D. (2013). Modulation of the endocannabinoid system: Vulnerability factor and new treatment target for stimulant addiction. *Frontiers in Psychiatry*, *4*, 109. https://doi.org/10.3389/fpsyt.2013.00109.

O'Malley, B. W., & Tsai, M.-J. (1992). Molecular pathways of steroid receptor action. *Biology of Reproduction*, *46*(2), 163–167. https://doi.org/10.1095/biolreprod46.2.163.

Opitz, B. (2014). Memory function and the hippocampus. *Frontiers of Neurology and Neuroscience*, *34*, 51–59. https://doi.org/10.1159/000356422.

Page, G., Khidir, F. A. L., Pain, S., Barrier, L., Fauconneau, B., Guillard, O., et al. (2006). Group I metabotropic glutamate receptors activate the p70S6 kinase via both mammalian target of rapamycin (mTOR) and extracellular signal-regulated kinase (ERK 1/2) signaling pathways in rat striatal and hippocampal synaptoneurosomes. *Neurochemistry International*, *49*(4), 413–421. https://doi.org/10.1016/j.neuint.2006.01.020.

Palomero-Gallagher, N., Bidmon, H., & Zilles, K. (2003). AMPA, kainate, and NMDA receptor densities in the hippocampus of untreated male rats and females in estrus and diestrus. *Journal of Comparative Neurology*, *459*(4), 468–474. https://doi.org/10.1002/cne.10638.

Pawlak, J., Karolczak, M., Krust, A., Chambon, P., & Beyer, C. (2005). Estrogen receptor-α is associated with the plasma membrane of astrocytes and coupled to the MAP/Src-kinase pathway. *Glia*, *50*(3), 270–275. https://doi.org/10.1002/glia.20162.

Pedram, A., Razandi, M., Deschenes, R. J., & Levin, E. R. (2012). DHHC-7 and -21 are palmitoylacyltransferases for sex steroid receptors. *Molecular Biology of the Cell*, *23*(1), 188–199. https://doi.org/10.1091/mbc.e11-07-0638.

Peris, J., Decambre, N., Coleman-Hardee, M. L., & Simpkins, J. W. (1991). Estradiol enhances behavioral sensitization to cocaine and amphetamine-stimulated striatal [3H] dopamine release. *Brain Research*, *566*(1–2), 255–264. https://doi.org/10.1016/0006-8993(91)91706-7.

Perry, A. N., Westenbroek, C., & Becker, J. B. (2013). Impact of pubertal and adult estradiol treatments on cocaine self-administration. *Hormones and Behavior*, *64*(4), 573–578. https://doi.org/10.1016/j.yhbeh.2013.08.007.

Peters, J., Kalivas, P. W., & Quirk, G. J. (2009). Extinction circuits for fear and addiction overlap in prefrontal cortex. *Learning & Memory*, *16*(5), 279–288. https://doi.org/10.1101/lm.1041309.

Peterson, B. M., Martinez, L. A., Meisel, R. L., & Mermelstein, P. G. (2016). Estradiol impacts the endocannabinoid system in female rats to influence behavioral and structural responses to cocaine. *Neuropharmacology*, *110*(Pt A), 118–124. https://doi.org/10.1016/j.neuropharm.2016.06.002.

Peterson, B. M., Mermelstein, P. G., & Meisel, R. L. (2015). Estradiol mediates dendritic spine plasticity in the nucleus accumbens core through activation of mGluR5. *Brain Structure and Function*, *220*(4), 2415–2422. https://doi.org/10.1007/s00429-014-0794-9.

Pfaff, D., & Keiner, M. (1973). Atlas of estradiol-concentrating cells in the central nervous system of the female rat. *Journal of Comparative Neurology*, *151*(2), 121–157. https://doi.org/10.1002/cne.901510204.

Pfaff, D., Ogawa, S., Kia, K., Vasudevan, N., Krebs, C., Frohlich, J., et al. (2002). Hormones, brain and behavior. Part III: Cellular and molecular mechanisms of hormone actions on behavior. *Neuroendocrinology*, *712000*. https://doi.org/10.1016/b978-012532104-4/50049-4. 441–XXII.

Pfaus, J. G., & Phillips, A. G. (1991). Role of dopamine in anticipatory and consummatory aspects of sexual behavior in the male rat. *Behavioral Neuroscience*, *105*(5), 727–743. https://doi.org/10.1037/0735-7044.105.5.727.

Post, R. M., Lockfeld, A., Squillace, K. M., & Contel, N. R. (1981). Drug-environment interaction: Context dependency of cocaine-induced behavioral sensitization. *Life Sciences*, *28*(7), 755–760. https://doi.org/10.1016/0024-3205(81)90157-0.

Proaño, S. B., Morris, H. J., Kunz, L. M., Dorris, D. M., & Meitzen, J. (2018). Estrous cycle-induced sex differences in medium spiny neuron excitatory synaptic transmission and intrinsic excitability in adult rat nucleus accumbens core. *Journal of Neurophysiology*, *120*(3), 1356–1373. https://doi.org/10.1152/jn.00263.2018.

Prossnitz, E. R., Arterburn, J. B., & Sklar, L. A. (2007). GPR30: A G protein-coupled receptor for estrogen. *Molecular and Cellular Endocrinology*, *265*, 138–142. https://doi.org/10.1016/j.mce.2006.12.010.

Prossnitz, E. R., & Barton, M. (2011). The G-protein-coupled estrogen receptor GPER in health and disease. *Nature Reviews Endocrinology*, *7*(12), 715–726. https://doi.org/10.1038/nrendo.2011.122.

Quesada, A., Romeo, H. E., & Micevych, P. (2007). Distribution and localization patterns of estrogen receptor-β and insulin-like growth factor-1 receptors in neurons and glial cells of the female rat substantia nigra: Localization of ERβ and IGF-1R in substantia nigra. *Journal of Comparative Neurology*, *503*(1), 198–208. https://doi.org/10.1002/cne.21358.

Rai, D., Frolova, A., Frasor, J., Carpenter, A. E., & Katzenellenbogen, B. S. (2005). Distinctive actions of membrane-targeted versus nuclear localized estrogen receptors in breast cancer cells. *Molecular Endocrinology*, *19*(6), 1606–1617. https://doi.org/10.1210/me.2004-0468.

Ramôa, C. P., Doyle, S. E., Naim, D. W., & Lynch, W. J. (2013). Estradiol as a mechanism for sex differences in the development of an addicted phenotype following extended access cocaine self-administration. *Neuropsychopharmacology*, *38*(9), 1698–1705. https://doi.org/10.1038/npp.2013.68.

Razandi, M., Oh, P., Pedram, A., Schnitzer, J., & Levin, E. R. (2002). ERs associate with and regulate the production of caveolin: Implications for signaling and cellular actions. *Molecular Endocrinology*, *16*(1), 100–115. https://doi.org/10.1210/mend.16.1.0757.

Razandi, M., Pedram, A., Greene, G., & Levin, E. (1999). Cell membrane and nuclear estrogen receptors (ers) originate from a single transcript: Studies of ERα and ERβ expressed in Chinese hamster ovary cells. *Molecular Endocrinology*, *13*(2), 307–319. https://doi.org/10.1210/me.13.2.307.

Reid, A. G., Lingford-Hughes, A. R., Cancela, L. M., & Kalivas, P. W. (2012). Chapter 24, Substance abuse disorders. *Handbook of Clinical Neurology*, *106*, 419–431. https://doi.org/10.1016/b978-0-444-52002-9.00024-3.

Revankar, C. M., Cimino, D. F., Sklar, L. A., Arterburn, J. B., & Prossnitz, E. R. (2005). A transmembrane intracellular estrogen receptor mediates rapid cell signaling. *Science*, *307*(5715), 1625–1630. https://doi.org/10.1126/science.1106943.

Richard, J. E., López-Ferreras, L., Anderberg, R. H., Olandersson, K., & Skibicka, K. P. (2017). Estradiol is a critical regulator of food-reward behavior. *Psychoneuroendocrinology*, *78*, 193–202. https://doi.org/10.1016/j.psyneuen.2017.01.014.

Rivas, F., & Mir, D. (1991). Accumbens lesion in female rats increases mount rejection without modifying lordosis. *Revista Española de Fisiología*, *47*(1), 1–6. http://europepmc.org/abstract/MED/1871414.

Roberts, D. C. S., Bennett, S. A. L., & Vickers, G. J. (1989). The estrous cycle affects cocaine self-administration on a progressive ratio schedule in rats. *Psychopharmacology*, *98*(3), 408–411. https://doi.org/10.1007/bf00451696.

Robinson, T. E., & Berridge, K. C. (2003). Addiction. *Annual Review of Psychology*, *54*(1), 25–53. https://doi.org/10.1146/annurev.psych.54.101601.145237.

Roughley, S., & Killcross, S. (2021). The role of the infralimbic cortex in decision making processes. *Current Opinion in Behavioral Sciences*, *41*, 138–143. https://doi.org/10.1016/j.cobeha.2021.06.003.

Ruda-Kucerova, J., Amchova, P., Babinska, Z., Dusek, L., Micale, V., & Sulcova, A. (2015). Sex differences in the reinstatement of methamphetamine seeking after forced abstinence in Sprague-Dawley rats. *Frontiers in Psychiatry*, *6*, 91. https://doi.org/10.3389/fpsyt.2015.00091.

Sar, M., Sahu, A., Crowley, W. R., & Kalra, S. P. (1990). Localization of neuropeptide-Y immunoreactivity in estradiol-concentrating cells in the hypothalamus*. *Endocrinology*, *127*(6), 2752–2756. https://doi.org/10.1210/endo-127-6-2752.

Sazdanović, M., Mitrović, S., Živanović-Mačužić, I., Jeremić, D., Tanasković, I., Milosavljević, Z., et al. (2013). Sexual dimorphism of medium-sized neurons with spines in human nucleus accumbens. *Archives of Biological Sciences*, *65*(3), 1149–1155. https://doi.org/10.2298/abs1303149s.

Schlegel, A., Wang, C., Katzenellenbogen, B. S., Pestell, R. G., & Lisanti, M. P. (1999). Caveolin-1 potentiates estrogen receptor α (ERα) signaling: Caveolin-1 drives ligand-independent nuclear translocation and activation of ERα. *Journal of Biological Chemistry*, *274*(47), 33551–33556. https://doi.org/10.1074/jbc.274.47.33551.

Schlegel, A., Wang, C., Pestell, R. G., & Lisanti, M. P. (2001). Ligand-independent activation of oestrogen receptor α by caveolin-1. *Biochemical Journal*, *359*(1), 203–210. https://doi.org/10.1042/bj3590203.

Schultz, K. N., von Esenwein, S. A., Hu, M., Bennett, A. L., Kennedy, R. T., Musatov, S., et al. (2009). Viral vector-mediated overexpression of estrogen receptor-α in striatum enhances the estradiol-induced motor activity in female rats and estradiol-modulated GABA release. *Journal of Neuroscience*, *29*(6), 1897–1903. https://doi.org/10.1523/jneurosci.4647-08.2009.

Segarra, A. C., Agosto-Rivera, J. L., Febo, M., Lugo-Escobar, N., Menéndez-Delmestre, R., Puig-Ramos, A., et al. (2010). Estradiol: A key biological substrate mediating the response to cocaine in female rats. *Hormones and Behavior*, *58*(1), 33–43. https://doi.org/10.1016/j.yhbeh.2009.12.003.

Sell, S. L., Thomas, M. L., & Cunningham, K. A. (2002). Influence of estrous cycle and estradiol on behavioral sensitization to cocaine in female rats. *Drug and Alcohol Dependence*, *67*(3), 281–290. https://doi.org/10.1016/s0376-8716(02)00085-6.

Shivers, B. D., Harlan, R. E., Morrell, J. I., & Pfaff, D. W. (1983). Absence of oestradiol concentration in cell nuclei of LHRH-immunoreactive neurones. *Nature*, *304*(5924), 345–347. https://doi.org/10.1038/304345a0.

Shughrue, P. J., Komm, B., & Merchenthaler, I. (1996). The distribution of estrogen receptor-β mRNA in the rat hypothalamus. *Steroids*, *61*(12), 678–681. https://doi.org/10.1016/s0039-128x(96)00222-x.

Shughrue, P. J., Lane, M. V., & Merchenthaler, I. (1997). Comparative distribution of estrogen receptor-α and -β mRNA in the rat central nervous system. *Journal of Comparative Neurology*, *388*(4), 507–525. https://doi.org/10.1002/(sici)1096-9861(19971201)388:4<507::aid-cne1>3.0.co;2-6.

Simerly, R. B., & Swanson, L. W. (1988). Projections of the medial preoptic nucleus: A Phaseolus vulgaris leucoagglutinin anterograde tract-tracing study in the rat. *Journal of Comparative Neurology*, *270*(2), 209–242. https://doi.org/10.1002/cne.902700205.

Simerly, R. B., Swanson, L. W., Chang, C., & Muramatsu, M. (1990). Distribution of androgen and estrogen receptor mRNA-containing cells in the rat brain: An in situ hybridization study. *The Journal of Comparative Neurology*, *294*(1), 76–95. https://doi.org/10.1002/cne.902940107.

Simpson, E. H., & Balsam, P. D. (2015). Behavioral neuroscience of motivation. *Current Topics in Behavioral Neurosciences*, *27*, 1–12. https://doi.org/10.1007/7854_2015_402.

Sinchak, K., Dewing, P., Ponce, L., Gomez, L., Christensen, A., Berger, M., et al. (2013). Modulation of the arcuate nucleus–medial preoptic nucleus lordosis regulating circuit: A role for GABAB receptors. *Hormones and Behavior*, *64*(1), 136–143. https://doi.org/10.1016/j.yhbeh.2013.06.001.

Sinchak, K., & Micevych, P. E. (2001). Progesterone blockade of estrogen activation of μ-opioid receptors regulates reproductive behavior. *The Journal of Neuroscience*, *21*(15), 5723–5729. https://doi.org/10.1523/jneurosci.21-15-05723.2001.

Sinchak, K., & Micevych, P. (2003). Visualizing activation of opioid circuits by internalization of G protein-coupled receptors. *Molecular Neurobiology*, *27*(2), 197–222. https://doi.org/10.1385/mn:27:2:197.

Sinchak, K., Mills, R. H., Tao, L., LaPolt, P., Lu, J. K. H., & Micevych, P. (2003). Estrogen induces de novo progesterone synthesis in astrocytes. *Developmental Neuroscience*, *25*(5), 343–348. https://doi.org/10.1159/000073511.

Sinchak, K., Mohr, M. A., & Micevych, P. E. (2020). Hypothalamic astrocyte development and physiology for neuroprogesterone induction of the luteinizing hormone surge. *Frontiers in Endocrinology*, *11*, 420. https://doi.org/10.3389/fendo.2020.00420.

Sircar, R., & Kim, D. (1999). Female gonadal hormones differentially modulate cocaine-induced behavioral sensitization in Fischer, Lewis, and Sprague-Dawley rats. *Journal of Pharmacology and Experimental Therapeutics*, *289*(1), 54–65. https://jpet.aspetjournals.org/content/289/1/54.

Smith, M. S., Freeman, M. E., & Neill, J. D. (1975). The control of progesterone secretion during the estrous cycle and early pseudopregnancy in the rat: Prolactin, gonadotropin and steroid levels associated with rescue of the corpus luteum of pseudopregnancy. *Endocrinology*, *96*(1), 219–226. https://doi.org/10.1210/endo-96-1-219.

Smith, R. J., Lobo, M. K., Spencer, S., & Kalivas, P. W. (2013). Cocaine-induced adaptations in D1 and D2 accumbens projection neurons (a dichotomy not necessarily synonymous with direct and indirect pathways). *Current Opinion in Neurobiology*, *23*(4), 546–552. https://doi.org/10.1016/j.conb.2013.01.026.

Song, Z., Yang, H., Peckham, E. M., & Becker, J. B. (2019). Estradiol-induced potentiation of dopamine release in dorsal striatum following amphetamine administration requires estradiol receptors and mGlu5. *ENeuro*, *6*(1). https://doi.org/10.1523/eneuro.0446-18.2019. ENEURO.0446-18.

Spampinato, S. F., Molinaro, G., Merlo, S., Iacovelli, L., Caraci, F., Battaglia, G., et al. (2012). Estrogen receptors and type 1 metabotropic glutamate receptors are interdependent in protecting cortical neurons against β-amyloid toxicity. *Molecular Pharmacology*, *81*(1), 12–20. https://doi.org/10.1124/mol.111.074021.

Spreafico, E., Bettini, E., Pollio, G., & Maggi, A. (1992). Nucleotide sequence of estrogen receptor cDNA from Sprague-Dawley rat. *European Journal of Pharmacology: Molecular Pharmacology*, *227*(3), 353–356. https://doi.org/10.1016/0922-4106(92)90016-o.

Staffend, N. A., Loftus, C. M., & Meisel, R. L. (2011). Estradiol reduces dendritic spine density in the ventral striatum of female Syrian hamsters. *Brain Structure and Function*, *215*(3–4), 187–194. https://doi.org/10.1007/s00429-010-0284-7.

Steinberg, E. E., Boivin, J. R., Saunders, B. T., Witten, I. B., Deisseroth, K., & Janak, P. H. (2014). Positive reinforcement mediated by midbrain dopamine neurons requires D1 and D2 receptor activation in the nucleus accumbens. *PLoS One*, *9*(4), e94771. https://doi.org/10.1371/journal.pone.0094771.

Stevis, P. E., Deecher, D. C., Suhadolnik, L., Mallis, L. M., & Frail, D. E. (1999). Differential effects of estradiol and estradiol-BSA conjugates. *Endocrinology*, *140*(11), 5455–5458.

Strakowski, S. M., Sax, K. W., Setters, M. J., & Keck, P. E. (1996). Enhanced response to repeated d-amphetamine challenge: Evidence for behavioral sensitization in humans. *Biological Psychiatry*, *40*(9), 872–880. https://doi.org/10.1016/0006-3223(95)00497-1.

Swalve, N., Smethells, J. R., Zlebnik, N. E., & Carroll, M. E. (2016). Sex differences in reinstatement of cocaine-seeking with combination treatments of progesterone and atomoxetine. *Pharmacology Biochemistry and Behavior*, *145*, 17–23. https://doi.org/10.1016/j.pbb.2016.03.008.

Sweis, B. M., Larson, E. B., Redish, A. D., & Thomas, M. J. (2018). Altering gain of the infralimbic-to-accumbens shell circuit alters economically dissociable decision-making algorithms. *Proceedings of the National Academy of Sciences*, *115*(27), 201803084. https://doi.org/10.1073/pnas.1803084115.

Szego, E. M., Barabás, K., Balog, J., Szilágyi, N., Korach, K. S., Juhász, G., et al. (2006). Estrogen induces estrogen receptor α-dependent cAMP response element-binding protein phosphorylation via mitogen activated protein kinase pathway in basal forebrain cholinergic neurons in vivo. *Journal of Neuroscience, 26*(15), 4104–4110. https://doi.org/10.1523/jneurosci.0222-06.2006.

Szego, C. M., & Davis, J. S. (1967). Adenosine 3′,5′-monophosphate in rat uterus: Acute elevation by estrogen. *Proceedings of the National Academy of Sciences, 58*(4), 1711–1718. https://doi.org/10.1073/pnas.58.4.1711.

Tanimura, S. M., & Watts, A. G. (1998). Corticosterone can facilitate as well as inhibit corticotropin-releasing hormone gene expression in the rat hypothalamic paraventricular nucleus. *Endocrinology, 139*(9), 3830–3836. https://doi.org/10.1210/endo.139.9.6192.

Tavares, H., Martins, S. S., Lobo, D. S. S., Silveira, C. M., Gentil, V., & Hodgins, D. C. (2003). Factors at play in faster progression for female pathological gamblers: An exploratory analysis. *The Journal of Clinical Psychiatry, 64*(4), 433–438. https://doi.org/10.4088/jcp.v64n0413.

Thomas, P., Pang, Y., Filardo, E. J., & Dong, J. (2005). Identity of an estrogen membrane receptor coupled to a G protein in human breast cancer cells. *Endocrinology, 146*(2), 624–632. https://doi.org/10.1210/en.2004-1064.

Thompson, T. L., & Moss, R. L. (1994). Estrogen regulation of dopamine release in the nucleus accumbens: Genomic-and nongenomic-mediated effects. *Journal of Neurochemistry, 62*(5), 1750–1756. https://doi.org/10.1046/j.1471-4159.1994.62051750.x.

Tobiansky, D. J., Will, R. G., Lominac, K. D., Turner, J. M., Hattori, T., Krishnan, K., et al. (2016). Estradiol in the preoptic area regulates the dopaminergic response to cocaine in the nucleus accumbens. *Neuropsychopharmacology, 41*(7), 1897–1906. https://doi.org/10.1038/npp.2015.360.

Trafton, J. A., Abbadie, C., Marek, K., & Basbaum, A. I. (2000). Postsynaptic signaling via the μ-opioid receptor: Responses of dorsal horn neurons to exogenous opioids and noxious stimulation. *The Journal of Neuroscience, 20*(23), 8578–8584. https://doi.org/10.1523/jneurosci.20-23-08578.2000.

Voss, J. L., Bridge, D. J., Cohen, N. J., & Walker, J. A. (2017). A closer look at the hippocampus and memory. *Trends in Cognitive Sciences, 21*(8), 577–588. https://doi.org/10.1016/j.tics.2017.05.008.

Vrtačnik, P., Ostanek, B., Mencej-Bedrač, S., & Marc, J. (2014). The many faces of estrogen signaling. *Biochemia Medica, 24*(3), 329–342. https://doi.org/10.11613/bm.2014.035.

Wade, C. B., & Dorsa, D. M. (2003). Estrogen activation of cyclic adenosine 5′-monophosphate response element-mediated transcription requires the extracellularly regulated kinase/mitogen-activated protein kinase pathway. *Endocrinology, 144*(3), 832–838. https://doi.org/10.1210/en.2002-220899.

Wakabayashi, Y., Nakada, T., Murata, K., Ohkura, S., Mogi, K., Navarro, V. M., et al. (2010). Neurokinin B and dynorphin A in kisspeptin neurons of the arcuate nucleus participate in generation of periodic oscillation of neural activity driving pulsatile gonadotropin-releasing hormone secretion in the goat. *Journal of Neuroscience, 30*(8), 3124–3132. https://doi.org/10.1523/jneurosci.5848-09.2010.

Wang, X., Moussawi, K., Knackstedt, L., Shen, H., & Kalivas, P. W. (2013). Role of mGluR5 neurotransmission in reinstated cocaine-seeking. *Addiction Biology, 18*(1), 40–49. https://doi.org/10.1111/j.1369-1600.2011.00432.x.

Warwick, H. K., Nahorski, S. R., & Challiss, R. A. J. (2005). Group I metabotropic glutamate receptors, mGlu1a and mGlu5a, couple to cyclic AMP response element binding protein (CREB) through a common Ca2+- and protein kinase C-dependent pathway. *Journal of Neurochemistry, 93*(1), 232–245. https://doi.org/10.1111/j.1471-4159.2005.03012.x.

Waselus, M., Flagel, S. B., Jedynak, J. P., Akil, H., Robinson, T. E., & Watson, S. J. (2013). Long-term effects of cocaine experience on neuroplasticity in the nucleus accumbens core of addiction-prone rats. *Neuroscience, 248*, 571–584. https://doi.org/10.1016/j.neuroscience.2013.06.042.

Watts, A. G., Kanoski, S. E., Sanchez-Watts, G., & Langhans, W. (2022). The physiological control of eating: Signals, neurons, and networks. *Physiological Reviews, 102*(2), 689–813. https://doi.org/10.1152/physrev.00028.2020.

Watts, A. G., & Swanson, L. W. (2002). Anatomy of motivation. In R. Gallistel, & H. Pashler (Eds.), *Vol. 3. Steven's handbook of experimental psychology: Learning, motivation, and emotion* (3rd ed.). John Wiley and Sons, Inc.

Weinberger, A. H., Smith, P. H., Allen, S. S., Cosgrove, K. P., Saladin, M. E., Gray, K. M., et al. (2015). Systematic and meta-analytic review of research examining the impact of menstrual cycle phase and ovarian hormones on smoking and cessation. *Nicotine & Tobacco Research, 17*(4), 407–421. https://doi.org/10.1093/ntr/ntu249.

Williams, T. M., & Lisanti, M. P. (2004). The caveolin proteins. *Genome Biology, 5*(3), 214. https://doi.org/10.1186/gb-2004-5-3-214.

Wilson, R. I., & Nicoll, R. A. (2002). Endocannabinoid signaling in the brain. *Science, 296*(5568), 678–682. https://doi.org/10.1126/science.1063545.

Wintermantel, T. M., Campbell, R. E., Porteous, R., Bock, D., Gröne, H.-J., Todman, M. G., et al. (2006). Definition of estrogen receptor pathway critical for estrogen positive feedback to gonadotropin-releasing hormone neurons and fertility. *Neuron, 52*(2), 271–280. https://doi.org/10.1016/j.neuron.2006.07.023.

Wise, R. A. (2004). Dopamine, learning and motivation. *Nature Reviews Neuroscience, 5*(6), 483–494. https://doi.org/10.1038/nrn1406.

Wong, A. M., Abrams, M. C., & Micevych, P. E. (2015). β-Arrestin regulates estradiol membrane-initiated signaling in hypothalamic neurons. *PLoS One, 10*(3), e0120530. https://doi.org/10.1371/journal.pone.0120530.

Wong, A. M., Scott, A. K., Johnson, C. S., Mohr, M. A., Mittelman-Smith, M., & Micevych, P. E. (2019). ERαΔ4, an ERα splice variant missing exon 4, interacts with caveolin-3 and mGluR2/3. *Journal of Neuroendocrinology, 31*(6), e12725. https://doi.org/10.1111/jne.12725.

Woolley, C., & McEwen, B. (1992). Estradiol mediates fluctuation in hippocampal synapse density during the estrous cycle in the adult rat [published erratum appears in J Neurosci 1992 Oct; 12(10):following table of contents]. *The Journal of Neuroscience, 12*(7), 2549–2554. https://doi.org/10.1523/jneurosci.12-07-02549.1992.

Wright, C. I., & Groenewegen, H. J. (1996). Patterns of overlap and segregation between insular cortical, intermediodorsal thalamic and basal amygdaloid afferents in the nucleus accumbens of the rat. *Neuroscience, 73*(2), 359–373. https://doi.org/10.1016/0306-4522(95)00592-7.

Wu, M. V., Manoli, D. S., Fraser, E. J., Coats, J. K., Tollkuhn, J., Honda, S.-I., et al. (2009). Estrogen masculinizes neural pathways and sex-specific behaviors. *Cell, 139*(1), 61–72. https://doi.org/10.1016/j.cell.2009.07.036.

Xiao, L., & Becker, J. B. (1997). Hormonal activation of the striatum and the nucleus accumbens modulates paced mating behavior in the female rat. *Hormones and Behavior, 32*(2), 114–124. https://doi.org/10.1006/hbeh.1997.1412.

Yager, L. M., Garcia, A. F., Wunsch, A. M., & Ferguson, S. M. (2015). The ins and outs of the striatum: Role in drug addiction. *Neuroscience, 301*, 529–541. https://doi.org/10.1016/j.neuroscience.2015.06.033.

Yin, H. H., Knowlton, B. J., & Balleine, B. W. (2005). Blockade of NMDA receptors in the dorsomedial striatum prevents action–outcome learning in instrumental conditioning. *European Journal of Neuroscience, 22*(2), 505–512. https://doi.org/10.1111/j.1460-9568.2005.04219.x.

Yin, H. H., Knowlton, B. J., & Balleine, B. W. (2006). Inactivation of dorsolateral striatum enhances sensitivity to changes in the action–outcome contingency in instrumental conditioning. *Behavioural Brain Research, 166*(2), 189–196. https://doi.org/10.1016/j.bbr.2005.07.012.

Yoest, K. E., Cummings, J. A., & Becker, J. B. (2015). Estradiol, dopamine and motivation. *Central Nervous System Agents in Medicinal Chemistry, 14*(2), 83–89. https://doi.org/10.2174/1871524914666141226103135.

Yoest, K. E., Cummings, J. A., & Becker, J. B. (2019). Ovarian hormones mediate changes in adaptive choice and motivation in female rats. *Frontiers in Behavioral Neuroscience, 13*, 250. https://doi.org/10.3389/fnbeh.2019.00250.

Yoest, K. E., Quigley, J. A., & Becker, J. B. (2018). Rapid effects of ovarian hormones in dorsal striatum and nucleus accumbens. *Hormones and Behavior, 104*, 119–129. https://doi.org/10.1016/j.yhbeh.2018.04.002.

Zahm, D. S., & Brog, J. S. (1992). On the significance of subterritories in the "accumbens" part of the rat ventral striatum. *Neuroscience, 50*(4), 751–767. https://doi.org/10.1016/0306-4522(92)90202-d.

Zakharova, E., Wade, D., & Izenwasser, S. (2009). Sensitivity to cocaine conditioned reward depends on sex and age. *Pharmacology Biochemistry and Behavior, 92*(1), 131–134. https://doi.org/10.1016/j.pbb.2008.11.002.

Zhang, D., Yang, S., Yang, C., Jin, G., & Zhen, X. (2008). Estrogen regulates responses of dopamine neurons in the ventral tegmental area to cocaine. *Psychopharmacology, 199*(4), 625–635. https://doi.org/10.1007/s00213-008-1188-6.

Zheng, J., & Ramirez, V. D. (1997). Demonstration of membrane estrogen binding proteins in rat brain by ligand blotting using a 17β-estradiol-[125I]bovine serum albumin conjugate. *The Journal of Steroid Biochemistry and Molecular Biology, 62*(4), 327–336. https://doi.org/10.1016/s0960-0760(97)00037-x.

Zwain, I. H., & Yen, S. S. C. (1999). Neurosteroidogenesis in astrocytes, oligodendrocytes, and neurons of cerebral cortex of rat brain. *Endocrinology, 140*(8), 3843–3852. https://doi.org/10.1210/endo.140.8.6907.

CHAPTER THREE

Metabotropic glutamate receptor function and regulation of sleep-wake cycles

Kimberly M. Holter, Bethany E. Pierce, and Robert W. Gould*

Department of Physiology and Pharmacology, Wake Forest University School of Medicine, Winston-Salem, NC, United States
*Corresponding author: e-mail address: rgould@wakehealth.edu

Contents

1. Introduction	94
2. Overview of metabotropic glutamate receptors (mGlu receptors)	95
2.1 Group I mGlu receptor distribution and function	96
2.2 Group II mGlu receptor distribution and function	97
2.3 Group III distribution and function	98
3. Sleep	99
3.1 Sleep stages and characteristics	100
3.2 Glutamate and the sleep-wake cycle	101
3.3 mGlu receptors and sleep	105
3.4 mGlu receptors, neuropsychiatric disorders and sleep	114
4. Discussion	137
References	144

Abstract

Metabotropic glutamate (mGlu) receptors are the most abundant family of G-protein coupled receptors and are widely expressed throughout the central nervous system (CNS). Alterations in glutamate homeostasis, including dysregulations in mGlu receptor function, have been indicated as key contributors to multiple CNS disorders. Fluctuations in mGlu receptor expression and function also occur across diurnal sleep-wake cycles. Sleep disturbances including insomnia are frequently comorbid with neuropsychiatric, neurodevelopmental, and neurodegenerative conditions. These often precede behavioral symptoms and/or correlate with symptom severity and relapse. Chronic sleep disturbances may also be a consequence of primary symptom progression and can exacerbate neurodegeneration in disorders including Alzheimer's disease (AD). Thus, there is a bidirectional relationship between sleep disturbances and CNS disorders; disrupted sleep may serve as both a cause and a consequence of the disorder. Importantly, comorbid sleep disturbances are rarely a direct target of primary pharmacological treatments for neuropsychiatric disorders even though improving sleep can positively impact other symptom clusters. This chapter details known roles of mGlu

International Review of Neurobiology, Volume 168
ISSN 0074-7742
https://doi.org/10.1016/bs.irn.2022.11.002

Copyright © 2023 Elsevier Inc.
All rights reserved.

receptor subtypes in both sleep-wake regulation and CNS disorders focusing on schizophrenia, major depressive disorder, post-traumatic stress disorder, AD, and substance use disorder (cocaine and opioid). In this chapter, preclinical electrophysiological, genetic, and pharmacological studies are described, and, when possible, human genetic, imaging, and post-mortem studies are also discussed. In addition to reviewing the important relationships between sleep, mGlu receptors, and CNS disorders, this chapter highlights the development of selective mGlu receptor ligands that hold promise for improving both primary symptoms and sleep disturbances.

1. Introduction

The word "sleep" holds different meaning and emotional weight for different individuals. While many people undergo an easy transition from waking to a peaceful state of altered consciousness, initiating and maintaining sleep is a fitful chore for others. Though there have been substantial advancements in understanding the underlying functions of homeostatic sleep, including its role in synaptic remodeling critical for memory consolidation (Manoach & Stickgold, 2019; Martin, Monroe, & Diering, 2019), many purposes remain unresolved. Although sleeping and waking states may be perceived as dichotomous processes, they are intimately intertwined. Acute sleep disruptions lead to decreased daytime arousal and cognitive impairments as well as increased fatigue or daytime somnolence. Further, chronic sleep disruptions including insomnia are often comorbid with neuropsychiatric, neurodevelopmental, and neurodegenerative disorders (Benca & Buysse, 2018; Benca, William, Thisted, & Gillin, 1992; Riemann, Krone, Wulff, & Nissen, 2020; Sprecher, Ferrarelli, & Benca, 2015; Wulff, Gatti, Wettstein, & Foster, 2010). Increasing evidence suggests a direct relationship between sleep disruptions and symptom onset and/or severity in central nervous system (CNS) disorders (Reeve, Sheaves, & Freeman, 2015; Riemann et al., 2020; Veatch et al., 2017; Wulff et al., 2010). Importantly, the *Diagnostics and Statistic Manual, 5th Edition* (DSM-5) delineated insomnia, defined by difficulty falling asleep and/or staying asleep, as a separate diagnosis from other mental disorders as opposed to a primary or secondary symptom, and, thus, sleep disturbances should be a primary target of pharmacotherapies (APA, 2013; Benca & Buysse, 2018).

Sleep is an intricately regulated process involving multiple circuits and neurotransmitter systems. These systems work in concert to actively inhibit arousal–related processes and shift from asynchronous electrical activity during wake to synchronous, slow oscillatory activity during sleep (Scammell,

Arrigoni, & Lipton, 2017). Seminal studies employing region-specific brain lesions and genetic manipulations in murine models as well as pharmacological studies have identified roles of noradrenergic, dopaminergic, cholinergic, histaminergic, serotonergic and orexinergic systems in sleep and arousal (for review, see España & Scammell, 2011; Weber & Dan, 2016). More recent developments suggest that the glutamatergic system is also extensively involved in sleep-wake processes (Jones, 2020; Martin et al., 2019; Saper & Fuller, 2017; Tononi & Cirelli, 2014). Although a critical regulator of several neurobiological processes including memory consolidation, the role of glutamate in sleep-specific processes is not as rigorously understood. With lines of evidence supporting a role of glutamatergic dysregulation in both CNS disorders and sleep impairments, examining the relationship between glutamate, sleep and CNS disorders is timely.

This chapter highlights research examining metabotropic glutamate (mGlu) receptor involvement in the regulation of sleep alone and in the context of CNS disorders. We focus on schizophrenia, major depressive disorder (MDD), post-traumatic stress disorder (PTSD), Alzheimer's disease (AD), and substance use disorder (SUD). It is important to note that these disorders are selected examples and that sleep disturbances and disrupted mGlu function are associated with many other CNS disorders. In this chapter, we review preclinical electrophysiological, genetic, and pharmacological studies contributing to our understanding of how various mGlu receptors impact sleep architecture and duration. When possible, we also discuss related human genetic, imaging, and post-mortem studies. Ultimately, we work to shed light on the underappreciated contribution of the glutamate system to sleep and, thereby, comorbid sleep disruptions in CNS disorders. As remediation of sleep disruptions should be a complementary focus of drug development efforts, this chapter highlights novel pharmacotherapeutic approaches targeting mGlu receptors that may effectively address both sleep disturbances and primary symptoms of the neuropsychiatric disorder.

2. Overview of metabotropic glutamate receptors (mGlu receptors)

Glutamate, the major excitatory neurotransmitter in the CNS, regulates rapid excitatory neurotransmission via ionotropic glutamate (iGlu) receptors and slower neuromodulatory effects via mGlu receptors. iGlu receptors are ligand-gated ion channels subdivided into four functional classes: α-amino-3-hydroxy-5-methyl-iso-xazolepropionic acid (AMPA) receptors, N-methyl-D-aspartate (NMDA) receptors, kainate receptors, and GluD receptors (Hansen et al., 2021). mGlu receptors are G-protein coupled receptors (GPCRs) and are the most abundant receptor family in the CNS with 8 distinct subtypes (mGlu$_1$-mGlu$_8$). These class C GPCRs are widely distributed, located on both neurons and glial cells influencing glutamatergic, GABAergic, and other neuromodulatory synaptic transmission (Luessen & Conn, 2022; Maksymetz, Moran, & Conn, 2017; Niswender & Conn, 2010). mGlu receptors are subdivided into three groups (Group I: mGlu$_1$ and mGlu$_5$; Group II: mGlu$_2$ and mGlu$_3$; Group III: mGlu$_4$, mGlu$_6$, mGlu$_7$, and mGlu$_8$) based on sequence homology, G-protein coupling, and agonist selectivity (Conn & Pin, 1997; Niswender & Conn, 2010; Wang & Zhuo, 2012). Acknowledging the extensive distribution and neuromodulatory role of mGlu receptors throughout the CNS, it is not surprising that preclinical and clinical studies suggest altered mGlu receptor expression and function contributes to many neuropsychiatric, neuro-developmental, and neurodegenerative disorders. In the following section, we provide a brief overview of the general functional and behavioral roles of the different mGlu receptor subtypes.

2.1 Group I mGlu receptor distribution and function

Group I mGlu receptors (mGlu$_{1/5}$) are widely distributed throughout the CNS in brain regions particularly relevant for learning, memory and moti-vation. mGlu$_5$ receptors are densely present in the cortex, striatum, and hippocampus whereas mGlu$_1$ receptors are found primarily in the cerebel-lum, thalamus, and hypothalamus (Niswender & Conn, 2010; Olive, 2010; Romano et al., 1995; Shigemoto et al., 1993; Shigemoto, Nakanishi, & Mizuno, 1992). These predominantly postsynaptic receptors are G$_{q/11}$-coupled and localized on glutamatergic and GABAergic terminals (Luessen & Conn, 2022; Niswender & Conn, 2010). Activation of G$_{q/11}$-coupled receptors canonically stimulates phospholipase C (PLC) and phosphoinositide hydrolysis leading to intracellular calcium mobiliza-tion. However, mGlu$_{1/5}$ receptors also activate several additional pathways including downstream effectors in the mitogen–activate protein kinase/extracellular protein kinase (MAPK/ERK) pathway (Luessen & Conn, 2022; Niswender & Conn, 2010; Page et al., 2006). Furthermore, mGlu$_{1/5}$ receptors are critical regulators of activity-dependent synaptic plasticity, influencing both long-term potentiation (LTP) and long-term depression (LTD), and this is in part attributed to their tight coupling with

NMDARs (Anwyl, 1999; Joffe, Centanni, Jaramillo, Winder, & Conn, 2018; Lüscher & Huber, 2010; Lutzu & Castillo, 2021; Lu et al., 1997). As these functions are outside the scope of this chapter, readers are referred to several excellent reviews (Luessen & Conn, 2022; Niswender & Conn, 2010; Olive, 2010; Shigemoto et al., 1993). Given the role of group I mGlu receptors in regulating synaptic plasticity and, thereby, learning and memory (for review, see Olive, 2010), it is not surprising that they have been implicated in neurodegenerative and neuropsychiatric disorders associated with cognitive disruptions including AD (Bruno et al., 2000; Kumar, Dhull, & Mishra, 2015), schizophrenia (Kinney et al., 2003; Maksymetz et al., 2017; Nicoletti et al., 2019), Parkinson's disease (Morin, Grégoire, Gomez-Mancilla, Gasparini, & di Paolo, 2010), MDD (Lindemann et al., 2015), and SUD (Gould et al., 2016; McGeehan & Olive, 2003; Veeneman et al., 2011). As will be described below, dysregulation of either mGlu$_1$ or mGlu$_5$ receptors also affects sleep which may further contribute to primary symptoms associated with these disorders (Aguilar, Strecker, Basheer, & Mcnally, 2020; Cavas, Scesa, & Navarro, 2013b; Holter et al., 2021).

2.2 Group II mGlu receptor distribution and function

Group II mGlu receptors (mGlu$_2$ and mGlu$_3$) are localized on both presynaptic and postsynaptic membranes and, similar to group I, are abundantly present in regions important for learning, memory, and motivation including the prefrontal cortex (PFC), hippocampus, striatum, and amygdala (Crupi, Impellizzeri, & Cuzzocrea, 2019; Luessen & Conn, 2022; Maksymetz et al., 2017). mGlu$_3$ receptors are also expressed on astrocytes, and this contributes to anti-inflammatory properties that may be neuroprotective (Bruno et al., 1998; Nicoletti et al., 2011). Group II mGlu receptors are G$_{i/o}$ coupled and inhibit adenylyl cyclase and phosphatidyl inositol 3-kinase which reduces intracellular calcium mobilization. These receptors also inhibit MAPK/ERK pathways (Luessen & Conn, 2022; Nicoletti et al., 2011; Niswender & Conn, 2010). Importantly, the primary function of presynaptic mGlu$_{2/3}$ receptors is to work as autoreceptors and maintain glutamate homeostasis via inhibition of release at glutamatergic and GABAergic synapses following periods of glutamate efflux (Conn & Jones, 2009; Luessen & Conn, 2022; Maksymetz et al., 2017; Mazzitelli, Palazzo, Maione, & Neugebauer, 2018; Moussawi & Kalivas, 2010; Wright, Arnold, Wheeler, Ornstein, & Schoepp, 2001). In general, group II mGlu receptors contribute to synaptic plasticity by reducing synaptic excitability; receptor activation leads to decreased EPSC amplitude and induction of postsynaptic hippocampal LTD (Altinbilek & Manahan-Vaughan, 2009; Grueter & Winder, 2005; Joffe et al., 2020; Walker et al., 2015; Yokoi et al., 1996). Behaviorally, mGlu$_{2/3}$ receptors have been shown to influence spatial learning and working memory (de Filippis et al., 2015; Lyon et al., 2011). Findings indicate pharmacological activation of mGlu$_{2/3}$ receptors can improve attention and working memory in acute pharmacological models of cognitive disruptions (e.g. NMDAR antagonists) in both humans and animals (Greco, Invernizzi, & Carli, 2005; Griebel et al., 2016; Krystal et al., 2005; Moghaddam & Adams, 1998). Activation may also produce anxiolytic- and antidepressant-like effects in rodent models (Dogra & Conn, 2021). As such, group II mGlu receptors may be promising pharmacotherapeutic targets for multiple disorders including schizophrenia (Hackler et al., 2010; Hiyoshi, Hikichi, Karasawa, & Chaki, 2014; Maksymetz et al., 2017; Moghaddam & Adams, 1998; Sokolenko, Hudson, Nithianantharajah, & Jones, 2019), AD (Lee et al., 2004; Richards et al., 2010), MDD (Chaki, 2017), pain (Davidson et al., 2017; Jones, Eberle, Peters, Monn, & Shannon, 2005; Mazzitelli et al., 2018), and SUD (Cleva & Olive, 2012; Hao, Martin-Fardon, & Weiss, 2010; Rodd et al., 2006).

2.3 Group III distribution and function

Group III mGlu receptors (mGlu$_4$, mGlu$_6$, mGlu$_7$, mGlu$_8$) are localized pre- and post-synaptically on glutamatergic and GABAergic terminals with differential expression throughout the CNS (Luessen & Conn, 2022; Niswender & Conn, 2010). mGlu$_4$ and mGlu$_8$ receptors are located presynaptically with regionally-restricted expression in the cerebellum and hippocampus (Luessen & Conn, 2022; Niswender & Conn, 2010a; Zhai et al., 2002). mGlu$_6$ receptors, unlike others, have strict localization to the retina, particularly on ON bipolar cells, and, thus, are not relevant for this chapter (Crupi et al., 2019; Luessen & Conn, 2022; Niswender & Conn, 2010). Lastly, mGlu$_7$ receptors are presynaptic and widely distributed throughout the CNS with expression in regions including the hippocampus, thalamus, hypothalamus and amygdala. mGlu$_7$ is unique in that it has a very low affinity for glutamate and, thus, it serves to only function in times of high synaptic activity to fine-tune and prevent excessive glutamate levels (Niswender & Conn, 2010). In general, group III mGlu receptors contribute to spontaneous glutamate transmission and hippocampal-mediated long-term synaptic plasticity (Altinbilek & Manahan-Vaughan, 2007). For example, $mGlu_4$ knockout mice had enhanced hippocampal LTP, though the PFC was unaffected (Iscru et al., 2013). Furthermore, bath application of the $mGlu_4$ PAM foliglurax to corticostriatal slices reduced glutamatergic transmission, as shown through a reduction in spontaneous EPSC frequency (Calabrese et al., 2022). Pharmacological activation of $mGlu_7$ induced LTD in mossy fiber inputs to stratum lucidum interneurons (SLINs) (Pelkey, Lavezzari, Racca, Roche, & McBain, 2005). In Schaffer collateral-CA1 synapses, activation of $mGlu_7$ potentiated a submaximal level of LTP and inhibition of $mGlu_7$ prevented the induction of LTP in these synapses (Kalinichev et al., 2013; Klar et al., 2015). Lastly, activation of $mGlu_8$ reduced field EPSPs induced by lateral perforant path (LPP) afferents in hippocampal slices (Zhai et al., 2002). In line with their regions of expression and preclinical findings, $mGlu_4$ receptors are being explored for their pharmacotherapeutic potential in Parkinson's disease and schizophrenia (Battaglia et al., 2006; Calabrese et al., 2022; Charvin, 2018; Luessen & Conn, 2022; Niswender & Conn, 2010). The $mGlu_7$ receptor has been a suggested target for disorders including AD (Crupi et al., 2019), neuropathic and inflammatory pain (Marabese et al., 2007; Palazzo, Fu, Ji, Maione, & Neugebauer, 2008), SUD (Li, Xi, & Markou, 2013), MDD and anxiety disorders (Bradley et al., 2012; O'Connor et al., 2013; Palucha & Pilc, 2007; Pałucha-Poniewiera & Pilc, 2013), and epilepsy (Girard et al., 2019; Sansig et al., 2001). Lastly, activation of $mGlu_8$ may be a promising therapeutic approach for anxiety disorders (Duvoisin et al., 2010; Linden et al., 2002; Niswender & Conn, 2010) and additional research supports $mGlu_8$ as a target for pain (Crupi et al., 2019; Marabese et al., 2007; Palazzo et al., 2008).

3. Sleep

Sleep is defined as an easily reversible state of reduced consciousness. However, from a neurobiological perspective, sleep is not passive. Rather, sleep is actively regulated by inhibition of arousal circuits which contrast the cortical activation and arousal that occurs during wake. The two-process model of homeostatic sleep regulation posits that sleep is driven by an interaction of sleep debt(Process S) with the circadian pacemaker (Process C). Process S increases during extended waking periods and decreases following periods of sleep, and the circadian pacemaker follows a similar oscillatory pattern, regulating sleep to specific times of day (Borbély, Daan, Wirz-Justice, & Deboer, 2016; Borbély & Wirz-Justice, 1982; Daan,

Beersma, & Borbély, 1984). Circadian factors influencing sleep regulation are outside the scope of this chapter but are thoroughly detailed in both Logan and McClung (2019) and Rosenwasser and Turek (2015). There are many neuroanatomical and neurochemical systems that contribute to sleep-wake regulation, and we refer the reader to many excellent reviews (Brown, Basheer, McKenna, Strecker, & McCarley, 2012; España & Scammell, 2011; Saper, Fuller, Pedersen, Lu, & Scammell, 2010; Saper, Scammell, & Lu, 2005). The following sections provide a brief overview of human and nonhuman sleep and the neurochemical systems regulating sleep-wake cycles with a specific focus on glutamatergic involvement.

3.1 Sleep stages and characteristics

On average, healthy adults sleep 6–8 h and undergo 4–5 sleep cycles per night (Hor & Tafti, 2009). Each cycle is comprised of non-rapid eye movement (NREM) and rapid eye movement (REM) sleep and lasts approximately 90 min in duration (le Bon, Lanquart, Hein, & Loas, 2019). In humans, NREM sleep is divided into three stages, termed N1, N2, and N3. The American Academy of Sleep Medicine (AASM) defines N1 as the lightest stage of sleep, representing a transition between wake and sleep when external stimuli can still be processed (Iber, Ancoli-Israel, & Chesson, 2007). N2 sleep is classified as light sleep and N3 is considered deep sleep, or slow wave sleep (SWS). Overall, in one night of sleep, the percentage of time in N1, N2, and N3 is ~5%, 50% and 20% with the remaining 25% being REM sleep (Iber et al., 2007). As a result of increased sleep debt throughout the day, N3 stage sleep (and duration of bouts) is generally more prevalent in the first half of the night whereas duration of REM bouts are generally longer in the second half of the night. However, the actual duration and percentage of time in each stage changes across development and normal aging in healthy populations (Ohayon, Carskadon, Guilleminault, & Vitiello, 2004).

Sleep stages and transitions are identified using electroencephalography (EEG) to measure and determine changes in brain oscillatory activity. The transition from wake to N1 sleep is associated with a shift from high/mixed frequency, lower amplitude EEG patterns to predominately lower frequency, higher amplitude oscillations. This includes a decrease in alpha band [~8–13 Hz] activity compared to waking states. N2 sleep is associated with sleep spindles (periodic synchronous burst firing in the sigma [~12–15 Hz] range) and K-complexes (brief waveforms characterized by high-amplitude spike and rebound patterns). N3 sleep, or SWS, is associated with high amplitude slow wave activity (SWA, delta band, [~0–4 Hz]). Lastly, REM sleep is associated with an increase in theta [~4–8 Hz] activity relative to delta activity as well as an increase in beta [18–30 Hz] and gamma [>30 Hz] frequencies. Historically, REM sleep was termed paradoxical sleep as the EEG waveforms paradoxically resemble EEG activity during waking states (Boissard et al., 2002; España & Scammell, 2011).

Humans and nonhuman primates are monophasic sleepers whereas rodents are polyphasic sleepers, rapidly cycling through sleep-wake states over a 24-h period. Despite these differences, from a neurophysiological perspective, mechanisms regulating sleep-wake cycles are evolutionarily well-conserved. Rodents are nocturnal and spend roughly 80% of their time sleeping during the light (inactive) phase and 20% of their time sleeping during the dark (active) phase when under a typical 12 h:12 h light:dark cycle. Due to the more rapid sleep cycle (~8–15 min/cycle), sleep is typically classified into REM and NREM sleep. Although EEG activity in rodent sleep stages are comparable to humans, electromyography (EMG; muscle activity) is often used in rodents to accurately distinguish REM sleep, which demonstrates a distinct muscle atonia, from waking periods. Research in animals using lesions and genetic and pharmacological manipulations has provided extensive insight into the mechanisms regulating sleep-wake transitions. These studies have also elucidated roles of different sleep variables in maintaining healthy daily function and how irregularities may contribute to various CNS disorders.

3.2 Glutamate and the sleep-wake cycle

Historically, the ascending reticular activation system (ARAS) was solely implicated in regulating waking and arousal. Brainstem nuclei, that were primarily thought to be monoaminergic (dopamine, norepinephrine, serotonin, histamine) and cholinergic, innervate thalamic projections to the cortex. Additional projections to the hypothalamus and basal forebrain innervate orexin/hypocretin neurons and forebrain cholinergic neurons, respectively, which then further project to the cortex (Saper & Fuller, 2017). Early seminal studies by Moruzzi and Magoun (1949) demonstrated that inactivation of the reticular formation and ARAS produced a passive state resembling sleep. However, more recent research demonstrated that lesions to different components of the ARAS had a minimal, if any, effect on sleep-wake durations in rats, suggesting there must be additional contributors (for detailed review see Saper & Fuller, 2017). Glutamatergic projections from the brainstem parabrachial and pedunculopontine nuclei to the basal forebrain as well as projections from the hypothalamic supramammillary area to the cortex have since been shown to contribute substantially to arousal and wake-promoting effects (see Saper & Fuller, 2017). Thus, multiple neurons, including glutamatergic neurons, in the brainstem, hypothalamus and basal forebrain regulate cortical arousal and waking.

Sleep promotion occurs via GABAergic inhibition of each of these wake-promoting nuclei. GABA neurons within the ventrolateral (VLPO) and median preoptic (MNPO) nuclei of the hypothalamus innervate all wake-regulating monoaminergic and cholinergic brainstem nuclei (Sherin, Elmquist, Torrealba, & Saper, 1998; Suntsova, Szymusiak, Alam, Guzman-Marin, & McGinty, 2002). Additional GABA neurons in the brainstem reticular formation inhibit glutamate neurons in the parabrachial nucleus (Anaclet et al., 2014). These GABA neurons are almost entirely inactive during waking periods, and NREM sleep is initiated when these neurons are activated to inhibit cortical activity. This shift in inhibitory balance from wake to sleep is in part driven by several metabolic and circadian influences including adenosine, prostaglandins, cytokines, and growth hormone-releasing hormone (Obal & Krueger, 2003). For example, extracellular concentrations of adenosine increase following prolonged periods of waking as active neurons hydrolyze ATP, and concentrations dissipate following sleep (Porkka-Heiskanen et al., 1997). Importantly, adenosine acts as a neuromodulator to inhibit arousal circuits and disinhibit VLPO GABA neurons thereby exerting sleep-promoting effects (Obal & Krueger, 2003).

There are multiple neuronal populations that contribute specifically to REM sleep. These include subsets of cholinergic neurons in the laterodorsal and pedunculopontine tegmentum (LTD/PPT) as well as the basal forebrain which promote cortical activation through acetylcholine release (Marrosu et al., 1995; Williams, Comisarow, Day, Fibiger, & Reiner, 1994). Additionally, neurons containing melanin-concentrating hormone and GABA neurons in the hypothalamus are active during REM sleep demonstrating inhibitory effects on wake-promoting nuclei (Verret et al., 2003). Lastly, a subset of glutamate neurons active in the sublaterodorsal (SLD) nucleus that project to the inhibitory neurons in the medulla and spinal cord are thought to contribute to muscle atonia during REM sleep (Boissard et al., 2002).

The distinct patterns of EEG activity that occur throughout sleep-wake cycles are derived from interactions between the subcortical systems described above (brainstem, basal forebrain, and hypothalamus) and the thalamus and cortex. Activity in the thalamocortical (TC) circuit also undergoes dynamic changes across sleep-wake cycles. The TC circuit is comprised of two groups of neurons. TC projection neurons are predominantly glutamatergic and relay sensory and motor information to the cortex. Thalamic relay neurons (TRNs) are predominantly GABAergic and are innervated by and can inhibit TC neurons (España & Scammell, 2011; Huguenard & McCormick, 2007). During wake and REM sleep, monoaminergic and cholinergic neurotransmitter release depolarizes thalamic neurons resulting in desynchronized fast, low-amplitude cortical oscillations and increased sensitivity to incoming stimuli (Aston-Jones, Smith, Moorman, & Richardson, 2009; Hu, Steriade, & Deschênes, 1989). In contrast, the firing patterns shift during NREM sleep as these neurons become hyperpolarized and undergo synchronized burst firing, reducing responsivity to external stimuli (Brown et al., 2012; España & Scammell, 2011; Livingstone & Hubel, 1981; Llinás & Steriade, 2006; Mccormick & Bal, 1997; Steriade, Iosif, & Apostol, 1968). These reciprocally acting projections in part drive cortical delta waves and sleep spindles during NREM sleep (España & Scammell, 2011; Steriade, Domich, Oakson, & Deschfines, 1987). Additional inhibitory feedback loops between cortical glutamate neurons and thalamic GABA neurons further contribute to cortical oscillatory rhythms and sleep spindle generation (Steriade et al., 1987). Lastly, Yu et al. (2019) recently found that activation of glutamate neurons in the VTA projecting to both the nucleus accumbens (NAc) and the lateral hypothalamus increased time awake, whereas inhibition decreased waking bout duration and increased frequency of wake: NREM sleep transitions. In short, there are many overlapping circuits and neurotransmitter systems involved in the transition and maintenance of sleep-wake cycles and the associated oscillatory functions, and the glutamate system is a strong contributor.

Preclinical and clinical studies support a strong relationship between extracellular glutamate and the different sleep stages. EEG studies paired with simultaneous *in vivo* microdialysis recordings in the orbitofrontal cortex found extracellular glutamate concentrations to be highest during REM sleep, modest during waking periods and lowest during NREM sleep in rats (Lopez-Rodriguez, Medina-Ceja, Wilson, Jhung, & Morales-Villagran, 2007). Amperometric detection of glutamate in the cortex showed rapid increases in glutamate at the beginning of wake and REM sleep and rapid decreases during NREM sleep (Dash, Douglas, Vyazovskiy, Cirelli, & Tononi, 2009). Additional microdialysis studies revealed similar elevations in glutamate levels during wake in the pontine reticular formation (Watson, Lydic, & Baghdoyan, 2011), though glutamate concentrations detected in the thalamus were contrastingly highest during NREM sleep (Kékesi, Dobolyi, Salfay, Nyitrai, & Juhász, 1997). These changes in glutamate concentrations likely align with the shift from asynchronous cortical excitation during wake to synchronous TC regulation during NREM (España & Scammell, 2011). Additionally, evidence from rodent amperometry studies and human studies using proton magnetic resonance spectroscopy (^{1}H-MRS) indicates that glutamate levels and receptor availability are not only state-dependent but also sensitive to additional factors including general sleep history and disruptions in sleep homeostasis that occur following periods of extended waking (Dash et al., 2009; Weigend et al., 2019).

Lastly, the glutamate system has a major role in regulating state-dependent neurobiological changes in synaptic plasticity. The sleep homeostasis hypothesis suggests that a key role of sleep is to maintain a balance in synaptic strength. Notably, the number of synapses undergoing LTP and LTD shift between sleep and wake (Tononi & Cirelli, 2014). During time awake, as daily learning occurs, synapses progressively strengthen and undergo LTP-like potentiation. This is followed by synaptic weakening and LTD during sleep (Tononi & Cirelli, 2014; Vyazovskiy, Cirelli, Pfister-Genskow, Faraguna, & Tononi, 2008). This synaptic weakening functions predominantly to prevent saturation of synapses, which would result in an inability to form new connections; this synaptic weakening is also an important process in memory consolidation (Diering et al., 2017; Martin et al., 2019; Tononi & Cirelli, 2014). Supporting evidence for the morphological changes that occur in sleep and wake has emerged from studies using Drosophila. During time awake, an increase in synapse number as well as in scaffold and post-synaptic proteins Bruchpilot (BRP) and Discs-large (DLG), both key regulators of glutamate release, were reported. These changes were exacerbated in periods of extended wake as a result of sleep deprivation and reduced during periods of sleep (Bushey, Tononi, & Cirelli, 2011; Gilestro, Tononi, & Cirelli, 2009). Importantly, in rodents, both iGlu (GluA1-containing AMPARs) and mGlu receptors, (primarily group I mGlu receptors, detailed below), have been found to orchestrate these synaptic changes (Diering et al., 2017; Vyazovskiy et al., 2008). Briefly, synaptic expression of cortical and hippocampal GluA1-containing AMPARs is highest during wake and lowest during NREM sleep, which corresponds with electrophysiological findings of synaptic potentiation (Vyazovskiy et al., 2008). Cognitive impairments, a primary symptom of many CNS disorders, may be a direct result of aberrant regulation of these synaptic plasticity changes.

3.3 mGlu receptors and sleep

Many studies spanning from Drosophila to humans highlight important contributions of the different mGlu receptor subtypes in sleep–wake regulation. On the most basic level, genetic knockdown of the only mGlu receptor in Drosophila (DmGluRA) produced drastic changes in sleep–wake behavior, notably increasing time spent asleep during the day and decreasing time spent asleep at night (Ly, Lee, Strus, Prober, & Naidoo, 2020). Additional support for mGlu involvement in sleep regulation is derived from EEG studies in mice with a constitutive knockout of a single mGlu receptor subtype or altered function of a downstream signaling protein (see Table 1). Recent development of pharmacological tool compounds with mGlu receptor subtype specificity has also added to this literature base (see Table 2). The following section describes literature examining genetic and pharmacological manipulations in rodents, a majority of which involves the group I and II mGlu receptor classes. Although less well characterized, some evidence supports the role of group III mGlu receptors in sleep–wake regulation. Lastly, when available, relevant human postmortem, GWAS, and neuroimaging studies are also included.

3.3.1 Group I mGlu receptors

Perhaps one of the most relevant contributions of $mGlu_{1/5}$ receptors is to maintain homeostasis in synaptic size and strength throughout the sleep–wake cycle via dynamic changes in the $mGlu_{1/5}$/Homer protein complex. Homer family proteins (Homer1, Homer2, and Homer3) are adaptor proteins that each contain several isoforms as a product of alternative splicing (Shiraishi-Yamaguchi & Furuichi, 2007). Homer protein variants are differentially recruited to the synapse during sleep and wake and, thereby, influence downstream effector systems of $mGlu_{1/5}$. During wake, $mGlu_{1/5}$ receptors are anchored to the excitatory post synaptic density via long-form Homer proteins, which contain both an Ena/VASP (EVP) domain and a coiled-coil domain. These long-form Homer proteins link $mGlu_{1/5}$ receptors to their primary signaling effector, the inositol triphosphate (IP3) receptor, in the endoplasmic reticulum as well as Shank1–3 proteins, facilitating cell excitability and increased GluA1-containing AMPAR levels during arousal states (Diering et al., 2017; Martin et al., 2019). Following prolonged waking periods, the activity-dependent immediate early gene Homer1a, a short-form protein containing only an EVP domain, is targeted to the synapse (Martin et al., 2019; Shiraishi-Yamaguchi & Furuichi, 2007). Homer1a uncouples the $mGlu_{1/5}$ receptor from the IP3 receptor, resulting in signaling through

Table 1 Effects of subtype-specific mGlu receptor genetic alterations on sleep in rodents.

Condition	Phase	Awake Duration	Sleep Total duration	NREM Duration	Bouts	Bout duration	Delta	REM Duration	Bouts	Bout duration	Theta
*PLCβ4[−/−] [1,2]	Light	↓	—	n.s., ↑	n.s.	↑	↑	n.s.	↓	↑	↓
	Dark	↓	—	n.s., ↑	↑	↓	↑	↑	n.s.	↑	↓
TC-restricted PLCβ4[−/−] [1]	Light	n.s.	—	↑	—	—	↑	n.s.	—	—	↓
	Dark	n.s.	—	n.s.	—	—	n.s.	↓	—	—	n.s.
*mGlu₁ Mutation[3]	24 hr	—	↓	↓	—	—	—	n.s.	—	—	—
mGlu₂/₃ [4]	Light	—	↑	—	—	—	—	—	—	—	—
	Dark	—	n.s.	—	—	—	—	—	—	—	—
mGlu₅ [5,6,7]	Light	n.s., ↓	—	n.s.	n.s.,↑	n.s.	n.s.,↓	↓	n.s.	n.s., ↓	↓
	Dark	n.s., ↑	—	n.s.	n.s.,↑	n.s., ↓	↓	n.s.	n.s.	n.s., ↓	↓
	24 hr	↓	—	↑	—	—	—	n.s.	—	—	—
mGlu₇ [8]	Light	n.s.	—	n.s.	—	—	n.s.	n.s.	n.s.	↓	n.s.
	Dark	↑	—	↓	—	—	n.s.	n.s.	—	—	n.s.

*PLCβ4[−/−] is a downstream effector in the mGlu₁ pathway; mGlu₁ mutation is loss of function.
n.s., not significant; —, not determined.
[1]Hong et al. (2016), [2]Ikeda et al. (2009), [3]Shi et al. (2021), [4]Pritchett et al. (2012), [5]Aguilar et al. (2020), [6]Ahnaou, Raeymaekers, Steckler and Drinkenbrug (2015), [7]Holst et al. (2017), [8]Fisher et al. (2020).

Table 2 Pharmacological studies examining subtype-specific mGlu receptor compounds on sleep in rodents.

Compound	Awake Duration	NREM Duration	NREM Latency	NREM Bouts	NREM Bout duration	REM Duration	REM Latency	REM Bouts	REM Bout duration
mGlu$_5$ receptor NAMS (within 4–6h post-dosing)									
MPEP[1,2] (1–20 mg/kg, IP)	↑	↑	n.s.	↓	↑	↓	↑	↓	↓
MTEP[1] (1–10 mg/kg, IP)	n.s.	↑	n.s.	↓	↑	↓	↑	n.s.	↓
GRN-529[3] (0.32–1 mg/kg, SC)	↑	↓	—	—	—	↓	—	—	—
Mavoglurant[3] (1–3.2 mg/kg, SC)	↑	↓	—	—	—	↓	—	—	—
Basimglurant[4] (0.03–0.3 mg/kg, PO)	↑	↓	↑	—	—	↓	↑	—	—
VU0424238[5] (1–30 mg/kg, IP)	↑	↓	n.s.	—	—	↓	↑	—	—
M-5MPEP (partial NAM)[5] (18–56.6 mg/kg, IP)	n.s.	n.s.	n.s.	—	—	↓	↑	—	—
mGlu$_5$ Receptor PAMS (within 4–6h post-dosing)									
LSN2814617[1,6] (2.5–10 mg/kg, PO)	↑	↓	↑	↓	↓	↓	↑	↓	↓
LSN2463359[6] (0.3–3 mg/kg, PO)	↑	↓	↑	—	—	↓	—	—	—
ADX47273[1,6] (10–300 mg/kg, PO)	↑	↓	↑	n.s.	↓	↓	↑	↓	n.s.
CDPPB[6,7] (30 mg/kg, IP)	↑ ; n.s.	↓	n.s.	—	—	↓	—	—	—
mGlu$_{2/3}$ Receptor Antagonists/NAMS (within 4–6h post-dosing)									
LY341495 (antagonist)[8,9] (1–10 mg/kg, SC)	↑	↓	↑	—	—	↓	↑	—	—
LY3020371 (antagonist)[10] (3–10 mg/kg, IP)	↑	↓	—	—	—	↓	—	—	—
Ro4491533 (NAM)[8] (2.5–40 mg/kg, PO)	↑	↓	↑	—	—	↓	↑	—	—

Continued

Table 2 Pharmacological studies examining subtype-specific mGlu receptor compounds on sleep in rodents.—cont'd

	Awake	NREM				REM			
Compound	Duration	Duration	Latency	Bouts	Bout duration	Duration	Latency	Bouts	Bout duration
mGlu_{2/3} Receptor Agonists/PAMS (within 4–6 h post-dosing)									
LY379268 (agonist)[10,11] (10 mg/kg, PO[10]; 0.25–1 mg/kg SC[11])	n.s. (dark) ↑ (light)	n.s. (dark) ↓ (light)	—	—	—	↓	—	—	—
LY354740 (agonist)[12] (1–10 mg/kg, SC)	n.s.	n.s.	n.s.	n.s.	n.s.	↓	↑	↓	↓
BINA (PAM)[12] (1–40 mg/kg, SC)	n.s.	n.s.	—	↑	↓	↓	↑	↓	↓
JNJ-42153605 (PAM)[13] (3 mg/kg, PO)	n.s.	n.s.	—	—	—	↓	n.s.	—	—
JNJ-40411813 (PAM)[14] (3–30 mg/kg, PO)	↓	↑	—	—	—	↓	↑	—	—
JNJ-40068782 (PAM)[15] (3–30 mg/kg, PO)	n.s.	n.s.	—	—	—	↓	↑	—	—
THIIC (PAM; 12 h post-dosing)[16] (10–30 mg/kg, PO)	—	↑	—	↑	—	↓	—	—	—
mGlu_{2/3} receptor agonists/PAMS (within 4–6 h post-dosing)									
(S)-3,4-dicarboxyphenylglycine[17] (5–20 mg/kg, IP)	n.s.	n.s.	n.s.	n.s.	n.s.	n.s.	n.s.	n.s.	n.s.

n.s., not significant; —, not determined; IP, intraperitoneal; SC, subcutaneous; PO, *per os* (oral).

[1]Ahnaou, Langlois, Steckler, Bartolome-Nebreda and Drinkenburg (2015), [2]Cavas et al. (2013b), [3]Harvey et al. (2013), [4]Lindemann et al. (2015), [5]Holter et al. (2021), [6] Gilmour et al. (2013), [7]Parmentier-Batteur et al. (2012), [8]Ahnaou, ver Donck, and Drinkenburg (2014), [9]Feinberg, Schoepp, Hsieh, Darchia, and Campbell (2005), [10]Wood et al. (2018), [11]Feinberg, Campbell, Schoepp, and Anderson (2002), [12]Ahnaou et al. (2009), [13]Cid et al. (2012), [14]Lavreysen et al. (2015), [15]Lavreysen et al. (2013), [16]Fell et al. (2011), [17]Cavas, Scesa, Martín-López, and Navarro (2017).

alternate pathways and reduced synaptic GluA1 (Kammermeier & Worley, 2007; Ronesi & Huber, 2008). While the specific downstream pathways of the $mGlu_{1/5}$/Homer1a protein complex are not yet extensively studied, there is support for increased MAPK/ERK signaling (Diering et al., 2017; Martin et al., 2019). Importantly, regulation of synaptic levels of Homer1a has been found to be mediated in part by neuromodulators including noradrenaline and adenosine which suppress and promote Homer1a synapse targeting, respectively (Diering et al., 2017). Although Homer1a mRNA is highest during wake, cortical synaptic levels are highest during sleep and states of increased sleep drive suggesting Homer1a levels are activity-dependent and contribute to sleep-maintained synaptic homeostasis (Diering et al., 2017; Maret et al., 2007). Overall, this process, termed homeostatic scaling-down, yields restorative benefits, increasing memory consolidation and opportunities for future learning (de Vivo et al., 2017; Diering et al., 2017; Martin et al., 2019; Tononi & Cirelli, 2014).

3.3.1.1 mGlu$_1$ receptors and sleep

In general, $mGlu_1$ has been implicated in modulating sleep duration. Recently, two loss-of-function mutations of GRM1, the gene encoding $mGlu_1$, were identified in two families diagnosed with familial natural short sleep (FNSS) (Shi et al., 2021). Individuals with FNSS can sleep less than 6 h a day (compared to the average 8 h of sleep) with no known adverse consequences (Aeschbach et al., 2003). When these mutations were introduced into mice, significant decreases in total and NREM sleep time were found with no effects on REM sleep time (Shi et al., 2021). No other studies to date have directly examined region-or circuit specific GRM1 genetic knockouts to further elucidate their involvement in sleep regulation. However, immunohistochemistry studies have shown that $mGlu_1$ expression was higher in rodents following a 12-h sleep deprivation period (Tadavarty, Rajput, Wong, Kumar, & Sastry, 2011), suggesting that $mGlu_1$ receptors may also serve a role in homeostatic regulation of sleep.

Additional studies examining downstream signaling pathways of $mGlu_1$ have elucidated the important role of this receptor in sleep via expression on TC neurons (Shigemoto et al., 1992). Activation of $mGlu_1$ leads to downstream PLC signaling, specifically the PLCβ4 isoform in TC neurons (Watanabe et al., 1998). Thus, to investigate the function of $mGlu_1$ in sleep, studies have examined sleep in PLCβ4-deficient (PLCβ4$^{-/-}$) mice. Both global PLCβ4$^{-/-}$ and TC-restricted PLCβ4$^{-/-}$ mice displayed longer total NREM sleep time as well as longer NREM and REM episodes.

Delta power was also increased during NREM sleep (Hong et al., 2016). Alternately, a separate study reported no differences in NREM sleep in $PLC\beta4^{-/-}$ mice compared to WT mice but did find increased total REM duration, abnormal wake-to-REM transitions, and altered ultradian body temperature rhythms (Ikeda et al., 2009). Altogether, though the GRM1 mutation studies and studies using $PLC\beta4^{-/-}$-deficient mice suggest different roles of $mGlu_1$ in sleep, it is clear that expression of $mGlu_1$ and the associated signaling pathways in TC circuits is likely involved in the transitions and maintenance of wake and sleep. Pharmacological studies examining receptor subtype-selective $mGlu_1$ compounds on sleep-wake architecture are also lacking, in large part due to the unavailability of selective ligands until recently (Luessen & Conn, 2022).

3.3.1.2 mGlu₅ receptors and sleep

Similar to the $mGlu_1$ receptor, human and animal studies suggest the $mGlu_5$ receptor serves a role in homeostatic regulation of sleep, with several lines of evidence supporting a role in sleep need/drive. Positron Emission Tomography (PET) imaging studies in healthy humans reported global increases in $mGlu_5$ receptor availability following a sleep deprivation period (Hefti et al., 2013; Holst et al., 2017; Weigend et al., 2019). $mGlu_5$ receptor availability was positively correlated with both subjective and objective measures of sleepiness as well as delta power during NREM sleep (Holst et al., 2017). Multiple EEG studies comparing $mGlu_5$ knockout (KO) mice with wild type (WT) mice corroborate these findings. First, $mGlu_5$ receptor KO mice showed less time awake and greater duration of NREM sleep during the active phase, though sleep was reportedly more fragmented (Aguilar et al., 2020). Alterations in spectral power were also found in $mGlu_5$ KO mice including decreased delta power during NREM, sleep, theta power during REM sleep, and alpha power during waking periods. These spectral changes could have implications in sleep quality and sleep homeostasis, sleep-dependent memory consolidation, and TRN function, respectively (Aguilar et al., 2020). Interestingly, some (Ahnaou, Langlois, et al., 2015; Ahnaou, Raeymaekers, et al., 2015; Holst et al., 2017) but not all (Aguilar et al., 2020) studies reported a lack of build-up in delta power following increased time awake across the dark (active) phase and following sleep deprivation, suggesting altered homeostatic sleep drive in KO animals. Additionally, one study reported extended waking periods following sleep deprivation in KO animals in contrast to the expected rebound sleep found in WT animals (Holst et al., 2017). However, a separate study reported a more rapid sleep rebound in mGlu$_5$ KO mice (Aguilar et al., 2020). Lastly, mGlu$_5$ KO mice showed higher basal gamma power during wake in both the light and dark phase (Ahnaou, Langlois, et al., 2015; Ahnaou, Raeymaekers, et al., 2015) as well as during NREM and REM sleep (Aguilar et al., 2020) suggesting heightened arousal and inability to shift from asynchronous to synchronous oscillatory activity.

Pharmacological studies using subtype-selective compounds have further aided in our understanding of the contributions of the mGlu$_5$ receptor to sleep regulation and underlying brain oscillatory activity (see Table 2 summarizing effects of mGlu selective compounds on sleep). Most highly selective compounds targeting mGlu$_5$ receptors include negative and positive allosteric modulators (NAMs and PAMs, respectively). NAMs and PAMs bind to an allosteric site and either reduce or enhance receptor function, respectively when a ligand binds to the orthosteric binding site, (Niswender & Conn, 2010). In rodent EEG studies, traditional mGlu$_5$ NAMs administered in the light cycle including MPEP and MTEP dose-dependently increased NREM sleep time and bout duration as well as REM latency and decreased REM duration within the 4–6h post-administration (Ahnaou, Langlois, et al., 2015; Ahnaou, Raeymaekers, et al., 2015; Cavas et al., 2013b; Harvey et al., 2013). Increased NREM duration, but not decreased REM duration, is consistent with results from mGlu$_5$ KO studies. Novel, more selective mGlu$_5$ compounds have also been examined. In contrast to MTEP and MPEP, basimglurant (RO4917523, RG7090; administered 2h into the dark cycle) and VU0424238 (administered 2h into the light cycle), both demonstrated initial but transient wake-promoting effects (Holter et al., 2021; Lindemann et al., 2015). However, as the wake-promoting effects of VU0424238 dissipated, selective reductions in REM sleep persisted for 8–10h following administration (Holter et al., 2021). Furthermore, all mGlu$_5$ NAMs mentioned above modulated quantitative EEG by producing increases in NREM delta power, which is possibly reflective of sleep drive, though this finding is inconsistent with studies in mGlu$_5$ KO mice (Harvey et al., 2013; Holter et al., 2021; Lindemann et al., 2015). However, similar to KO mice, multiple mGlu$_5$ NAMs increased high frequency gamma power during waking periods, which may be reflective of increased arousal (Ahnaou, Langlois, et al., 2015; Ahnaou, Raeymaekers, et al., 2015; Holter et al., 2021). Interestingly, M-5MPEP, a partial mGlu$_5$ NAM (at full receptor occupancy, there is only 50% functional inhibition), selectively decreased REM sleep without affecting quantitative EEG (Holter et al., 2021; Rodriguez et al., 2005). Lastly, in contrast, to mGlu$_5$ NAMs, mGlu$_5$ PAMs administered in the light phase including LSN2814617, CDPPB and ADX47273 demonstrated wake-promoting effects for up to 7h following administration (Ahnaou, Langlois, et al., 2015; Ahnaou, Raeymaekers, et al., 2015; Gilmour et al., 2013; Parmentier-Batteur et al., 2012). In the case of LSN2814617 (3mg/kg), while wake and REM sleep effects normalized at around 6h post-administration, decreases in deep sleep duration did not normalize until the onset of the dark cycle at 10h post-administration (Ahnaou, Langlois, et al., 2015; Ahnaou, Raeymaekers, et al., 2015). It is important to note tolerance developed to the wake-promoting effects of CDPPB, such that sleep-wake durations and latencies were not different from baseline following 7 days of repeated administration (Parmentier-Batteur et al., 2012).

Overall, mGlu$_5$ receptor activity influences multiple stages of the sleep-wake cycle. Human studies support a relationship between mGlu$_5$ receptor availability and sleep homeostasis/sleep need. In rodent models, both mGlu$_5$ KO mice and pharmacological studies confirm altered sleep-wake patterns. Of note, differences in selectivity, potency, and pharmacokinetics of mGlu$_5$ compounds need to be considered as these may contribute to a state-specific effect (e.g. selective REM sleep reduction) or initial wake-promotion that may artificially reduce both NREM and REM sleep; off-target effects of some compounds, especially MTEP and MPEP, may also be of influence (Montana et al., 2009).

3.3.2 Group II mGlu receptors and sleep

Similar to group I mGlu receptors, substantial preclinical literature describes the modulatory role of group II mGlu receptors on sleep. Using video analysis to determine sleep vs waking periods, Grm2/3$^{-/-}$ mice had significant reductions in sleep time during both the light and dark phase as well as fragmented sleep (increased number of sleep bouts and decreased bout duration) (Pritchett et al., 2015). However, two additional studies using EEG reported no marked differences between Grm2/3$^{-/-}$ and Grm2$^{-/-}$ mice in waking, NREM and REM durations compared to WT mice (Ahnaou et al., 2009; Wood et al., 2018).

Pharmacological modulation of mGlu$_{2/3}$ receptors, however, has produced robust changes in sleep/wake patterns. Activation of mGlu$_{2/3}$ during the light cycle via agonists including LY379268 and LY354740 selectively increased REM sleep latency and reduced REM sleep duration without impacting wake or NREM sleep for 7–12h following administration (Ahnaou et al., 2009; Feinberg et al., 2002). mGlu$_{2/3}$ agonists also reduced high frequency oscillatory activity in the 10–50 Hz range indicating that, in addition to selectively reducing REM sleep, these compounds also reduce cortical arousal during wake (Ahnaou et al., 2009; Feinberg et al., 2002). Importantly, these effects were absent in $Grm2^{-/-}$ mice suggesting that a reduction in REM sleep is specifically attributed to activation of $mGlu_2$ and not $mGlu_3$ receptors (Ahnaou et al., 2009; Wood et al., 2018). $mGlu_{2/3}$ agonists have similar effects on sleep when compared to $mGlu_{2/3}$ PAMs including BINA, JNJ-42153605, THIIC and JNJ-40411813, although the latter two compounds also increased sleep duration and bout length for 8–12 h following administration (Ahnaou et al., 2009; Cid et al., 2012; Fell et al., 2011; Lavreysen et al., 2013). Effects of $mGlu_{2/3}$ PAMs were also absent in $Grm2^{-/-}$ mice (Ahnaou et al., 2009; Wood et al., 2018). In contrast, $mGlu_{2/3}$ antagonists (LY341495 and LY3020371) administered in both the light and dark phase and NAM (Ro-4491533) increased wake duration (both active and passive) and non-selectively decreased total sleep time. These wake-promoting effects were long-lasting, such that administration in the light cycle resulted in sustained changes through the beginning of the dark cycle (Ahnaou et al., 2014). Likely tied to increased arousal, increased gamma power was reported following administration, and this opposes effects to receptor activation (Ahnaou et al., 2014; Feinberg et al., 2005; Wood et al., 2018). Interestingly, reduced, yet still measurable, effects of the $mGlu_{2/3}$ antagonist LY3020371 and NAM Ro-4491533 were present in $mGlu2^{-/-}$ rats suggesting that both $mGlu_2$ and $mGlu_3$ contribute to wake promotion (Wood et al., 2018). As noted above, acute pharmacological studies recapitulate some, but not all, findings from knockout mice.

3.3.3 Group III mGlu receptors and sleep

Unlike group I and II mGlu receptors, the influence of group III mGlu receptors on sleep-wake regulation are understudied. To our knowledge, no study has examined the influence of $mGlu_4$ receptors on sleep-wake architecture in KO mice or with selective ligands despite an extensive catalogue of pharmacological compounds and widespread interest in $mGlu_4$ receptors with regard to Parkinson's Disease (Hopkins, Lindsley, & Niswender, 2009). Furthermore, research examining the influence $mGlu_7$ and $mGlu_8$ receptors on sleep regulation is scarce. A single study reported increased time awake and decreased NREM sleep duration in mice with a GRM7 deletion, the gene that encodes $mGlu_7$, compared to WT mice. Additionally, though these $Grm7^{-/-}$ mice displayed no change in overall REM sleep duration, they demonstrated decreased average REM bout duration and increased bout numbers, suggesting changes in the maintenance of REM sleep (Fisher et al., 2020). In part attributed to the scarcity of selective compounds, pharmacological studies are also limited. The mGlu7 PAM AMN082 produced differential effects on sleep in healthy rats. Acute administration of lower doses (5 and 10 mg/kg) during the light phase increased NREM sleep and decreased time awake whereas a high dose (20 mg/kg) increased time awake and decreased both NREM and REM sleep (Cavas, Scesa, & Navarro, 2013a). However, similar effects were present when examined in Grm7$^{-/-}$ mice (Ahnaou, Raeyemaekers, Huysmans, & Drinkenburg, 2016) suggesting off-target effects. Thus, there is still a need to examine more selective compounds to better characterize the role for mGlu7 in sleep regulation. Lastly, a single study reports that the mGlu8 agonist (S)-3,4–DCPG did not affect any sleep parameter in healthy rats (Cavas et al., 2017).

Overall, the individual mGlu receptors have differential roles in sleep/wake regulation. In general, both pharmacological and knockout models examining group I receptors identified contributions to both NREM and REM sleep time. Further, in conjunction with human findings, several rodent studies identified changes in spectral activity that reflected contributions to homeostatic sleep drive. In contrast, group II receptors appear to have less profound contributions in EEG knockout studies, and, in pharmacological studies, many compounds appear to produce REM-selective inhibition (agonists/PAMs) or robust wake-promotion (antagonists/NAMs). Lastly, while some roles of group III receptors have been reported, the literature is still in early stages with genetic and pharmacological advancements necessary to further our understanding of their role (or lack thereof) in regulating sleep.

3.4 mGlu receptors, neuropsychiatric disorders and sleep

Comorbid sleep disturbances including insomnia have been found in the majority of CNS disorders (Benca & Buysse, 2018; Winkelman & de Lecea, 2020; Wulff et al., 2010). Although primarily diagnosed subjectively and defined as difficulty initiating or maintaining sleep, insomnia is also associated with aberrant quantitative EEG characteristics including increased high frequency (beta/gamma) activity at sleep onset and during NREM sleep (for review, see Zhao et al., 2021). Importantly, the DSM-5 delineated insomnia as a separate diagnosis from other mental disorders as opposed to a primary or secondary symptom (APA, 2013). Thus, as a separate diagnosis, insomnia should also be a primary target of pharmacotherapies (Benca & Buysse, 2018). Furthermore, the DSM-5 emphasizes the bidirectional relationship between insomnia and neuropsychiatric illness, where both disorders can serve as a risk factor for the development of one another (Benca & Buysse, 2018; Winkelman & de Lecea, 2020). Support for insomnia as a risk factor stemmed from data collected from the National Institute of Mental Health's Epidemiologic Catchment Area Survey. This survey collected comprehensive information on mental disorders in the United States and included questions about sleep disturbances in addition to current psychiatric symptoms. Findings from this survey suggested that insomnia increased the risk of first episode diagnosis of MDD (Ford & Kamerow, 1989). As discussed below, preexisting insomnia has also been implicated in the onset of SUD, PTSD, schizophrenia, and AD (Irwin & Vitiello, 2019; Maher, Rego, & Asnis, 2006; Manoach & Stickgold, 2009; Neylan et al., 2021; Riemann, Berger, & Voderholzer, 2001; Roehrs, Sibai, & Roth, 2021; Zhang et al., 2022).

It is important to understand how the glutamate system may contribute to the comorbid and bidirectional relationship between sleep and CNS disorders. As there are ongoing efforts to develop mGlu receptor subtype-selective treatments for these conditions, current and future research should work to determine if mGlu receptor compounds can target both sleep disturbances and the comorbid neuropsychiatric disorders. In the following sections, we present preclinical and clinical literature detailing known alterations in glutamatergic function associated with schizophrenia, MDD, PTSD, AD, and SUD as well as the comorbid sleep disruptions frequently linked with these conditions. As the field is still expanding, we build on the predominately preclinical pharmacological data above and speculate on possible mGlu receptor manipulations/strategies to improve sleep disturbances in addition to other symptoms of the CNS disorder.

3.4.1 Schizophrenia

Schizophrenia is a neuropsychiatric disorder that affects around 1% of the human population and is characterized by three primary symptoms clusters: positive, negative, and cognitive (Jones, Watson, & Fone, 2011; Manoach & Stickgold, 2019; Winship et al., 2019). Sleep disruptions represent an additional underappreciated symptom cluster (Manoach & Stickgold, 2019). Sleep disturbances are reported in both medicated and drug-naïve patients, affecting ~80% of all patients, and are linked with symptom prevalence and severity (Chen et al., 2020; Cohrs, 2008; Korenic et al., 2020; Manoach &

Stickgold, 2019; Sprecher et al., 2015). Patients commonly experience insomnia and related increases in NREM sleep latency as well as reduced sleep quality (e.g. reduced delta power in N3 sleep) and sleep efficiency (defined as the ratio of time spent asleep when in bed) (Benca & Buysse, 2018; Manoach & Stickgold, 2019; Sprecher et al., 2015). Decreased REM sleep latency has also been found during acute psychosis, though reported changes in REM sleep duration in patients with schizophrenia are inconsistent throughout literature (Pritchett et al., 2012; Sprecher et al., 2015). Importantly, sleep disturbances can precede the onset of schizophrenia and associated symptoms. Insomnia has been found to precede the onset of positive symptoms including hallucinations and paranoia (for review, see Reeve et al., 2015), and, similarly, decreased NREM sleep duration and quality has been associated with the onset of negative and cognitive symptoms (Göder et al., 2004; Göder et al., 2006; Keshavan et al., 1995). Sleep spindles, which are important for memory consolidation and cognition, also occur less frequently in medicated patients with schizophrenia relative to healthy humans (Göder et al., 2006, 2015; Manoach & Stickgold, 2019; Wamsley et al., 2012). Together, altered sleep duration and architecture are directly associated with all primary symptom clusters of schizophrenia.

Current antipsychotic medications, which exert therapeutic effects via blockade of dopamine D2 and serotonin 5-HT2A receptors, affect sleep and can contribute to results of sleep studies conducted in medicated patients. Both first- and second-generation antipsychotics increase total sleep time in healthy humans and in patients with schizophrenia in part by decreasing monoaminergic mediated arousal circuits, and these sedative effects may persist during wake (Göder et al., 2004; Manoach et al., 2010; Manoach & Stickgold, 2009). Although literature is mixed on specific changes, second generation antipsychotics have commonly been found to increase both NREM and REM sleep time in patients with schizophrenia (for review and more detailed descriptions, see Cohrs, 2008). Importantly, antipsychotic medications are used as off-label agents to treat insomnia (Benca & Buysse, 2018). Thus, although many of the current FDA-approved medications can increase sleep in patients with schizophrenia, novel, more effective pharmacotherapies that target cognitive impairments and have a lower adverse effect risk than current antipsychotics are needed.

Although antipsychotics were initially developed on the premise that excessive dopamine in the striatum contributed to the positive symptoms, growing evidence supports glutamatergic dysregulation, specifically

N-methyl-D-aspartate receptor (NMDAR) hypofunction, as a key contributor to multiple symptoms of schizophrenia (Coyle, 2006; Javitt, 2007; Jones et al., 2011; Kantrowitz & Javitt, 2012). This hypothesis stemmed from early studies that reported psychotic-like behavior in healthy humans following administration of NMDAR antagonists (Davies & Beech, 1960; Luby, Cohen, Rosenbaum, Gottlieb, & Kelley, 1959). Since then, multiple studies have reported similarities in brain function, psychotic-like behaviors, and cognitive impairments between healthy humans administered NMDAR-antagonists and patients with schizophrenia, and NMDAR antagonists have been integrated into many preclinical studies to model all symptom clusters (for review, see Javitt, 2007; Jones et al., 2011). Genetic, functional imaging, and post-mortem brain tissue studies have provided substantial evidence of neurobiological alterations in the glutamate system in patients with schizophrenia (Hu, MacDonald, Elswick, & Sweet, 2015; Korenic et al., 2020). For example, altered pyramidal cell morphology was found in the cortex of post-mortem brain tissue, including decreased dendritic spine density, dendritic length, and somal size (Black et al., 2004; Glantz & Lewis, 2000; Kalus, Müller, Zuschratter, & Senitz, 2000; Pierri, Volk, Auh, Sampson, & Lewis, 2001, 2003). Changes in enzymes that have a role in glutamate synthesis and metabolism have also been reported. Protein expression of glutamine synthetase was lower and activity of phosphate-activated glutaminase (PAG) was higher in the PFC. These enzymes are responsible for converting glutamate to glutamine and vice versa suggesting impairment in glutamate metabolism in patients with schizophrenia compared to healthy controls (Burbaeva et al., 2003; Gluck, Thomas, Davis, & Haroutunian, 2002). Importantly, studies have found that alterations in glutamate neurotransmission correspond with associated sleep impairments. Using $[^1H]$-MRS, Korenic et al., 2020 found poor sleep quality in patients with schizophrenia was associated with higher Glx (glutamate and glutamine) levels in the hippocampus and lower levels of glutamate in the anterior cingulate cortex. The glutamate levels in the anterior cingulate cortex correlated with lower total scores on the Brief Psychiatric Rating Scale (BPRS). Moreover, poor sleep quality corresponded with heightened severity of positive symptoms. Taken together, these clinical studies firmly support a relationship between glutamate dysregulation, schizophrenia, and sleep.

Alterations in mGlu receptors have also been identified. Post-mortem tissue from patients with schizophrenia revealed higher protein expression of $mGlu_5$ and several $mGlu_5$ trafficking proteins in the hippocampus

(Matosin et al., 2015). Conversely, a PET imaging study reported lower cortical mGlu$_5$ binding potential in patients with schizophrenia compared to healthy controls and lower binding was associated with an increase in negative symptoms and worse cognitive performance (Régio Brambilla et al., 2020). Lastly, polymorphisms of GRM5, the gene encoding the mGlu$_5$ receptor, have also been reported in patients with schizophrenia (Matosin et al., 2018). This relationship between mGlu$_5$ and symptoms of schizophrenia is further supported by pharmacological studies. Some mGlu$_5$ NAMs, including fenobam, induced psychotomimetic-like effects in both humans and animals. These compounds were also associated with increased resting-state high frequency gamma power, which is commonly found in patients with schizophrenia (Ahnaou, Langlois, et al., 2015; Ahnaou, Raeymaekers, et al., 2015; Holter et al., 2021; Homayoun, Stefani, Adams, Tamagan, & Moghaddam, 2004; Jacob et al., 2009). The mGlu$_5$ receptor is coupled to the NMDAR (Niswender & Conn, 2010; Shigemoto et al., 1993), and, thus, one hypothesized mechanism through which mGlu$_5$ NAMs produce these psychotomimetic-like effects is via downstream NMDAR inhibition (Sengmany & Gregory, 2016). Furthermore, as described above, global knockout of mGlu$_5$ in mice recapitulated some sleep deficits similar to those in patients with schizophrenia (Aguilar et al., 2020). Taken together, these preclinical and clinical studies support an underlying role of mGlu$_5$ hypofunction in schizophrenia symptomology.

While overwhelming evidence suggests NMDAR dysregulation may underlie all symptom clusters of schizophrenia, directly targeting NMDARs as a therapeutic approach has been met with adverse effects including seizures. Thus, modulating mGlu receptors represents a more tolerable approach than direct NMDAR activation, and all groups of mGlu receptors have gained preclinical support as potential therapeutic targets (Dogra & Conn, 2022; Luessen & Conn, 2022). Activation/allosteric modulation of group I mGlu receptors has shown promise through increasing presynaptic glutamate release and postsynaptic NMDAR function. Support for the use of both mGlu$_1$ and mGlu$_5$ PAMs spans multiple preclinical models of all primary symptom clusters. For example, compounds targeting either receptor have reversed amphetamine and NMDAR antagonist-induced hyperlocomotion, deficits in paired pulse inhibition, and deficits in multiple cognitive assessments (for review, seeDogra & Conn, 2022; Maksymetz et al., 2017). Although mGlu$_5$ PAMs increase time awake and reduce sleep in healthy rodents, the potential for these compounds to modulate sleep spindles still warrants examination of the sleep-altering effects in animal models of schizophrenia. As noted above, patients with schizophrenia and mGlu$_5$ KO mice showed lower sleep spindle frequencies (Aguilar et al., 2020; Göder et al., 2006, 2015; Manoach & Stickgold, 2019; Wamsley et al., 2012). Sleep spindles are generated in the TRN (Bazhenov, Timofeev, Steriade, & Sejnowski, 2000; Ferrarelli & Tononi, 2017) and are in part mediated by group I mGlu receptor activation (Sun et al., 2016). Thus, both mGlu$_1$ and mGlu$_5$ PAMs may enhance sleep spindle activity and, thereby, memory consolidation. Patients with schizophrenia also report excessive daytime sleepiness which may be a product of insomnia, reduced sleep quality, and/or the sedative effects of current antipsychotics (Cohrs, 2008; Miller, 2004; Sharma, 2016). Increasing arousal/time awake during the day could aid with sleep consolidation at night, similar to the primary goal of the cognitive behavioral therapy approach for insomnia (Haynes, Talbert, Fox, & Close, 2018). These can also be employed as an adjunct treatment to counteract the sedative effects of many antipsychotics. mGlu$_5$ PAMs have not yet been examined on sleep-wake activity in preclinical models of schizophrenia, though these studies are critical to determine potential of these compounds to normalize sleep disturbances. It is also important to consider time of dosing in future studies to mitigate undesired wake promoting/sleep-disrupting effects.

Group II mGlu receptors are also promising pharmacotherapeutic targets for schizophrenia. NMDAR hypofunction induced by ketamine and PCP ultimately leads to increased levels of synaptic glutamate in regions including the PFC (Lorrain, Baccei, Bristow, Anderson, & Varney, 2003; Maksymetz et al., 2017; Moghaddam & Jackson, 2003). Activation of mGlu$_{2/3}$ presynaptic autoreceptors causes a reduction in synaptic glutamate concentrations (Dogra & Conn, 2022; Maksymetz et al., 2017) and, thus, may be a viable mechanism to attenuate this aberrant glutamate efflux. Similar to group I PAMs, mGlu$_{2/3}$ agonists and PAMs have shown efficacy in preclinical models of all primary symptom clusters of schizophrenia (Benneyworth et al., 2007; Galici et al., 2006; Greco et al., 2005; Griebel et al., 2016; Hackler et al., 2010; Kawaura, Karasawa, & Hikichi, 2016; Maksymetz et al., 2017; Moghaddam & Adams, 1998). Following extensive support from preclinical studies, the mGlu$_{2/3}$ agonist LY2140023 advanced into clinical trials for treatment of schizophrenia but had mixed results in phase II studies. However, significant improvements in positive and negative symptoms were reported in some trials supporting activation of mGlu$_{2/3}$ as a mechanism for further pursuit (for review, see Maksymetz et al., 2017). Unlike mGlu$_5$ PAMs, mGlu$_{2/3}$ PAMs may influence sleep directly by reducing extracellular glutamate levels. Newer, more selective compounds increased NREM sleep time and bout length (Fell et al., 2011; Lavreysen et al., 2013) and, thus, may be of benefit to patients experiencing insomnia. Furthermore, although not directly investigated, $mGlu_2$ receptors are also located in the TRN and regulate signaling (Crabtree, Lodge, Bashir, & Isaac, 2013) suggesting a possible, but not yet studied, influence on sleep spindle activity.

In summary, both $mGlu_5$ and $mGlu_{2/3}$ PAMs represent promising therapeutic approaches for multiple symptoms of schizophrenia. While speculative, these compounds may appropriately address associated sleep disturbances. It is important to note that preclinical studies also support targeting group III mGlu receptors, specifically with $mGlu_4$ PAMs and $mGlu_7$ NAMs, for all primary symptom clusters of schizophrenia (for review, see Dogra & Conn, 2022 and Luessen & Conn, 2022). However, as reiterated throughout this chapter, further research is needed to understand the potential of group III compounds to target schizophrenia-associated sleep disturbances.

3.4.2 Major depressive disorder

Major depressive disorder (MDD) is characterized by the DSM-5 as having at least five symptoms including depressed mood, loss of interest in activities, fatigue, and sleep disturbances for a 2-week period (APA, 2013). Sleep disturbances are reported in around 90% of patients with MDD, and, while the majority of these patients report insomnia, there is a smaller subset of patients (15–35%) who report hypersomnia (Armitage, 2007; Reynolds & Kupfer, 1987; Steiger & Pawlowski, 2019). Further, there appears to be a strong bidirectional relationship between depression and insomnia. First, in an epidemiological study examining an adult sample from the UK population, insomnia was linked to preexisting depression (Morphy, Dunn, Lewis, Boardman, & Croft, 2007). Multiple epidemiological studies have also found preexisting insomnia was associated with the development of depression (Fang, Tu, Sheng, & Shao, 2019; Ford & Kamerow, 1989; Morphy et al., 2007; Riemann et al., 2020). However, it is important to note that while the majority of patients with depression have comorbid insomnia, only around 20% of patients with insomnia are diagnosed with depression (Breslau, Roth, Rosenthal, & Andreski, 1996). Polysomnography studies reveal that, in general, patients with depression experience a reduction in overall sleep time and display increased NREM sleep latency alongside decreased NREM sleep duration and increased wake after sleep onset (WASO) (Benca et al., 1992; Steiger & Kimura, 2010; Wichniak,

Wierzbicka, & Jernajczyk, 2013; Wichniak, Wierzbicka, Walęcka, & Jernajczyk, 2017). This is paired with reduced delta power during NREM sleep (Armitage, 1995; Kupfer, Reynolds, Ulrich, & Grochocinski, 1986; Steiger & Kimura, 2010). Interestingly, in contrast to NREM sleep, patients with MDD concurrently experience an increase in REM sleep duration and density and a reduction in REM sleep latency (Foster, Kupfer, Coble, & McPartland, 1976; Steiger & Pawlowski, 2019; Wichniak et al., 2013). Although variable, most antidepressant medications selectively decrease REM sleep and increase REM sleep latency without producing concomitant decreases in NREM sleep (Steiger & Pawlowski, 2019; Wichniak et al., 2017). Antidepressants have also been found to increase NREM sleep delta power (Kupfer, Ehlers, Pollock, Swami Nathan, & Perel, 1989; Steiger & Kimura, 2010). Importantly, the ability of medications to mitigate REM sleep abnormalities has been shown to correspond with treatment efficacy, reiterating the importance of improving sleep for overall treatment outcome (Riemann et al., 2001).

There is substantial evidence for glutamatergic contributions to the pathology of MDD and some even suggest there should be a glutamate hypothesis of depression (Sanacora, Treccani, & Popoli, 2012). Several studies in clinical populations have shown differences in glutamate levels in the CSF, plasma and platelets, and post-mortem brain tissue between individuals with MDD and healthy individuals. For example, increased glutamate levels in the plasma have been found in patients with MDD relative to a healthy control group (Altamura et al., 1993; Kim, Schmid-Burgk, Claus, & Kornhuber, 1982; Mauri et al., 1998). In contrast, decreased glutamate levels were reported in the CSF in patients with refractory affective disorder (includes both unipolar and bipolar patients) (Frye, Tsai, Huggins, Coyle, & Post, 2007), although earlier studies examining glutamine levels in the CSF found increases in patients with unipolar and bipolar depression (Levine et al., 2000). In studies examining post-mortem tissue, increased glutamate levels were found in the frontal cortex (Hashimoto, Sawa, & Iyo, 2007) and dorsolateral prefrontal cortex (Lan et al., 2009). Lastly, several studies have examined glutamate levels in patients with MDD using ^{1}H-MRS. Collectively, a meta-analysis reported consistent decreases in concentrations of glutamate both globally and specifically in the anterior cingulate cortex (Luykx et al., 2012). In short, these clinical findings have pointed to clear dysregulations in glutamate levels in patients with MDD, and the sampling methodology, brain-region, and/or heterogeneity associated with MDD may have influenced varying findings across studies.

Targeting the glutamate system through both iGlu and mGlu receptors has been explored for depression. Early findings suggest that ketamine, an NMDAR antagonist, produced rapid-acting antidepressant effects, and NMDAR antagonism remains a promising therapeutic approach (Berman et al., 2000; Raab-Graham, Workman, Namjoshi, & Niere, 2016; Trullas & Skolnick, 1990; Zarate et al., 2006). Of relevance, ketamine administration to individuals with MDD decreased mGlu5 receptor availability in several brain regions and hippocampal mGlu5 availability correlated with Montgomery-Åsberg Depression Rating Scale (MADRS) total score (Esterlis et al., 2018). Thus, long-standing research also supports targeting mGlu receptors as a treatment approach for MDD. Functional antagonism of mGlu5 displayed therapeutic potential in preclinical models that have predictive validity for antidepressant medications including the forced swim test (FST) and tail suspension test (TST) (Felts et al., 2017; Gould et al., 2016; Hughes et al., 2013; Kato et al., 2015; Lindemann et al., 2015). Importantly, extensive preclinical work suggests mGlu5 NAMs may be beneficial for treating sleep disturbances associated with MDD. For example, the mGlu5 NAM basimglurant significantly increased REM latency and decreased REM sleep and the REM/NREM ratio in healthy rats when administered during the active phase. Furthermore, basimglurant increased delta power during NREM sleep. This profile is similar to other mGlu5 NAMs (e.g. Holter et al., 2021) and consistent with several clinically prescribed antidepressants (Steiger & Pawlowski, 2019; Wichniak et al., 2013, 2017). Furthermore, some mGlu5 NAMs have progressed into clinical trials for MDD, though historically with negative outcomes due to the adverse effects noted above (e.g. fenobam, Pecknold, McClure, Appeltauer, Wrzesinski, & Allan, 1982). However, basimglurant advanced to phase II clinical trials as an adjunctive therapy for depression. Although the primary endpoint criteria, change in MADRS total score, was not met with this compound, significant results were found in secondary endpoint criteria including other self-report measures of depressive symptoms (Quiroz et al., 2016). It is critical to note that sleep was not evaluated in this clinical trial.

Clinical evidence, although mixed, also supports a role of group II mGlu receptors in MDD. One study reported increased mGlu2/3 protein levels in the PFC of postmortem tissue (Feyissa et al., 2010) while a PET imaging study reported significantly lower binding in the anterior cingulate cortex in patients with MDD suggesting the relationship between mGlu2/3 and MDD is complex (Mcomish et al., 2016). Both agonists and antagonists of group II mGlu receptors have been pursued in preclinical and clinical studies for their antidepressant-like potential. Antagonism of the $mGlu_{2/3}$ receptor has been most heavily pursued because these compounds have similar mechanisms to ketamine. Ultimately, similar to subanesthetic doses of ketamine, administration of an $mGlu_{2/3}$ antagonist should lead to increased glutamate release because presynaptic $mGlu_{2/3}$ receptors function as autoreceptors (Chaki, Koike, & Fukumoto, 2019). Aside from sleep, preclinical studies showed $mGlu_{2/3}$ antagonists and NAMs display antidepressant-like effects in preclinical models including the FST and TST (Chaki et al., 2004; Fukumoto, Iijima, & Chaki, 2016; Witkin et al., 2016). Interestingly, both $mGlu_2$ and $mGlu_3$ contribute to this effect because selective compounds VU6001966 ($mGlu_2$ NAM) and VU650786 ($mGlu_3$ NAM) both decreased immobility time in the FST (Joffe et al., 2020). Additionally, the $mGlu_{2/3}$ NAM decoglurant advanced to clinical trials as an adjunctive treatment to SSRIs or SNRIs but was unsuccessful in meeting primary or secondary endpoint criteria. However, complexities with the clinical trial including a high placebo response made it difficult to adequately evaluate the compound (Umbricht et al., 2020). Importantly, $mGlu_{2/3}$ antagonists may be relevant to target sleep disturbances in patients with depression because of their ability to reduce REM sleep. However, applicability of $mGlu_{2/3}$ antagonists may be population-dependent, perhaps better suited for the sub-population presenting with hypersomnia, as $mGlu_{2/3}$ antagonists also have wake-promoting effects.

$mGlu_{2/3}$ agonists and PAMs may have greater potential for targeting sleep disturbances in MDD. Several of these compounds have also displayed antidepressant-like effects in preclinical screens including the FST (Fell et al., 2011). Furthermore, the $mGlu_{2/3}$ agonist LY379268 demonstrated promising antidepressant-like effects when administered as an adjunctive treatment to fluoxetine and may help bridge the gap between treatment initiation and delayed onset of the therapeutic effects of classic antidepressants (Matrisciano et al., 2007). Importantly, similar to $mGlu_{2/3}$ antagonists, these compounds also significantly decrease REM sleep time and increase REM latency, but, unlike $mGlu_{2/3}$ antagonists, most do not promote wake (Ahnaou et al., 2009; Cid et al., 2012). Thus, this approach may be more beneficial to those who are concurrently experiencing insomnia. Because elevated extracellular glutamate concentrations were found in the orbitofrontal cortex during REM sleep in the rat (Lopez-Rodriguez et al., 2007), activation of $mGlu_{2/3}$ agonists may be a more suitable approach to correct REM sleep abnormalities in MDD by reducing overall synaptic glutamate levels and selectively suppressing REM sleep (Ahnaou et al., 2009).

In summary, reducing glutamate function via mGlu$_5$ NAMs or mGlu$_{2/3}$ PAMs may be a viable approach to target primary symptoms and sleep disruptions in MDD. As evaluated with basimglurant, the ability of new compounds to normalize sleep disruptions associated with MDD should be integrated into future clinical trial design.

3.4.3 Post-traumatic stress disorder

Post-traumatic stress disorder (PTSD) affects roughly 4% of the population and is associated with fear, anxiety, avoidance, negative mood, intrusive thoughts, hyperarousal, and sleep disturbances, including insomnia, that persist following a traumatic event (Kessler, Wai, Demler, & Walters, 2005; Kilpatrick et al., 2013). While objective measures of sleep within the PTSD population are inconclusive, perhaps due to the heterogeneity of individuals and their associated trauma, many studies report increased light sleep, decreased SWS, increased REM density, and decreased REM bout length (Kobayashi, Boarts, & Delahanty, 2007; Kobayashi, Huntley, Lavela, & Mellman, 2012; Lewis et al., 2020; Woodward, Leskin, & Sheikh, 2002). Subjectively, patients with PTSD report sleep disturbances including nightmares, sleep terrors, limb movements, and dream enactment (Baird et al., 2018). Additionally, the prevalence of sleep disordered breathing (e.g. sleep apnea) and sleep movement disorders, including periodic limb movement disorder and REM movement disorders, are higher among those with PTSD (Baird et al., 2018; Maher et al., 2006; Neylan et al., 2021). Sleep disturbances have also been found to correlate with prevalence and severity of other symptoms. For example, insomnia severity correlated with intrusion and hyperarousal (Lauterbach, Behnke, & McSweeney, 2011). The frequency and severity of nightmares was associated with a higher prevalence of hopelessness and suicidal behaviors (Littlewood, Gooding, Panagioti, & Kyle, 2016). Further, in support of a bidirectional relationship, it is well-established that preexisting sleep disturbances influence the development of PTSD. Insomnia and stress reactivity prior to a motor vehicle collision predicted subsequent PTSD diagnosis (Neylan et al., 2021). Additionally, some studies evaluating self-reported sleep measures including nightmares and insomnia prior to deployment corresponded to subsequent PTSD development in combat veterans (van Liempt, van Zuiden, Westenberg, Super, & Vermetten, 2013; Wang et al., 2019). Given the important role of sleep in regulating stress, emotion, and learning and memory, it is not surprising that dysregulated sleep may exacerbate the development and progression of PTSD.

Glutamatergic function has been shown to be altered in individuals with PTSD which may directly contribute to PTSD symptoms including altered fear reactivity, hyperarousal, and dysregulated sleep. Using ^{13}C-acetate MRS, patients with PTSD showed reduced glutamatergic synaptic strength in the PFC compared to healthy controls (Averill et al., 2022). Similarly, in rodents, the single prolonged stress model reduced glutamatergic excitation of pyramidal neurons in the infralimbic medial PFC (Nawreen, Baccei, & Herman, 2021). Another study using ^{1}H-MRS imaging in humans reported lower GABA and increased glutamate levels in the cortex in patients with PTSD, and these findings correlated with higher subjective ratings of insomnia (Meyerhoff, Mon, Metzler, & Neylan, 2014). In further support of research that suggests sleep disturbances may exacerbate neurobiological factors in PTSD, studies conducted in healthy humans found a single night of sleep deprivation altered functional connectivity between the amygdala and PFC and was associated with aberrant amygdalar reactivity (Killgore, 2013; Yoo, Gujar, Hu, Jolesz, & Walker, 2007). These data support the prevailing theory that inadequate top-down PFC inhibitory control leads to hyperactivity in the amygdala and over-reactivity to stressful stimuli (Rauch, Shin, & Phelps, 2006). Furthermore, as many of the sleep disturbances and other symptoms derived from hyperarousal occur in part because of increased glutamate activation of the hippocampus and amygdala, putative pharmacotherapeutics should attenuate glutamate to restore excitatory:inhibitory balance.

Current treatments for PTSD include cognitive behavioral therapy and exposure therapy, which are aimed to help individuals re-process traumatic events and re-structure responses in an effort to attenuate hyperarousal and responsivity to environmental stimuli (American Psychological Association, 2017). Treatment of the emotion-related symptoms of PTSD have been found to positively impact sleep. For example, subjective and objective measures of sleep improved in patients with PTSD following successful outcomes of trauma-targeted psychotherapies (e.g. remission of emotional symptoms) (Rousseau et al., 2021). While severity of sleep disturbances may be reduced following behavioral treatment approaches, they are not fully resolved (Taylor et al., 2020). There are only two FDA-approved pharmacological treatments for PTSD, the antidepressant medications sertraline and paroxetine (Brady et al., 2000; Krystal et al., 2017; Marshall, Beebe, Oldham, & Zaninelli, 2001). However, both antidepressants, in addition to commonly used anxiolytics, target depressed mood and anxiety symptoms rather than directly targeting hyperarousal or sleep disturbances. Since

NMDAR hyperactivity has been associated with this hyperarousal and PTSD formation after trauma, NMDAR antagonists including ketamine have been proposed to both target this hyperarousal and provide rapid antidepressant effects (Liriano, Hatten, & Schwartz, 2019). A single dose of ketamine decreased subjective measures of intrusion, avoidance, and hyperarousal in patients with PTSD (Feder et al., 2014). Additionally, ketamine reduced the likelihood of developing PTSD after burn injury (McGhee, Maani, Garza, Gaylord, & Black, 2008). The partial NMDAR agonist D-cycloserine has also been investigated and was administered alongside exposure therapy to enhance extinction learning (Baker, Cates, & Luthin, 2017; George et al., 2018). Lastly, while sleep aids such as benzodiazepines, zolpidem, buspirone, and gabapentin have been tested in addition to cognitive behavioral therapy for PTSD-induced sleep disturbances, results are variable (Maher et al., 2006). Thus, as reiterated throughout, novel, disorder-specific medications that target sleep are still needed.

Emerging evidence supports targeting both group I and group II mGlu receptors for symptoms of PTSD and associated sleep disturbances. Importantly, unlike clinical and preclinical studies for other CNS disorders, mGlu receptor compounds have been assessed for their ability to improve sleep disturbances in the context of animal models of traumatic stress. First, functional antagonism of the $mGlu_5$ receptor may be effective in treating both sleep disruptions and hyperarousal associated with PTSD. In vivo PET imaging showed that patients with PTSD have higher $mGlu_5$ receptor availability in the hippocampus relative to healthy controls, and $mGlu_5$ receptor availability correlated with heightened avoidance symptoms and suicidal ideations (Davis et al., 2019; Esterlis, Holmes, Sharma, Krystal, & Delorenzo, 2018; Holmes et al., 2017). In preclinical rodent models, treatments targeting $mGlu_5$ receptors also show promise in treating PTSD due to their role in fear, avoidance, and extinction learning (Riedel, Casabona, Platt, MacPhail, & Nicoletti, 2000). As previously discussed, $mGlu_5$ NAMs inhibit arousal and have been shown to selectively reduce REM sleep, and may modulate sleep disruptions induced by single prolonged stress in rodents (Nedelcovych et al., 2015). Given that the hippocampus influences memory consolidation and REM sleep and the amygdala regulates emotional processing (Rasch & Born, 2013; Wang, Liu, Li, Qu, & Huang, 2021), it is theorized that hippocampal and amygdalar hyperactivity during REM sleep decreases extinction learning, increases emotional response to trauma, and increases PTSD symptom manifestation (Bottary et al., 2020; Cominski, Jiao, Catuzzi, Stewart, & Pang, 2014; Pace-Schott,

Germain, & Milad, 2015; Straus et al., 2018). Thus, mGlu$_5$ NAMs may be a two-fold treatment approach that facilitates extinction learning and targets the sleep disturbances associated with PTSD.

To date, most work has focused on mGlu$_5$ as a potential treatment for PTSD and associated sleep impairments. However, activation of group II receptors also produces selective decreases in REM sleep and preclinical studies support group II receptors as a therapeutic target for PTSD. Given their expression in regions including the amygdala, both mGlu$_2$ and mGlu$_3$ receptors have been implicated in fear memory and extinction learning (Kim et al., 2015; Linden et al., 2005; Potter, Zanos, & Gould, 2020; Sweeten, Adkins, Wellman, & Sanford, 2021). Of relevance, Sweeten et al., 2021 explored the therapeutic potential of mGlu$_{2/3}$ agonist LY379268 for sleep disturbances associated with fear extinction and found that a microinjection of LY379268 to the basolateral amygdala normalized the increased REM response associated with context re-exposure. Thus, group II receptors remain an exciting area of future research and should be further pursued. Lastly, it is important to note group III receptors (notably mGlu$_7$ and mGlu$_8$) also modulate learning and memory, including aversive and extinction learning, though their effects on sleep disturbances have not yet been studied (Fendt et al., 2013; O'Connor et al., 2019).

3.4.4 Alzheimer's disease

As the most common form of dementia, Alzheimer's Disease (AD) is characterized by neurodegeneration, aggregation of extracellular amyloid β (Aβ) into amyloid plaques, and the aggregation of intracellular tau into neurofibrillary tangles (NFTs), which begin to accumulate approximately 15 years before cognitive decline (Jack et al., 2010; Jack et al., 2016). Importantly, sleep disruptions are common in both clinical populations (even in prodromal stages) and preclinical models, and these have correlated with cognitive decline (Lim, Kowgier, Yu, Buchman, & Bennett, 2013; Potvin et al., 2012; Romanella et al., 2021). In clinical populations, total sleep time and SWS percentage were decreased and REM sleep latency and waking after sleep onset (WASO) were increased in patients with AD when compared to healthy controls (Bliwise et al., 1989; Zhang et al., 2022). Furthermore, a recent meta-analysis examining polysomnographic sleep changes in patients with AD found a significant positive relationship between SWS and REM sleep duration and Mini-Mental Sate Examination (MMSE) scores, such that reduced sleep duration correlated with lower MMSE scores (Zhang et al., 2022).

Electroencephalography (EEG) studies in patients with AD have reported several age-related differences, especially in spectral waveforms, that may directly contribute to cognitive dysfunction and AD pathology (Lyashenko, Poluektov, Levin, & Pchelina, 2016; Scullin & Bliwise, 2015a, 2015b; Song et al., 2015; Xie et al., 2013; Zhang et al., 2022). Patients with AD demonstrated increased power in low frequency (delta/theta) bands and decreased alpha power when awake compared to healthy aged or young cohorts (Czigler et al., 2008; Rondina et al., 2016). These waveforms each have distinct roles in cognitive processes and, thus, may correspond with AD-associated cognitive decline. Additionally, findings of reduced delta power (0.5–4 Hz) during NREM sleep, as well as sleep spindle abnormalities including reduced density, amplitude, and duration, have been reported in patients with AD; these further correlated with impaired cognitive performance (Scullin & Bliwise, 2015b; Weng, Lei, & Yu, 2020).

Importantly, disrupted sleep has been implicated as both a cause and a consequence of AD pathology. Homeostatic sleep is important for Aβ clearance, and, in healthy adult participants, disrupted slow wave activity was found to increase AD pathology and Aβ levels (Ju et al., 2017; Lucey et al., 2019). Furthermore, healthy individuals with amyloid deposition demonstrated worsened sleep efficiency compared to individuals without amyloid deposition (Ju et al., 2013). Murine models of AD-like pathology also support a relationship between sleep and pathological development. For example, in transgenic mice expressing the human amyloid precursor protein (APP), Aβ accumulation was observed in brain interstitial fluid following sleep deprivation (Kang et al., 2009). Additionally, sleep disturbances associated with AD may also contribute to sundown syndrome, a phenomenon in which neuropsychiatric symptoms such as delusions, irritation, and depression are exacerbated during the late afternoon and night hours (Canevelli et al., 2016). Although the underlying causes of these are not well understood, possible contributing factors include abnormal sleep schedules (e.g. reduced nighttime sleep and increased daytime napping) as well as circadian disruptions (e.g. degeneration of the SCN and decreased melatonin release) (Canevelli et al., 2016; Srinivasan, Pandi-Perumal, Cardinali, Poeggeler, & Hardeland, 2006; Stopa et al., 1999). Taken together, identifying at risk individuals with sleep disturbances early in life may present a relevant treatment window to slow or prevent pathological progression.

Recent evidence points to a glutamate hypothesis of AD, which is supported by the increased release and decreased clearance of synaptic glutamate (Revett, Baker, Jhamandas, & Kar, 2013; Tok, Ahnaou, &

Drinkenburg, 2021). These alterations in glutamate levels may contribute to network hyperexcitability and, thereby, pathological development (Tok et al., 2021). Glutamate receptors have also been targeted for the treatment of AD. Aβ-mediated neurotoxicity has been hypothesized to occur via an NMDAR-activity dependent mechanism, and, thus, memantine, a non-competitive NMDAR antagonist, has been FDA-approved as a treatment for moderate to severe AD (Kumar et al., 2015; Miguel-Hidalgo, Alvarez, Cacabelos, & Quack, 2002; Rogawski & Wenk, 2003; Shankar et al., 2007). However, memantine is limited in that it only relieves symptoms and does not improve existing pathology (Matsunaga et al., 2018).

Several studies support targeting both group I and group II mGlu receptors for AD. Because $mGlu_5$ is coupled to the NMDAR (Niswender & Conn, 2010; Shigemoto et al., 1993), reducing $mGlu_5$ function may exert neuroprotective effects by blocking Aβ-mediated neurotoxicity, similar to memantine, with lesser adverse effects (Kumar et al., 2015). Furthermore, $mGlu_5$ has been shown to directly engage with Aβ oligomers (Aβo). One possible contributor to AD pathology is Aβo binding to cellular prion protein (PrP^c) which was shown in neuronal cultures to ultimately induce synaptic dysfunction (Laurén, Gimbel, Nygaard, Gilbert, & Strittmatter, 2009). Interaction of the Aβo-PrP^c complex with $mGlu_5$ induces downstream calcium mobilization and increases Aβ formation, and these effects may be associated with deficits in learning and memory (Hamilton, Esseltine, Devries, Cregan, & Ferguson, 2014; Um et al., 2013). In support of this interaction, both deletion of $mGlu_5$ and blockade of $mGlu_5$ with NAMs including MTEP in a mouse model of AD (APPswe/PS1ΔE9 mice) reversed deficits in spatial learning and working memory and prevented Aβo-associated changes in synaptic density (Budgett, Bakker, Sergeev, Bennett, & Bradley, 2022; Hamilton et al., 2014, 2016; Um et al., 2013). Lastly, there is preclinical support for both inhibition and activation of group II mGlu receptors. Support for inhibition as a therapeutic approach stems from findings that $mGlu_2$ receptor activation increased accumulation of Aβ42 in cortical synaptoneurosomes (Kim et al., 2010). Pharmacological activation of $mGlu_2$ receptors also contributed to microglial neurotoxicity, whereas blockade of $mGlu_2$ receptors reduced microglia apoptosis (Taylor, Diemel, Cuzner, & Pocock, 2002). Importantly, increased $mGlu_2$ receptor expression was found in hippocampal neurons in post-mortem tissue from patients with AD when compared to age-matched controls suggesting a close relationship between $mGlu_2$ and AD pathogenesis (Lee et al., 2004). Conversely, there is some support for $mGlu_{2/3}$ receptor activation for the treatment of AD. Caraci et al., 2011 found activation of both $mGlu_2$ and $mGlu_3$ receptors with LY379268 exerted neuroprotective effects, reducing Aβ-mediated neuronal death. Further, neuroprotective effects of LY379268 did not occur in cells lacking $mGlu_3$ receptors and selective activation of $mGlu_2$ receptors increased Aβ-mediated neurotoxicity suggesting a role for $mGlu_3$ receptor activation in these neuroprotective effects (Caraci et al., 2011; Caraci, Nicoletti, & Copani, 2018).

$mGlu_5$ NAMs may also hold promise for targeting sleep disturbances associated with AD. Aforementioned effects of $mGlu_5$ NAMs on sleep, primarily increases in NREM sleep and NREM sleep quality, could mitigate the accelerated Aβ deposition caused by sleep disruptions and, thereby, slow cognitive impairment and behavioral effects (Ahnaou, Langlois, et al., 2015; Ahnaou, Raeymaekers, et al., 2015; Cavas et al., 2013b). Furthermore, while speculative, arousal enhancing effects of some $mGlu_5$ NAMs may be favorable to target sundowning effects if administered at the correct time of day. For example, VU042438 has initial wake-promoting effects followed by compensatory increases in NREM sleep and increased NREM sleep quality (Holter et al., 2021). This profile could normalize sleep patterns that are perceived contributors to sundowning pathophysiology, serving to prevent daytime napping and increase nighttime sleep. However, further research is needed to determine if effects of $mGlu_5$ NAMs on sleep in healthy rodents will translate similarly to rodents in models of AD-pathology and humans with and without AD. Lastly, while both inhibition and activation of $mGlu_{2/3}$ receptors has been explored for the possible benefits in targeting AD pathology, the effects of these compounds on sleep in healthy animals (both decreased REM sleep) appear to counteract the desired effects in patients with Alzheimer's disease. Nonetheless, examining these compounds on sleep in the context of AD is still important to acknowledge and address in future studies.

3.4.5 Substance use disorders

Current research suggests a bidirectional relationship between sleep and substance use/misuse. On one hand, sleep disturbances increase the risk of licit and illicit drug use which contributes to the subsequent development of substance use disorders (SUDs). For example, reduced sleep duration or quality as well as social jetlag (reduced sleep time on weekends) during adolescence is associated with increased problematic drinking and continued illicit substance use in adulthood (Dolsen & Harvey, 2017; Hasler & Pedersen, 2020; Roehrs et al., 2021; Troxel et al., 2021). Conversely, substance use also induces both acute and long-lasting sleep disruptions (Koob & Colrain, 2020; Roehrs et al., 2021; Valentino & Volkow, 2020). Although all illicit substances, as well as alcohol, cannabis, and nicotine alter sleep, specific disturbances and severity vary depending on a number of factors including, but not limited to, substance, duration, and frequency/typography of use. Chronic drug exposure associated with SUD often leads to long-lasting sleep disruptions including insomnia that last for weeks to months into abstinence. Importantly, sleep disturbances during abstinence have been correlated with increased craving and risk of relapse (Brower, 2003; Dolsen & Harvey, 2017; Lydon-Staley et al., 2017). Thus, as with other CNS disorders, targeting the sleep disruptions associated with SUD represents a promising treatment approach to not only prevent progression from recreational use to SUD but also to reduce relapse following abstinence. While most substances affect sleep (Hser et al., 2017; Jaehne, Loessl, Bárkai, Riemann, & Hornyak, 2009; Mondino et al., 2021), the following sections focus on relevant clinical and preclinical literature describing effects of cocaine and opioids on sleep.

3.4.5.1 Cocaine use disorder (CUD)

There is growing preclinical and clinical evidence for the presence of sleep disturbances throughout various stages of cocaine use disorder (CUD). As a stimulant, acute use of cocaine is associated with increased arousal and time awake and decreased total sleep time across species (Bjorness & Greene, 2018, 2021; Post, Gillin, Wyatt, & Goodwin, 1974; Schierenbeck, Riemann, Berger, & Hornyak, 2008). However, abstinence following chronic use is associated with persistent alterations in sleep. In general, polysomnography studies show a decrease in REM sleep on the first day of abstinence (Schierenbeck et al., 2008). Sleep disturbances further into abstinence include decreased total and SWS time, decreased REM latency, and increased REM duration, which may be a REM rebound effect (Bjorness & Greene, 2022; Schierenbeck et al., 2008). During early abstinence (<1 week), objective polysomnography and subjective self-report measures of sleep concurrently indicate decreased sleep time, increased sleep latency, and decreased sleep quality. Interestingly, although most patients continue to display insomnia-like symptoms on objective measures during later abstinence periods (>2 weeks), subjective reports do not align, as patients report either unchanged or improved sleep (Morgan et al., 2006; Morgan, Perry, Cho, Krystal, & D'Souza, 2006; Morgan & Malison, 2007; Pace-Schott et al., 2005). Importantly, the DSM-5 characterizes these sleep disturbances as a subset of stimulant-induced disorders, which are symptoms that last months to years into abstinence (APA, 2013).

Surprisingly, few studies to date have examined cocaine-associated sleep alterations in animal models of CUD. One nonhuman primate study using actigraphy-based assessments reported sleep fragmentation and decreased NREM sleep during abstinence following cocaine self-administration (Cortes et al., 2016). Further, only two rodent studies have used polysomnography to examine effects of repeated cocaine exposure on sleep. Compared to a drug-naïve baseline, Chen and colleagues reported sleep fragmentation and, in contrast to human reports, both decreased NREM and REM sleep following one week of cocaine self-administration (Chen et al., 2015). These effects were only in part recapitulated in a study examining sleep following one week of repeated, non-contingent administration of cocaine in mice (Bjorness & Greene, 2022). Thus, this suggests frequency, dose, duration, and route of administration (contingent vs non-contingent) may be influential factors. Lastly, reduced REM sleep was correlated with increased operant responding to a cocaine cue (incubation paradigm; a preclinical model of drug seeking) and restoring REM sleep reduced cue-associated responding (Chen et al., 2015). Thus, this study established a direct relationship between cocaine induced sleep-disruptions and future drug-related behaviors.

Impaired glutamate homeostasis has been widely implicated in CUD. The primary reinforcing effects of cocaine are attributed to the blockade of the dopamine transporter in the mesocorticolimbic dopamine pathway (Di Chiara & Imperato, 1988), and chronic cocaine self-administration is associated with a hypodopaminergic state (Goldstein & Volkow, 2002; Gould, Porrino, & Nader, 2012; Mateo, Lack, Morgan, Roberts, & Jones, 2005; Siciliano, Ferris, & Jones, 2016). However, cocaine self-administration also induces long-lasting glutamate-mediated increases in synaptic plasticity in mesocorticolimbic regions including the VTA, NAc, and PFC (Niedzielska-Andres et al., 2021; van Huijstee & Mansvelder, 2014). During abstinence, exposure to cocaine-paired cues increases cortical glutamate concentrations in the NAc, and this excessive glutamate-mediated response drives increased DA release (Korpi et al., 2015; Shin, Gadzhanova, Roughead, Ward, & Pont, 2016), a purported mechanism contributing to craving. Excessive cue-reactivity following extended periods of abstinence is associated with an accumulation of calcium-permeable AMPA receptors (CP-AMPARs; for review, see Wolf, 2016), and this process has been shown to be regulated via $mGlu_1$ receptor activity (McCutcheon et al., 2011).

Importantly, Chen and colleagues (2015) demonstrated that altering sleep impacted CP-AMPARs and subsequent cocaine-related behaviors in rats. Specifically, cocaine-induced REM sleep disruptions correlated with the accumulation of CP-AMPARs and restoring REM sleep during abstinence reduced CP-AMPARs and cocaine cue-reactivity (Chen et al., 2015). Alterations in mGlu receptors are also present during abstinence following cocaine self-administration. Rats showed lower $mGlu_{2/3}$ receptor expression (Logan et al., 2020), density (Pomierny-Chamiolo, Miszkiel, Frankowska, & Mizera, 2017), and function (Moussawi et al., 2011) in the NAc as well as lower $mGlu_5$ receptor expression in regions including the NAc and striatum (Hao et al., 2010; Knackstedt, Trantham-Davidson, & Schwendt, 2014; Swanson, Baker, Carson, Worley, & Kalivas, 2001).

3.4.5.2 Opioid use disorder (OUD)

Understanding the direct effects of opioids on human sleep is complicated by a number of factors including the type/frequency of use (prescription use, misuse, OUD), the presence or absence of pain, and whether studies use subjective or objective measures. Studies conducted in healthy, pain-free humans reported that opioid administration increased lighter NREM sleep stages (N1 and N2) and decreased N3 and REM sleep (Dimsdale, Norman, DeJardin, & Wallace, 2007; Shaw, Lavigne, & Mayer, 2006). Prescription opioid-dependent users with chronic pain experience subjective and objective (actigraphy-based) decreases in sleep duration and sleep quality (Hartwell, Pfeifer, McCauley, Moran-Santa Maria, & Back, 2014), and, in some instances, prescription opioids have been shown to further exacerbate sleep disturbances associated with chronic pain (Robertson et al., 2016). Moreover, illicit opioid use may affect sleep differently than FDA-approved opioids for the treatment of pain and OUD based on differences in potency/efficacy, route of administration, and related pharmacokinetic factors. In studies that used the Pittsburg Sleep Quality Index (PQSI) to assess sleep, 70–84% of sampled participants in methadone maintenance therapy were poor sleepers, and PQSI scores correlated with methadone dosage (Hsu et al., 2012; Stein et al., 2004). These studies are not exhaustive but highlight the complexity of interpreting opioid-related effects on sleep and reiterate that sleep is negatively impacted across conditions.

Specifically, within the context of OUD, sleep impairments including insomnia are some of the most common complaints during acute (days to weeks) and protracted (weeks to months) withdrawal (Dunn, Huhn, Bergeria, Gipson, & Weerts, 2019). Other common symptoms reported during abstinence periods in non-treatment seeking individuals with OUD include restless sleep, fatigue, and unusual dreaming (Hartwell et al., 2014; Tripathi, Rao, Dhawan, Jain, & Sinha, 2020). For example, increased N1 and reduced N2 sleep was associated with relapse in heroin detoxification patients at a 6-month follow-up assessment (Rady, Mekky, Moulokheya, & Elsheshai, 2020). Sleep disruption can also lead to decreased pain tolerance, which may influence subsequent opioid use (Eacret, Veasey, & Blendy, 2020). Together, similar to cocaine, chronic opioid use is associated with long-lasting sleep alterations and influences the likelihood of relapse.

Animal studies provide the opportunity to directly assess the effects of opioids on sleep, controlling for a number of the aforementioned confounding variables associated with human studies. In rats, acute morphine administration reduced NREM and REM sleep, yet NREM, but not REM, sleep duration returned to normal within three consecutive days of administration (Khazan, Weeks, & Schroeder, 1967). Acute effects of opioids are largely consistent between humans and rodents, except that opioids increased delta power during NREM sleep in rodents whereas opioids decreased delta power during N3 sleep in humans (Gauthier, Guzick, Brummett, Baghdoyan, & Lydic, 2011; Shaw et al., 2006). Additionally, actigraphy-based assessments in non-human primates were used to examine acute effects of the FDA-approved treatments for OUD on sleep-like parameters including sleep latency and efficiency (Berro, Zamarripa, Talley, Freeman, & Rowlett, 2022). Interestingly, while methadone (full mu opioid receptor agonist) and buprenorphine (partial mu opioid receptor agonist) both increased and decreased sleep latency and efficiency, respectively, disruptive effects of buprenorphine were longer lasting. In contrast, naltrexone (mu opioid receptor antagonist) reduced sleep latency and increased efficiency (Berro et al., 2022). While these studies were conducted in non-opioid dependent monkeys and did not employ polysomnography, they elegantly distinguished effects of activating or inhibiting opioid receptors on sleep-like parameters.

Similar to CUD, few preclinical studies have examined effects of chronic opioid self-administration and abstinence on sleep/wake cycles. In the only study to our knowledge, rats given 6-hr access to self-administer heroin during the light cycle for 2 weeks demonstrated a complete reversal in sleep pattern compared to a saline control group. REM and NREM sleep were increased during the dark period but decreased during the light period (Coffey, Guan, Grigson, & Fang, 2016). While NREM sleep duration returned to baseline levels in both the light and dark period within three days of abstinence, REM sleep impairments in the dark period persisted. These findings suggest altered recovery duration and possible underlying neurobiological alterations following heroin self-administration. Future preclinical research is needed to understand, in a well-controlled setting, how chronic opioid self-administration affects sleep and how FDA-approved opioid treatments affect sleep during abstinence.

There are many known and posited mechanisms through which opioids affect sleep-wake regulation. Reinforcing effects (e.g. abuse liability) of opioids are attributed to activation of mu-opioid receptors (MORs). MORs are broadly distributed throughout the CNS, influencing reward, pain, stress reactivity, mood, cognition and sleep (for review, see Welsch, Bailly, Darcq, & Kieffer, 2020). Due to the broad receptor distribution, opioids likely affect sleep-wake regulation via multiple circuits including excitation of the mesolimbic dopamine system and direct activation of MORs on hypocretin-containing hypothalamic neurons (Eacret et al., 2020). Additionally, opioids influence the suprachiasmatic nucleus affecting arousal and circadian rhythm and disrupt sleep by inhibiting respiratory activity and contributing to sleep apnea (Bergum, Berezin, & Vigh, 2022; Rosen et al., 2019; Wang & Teichtahl, 2007). While these are noted mechanisms through which acute exposure may alter sleep-wake regulation, similar to CUD, neurobiological alterations spanning many of these circuits may contribute to the long-lasting sleep disturbances associated with OUD.

Extensive evidence implicates excessive glutamate associated with OUD as a contributor to opioid-associated withdrawal symptoms and risk for relapse. In fact, hyperglutamatergic function may be a critical component of withdrawal. Glutamate concentrations within the locus coeruleus (LC), a noradrenergic nucleus in the brainstem, have been shown to increase during naloxone-precipitated withdrawal in opioid-dependent rats (Feng, Zhang, Rockhold, & Ho, 1995) and acute injections of glutamate into the LC induced behavioral signs of withdrawal similar to naloxone in morphine-dependent animals (Medrano, Mendiguren, & Pineda, 2015; Sekiya, Nakagawa, Ozawa, Minami, & Satoh, 2004; Tokuyama, Wakabayashi, & Ho, 1995; Zhu, Rockhold, & Ho, 1998). Given that the LC is largely implicated in OUD and associated withdrawal, as well as a key contributor for wake-promoting circuits, reducing glutamatergic function may be a viable therapeutic approach for multiple symptoms of OUD during abstinence including sleep disturbances. Additionally, like cocaine, long-lasting glutamatergic-mediated synaptic plasticity changes likely contribute to drug craving and risk of relapse (van Huijstee &

Mansvelder, 2014). While less well characterized, recent evidence shows increased CP-AMPARs in the NAc that contribute to drug seeking during abstinence following oxycodone self-administration, which is a finding similar to cocaine studies (Wong, Zimbelman, Milovanovic, Wolf, & Stefanik, 2022).

3.4.5.3 Modulating mGlu receptors as treatments for SUD

There are currently 3 FDA-approved pharmacological treatments for OUD and no pharmacological treatments for CUD. While each is successful in some regard, all have adverse risk factors and none directly target sleep disturbances. Decades of research have examined compounds targeting various mGlu receptor subtypes as treatments for SUD. For example, mGlu$_5$ antagonists/NAMs as well as mGlu$_{2/3}$ PAMs have shown promise in reducing cocaine and opioid-related behaviors in rodent and nonhuman primate models of drug taking or seeking (Amato et al., 2013; Gould et al., 2016; Gould, Felts, & Jones, 2017; Johnson & Lovinger, 2020; Keck et al., 2013; Kumaresan et al., 2009; Lee, Platt, Rowlett, Adewale, & Spealman, 2005; Lou, Chen, Liu, Ruan, & Zhou, 2014; Salling, Grassetti, Ferrera, Martinez, & Foltin, 2021; van der Kam, de Vry, & Tzschentke, 2007). Moreover, many of these compounds have demonstrated anxiolytic-like and antidepressant-like activity suggesting potential to alleviate anxiety/ depressive symptoms often present during withdrawal (Gould et al., 2016). Lastly, cocaine abstinence-induced accumulation of CP-AMPARs can be restored followed administration of an mGlu$_1$ receptor agonist (McCutcheon et al., 2011), although this has not been examined in the context of sleep, nor extended to studies involving opioids.

A causal relationship has not yet been established between sleep disruptions and altered mGlu receptor function during abstinence following cocaine or opioid self-administration. However, based on KO and pharmacological studies in drug-naïve rodents (Tables 1 and 2), we can speculate on possible treatment approaches. Results from several studies suggest targeting mGlu$_1$ and mGlu$_{2/3}$ receptors for CUD and OUD may benefit both abstinence-related behaviors and abstinence-induced sleep disturbances including insomnia and decreases in REM sleep. For example, mGlu$_1$ agonists may alter CP-AMPARs during abstinence in part by altering sleep. CP-AMPARs have a strong role in regulating sleep and sleep-maintained synaptic homeostasis and, thus, balancing these may normalize the restorative benefits of sleep including memory consolidation (for review, see Shepherd, 2012 and Martin et al., 2019). Additionally, some novel mGlu$_{2/3}$ PAMs may be able to target aberrant decreases in NREM sleep associated with abstinence. While mGlu$_5$ NAMs are promising for reducing drug taking/seeking, these compounds may not produce the desired effects on sleep as it relates to findings in some, but not all, rodent and human studies (i.e. mGlu$_5$ NAMs reduce NREM and REM sleep when increased NREM and REM sleep is needed). However, conflicting evidence between clinical and preclinical studies that show increased and decreased REM sleep during abstinence, respectively, indicates the appropriate way to target REM sleep still needs further investigation. Additionally, this does not discount the ability of these compounds to normalize possible sleep disturbances associated with stimulant-induced disorders including anxiety and depression. Lastly, though there is minimal literature surrounding group III mGlu receptors and sleep, it is important to note that activating mGlu$_7$ and mGlu$_8$ receptors was effective in attenuating self-administration, reinstatement, and withdrawal as it relates to cocaine and opioids (Hajasova, Canestrelli, Acher, Noble, & Marie, 2018; Kahvandi et al., 2021; Li et al., 2013; Luessen & Conn, 2022; Pałucha-Poniewiera & Pilc, 2013; Salling et al., 2021). Together, these studies provide compelling rationale to further examine group I-III mGlu receptor compounds in multiple preclinical models of SUD, with a particular need to examine associated impacts on sleep disturbances.

4. Discussion

As described in this chapter and summarized in Tables 3 and 4, sleep disturbances including insomnia are prevalent in many neuropsychiatric disorders, and there is a clear bidirectional relationship between the two. Sleep disturbances can precede traditional symptom onset and/or correlate with symptom severity or relapse (e.g. schizophrenia, MDD, SUD, AD). Sleep disturbances may also contribute to or exacerbate neuropathology associated with disorders including AD. Although disrupted sleep is common in both subjective and objective reports, primary pharmacological treatments for neuropsychiatric disorders have historically disregarded sleep as a modifiable symptom. Importantly, the DSM-5 distinguished CNS disorders and sleep disturbances as separate diagnoses. While common sleep medications including benzodiazepines and z-drugs are often prescribed for many neuropsychiatric disorders, they all decrease arousal by increasing inhibitory GABA activity (Benca & Buysse, 2018). Thus, depending on the disorder, these compounds may not directly affect the disrupted neurobiology/

Table 3 Objective measures of sleep disruptions in neuropsychiatric disorders.

Neuropsychiatric condition	NREM		REM					Awakenings		qEEG	Sleep spindles		Comorbidity
	Duration	Latency	Duration	latency	Density	Bout Length	%	WASO	Sleep fragmentation	NREM Delta	Number	Density	Sleep disorders
Schizophrenia 1,4–13	↓	↑	n.s	↓	↑	—	n.s	↑	↑	↓	↓	↓	↑
Major Depressive Disorder 1,14–24	↓	↑	↑	↓	↑	↑	↑	↑	↑	↕	↓	↓	↑
Post-traumatic stress disorder 25–39	↓ ; n.s	↑	↕ ; n.s	n.s	↑	↓	n.s	↑ ; n.s	—	↓	—	↑	↑
Alzheimer's Disease 1,40–42	↓	↑	↓ ; n.s	↑	↓	—	↓	↑	↑	↓	↓	—	↑

Substance Use Disorders: Withdrawal and Abstinence															
	Cocaine (Early Abstinence) 2,3,43-50	↑	↓	↑	↓	—	— ·	↑	—	—	—	—	—	—	-
	Cocaine (Later Abstinence) 2,3,43-49	↓	↑	↓	↑ ; n.s	—	—	↕ ↑	↑		↓	—	—	-
	Opioid (w/o medication assistance) 2,3,50-57	↓	↑	↓	↑	—	—	— —	—		↓	—	—	↑

n.s., not significant; —, not determined; qEEG, quantitative EEG; REM Density, number of REM bouts per minute of sleep; % REM, duration of REM out of total sleep time; WASO, time awake after sleep onset until final awakening; Early abstinence, <1 week; Later abstinence, >2 weeks.

[1]Benca et al. (1992), [2]Angarita, Emadi, Hodges, and Morgan (2016), [3]Dolsen and Harvey (2017), [4]Cohrs (2008), [5]Gerstenberg et al. (2020), [6]Göder et al. (2006), [7]Hodges, Pittman, and Morgan (2017), [8]Kaskie, Graziano, and Ferrarelli (2017), [9]Keshavan et al. (1998), [10]Manoach et al. (2004), [11]Manoach and Stickgold (2009), [12]Manoach and Stickgold (2019), [13]Sprecher et al. (2015), [14]Armitage, Hudson, Trivedi, and Rush (1995), [15]Armitage, Hoffmann, Fitch, Morel, and Bonato (1995), [16]Bovy et al. (2021), [17]Coble, Foster, and Kupfer (1976), [18]Steiger, Pawlowski, and Kimura (2015), [19]Steiger and Kimura (2010), [20]Steiger and Pawlowski (2019), [21]Steiger et al. (2015), [22]Tesler et al. (2016), [23]Wichniak, Antczak, Wierzbicka, and Jernajczyk (2002), [24]Wichniak et al. (2013), [25]Baird et al. (2018), [26]Khazaie and Masoudi (2016), [27]Kobayashi et al. (2007), Kobayashi et al. (2012), [29]Krakow, Ulibarri, Moore, and McIver (2015), [30]Laniepce et al. (2020), [31]Le Bon et al. (1997), [32]Leeman-markowski et al. (2020), [33]Lewis et al. (2020), [34]Mellman, Bustamante, Fins, Pigeon, and Nolan (2002), [35]Mellman, Pigeon, Nowell, and Nolan (2007), [36]Mellman, Kobayashi, Lavela, Wilson, and Hall Brown (2014), [37]Neylan et al. (2021), [38]Swift (2020), [39]Woodward et al. (2002), [40]Bliwise (1993), [41]Scullin and Bliwise (2015a, 2015b), [42]Zhang et al. (2022), [43]Bjorness and Greene (2021), [44]Chen et al. (2015), [45]Johanson, Roehrs, Schuh, and Warbasse (1999), [46]Matuskey, Pittman, Forselius, Malison, and Morgan (2011), [47]Morgan, Pace-Schott, et al. (2006), Morgan, Perry, et al., 2006, [48]Schierenbeck et al. (2008), [49]Valladares and Irwin (2007), [50]Chen, Ting, Wu, Lin, and Gossop (2017), [51]Eckert and Yaggi (2022), [52]Gauthier et al. (2011), [53]Kay (1975), [54]Khazan and Colasanti (1972), [55]Rosen et al. (2019), [56]Shaw et al. (2006), [57]Wang and Teichtahl (2007).

Table 4 Subjective assessments of sleep disruptions in neuropsychiatric disorders.

Neuropsychiatric condition		Sleep				Nightmares	Limb movement	Daytime napping
		Duration	Latency	Quality	WASO			
Schizophrenia 2–7		↕	↑	↓	↑	↑	—	↑
Major depressive disorder 8–11		↕	↑	↓	↑	—	—	↑
Post-traumatic stress disorder 12–16		↓	↑	↓	↑	↑	↑	—
Alzheimer's disease 17–22		↓	↑	↓	↑	—	—	↑
Substance use disorders: Withdrawal and abstinence	Cocaine (Early abstinence) [1,23-27]	↓	↑	↓	—	—	—	—
	Cocaine (Later abstinence) [1,23-27]	↑	↓	↑	—	—	↑	—
	Opioid (w/o medication assistance) 1,28-30	↓	↑	↓	↑	↑	↑	↑

Abbreviations: n.s., not significant; —, not determined. WASO, time awake after sleep onset until final awakening, Early abstinence, <1 week; Later abstinence, >2 weeks.

[1]Angarita et al. (2016), [2]Alam and Chengappa (2011), [3]Chiu, Ree, Janca, and Waters (2016), [4]Chouinard, Poulin, Stip, and Godbout (2004), [5]Dule, Ahmed, Tessema, and Soboka (2020), [6]Göder et al. (2006), [7]Manoach and Stickgold (2019), [8]Ağargün, Kara, and Solmaz (1997), [9]Lopez, Hoffmann, Armitage, and Are (2010), [10]Maglione et al. (2014), [11]Steiger and Pawlowski (2019), [12]Denis et al. (2021), [13]Kobayashi et al. (2012), [14]Westermeyer et al. (2010), [15]Woodward et al. (2002), [16]Yeh et al. (2021), [17]Borges et al. (2021), [18]Kim, Ahn, Min, Yoo, and Choi (2021), [19]Kuo, Hsiao, Lo, and Nikolai (2021), [20]Li et al. (2022), [21]Scullin and Bliwise (2015a, 2015b), [22]Zhang et al. (2022), [23]Hodges et al. (2017), [24]Matuskey et al. (2011), [25]Morgan, Perry, et al. (2006), [26]Morgan and Malison (2007), [27]Schierenbeck et al. (2008), [28]Chen et al. (2017), [29]Gupta, Ali, and Ray (2018), [30]Sharkey et al. (2011).

circuits implicated in disrupted sleep–wake activity in a disease-specific manner. Given that many CNS disorders have seen a dearth of novel treatments, future advancements should investigate sleep modifying pharmacotherapies in parallel with strategies to mitigate other symptoms.

As highlighted throughout, glutamate is a key regulator of homeostatic sleep, and glutamate dysfunction is commonly associated with many CNS disorders. Thus, it is important to investigate how glutamate dysfunction may directly contribute to sleep disturbances associated with specific CNS disorders. This chapter provided an overview of known roles of glutamatergic and, more specifically, mGlu receptor regulation of the sleep-wake cycle. Key contributions of multiple mGlu receptor subtypes in sleep regulation are discussed alone and in context of multiple CNS disorders. This chapter also highlights current knowledge gaps and speculates on plausible mGlu receptor subtypes to target that mitigate both the primary symptoms of a specific disorder and the associated sleep disturbances. We argue that, while possible to identify a singular treatment to augment sleep across multiple conditions, given the heterogeneity of CNS disorders, unique treatments based on neurobiological alterations may have greater therapeutic efficacy across multiple symptoms.

As summarized in Table 1 and 2, much of our understanding of mGlu receptor contributions to sleep regulation stem from constitutive knockout studies or studies using acute pharmacological challenges with subtype selective (or preferring) compounds. While these studies are vital to our basic understanding, methodological considerations should be discussed to assist with accurate interpretation of findings. For example, the absence of altered baseline sleep between wild type mice and mice with genetically altered receptor expression/function should not entirely rule out the possible involvement of that receptor influencing sleep regulation. Given the highly conserved circuitry regulating the sleep–wake cycle, constitutive knockout may result in compensatory functional and circuit-level changes that mask subtle differences. Moreover, baseline sleep comparisons reflect normal sleep regulation but not the mechanisms regulating sleep following a perturbation (sleep restriction/deprivation, stressor, etc.). For example, $mGlu_5$ studies found disruptions in homeostatic sleep drive in KO animals following a sleep deprivation period (Ahnaou, Raeymaekers, et al., 2015; Holst et al., 2017). Importantly, while many studies have examined direct effects of mGlu receptor knockouts (notably $mGlu_{2/3}$, $mGlu_5$, $mGlu_7$) or altered downstream signaling (e.g. PLCβ4, a downstream effector of $mGlu_1$) as shown in Table 1, direct effects of $mGlu_1$, $mGlu_4$, and $mGlu_8$ have not yet been examined. Furthermore, recent advances in neuroscience methods afford the exciting opportunity to examine sleep-wake activity changes at baseline and in response to perturbations. These include studies with conditional genetic alterations in mice as well as circuit/cell-specific modulation via opto/chemogenetic techniques.

Similar methodological considerations should be reviewed when designing/interpreting pharmacological studies including time of dosing and selectivity of pharmacological effects. In rodent studies, compounds administered during the dark (active) phase can provide information regarding general sleep-promoting effects (or sedative effects based on quantitative assessments of spectral power distribution) whereas compound administration during the light (quiescent) phase may better align with investigating wake-promoting/sleep-disrupting effects, NREM:REM ratios/latencies, and delta power during NREM sleep. Given the known fluctuations in both neurotransmitters and receptor cycling across diurnal periods, compounds may have divergent profiles on sleep-wake duration and architecture depending on time of dosing. Examining pharmacological effects on rodent sleep-wake behavior in both periods is even more critical when considering that recently developed compounds with increased mGlu receptor subtype selectivity predominately involve allosteric modulators (for review, see Luessen & Conn, 2022). Unlike compounds that bind to the orthosteric site, that activate/inhibit receptor function regardless of endogenous physiological conditions, allosteric modulators require the presence of an endogenous agonist (e.g. glutamate) to augment intracellular signaling. In this regard, allosteric modulation may provide unprecedented temporal selectivity, effectively modulating signaling when endogenous glutamate is present at higher concentrations (wake and REM) without altering sleep/wake state when glutamate is lower (NREM). Importantly, most pharmacological studies included in this chapter examined acute but not chronic pharmacological effects on sleep-wake cycles. Many studies have also not examined effects of these compounds on sleep in higher order, gyrencephalic species with more analogous sleep-wake patterns to humans. Lastly, it must be noted that, given the absence of literature examining mGlu receptor modulation on sleep in preclinical models of a CNS disorder, we have speculated on most potential pharmacotherapeutic applications. Future studies are thus needed to expand on findings from pharmacological assessments in otherwise healthy animals and incorporate these methodologies into animal models of disease states and/or in clinical populations to confirm/refute these hypotheses.

In summary, normalizing sleep disturbances associated with CNS disorders represents a novel treatment approach. Improving sleep may serve as a critical component in the therapeutic outcome, and restoration of homeostatic sleep may have several downstream benefits for other primary symptoms of a CNS disorder. Recent development of mGlu receptor-subtype selective ligands presents an opportunity to further interrogate the role of group I, II and III mGlu receptor regulation of the sleep-wake cycle with separate or synergistic effects on other symptoms. In general, group I mGlu receptors appear to be involved in homeostatic sleep pressure in addition to influences in sleep duration (Aguilar et al., 2020; Martin et al., 2019) whereas group II receptors appear to be more so involved in cortical arousal and selective effects on REM sleep following activation (Ahnaou et al., 2009; Pritchett et al., 2015). Fewer studies have examined group III, though their role (or lack thereof) will likely be determined in future studies with the recent development of more subtype-selective ligands (Lin et al., 2022; Luessen & Conn, 2022).

Importantly, the potential for specific mGlu receptors and the associated pharmacological effect (activation or inhibition) to ameliorate sleep disturbances and primary symptoms varies across disorders. For example, where $mGlu_5$ agonists/PAMs may have pharmacotherapeutic relevance for sleep disturbances and symptoms in schizophrenia (Maksymetz et al., 2017), $mGlu_5$ antagonists/NAMs may be more relevant for MDD, PTSD, AD, and SUD (Hamilton et al., 2014; Luessen & Conn, 2022). Likewise, both activation and inhibition of $mGlu_{2/3}$ may be therapeutically relevant for MDD (Dogra & Conn, 2021) whereas activation may be most relevant with regard to schizophrenia (Maksymetz et al., 2017). The current, most effective therapies for both schizophrenia and depression influence sleep, predominantly through increasing sleep duration (antipsychotics; Cohrs, 2008) or selectively decreasing REM sleep (antidepressants; Steiger & Pawlowski, 2019; Wichniak et al., 2017). Thus, systematically examining sleep as a primary or secondary endpoint should be considered in the development of novel therapies. Recent preclinical studies investigating $mGlu_5$ NAMs and $mGlu_{2/3}$ PAMs have shown antidepressant-like effects, both in classic models and on sleep (Ahnaou et al., 2009; Gould et al., 2016; Holter et al., 2021; Lindemann et al., 2015). Several compounds targeting mGlu receptors have advanced to clinical trials for disorders including schizophrenia and MDD. For example, both $mGlu_5$ NAMs (Quiroz et al., 2016) and $mGlu_{2/3}$ antagonists (see Maksymetz et al., 2017 for discussion) have advanced to phase II for the treatment of MDD but were stalled due to a lack of significant effects on primary endpoints despite promising results on secondary endpoints. As multiple development programs progress forward with novel mGlu receptor subtype selective compounds for various indications, incorporating sleep measures into pre-clinical and clinical studies (although costly) should be considered. Lastly, while this chapter highlights the importance of mGlu receptor function on sleep in the context of schizophrenia, MDD, PTSD, SUD, and AD, there is impetus for targeting sleep disturbances as a treatment approach for numerous other neurodegenerative and neurodevelopmental disorders.

References

Aeschbach, D., Sher, L., Postolache, T. T., Matthews, J. R., Jackson, M. A., & Wehr, T. A. (2003). A longer biological night in long sleepers than in short sleepers. *The Journal of Clinical Endocrinology and Metabolism, 88*(1), 26–30. https://doi.org/10.1210/jc.2002-020827.

Ağargün, M. Y., Kara, H., & Solmaz, M. (1997). Subjective sleep quality and suicidality in patients with major depression. *Journal of Psychiatric Research, 31*(3), 377–381. https://doi.org/10.1016/S0022-3956(96)00037-4.

Aguilar, X. D. D., Strecker, R. E., Basheer, R., & Mcnally, J. M. (2020). Alterations in sleep, sleep spindle, and EEG power in mGluR5 knockout mice. *Journal of Neurophysiology, 123*(1), 22–33. https://doi.org/10.1152/jn.00532.2019.

Ahnaou, A., Dautzenberg, F. M., Geys, H., Imogai, H., Gibelin, A., Moechars, D., et al. (2009). Modulation of group II metabotropic glutamate receptor (mGlu2) elicits common changes in rat and mice sleep-wake architecture. *European Journal of Pharmacology, 603*(1–3), 62–72. https://doi.org/10.1016/j.ejphar.2008.11.018.

Ahnaou, A., Langlois, X., Steckler, T., Bartolome-Nebreda, J. M., & Drinkenburg, W. H. I. M. (2015). Negative versus positive allosteric modulation of metabotropic glutamate receptors (mGluR5): Indices for potential pro-cognitive drug properties based on EEG network oscillations and sleep-wake organization in rats. *Psychopharmacology, 232*(6), 1107–1122. https://doi.org/10.1007/s00213-014-3746-4.

Ahnaou, A., Raeyemaekers, L., Huysmans, H., & Drinkenburg, W. H. I. M. (2016). Off-target potential of AMN082 on sleep EEG and related physiological variables: Evidence from mGluR7 (-/-) mice. *Behavioural Brain Research, 311*, 287–297. https://doi.org/10.1016/j.bbr.2016.05.035.

Ahnaou, A., Raeymaekers, L., Steckler, T., & Drinkenbrug, W. H. I. M. (2015). Relevance of the metabotropic glutamate receptor (mGluR5) in the regulation of NREM-REM sleep cycle and homeostasis: Evidence from mGluR5 (-/-) mice. *Behavioural Brain Research, 282*, 218–226. https://doi.org/10.1016/j.bbr.2015.01.009.

Ahnaou, A., ver Donck, L., & Drinkenburg, W. H. I. M. (2014). Blockade of the metabotropic glutamate (mGluR2) modulates arousal through vigilance states transitions: Evidence from sleep-wake EEG in rodents. *Behavioural Brain Research, 270*, 56–67. https://doi.org/10.1016/j.bbr.2014.05.003.

Alam, A., & Chengappa, K. N. R. (2011). Obstructive sleep apnoea and schizophrenia: A primer for psychiatrists. *Acta Neuropsychiatrica, 23*(5), 201–209. https://doi.org/10.1111/j.1601-5215.2011.00589.x.

Altamura, C. A., Mauri, M. C., Ferrara, A., Moro, A. R., D'Andrea, G., & Zamberlan, F. (1993). Plasma and platelet excitatory amino acids in psychiatric disorders. *The American Journal of Psychiatry, 150*(11), 1731–1733. https://doi.org/10.1176/ajp.150.11.1731.

Altinbilek, B., & Manahan-Vaughan, D. (2007). Antagonism of group III metabotropic glutamate receptors results in impairment of LTD but not LTP in the hippocampal CA1 region, and prevents long-term spatial memory. *The European Journal of Neuroscience*, 26(5), 1166–1172. https://doi.org/10.1111/j.1460-9568.2007.05742.x.

Altinbilek, B., & Manahan-Vaughan, D. (2009). A specific role for group II metabotropic glutamate receptors in hippocampal long-term depression and spatial memory. *Neuroscience*, 158(1), 149–158. https://doi.org/10.1016/j.neuroscience.2008.07.045.

Amato, R. J., Felts, A. S., Rodriguez, A. L., Venable, D. F., Morrison, R. D., Byers, F. W., et al. (2013). Substituted 1-phenyl-3-(pyridin-2-yl)urea negative allosteric modulators of mGlu5: Discovery of a new tool compound VU0463841 with activity in rat models of cocaine addiction. *ACS Chemical Neuroscience*, 4(8), 1217–1228. https://doi.org/10.1021/cn400070k.

American Psychological Association. (2017). *Clinical Practice Guideline for the Treatment of Posttraumatic Stress Disorder (PTSD)*. *Vol. 139*. Washington, DC: APA, Guideline Development Panel for the Treatment of Posttraumatic Stress Disorder in Adults. https://doi.org/10.1162/jocn.

Anaclet, C., Ferrari, L., Arrigoni, E., Bass, C. E., Saper, C. B., Lu, J., et al. (2014). The GABAergic parafacial zone is a medullary slow wave sleep-promoting center. *Nature Neuroscience*, 17(9), 1217–1224. https://doi.org/10.1038/nn.3789.

Angarita, G. A., Emadi, N., Hodges, S., & Morgan, P. T. (2016). Sleep abnormalities associated with alcohol, cannabis, cocaine, and opiate use: A comprehensive review. *Addiction Science and Clinical Practice*, 11(1), 1–17. https://doi.org/10.1186/s13722-016-0056-7.

Anwyl, R. (1999). Metabotropic glutamate receptors: Electrophysiological properties and role in plasticity. *Brain Research. Brain Research Reviews*, 29(1), 83–120. https://doi.org/10.1016/s0165-0173(98)00050-2.

APA. (2013). *Diagnostic and statistical manual of mental disorders* (5th ed.).

Armitage, R. (1995). The distribution of EEG frequencies in REM and NREM sleep stages in healthy young adults. *Sleep*, 18(5), 334–341. https://doi.org/10.1093/sleep/18.5.334.

Armitage, R. (2007). Sleep and circadian rhythms in mood disorders. *Acta Psychiatrica Scandinavica*, 433, 104–115. https://doi.org/10.1111/j.1600-0447.2007.00968.x.

Armitage, R., Hoffmann, R., Fitch, T., Morel, C., & Bonato, R. (1995). A comparison of period amplitude and power spectral analysis of sleep EEG in normal adults and depressed outpatients. *Psychiatry Research*, 56(3), 245–256. https://doi.org/10.1016/0165-1781(95)02615-4.

Armitage, R., Hudson, A., Trivedi, M., & Rush, A. J. (1995). Sex differences in the distribution of EEG frequencies during sleep: Unipolar depressed outpatients. *Journal of Affective Disorders*, 34(2), 121–129. https://doi.org/10.1016/0165-0327(95)00009-C.

Aston-Jones, G., Smith, R. J., Moorman, D. E., & Richardson, K. A. (2009). Role of lateral hypothalamic orexin neurons in reward processing and addiction. *Neuropharmacology*, 56(Suppl. 1), 112–121. https://doi.org/10.1016/j.neuropharm.2008.06.060.

Averill, L. A., Jiang, L., Purohit, P., Coppoli, A., Averill, C. L., Roscoe, J., et al. (2022). Prefrontal glutamate neurotransmission in PTSD: A novel approach to estimate synaptic strength in vivo in humans. *Chronic Stress*, 6. https://doi.org/10.1177/24705470221092734.

Baird, T., Theal, R., Gleeson, S., McLeay, S., O'Sullivan, R., Initiative, P. T. S. D., et al. (2018). Detailed polysomnography in Australian Vietnam veterans with and without posttraumatic stress disorder. *Journal of Clinical Sleep Medicine*, 14(9), 1577–1586. https://doi.org/10.5664/jcsm.7340.

Baker, J. F., Cates, M. E., & Luthin, D. R. (2017). D-cycloserine in the treatment of posttraumatic stress disorder. *Mental Health Clinician*, 7(2), 88–94. https://doi.org/10.9740/mhc.2017.03.088.

Battaglia, G., Busceti, C. L., Molinaro, G., Biagioni, F., Traficante, A., Nicoletti, F., et al. (2006). Pharmacological activation of mGlu4 metabotropic glutamate receptors reduces nigrostriatal degeneration in mice treated with 1-methyl-4-phenyl-1,2,3,6-tetrahydropyridine. *The Journal of Neuroscience, 26*(27), 7222–7229. https://doi.org/10.1523/JNEUROSCI.1595-06.2006.

Bazhenov, M., Timofeev, I., Steriade, M., & Sejnowski, T. (2000). Spiking-bursting activity in the thalamic reticular nucleus initiates sequences of spindle oscillations in thalamic networks. *Journal of Neurophysiology, 84*(2), 1076–1087. https://doi.org/10.1152/jn.2000.84.2.1076.

Benca, R. M., & Buysse, D. J. (2018). Reconsidering insomnia as a disorder rather than just a symptom in psychiatric practice. *The Journal of Clinical Psychiatry, 79*(1). https://doi.org/10.4088/JCP.me17008ah1c.

Benca, R. M., William, H., Thisted, R. A., & Gillin, J. C. (1992). Sleep and psychiatric disorders: A meta-analysis. *Sleep and Psychiatric Disorders, Vol. 49*, 651. Benca RM, Obermeyer WH, Thisted RA, Gillin JC. https://doi.org/10.1001/archpsyc.1992.01820080059010.

Benneyworth, M. A., Xiang, Z., Smith, R. L., Garcia, E. E., Conn, P. J., & Sanders-Bush, E. (2007). A selective positive allosteric modulator of metabotropic glutamate receptor subtype 2 blocks a hallucinogenic drug model of psychosis. *Molecular Pharmacology, 72*(2), 477–484. https://doi.org/10.1124/mol.107.035170.

Bergum, N., Berezin, C.-T., & Vigh, J. (2022). A retinal contribution to opioid-induced sleep disorders? *Frontiers in Neuroscience, 16*, 981939. https://doi.org/10.3389/fnins.2022.981939.

Berman, R. M., Cappiello, A., Anand, A., Oren, D. A., Heninger, G. R., Charney, D. S., et al. (2000). Antidepressant effects of ketamine in depressed patients. *Biological Psychiatry, 47*(4), 351–354. https://doi.org/10.1016/S0006-3223(99)00230-9.

Berro, L. F., Zamarripa, C. A., Talley, J. T., Freeman, K. B., & Rowlett, J. K. (2022). Effects of methadone, buprenorphine, and naltrexone on actigraphy-based sleep-like parameters in male rhesus monkeys. *Addictive Behaviors, 135*, 107433. https://doi.org/10.1016/j.addbeh.2022.107433.

Bjorness, T. E., & Greene, R. W. (2018). Dose response of acute cocaine on sleep/waking behavior in mice. *Neurobiology of Sleep and Circadian Rhythms, 5*, 84–93. https://doi.org/10.1016/j.nbscr.2018.02.001.

Bjorness, T. E., & Greene, R. W. (2021). Interaction between cocaine use and sleep behavior: A comprehensive review of cocaine's disrupting influence on sleep behavior and sleep disruptions influence on reward seeking. *Pharmacology Biochemistry and Behavior, 206*(November 2020), 173194. https://doi.org/10.1016/j.pbb.2021.173194.

Bjorness, T. E., & Greene, R. W. (2022). Arousal-mediated sleep disturbance persists during cocaine abstinence in male mice. *Frontiers in Neuroscience, 16*, 868049. https://doi.org/10.3389/fnins.2022.868049.

Black, J. E., Kodish, I. M., Grossman, A. W., Klintsova, A. Y., Orlovskaya, D., Vostrikov, V., et al. (2004). Pathology of layer V pyramidal neurons in the prefrontal cortex of patients with schizophrenia. *The American Journal of Psychiatry, 161*(4), 742–744. https://doi.org/10.1176/appi.ajp.161.4.742.

Bliwise, D. L. (1993). Sleep in normal aging and dementia. *Sleep, 16*(1), 40–81. https://doi.org/10.1093/sleep/16.1.40.

Bliwise, D. L., Tinklenberg, J., Yesavage, J. A., Davies, H., Pursley, A. M., Petta, D. E., et al. (1989). REM latency in Alzheimer's disease. *Biological Psychiatry, 25*(3), 320–328. https://doi.org/10.1016/0006-3223(89)90179-0.

Boissard, R., Gervasoni, D., Schmidt, M. H., Barbagli, B., Fort, P., & Luppi, P. H. (2002). The rat ponto-medullary network responsible for paradoxical sleep onset and maintenance: A combined microinjection and functional neuroanatomical study. *European Journal of Neuroscience, 16*(10), 1959–1973. https://doi.org/10.1046/j.1460-9568.2002.02257.x.

Borbély, A. A., Daan, S., Wirz-Justice, A., & Deboer, T. (2016). The two-process model of sleep regulation: A reappraisal. *Journal of Sleep Research, 25*(2), 131–143. https://doi.org/10.1111/jsr.12371.

Borbély, A. A., & Wirz-Justice, A. (1982). Sleep, sleep deprivation and depression. A hypothesis derived from a model of sleep regulation. *Human Neurobiology, 1*(3), 205–210.

Borges, C. R., Piovezan, R. D., Poyares, D. R., Busatto Filho, G., Studart-Neto, A., Coutinho, A. M., et al. (2021). Subjective sleep parameters in prodromal Alzheimer's disease: A case-control study. *Brazilian Journal of Psychiatry, 43*(5), 510–513. https://doi.org/10.1590/1516-4446-2020-1503.

Bottary, R., Seo, J., Daffre, C., Gazecki, S., Moore, K. N., Kopotiyenko, K., et al. (2020). Fear extinction memory is negatively associated with REM sleep in insomnia disorder. *Sleep, 43*(7), 1–12. https://doi.org/10.1093/SLEEP/ZSAA007.

Bovy, L., Weber, F. D., Tendolkar, I., Fernández, G., Czisch, M., Steiger, A., et al. (2021). Non-REM sleep in major depressive disorder. *BioRxiv*. 2021.03.19.436132 https://www.biorxiv.org/content/10.1101/2021.03.19.436132v1%0Ahttps://www.biorxiv.org/content/10.1101/2021.03.19.436132v1.abstract.

Bradley, S. R., Uslaner, J. M., Flick, R. B., Lee, A., Groover, K. M., & Hutson, P. H. (2012). The mGluR7 allosteric agonist AMN082 produces antidepressant-like effects by modulating glutamatergic signaling. *Pharmacology, Biochemistry, and Behavior, 101*(1), 35–40. https://doi.org/10.1016/j.pbb.2011.11.006.

Brady, K., Pearlstein, T., Asnis, G. M., Baker, D., Rothbaum, B., Sikes, C. R., et al. (2000). Efficacy and safety of sertraline treatment of posttraumatic stress disorder: A randomized controlled trial. *JAMA, 283*(14), 1837–1844. https://doi.org/10.1001/jama.283.14.1837.

Breslau, N., Roth, T., Rosenthal, L., & Andreski, P. (1996). Sleep disturbance and psychiatric disorders: A longitudinal epidemiological study of young adults. *Biological Psychiatry, 39*(6), 411–418. https://doi.org/10.1016/0006-3223(95)00188-3.

Brower, K. J. (2003). Insomnia, alcoholism and relapse. *Sleep Medicine Reviews, 7*(6), 523–539. https://doi.org/10.1016/s1087-0792(03)90005-0.

Brown, R. E., Basheer, R., McKenna, J. T., Strecker, R. E., & McCarley, R. W. (2012). Control of sleep and wakefulness. *Physiological Reviews, 92*(3), 1087–1187. https://doi.org/10.1152/physrev.00032.2011.

Bruno, V., Battaglia, G., Casabona, G., Copani, A., Caciagli, F., & Nicoletti, F. (1998). Neuroprotection by glial metabotropic glutamate receptors is mediated by transforming growth factor-beta. *The Journal of Neuroscience: The Official Journal of the Society for Neuroscience, 18*(23), 9594–9600. https://doi.org/10.1523/JNEUROSCI.18-23-09594.1998.

Bruno, V., Ksiazek, I., Battaglia, G., Lukic, S., Leonhardt, T., Sauer, D., et al. (2000). Selective blockade of metabotropic glutamate receptor subtype 5 is neuroprotective. *Neuropharmacology, 39*(12), 2223–2230. https://doi.org/10.1016/S0028-3908(00)00079-4.

Budgett, R. F., Bakker, G., Sergeev, E., Bennett, K. A., & Bradley, S. J. (2022). Targeting the type 5 metabotropic glutamate receptor: A potential therapeutic strategy for neurodegenerative diseases? *Frontiers in Pharmacology, 13*, 1–20. https://doi.org/10.3389/fphar.2022.893422.

Burbaeva, G. S., Boksha, I. S., Turishcheva, M. S., Vorobyeva, E. A., Savushkina, O. K., & Tereshkina, E. B. (2003). Glutamine synthetase and glutamate dehydrogenase in the prefrontal cortex of patients with schizophrenia. *Progress in Neuro-Psychopharmacology & Biological Psychiatry, 27*(4), 675–680. https://doi.org/10.1016/s0278-5846(03)00078-2.

Bushey, D., Tononi, G., & Cirelli, C. (2011). Sleep and synaptic homeostasis: Structural evidence in Drosophila. *Science*, *332*(6037), 1576–1581. https://doi.org/10.1126/science. 1202839.

Calabrese, V., Picconi, B., Heck, N., Campanelli, F., Natale, G., Marino, G., et al. (2022). A positive allosteric modulator of mGlu4 receptors restores striatal plasticity in an animal model of l-Dopa-induced dyskinesia. *Neuropharmacology*, *218*, 109205. https://doi. org/10.1016/j.neuropharm.2022.109205.

Canevelli, M., Valletta, M., Trebbastoni, A., Sarli, G., D'Antonio, F., Tariciotti, L., et al. (2016). Sundowning in dementia: Clinical relevance, pathophysiological determinants, and therapeutic approaches. *Frontiers in Medicine*, *3*, 73. https://doi.org/10.3389/fmed. 2016.00073.

Caraci, F., Molinaro, G., Battaglia, G., Giuffrida, M. L., Riozzi, B., Traficante, A., et al. (2011). Targeting group II metabotropic glutamate (mGlu) receptors for the treatment of psychosis associated with Alzheimer's disease: Selective activation of mGlu2 receptors amplifies beta-amyloid toxicity in cultured neurons, whereas dual activation of mGlu2 and mGlu3 receptors is neuroprotective. *Molecular Pharmacology*, *79*(3), 618–626. https:// doi.org/10.1124/mol.110.067488.

Caraci, F., Nicoletti, F., & Copani, A. (2018). Metabotropic glutamate receptors: The potential for therapeutic applications in Alzheimer's disease. *Current Opinion in Pharmacology*, *38*, 1–7. https://doi.org/10.1016/j.coph.2017.12.001.

Cavas, M., Scesa, G., Martín-López, M., & Navarro, J. F. (2017). Selective agonism of mGlu8 receptors by (S)-3,4-dicarboxyphenylglycine does not affect sleep stages in the rat. *Pharmacological Reports*, *69*(1), 97–104. https://doi.org/10.1016/j.pharep.2016.09.019.

Cavas, M., Scesa, G., & Navarro, J. F. (2013a). Positive allosteric modulation of mGlu7 receptors by AMN082 affects sleep and wakefulness in the rat. *Pharmacology Biochemistry and Behavior*, *103*(4), 756–763. https://doi.org/10.1016/j.pbb.2012.12.008.

Cavas, M., Scesa, G., & Navarro, J. F. (2013b). Effects of MPEP, a selective metabotropic glutamate mGlu5 ligand, on sleep and wakefulness in the rat. *Progress in Neuro-Psychopharmacology and Biological Psychiatry*, *40*, 18–25. https://doi.org/10.1016/ J.PNPBP.2012.09.011.

Chaki, S. (2017). mGlu2/3 receptor antagonists as novel antidepressants. *Trends in Pharmacological Sciences*, *38*(6), 569–580. https://doi.org/10.1016/j.tips.2017.03.008.

Chaki, S., Koike, H., & Fukumoto, K. (2019). Targeting of metabotropic glutamate receptors for the development of novel antidepressants. *Chronic Stress*, *3*. https://doi.org/10. 1177/2470547019837712.

Chaki, S., Yoshikawa, R., Hirota, S., Shimazaki, T., Maeda, M., Kawashima, N., et al. (2004). MGS0039: A potent and selective group II metabotropic glutamate receptor antagonist with antidepressant-like activity. *Neuropharmacology*, *46*(4), 457–467. https://doi.org/10. 1016/j.neuropharm.2003.10.009.

Charvin, D. (2018). mGlu4 allosteric modulation for treating Parkinson's disease. *Neuropharmacology*, *135*, 308–315. https://doi.org/10.1016/j.neuropharm.2018. 03.027.

Chen, M. H., Korenic, S. A., Wickwire, E. M., Wijtenburg, S. A., Hong, L. E., & Rowland, L. M. (2020). Sex differences in subjective sleep quality patterns in Schizophrenia. *Behavioral Sleep Medicine*, *18*(5), 668–679. https://doi.org/10.1080/15402002.2019. 1660168.

Chen, V. C. H., Ting, H., Wu, M. H., Lin, T. Y., & Gossop, M. (2017). Sleep disturbance and its associations with severity of dependence, depression and quality of life among heroin-dependent patients: A cross-sectional descriptive study. *Substance Abuse: Treatment, Prevention, and Policy*, *12*(1), 1–9. https://doi.org/10.1186/s13011-017-0101-x.

Chen, B., Wang, Y., Liu, X., Liu, Z., Dong, Y., & Huang, Y. H. (2015). Sleep regulates incubation of cocaine craving. *J Neurosci*, *35*(39), 13300–13310. https://doi.org/10. 1523/jneurosci.1065-15.2015.

Chiu, V. W., Ree, M., Janca, A., & Waters, F. (2016). Sleep in Schizophrenia: Exploring subjective experiences of sleep problems, and implications for treatment. *Psychiatric Quarterly*, *87*(4), 633–648. https://doi.org/10.1007/s11126-015-9415-x.

Chouinard, S., Poulin, J., Stip, E., & Godbout, R. (2004). Sleep in untreated patients with schizophrenia: A meta-analysis. *Schizophrenia Bulletin*, *30*(4), 957–967. https://doi.org/10.1093/oxfordjournals.schbul.a007145.

Cid, J. M., Tresadern, G., Vega, J. A., de Lucas, A. I., Matesanz, E., Iturrino, L., et al. (2012). Discovery of 3-cyclopropylmethyl-7-(4-phenylpiperidin-1-yl)-8-trifluoromethyl[1,2,4]triazolo[4,3-a]pyridine (JNJ-42153605): A positive allosteric modulator of the metabotropic glutamate 2 receptor. *Journal of Medicinal Chemistry*, *55*(20), 8770–8789. https://doi.org/10.1021/jm3010724.

Cleva, R. M., & Olive, M. F. (2012). mGlu receptors and drug addiction. *Wiley Interdisciplinary Reviews. Membrane Transport and Signaling*, *1*(3), 281–295. https://doi.org/10.1002/wmts.18.

Coble, P., Foster, F. G., & Kupfer, D. J. (1976). Electroencephalographic sleep diagnosis of primary depression. *Archives of General Psychiatry*, *33*(9), 1124–1127. https://doi.org/10.1001/archpsyc.1976.01770090114012.

Coffey, A. A., Guan, Z., Grigson, P. S., & Fang, J. (2016). Reversal of the sleep–wake cycle by heroin self-administration in rats. *Brain Research Bulletin*, *123*, 33–46. https://doi.org/10.1016/j.brainresbull.2015.09.008.

Cohrs, S. (2008). Sleep disturbances in patients with schizophrenia: Impact and effect of antipsychotics. *CNS Drugs*. https://doi.org/10.2165/00023210-200822110-00004.

Cominski, T. P., Jiao, X., Catuzzi, J. E., Stewart, A. L., & Pang, K. C. H. (2014). The role of the hippocampus in avoidance learning and anxiety vulnerability. *Frontiers in Behavioral Neuroscience*, *8*, 1–10. https://doi.org/10.3389/fnbeh.2014.00273.

Conn, P. J., & Jones, C. K. (2009). Promise of mGluR2/3 activators in psychiatry. *Neuropsychopharmacology*, *34*(1), 248–249. https://doi.org/10.1038/npp.2008.156.

Conn, P. J., & Pin, J. P. (1997). Pharmacology and functions of metabotropic glutamate receptors. *Annual Review of Pharmacology and Toxicology*, *37*, 205–237. https://doi.org/10.1146/annurev.pharmtox.37.1.205.

Cortes, J. A., Gomez, G., Ehnerd, C., Gurnsey, K., Nicolazzo, J., Bradberry, C. W., et al. (2016). Altered activity-based sleep measures in rhesus monkeys following cocaine self-administration and abstinence. *Drug and Alcohol Dependence*, *163*, 202–208. https://doi.org/10.1016/j.drugalcdep.2016.04.014.

Coyle, J. T. (2006). Glutamate and schizophrenia: Beyond the dopamine hypothesis. *Cellular and Molecular Neurobiology*, *26*(4–6), 365–384. https://doi.org/10.1007/s10571-006-9062-8.

Crabtree, J. W., Lodge, D., Bashir, Z. I., & Isaac, J. T. R. (2013). GABAA, NMDA and mGlu2 receptors tonically regulate inhibition and excitation in the thalamic reticular nucleus. *European Journal of Neuroscience*, *37*(6), 850–859. https://doi.org/10.1111/ejn.12098.

Crupi, R., Impellizzeri, D., & Cuzzocrea, S. (2019). Role of metabotropic glutamate receptors in neurological disorders. *Frontiers in Molecular Neuroscience*, *12*, 1–11. https://doi.org/10.3389/fnmol.2019.00020.

Czigler, B., Csikós, D., Hidasi, Z., Anna Gaál, Z., Csibri, É., Kiss, É., et al. (2008). Quantitative EEG in early Alzheimer's disease patients—Power spectrum and complexity features. *International Journal of Psychophysiology*, *68*(1), 75–80. https://doi.org/10.1016/j.ijpsycho.2007.11.002.

Daan, S., Beersma, D. G., & Borbély, A. A. (1984). Timing of human sleep: Recovery process gated by a circadian pacemaker. *The American Journal of Physiology*, *246*(2 Pt. 2), R161–R183. https://doi.org/10.1152/ajpregu.1984.246.2.R161.

Dash, M. B., Douglas, C. L., Vyazovskiy, V. V., Cirelli, C., & Tononi, G. (2009). Long-term homeostasis of extracellular glutamate in the rat cerebral cortex across sleep and waking states. *Journal of Neuroscience*, *29*(3), 620–629. https://doi.org/10.1523/JNEUROSCI.5486-08.2009.

Davidson, S., Golden, J., Copits, B., Pradipta, R., Vogt, S., Valtcheva, M., et al. (2017). Group II mGluRs suppress hyperexcitability in mouse and human nociceptors. *Physiology & Behavior, 176*(5), 139–148. https://doi.org/10.1097/j.pain.0000000000000621.

Davies, B. M., & Beech, H. R. (1960). The effect of 1-arylcylohexylamine (sernyl) on twelve normal volunteers. *The Journal of Mental Science, 106*, 912–924. https://doi.org/10.1192/bjp.106.444.912.

Davis, M. T., Hillmer, A., Holmes, S. E., Pietrzak, R. H., DellaGioia, N., Nabulsi, N., et al. (2019). In vivo evidence for dysregulation of mGluR5 as a biomarker of suicidal ideation. *Proceedings of the National Academy of Sciences of the United States of America, 166*(23), 11490–11495. https://doi.org/10.1073/pnas.1818871116.

de Filippis, B., Lyon, L., Taylor, A., Lane, T., Burnet, P. W. J., Harrison, P. J., et al. (2015). The role of group II metabotropic glutamate receptors in cognition and anxiety: Comparative studies in GRM2−/−, GRM3−/− and GRM2/3−/− knockout mice. *Neuropharmacology, 89*, 19–32. https://doi.org/10.1016/j.neuropharm.2014.08.010.

de Vivo, L., Bellesi, M., Marshall, W., Bushong, E. A., Ellisman, M. H., Tononi, G., et al. (2017). Ultrastructural evidence for synaptic scaling across the wake/sleep cycle. *Science (New York, N.Y.), 355*(6324), 507–510. https://doi.org/10.1126/science.aah5982.

Denis, D., Bottary, R., Cunningham, T. J., Zeng, S., Daffre, C., Oliver, K. L., et al. (2021). Sleep power spectral density and spindles in PTSD and their relationship to symptom severity. *Frontiers in Psychiatry, 12*(November), 1–14. https://doi.org/10.3389/fpsyt.2021.766647.

Di Chiara, G., & Imperato, A. (1988). Drugs abused by humans preferentially increase synaptic dopamine concentrations in the mesolimbic system of freely moving rats. *Proceedings of the National Academy of Sciences of the United States of America, 85*(14), 5274–5278. https://doi.org/10.1073/pnas.85.14.5274.

Diering, G. H., Nirujogi, R. S., Roth, R. H., Worley, P. F., Pandey, A., & Huganir, R. L. (2017). Homer1a drives homeostatic scaling-down of excitatory synapses during sleep. *Science, 515*(February), 511–515. https://doi.org/10.1126/science.aai8355.

Dimsdale, J. E., Norman, D., DeJardin, D., & Wallace, M. S. (2007). The effect of opioids on sleep architecture. *Journal of Clinical Sleep Medicine, 3*(1), 33–36.

Dogra, S., & Conn, P. J. (2021). Targeting metabotropic glutamate receptors for the treatment of depression and other stress-related disorders. *Neuropharmacology, 196*, 108687. https://doi.org/10.1016/j.neuropharm.2021.108687.

Dogra, S., & Conn, P. J. (2022). Metabotropic glutamate receptors as emerging targets for the treatment of Schizophrenia. *Molecular Pharmacology, 101*(5), 275–285. https://doi.org/10.1124/molpharm.121.000460.

Dolsen, M. R., & Harvey, A. G. (2017). Life-time history of insomnia and hypersomnia symptoms as correlates of alcohol, cocaine and heroin use and relapse among adults seeking substance use treatment in the United States from 1991 to 1994. *Addiction, 112*(6), 1104–1111. https://doi.org/10.1111/add.13772.

Dule, A., Ahmed, G., Tessema, W., & Soboka, M. (2020). Sleep quality in Schizophrenia. *Journal of Mental Health & Clinical Psychology, 4*(4), 57–64. https://doi.org/10.29245/2578-2959/2020/4.1223.

Dunn, K. E., Huhn, A. S., Bergeria, C. L., Gipson, C. D., & Weerts, E. M. (2019). Nonopioid neurotransmitter systems that contribute to the opioid withdrawal syndrome: A review of preclinical and human evidence. *Journal of Pharmacology and Experimental Therapeutics, 371*(2), 422–452. https://doi.org/10.1124/jpet.119.258004.

Duvoisin, R. M., Pfankuch, T., Wilson, J. M., Grabell, J., Chhajlani, V., Brown, D. G., et al. (2010). Acute pharmacological modulation of mGluR8 reduces measures of anxiety. *Behavioural Brain Research, 212*(2), 130–134. https://doi.org/10.1037/12218-043.

Eacret, D., Veasey, S. C., & Blendy, J. A. (2020). Bidirectional relationship between opioids and disrupted sleep: Putative mechanisms. *Molecular Pharmacology, 98*(4), 445–453. https://doi.org/10.1124/mol.119.119107.

Eckert, D. J., & Yaggi, H. K. (2022). Opioid use disorder, sleep deficiency, and ventilatory control: Bidirectional mechanisms and therapeutic targets. *American Journal of Respiratory and Critical Care Medicine, 206*, 937–949. https://doi.org/10.1164/rccm.202108-2014ci.

España, R. A., & Scammell, T. E. (2011). Sleep neurobiology from a clinical perspective. *Sleep, 34*(7), 845–858. https://doi.org/10.5665/SLEEP.1112.

Esterlis, I., DellaGioia, N., Pietrzak, R. H., Matuskey, D., Nabulsi, N., Abdallah, C. G., et al. (2018). Ketamine-induced reduction in mGluR5 availability is associated with an antidepressant response: An [^{11}C]ABP688 and PET imaging study in depression. *Molecular Psychiatry, 23*(4), 824–832. https://doi.org/10.1038/mp.2017.58.

Esterlis, I., Holmes, S. E., Sharma, P., Krystal, J. H., & Delorenzo, C. (2018). mGluR5 and stress disorders: Knowledge gained from receptor imaging studies. *Biological Psychiatry, 84*(2), 95–105. https://doi.org/10.1016/j.biopsych.2017.08.025.

Fang, H., Tu, S., Sheng, J., & Shao, A. (2019). Depression in sleep disturbance: A review on a bidirectional relationship, mechanisms and treatment. *Journal of Cellular and Molecular Medicine, 23*(4), 2324–2332. https://doi.org/10.1111/jcmm.14170.

Feder, A., Parides, M. K., Murrough, J. W., Perez, A. M., Morgan, J. E., Saxena, S., et al. (2014). Efficacy of intravenous ketamine for treatment of chronic posttraumatic stress disorder: A randomized clinical trial. *JAMA Psychiatry, 71*(6), 681–688. https://doi.org/10.1001/jamapsychiatry.2014.62.

Feinberg, I., Campbell, I. G., Schoepp, D. D., & Anderson, K. (2002). The selective group mGlu2/3 receptor agonist LY379268 suppresses REM sleep and fast EEG in the rat. *Pharmacology Biochemistry and Behavior, 73*(2), 467–474. https://doi.org/10.1016/S0091-3057(02)00843-2.

Feinberg, I., Schoepp, D. D., Hsieh, K. C., Darchia, N., & Campbell, I. G. (2005). The metabotropic glutamate (mGLU)2/3 receptor antagonist LY341495 [2S-2-amino-2-(1S,2S-2-carboxycyclopropyl-1-yl)-3-(xanth-9-yl)propanoic acid] stimulates waking and fast electroencephalogram power and blocks the effects of the mGLU2/3 receptor agonist LY. *Journal of Pharmacology and Experimental Therapeutics, 312*(2), 826–833. https://doi.org/10.1124/jpet.104.076547.

Fell, M. J., Witkin, J. M., Falcone, J. F., Katner, J. S., Perry, K. W., Hart, J., et al. (2011). N-(4-((2-(trifluoromethyl)-3-hydroxy-4-(isobutyryl)phenoxy)methyl) benzyl)-1-methyl-1h-imidazole-4-carboxamide (THIIC), a novel metabotropic glutamate 2 potentiator with potential anxiolytic/antidepressant properties: In vivo profiling suggests a link bet. *Journal of Pharmacology and Experimental Therapeutics, 336*(1), 165–177. https://doi.org/10.1124/jpet.110.172957.

Felts, A. S., Rodriguez, A. L., Blobaum, A. L., Morrison, R. D., Bates, B. S., Thompson Gray, A., et al. (2017). Discovery of N-(5-fluoropyridin-2-yl)-6-methyl-4-(pyrimidin-5-yloxy)picolinamide (VU0424238): A novel negative allosteric modulator of metabotropic glutamate receptor subtype 5 selected for clinical evaluation. *Journal of Medicinal Chemistry, 60*(12), 5072–5085. https://doi.org/10.1021/acs.jmedchem.7b00410.

Fendt, M., Imobersteg, S., Peterlik, D., Chaperon, F., Mattes, C., Wittmann, C., et al. (2013). Differential roles of mGlu7 and mGlu8 in amygdala-dependent behavior and physiology. *Neuropharmacology, 72*, 215–223. https://doi.org/10.1016/j.neuropharm.2013.04.052.

Feng, Y. Z., Zhang, T., Rockhold, R. W., & Ho, I. K. (1995). Increased locus coeruleus glutamate levels are associated with naloxone-precipitated withdrawal from butorphanol in the rat. *Neurochemical Research, 20*(6), 745–751. https://doi.org/10.1007/BF01705544.

Ferrarelli, F., & Tononi, G. (2017). Reduced sleep spindle activity point to a TRN-MD thalamus-PFC circuit dysfunction in schizophrenia. *Schizophrenia Research, 180*, 36–43. https://doi.org/10.1016/j.schres.2016.05.023.

Feyissa, A. M., Woolverton, W. L., Miguel-Hidalgo, J. J., Wang, Z., Kyle, P. B., Hasler, G., et al. (2010). Elevated level of metabotropic glutamate receptor 2/3 in the prefrontal cortex in major depression. *Progress in Neuro-Psychopharmacology and Biological Psychiatry*, *34*(2), 279–283. https://doi.org/10.1016/j.pnpbp.2009.11.018.

Fisher, N. M., Gould, R. W., Gogliotti, R. G., McDonald, A. J., Badivuku, H., Chennareddy, S., et al. (2020). Phenotypic profiling of mGlu(7) knockout mice reveals new implications for neurodevelopmental disorders. *Genes, Brain, and Behavior, 19*(7), e12654. https://doi.org/10.1111/gbb.12654.

Ford, D. E., & Kamerow, D. B. (1989). Epidemiologic study of sleep disturbances and psychiatric disorders. An opportunity for prevention? *JAMA, 262*(11), 1479–1484. https://doi.org/10.1001/jama.262.11.1479.

Foster, F. G., Kupfer, D. J., Coble, P., & McPartland, R. J. (1976). Rapid eye movement sleep density. An objective indicator in severe medical-depressive syndromes. *Archives of General Psychiatry, 33*(9), 1119–1123. https://doi.org/10.1001/archpsyc.1976.01770090109011.

Frye, M. A., Tsai, G. E., Huggins, T., Coyle, J. T., & Post, R. M. (2007). Low cerebrospinal fluid glutamate and glycine in refractory affective disorder. *Biological Psychiatry, 61*(2), 162–166. https://doi.org/10.1016/j.biopsych.2006.01.024.

Fukumoto, K., Iijima, M., & Chaki, S. (2016). The antidepressant effects of an mGlu2/3 receptor antagonist and ketamine require AMPA receptor stimulation in the mPFC and subsequent activation of the 5-HT neurons in the DRN. *Neuropsychopharmacology, 41*(4), 1046–1056. https://doi.org/10.1038/npp.2015.233.

Galici, R., Jones, C. K., Hemstapat, K., Nong, Y., Echemendia, N. G., Williams, L. C., et al. (2006). Biphenyl-indanone A, a positive allosteric modulator of the metabotropic glutamate receptor subtype 2, has antipsychotic- and anxiolytic-like effects in mice. *Journal of Pharmacology and Experimental Therapeutics, 318*(1), 173–185. https://doi.org/10.1124/jpet.106.102046.

Gauthier, E. A., Guzick, S. E., Brummett, C. M., Baghdoyan, H. A., & Lydic, R. (2011). Buprenorphine disrupts sleep and decreases adenosine concentrations in sleep-regulating brain regions of sprague dawley rat. *Anesthesiology, 115*(4), 743–753. https://doi.org/10.1097/ALN.0b013e31822e9f85.

George, S. A., Rodriguez-Santiago, M., Riley, J., Abelson, J. L., Floresco, S. B., & Liberzon, I. (2018). D-Cycloserine facilitates reversal in an animal model of post-traumatic stress disorder. *Behavioural Brain Research, 347*, 332–338. https://doi.org/10.1016/j.bbr.2018.03.037.

Gerstenberg, M., Furrer, M., Tesler, N., Franscini, M., Walitza, S., & Huber, R. (2020). Reduced sleep spindle density in adolescent patients with early-onset schizophrenia compared to major depressive disorder and healthy controls. *Schizophrenia Research, 221*, 20–28. https://doi.org/10.1016/j.schres.2019.11.060.

Gilestro, G. F., Tononi, G., & Cirelli, C. (2009). Widespread changes in synaptic markers as a function of sleep and wakefulness in drosophila. *Science, 324*(5923), 109–112. https://doi.org/10.1126/science.1166673.

Gilmour, G., Broad, L. M., Wafford, K. A., Britton, T., Colvin, E. M., Fivush, A., et al. (2013). In vitro characterisation of the novel positive allosteric modulators of the mGlu5 receptor, LSN2463359 and LSN2814617, and their effects on sleep architecture and operant responding in the rat. *Neuropharmacology, 64*, 224–239. https://doi.org/10.1016/j.neuropharm.2012.07.030.

Girard, B., Tuduri, P., Moreno, M. P., Sakkaki, S., Barboux, C., Bouschet, T., et al. (2019). The mGlu7 receptor provides protective effects against epileptogenesis and epileptic seizures. *Neurobiology of Disease, 129*, 13–28. https://doi.org/10.1016/j.nbd.2019.04.016.

Glantz, L. A., & Lewis, D. A. (2000). Decreased dendritic spine density on prefrontal cortical pyramidal neurons in schizophrenia. *Archives of General Psychiatry, 57*(1), 65–73. https://doi.org/10.1001/archpsyc.57.1.65.

Gluck, M. R., Thomas, R. G., Davis, K. L., & Haroutunian, V. (2002). Implications for altered glutamate and GABA metabolism in the dorsolateral prefrontal cortex of aged schizophrenic patients. *The American Journal of Psychiatry*, *159*(7), 1165–1173. https://doi.org/10.1176/appi.ajp.159.7.1165.

Göder, R., Aldenhoff, J. B., Boigs, M., Braun, S., Koch, J., & Fritzer, G. (2006). Delta power in sleep in relation to neuropsychological performance in healthy subjects and schizophrenia patients. *The Journal of Neuropsychiatry and Clinical Neurosciences*, *18*(4), 529–535. https://doi.org/10.1176/jnp.2006.18.4.529.

Göder, R., Boigs, M., Braun, S., Friege, L., Fritzer, G., Aldenhoff, J. B., et al. (2004). Impairment of visuospatial memory is associated with decreased slow wave sleep in schizophrenia. *Journal of Psychiatric Research*, *38*(6), 591–599. https://doi.org/10.1016/j.jpsychires.2004.04.005.

Göder, R., Graf, A., Ballhausen, F., Weinhold, S., Baier, P. C., Junghanns, K., et al. (2015). Impairment of sleep-related memory consolidation in schizophrenia: Relevance of sleep spindles? *Sleep Medicine*, *16*(5), 564–569. https://doi.org/10.1016/j.sleep.2014.12.022.

Goldstein, R. Z., & Volkow, N. D. (2002). Drug addiction and its underlying neurobiological basis: Neuroimaging evidence for the involvement of the frontal cortex. *Am J Psychiatry*, *159*(10), 1642–1652. https://doi.org/10.1176/appi.ajp.159.10.1642.

Gould, R. W., Amato, R. J., Bubser, M., Joffe, M. E., Nedelcovych, M. T., Thompson, A. D., et al. (2016). Partial mGlu 5 negative allosteric modulators attenuate cocaine-mediated behaviors and lack psychotomimetic-like effects. *Neuropsychopharmacology*, *41*(4), 1166–1178. https://doi.org/10.1038/npp.2015.265.

Gould, R. W., Felts, A. S., & Jones, C. K. (2017). Negative allosteric modulators of the metabotropic glutamate receptor subtype 5 for the treatment of cocaine use disorder. In *The neuroscience of cocaine: Mechanisms and treatment*. https://doi.org/10.1016/B978-0-12-803750-8.00071-3.

Gould, R. W., Porrino, L. J., & Nader, M. A. (2012). Nonhuman primate models of addiction and PET imaging: Dopamine system dysregulation. *Current Topics in Behavioral Neurosciences*, *11*, 25–44. https://doi.org/10.1007/7854_2011_168.

Greco, B., Invernizzi, R. W., & Carli, M. (2005). Phencyclidine-induced impairment in attention and response control depends on the background genotype of mice: Reversal by the mGLU2/3 receptor agonist LY379268. *Psychopharmacology*, *179*(1), 68–76. https://doi.org/10.1007/s00213-004-2127-9.

Griebel, G., Pichat, P., Boulay, D., Naimoli, V., Potestio, L., Featherstone, R., et al. (2016). The mGluR2 positive allosteric modulator, SAR218645, improves memory and attention deficits in translational models of cognitive symptoms associated with schizophrenia. *Scientific Reports*, *6*, 1–23. https://doi.org/10.1038/srep35320.

Grueter, B. A., & Winder, D. G. (2005). Group II and III metabotropic glutamate receptors suppress excitatory synaptic transmission in the dorsolateral bed nucleus of the stria terminalis. *Neuropsychopharmacology*, *30*(7), 1302–1311. https://doi.org/10.1038/sj.npp.1300672.

Gupta, R., Ali, R., & Ray, R. (2018). Willis-Ekbom disease/restless legs syndrome in patients with opioid withdrawal. *Sleep Medicine*, *45*, 39–43. https://doi.org/10.1016/j.sleep.2017.09.028.

Hackler, E. A., Byun, N. E., Jones, C. K., Williams, J. M., Baheza, R., Sengupta, S., et al. (2010). Selective potentiation of the metabotropic glutamate receptor subtype 2 blocks phencyclidine-induced hyperlocomotion and brain activation. *Neuroscience*, *168*(1), 209–218. https://doi.org/10.1016/j.neuroscience.2010.02.057.

Hajasova, Z., Canestrelli, C., Acher, F., Noble, F., & Marie, N. (2018). Role of mGlu7 receptor in morphine rewarding effects is uncovered by a novel orthosteric agonist. *Neuropharmacology*, *131*, 424–430. https://doi.org/10.1016/j.neuropharm.2018.01.002.

Hamilton, A., Esseltine, J. L., Devries, R. A., Cregan, S. P., & Ferguson, S. S. G. (2014). Metabotropic glutamate receptor 5 knockout reduces cognitive impairment and pathogenesis in a mouse model of Alzheimer's disease. *Molecular Brain, 7*(1), 1–12. https://doi.org/10.1186/1756-6606-7-40.

Hamilton, A., Vasefi, M., vander Tuin, C., McQuaid, R. J., Anisman, H., & Ferguson, S. S. G. (2016). Chronic pharmacological mGluR5 inhibition prevents cognitive impairment and reduces pathogenesis in an Alzheimer disease mouse model. *Cell Reports, 15*(9), 1859–1865. https://doi.org/10.1016/j.celrep.2016.04.077.

Hansen, K. B., Wollmuth, L. P., Bowie, D., Furukawa, H., Menniti, F. S., Sobolevsky, A. I., et al. (2021). Structure, function, and pharmacology of glutamate receptor ion channels. *Pharmacological Reviews, 73*(4), 298–487. https://doi.org/10.1124/pharmrev.120.000131.

Hao, Y., Martin-Fardon, R., & Weiss, F. (2010). Behavioral and functional evidence of metabotropic glutamate receptor 2/3 and metabotropic glutamate receptor 5 dysregulation in cocaine-escalated rats: Factor in the transition to dependence. *Biological Psychiatry, 68*(3), 240–248. https://doi.org/10.1016/j.biopsych.2010.02.011.

Hartwell, E. E., Pfeifer, J. G., McCauley, J. L., Moran-Santa Maria, M., & Back, S. E. (2014). Sleep disturbances and pain among individuals with prescription opioid dependence. *Addictive Behaviors, 39*(10), 1537–1542. https://doi.org/10.1016/j.addbeh.2014.05.025.

Harvey, B. D., Siok, C. J., Kiss, T., Volfson, D., Grimwood, S., Shaffer, C. L., et al. (2013). Neurophysiological signals as potential translatable biomarkers for modulation of metabotropic glutamate 5 receptors. *Neuropharmacology, 75*, 19–30. https://doi.org/10.1016/j.neuropharm.2013.06.020.

Hashimoto, K., Sawa, A., & Iyo, M. (2007). Increased levels of glutamate in brains from patients with mood disorders. *Biological Psychiatry, 62*(11), 1310–1316. https://doi.org/10.1016/j.biopsych.2007.03.017.

Hasler, B. P., & Pedersen, S. L. (2020). Sleep and circadian risk factors for alcohol problems: A brief overview and proposed mechanisms. *Current Opinion in Psychology, 34*, 57–62. https://doi.org/10.1016/j.copsyc.2019.09.005.

Haynes, J., Talbert, M., Fox, S., & Close, E. (2018). Cognitive behavioral therapy in the treatment of insomnia. *Southern Medical Journal, 111*(2), 75–80. https://doi.org/10.14423/SMJ.0000000000000769.

Hefti, K., Holst, S. C., Sovago, J., Bachmann, V., Buck, A., Ametamey, S. M., et al. (2013). Increased metabotropic glutamate receptor subtype 5 availability in human brain after one night without sleep. *Biological Psychiatry, 73*(2), 161–168. https://doi.org/10.1016/j.biopsych.2012.07.030.

Hiyoshi, T., Hikichi, H., Karasawa, J. I., & Chaki, S. (2014). Metabotropic glutamate receptors regulate cortical gamma hyperactivities elicited by ketamine in rats. *Neuroscience Letters, 567*, 30–34. https://doi.org/10.1016/j.neulet.2014.03.025.

Hodges, S. E., Pittman, B., & Morgan, P. T. (2017). Sleep perception and misperception in chronic cocaine users during abstinence. *Sleep, 40*(3). https://doi.org/10.1093/sleep/zsw069.

Holmes, S. E., Girgenti, M. J., Davis, M. T., Pietrzak, R. H., Dellagioia, N., Nabulsi, N., et al. (2017). Altered metabotropic glutamate receptor 5 markers in PTSD: In vivo and postmortem evidence. *Proceedings of the National Academy of Sciences of the United States of America, 114*(31), 8390–8395. https://doi.org/10.1073/pnas.1701749114.

Holst, S. C., Sousek, A., Hefti, K., Saberi-Moghadam, S., Buck, A., Ametamey, S. M., et al. (2017). Cerebral mGluR5 availability contributes to elevated sleep need and behavioral adjustment after sleep deprivation. *ELife, 6*, 1–23. https://doi.org/10.7554/eLife.28751.001.

Holter, K. M., Lekander, A. D., LaValley, C. M., Bedingham, E. G., Pierce, B. E., Sands, L. P., et al. (2021). Partial mGlu5 negative allosteric modulator M-5MPEP demonstrates antidepressant-like effects on sleep without affecting cognition or quantitative EEG. *Frontiers in Neuroscience*, *15*, 700822. https://doi.org/10.3389/fnins.2021.700822.

Homayoun, H., Stefani, M. R., Adams, B. W., Tamagan, G. D., & Moghaddam, B. (2004). Functional interaction between NMDA and mGlu5 receptors: Effects on working memory, instrumental learning, motor behaviors, and dopamine release. *Neuropsychopharmacology*, *29*(7), 1259–1269. https://doi.org/10.1038/sj.npp.1300417.

Hong, J., Lee, J., Song, K., Ha, G. E., Yang, Y. R., Ma, J. S., et al. (2016). The thalamic mGluR1-PLCβ4 pathway is critical in sleep architecture. *Molecular Brain*, *9*(1), 1–12. https://doi.org/10.1186/s13041-016-0276-5.

Hopkins, C. R., Lindsley, C. W., & Niswender, C. M. (2009). mGluR4-positive allosteric modulation as potential treatment for Parkinson's disease. *Future Medicinal Chemistry*, *1*(3), 501–513. https://doi.org/10.4155/fmc.09.38.

Hor, H., & Tafti, M. (2009). How much sleep do we need? *Science*, *325*(5942), 825–826. https://doi.org/10.1126/science.1178713.

Hser, Y.-I., Mooney, L. J., Huang, D., Zhu, Y., Tomko, R. L., McClure, E., et al. (2017). Reductions in cannabis use are associated with improvements in anxiety, depression, and sleep quality, but not quality of life. *Journal of Substance Abuse Treatment*, *81*, 53–58. https://doi.org/10.1016/j.jsat.2017.07.012.

Hsu, W. Y., Chiu, N. Y., Liu, J. T., Wang, C. H., Chang, T. G., Liao, Y. C., et al. (2012). Sleep quality in heroin addicts under methadone maintenance treatment. *Acta Neuropsychiatrica*, *24*(6), 356–360. https://doi.org/10.1111/j.1601-5215.2011.00628.x.

Hu, W., MacDonald, M. L., Elswick, D. E., & Sweet, R. A. (2015). The glutamate hypothesis of schizophrenia: Evidence from human brain tissue studies. *Annals of the New York Academy of Sciences*, *1338*, 38–57. https://doi.org/10.1111/nyas.12547.

Hu, B., Steriade, M., & Deschênes, M. (1989). The cellular mechanism of thalamic pontogeniculo-occipital waves. *Neuroscience*, *31*(1), 25–35. https://doi.org/10.1016/0306-4522(89)90028-6.

Hughes, Z. A., Neal, S. J., Smith, D. L., Sukoff Rizzo, S. J., Pulicicchio, C. M., Lotarski, S., et al. (2013). Negative allosteric modulation of metabotropic glutamate receptor 5 results in broad spectrum activity relevant to treatment resistant depression. *Neuropharmacology*, *66*, 202–214. https://doi.org/10.1016/j.neuropharm.2012.04.007.

Huguenard, J. R., & McCormick, D. A. (2007). Thalamic synchrony and dynamic regulation of global forebrain oscillations. *Trends in Neurosciences*, *30*(7), 350–356. https://doi.org/10.1016/j.tins.2007.05.007.

Iber, C., Ancoli-Israel, S., & Chesson, A. (2007). *The AASM manual for the scoring of sleep and associated events: Rules, terminology, and technical specification* (1st ed.). American Academy of Sleep Medicine.

Ikeda, M., Hirono, M., Sugiyama, T., Moriya, T., Ikeda-Sagara, M., Eguchi, N., et al. (2009). Phospholipase C-β4 is essential for the progression of the normal sleep sequence and ultradian body temperature rhythms in mice. *PLoS One*, *4*(11). https://doi.org/10.1371/journal.pone.0007737.

Irwin, M. R., & Vitiello, M. V. (2019). Implications of sleep disturbance and inflammation for Alzheimer's disease dementia. *The Lancet. Neurology*, *18*(3), 296–306. https://doi.org/10.1016/S1474-4422(18)30450-2.

Iscru, E., Goddyn, H., Ahmed, T., Callaerts-Vegh, Z., D'Hooge, R., & Balschun, D. (2013). Improved spatial learning is associated with increased hippocampal but not prefrontal long-term potentiation in mGluR4 knockout mice. *Genes, Brain, and Behavior*, *12*(6), 615–625. https://doi.org/10.1111/gbb.12052.

Jack, C. R., Bennett, D. A., Blennow, K., Carrillo, M. C., Feldman, H. H., Frisoni, G. B., et al. (2016). A/T/N: An unbiased descriptive classification scheme for Alzheimer disease biomarkers. *Neurology*, *87*(5), 539–547. https://doi.org/10.1212/WNL.0000000000002923.

Jack, C. R., Knopman, D. S., Jagust, W. J., Shaw, L. M., Aisen, P. S., Weiner, M. W., et al. (2010). Hypothetical model of dynamic biomarkers of the Alzheimer's pathological cascade. *The Lancet. Neurology, 9*(1), 119–128. https://doi.org/10.1016/S1474-4422(09)70299-6.

Jacob, W., Gravius, A., Pietraszek, M., Nagel, J., Belozertseva, I., Shekunova, E., et al. (2009). The anxiolytic and analgesic properties of fenobam, a potent mGlu5 receptor antagonist, in relation to the impairment of learning. *Neuropharmacology, 57*(2), 97–108. https://doi.org/10.1016/J.NEUROPHARM.2009.04.011.

Jaehne, A., Loessl, B., Bárkai, Z., Riemann, D., & Hornyak, M. (2009). Effects of nicotine on sleep during consumption, withdrawal and replacement therapy. *Sleep Medicine Reviews, 13*(5), 363–377. https://doi.org/10.1016/j.smrv.2008.12.003.

Javitt, D. C. (2007). Glutamate and Schizophrenia: Phencyclidine, N-Methyl-d-aspartate receptors, and dopamine-glutamate interactions. *International Review of Neurobiology, 78*, 69–108. https://doi.org/10.1016/S0074-7742(06)78003-5.

Joffe, M. E., Centanni, S. W., Jaramillo, A. A., Winder, D. G., & Conn, P. J. (2018). Metabotropic glutamate receptors in alcohol use disorder: Physiology, plasticity, and promising pharmacotherapies. *ACS Chemical Neuroscience, 9*(9), 2188–2204. https://doi.org/10.1021/acschemneuro.8b00200.

Joffe, M. E., Santiago, C. I., Oliver, K. H., Maksymetz, J., Harris, N. A., Engers, J. L., et al. (2020). mGlu2 and mGlu3 negative allosteric modulators divergently enhance thalamocortical transmission and exert rapid antidepressant-like effects. *Neuron, 105*(1), 46–59.e3. https://doi.org/10.1016/j.neuron.2019.09.044.

Johanson, C. E., Roehrs, T., Schuh, K., & Warbasse, L. (1999). The effects of cocaine on mood and sleep in cocaine-dependent males. *Experimental and Clinical Psychopharmacology, 7*(4), 338–346. https://doi.org/10.1037/1064-1297.7.4.338.

Johnson, K. A., & Lovinger, D. M. (2020). Allosteric modulation of metabotropic glutamate receptors in alcohol use disorder: Insights from preclinical investigations. *Advances in Pharmacology, 88*, 193–232. https://doi.org/10.1016/bs.apha.2020.02.002.

Jones, B. E. (2020). Arousal and sleep circuits. *Neuropsychopharmacology, 45*(1), 6–20. https://doi.org/10.1038/s41386-019-0444-2.

Jones, C. K., Eberle, E. L., Peters, S. C., Monn, J. A., & Shannon, H. E. (2005). Analgesic effects of the selective group II (mGlu2/3) metabotropic glutamate receptor agonists LY379268 and LY389795 in persistent and inflammatory pain models after acute and repeated dosing. *Neuropharmacology, 49*(Suppl), 206–218. https://doi.org/10.1016/j.neuropharm.2005.05.008.

Jones, C., Watson, D., & Fone, K. (2011). Animal models of schizophrenia. *British Journal of Pharmacology, 164*(4), 1162–1194. https://doi.org/10.1111/j.1476-5381.2011.01386.x.

Ju, Y.-E. S., McLeland, J. S., Toedebusch, C. D., Xiong, C., Fagan, A. M., Duntley, S. P., et al. (2013). Sleep quality and preclinical Alzheimer disease. *JAMA Neurology, 70*(5), 587–593. https://doi.org/10.1001/jamaneurol.2013.2334.

Ju, Y.-E. S., Ooms, S. J., Sutphen, C., Macauley, S. L., Zangrilli, M. A., Jerome, G., et al. (2017). Slow wave sleep disruption increases cerebrospinal fluid amyloid-β levels. *Brain: A Journal of Neurology, 140*(8), 2104–2111. https://doi.org/10.1093/brain/awx148.

Kahvandi, N., Ebrahimi, Z., Karimi, S. A., Shahidi, S., Salehi, I., Naderishahab, M., et al. (2021). The effect of the mGlu8 receptor agonist, (S)-3,4-DCPG on acquisition and expression of morphine-induced conditioned place preference in male rats. *Behavioral and Brain Functions, 17*(1), 1–10. https://doi.org/10.1186/s12993-021-00174-0.

Kalinichev, M., Rouillier, M., Girard, F., Royer-Urios, I., Bournique, B., Finn, T., et al. (2013). ADX71743, a potent and selective negative allosteric modulator of metabotropic glutamate receptor 7: In vitro and in vivo characterization. *The Journal of Pharmacology and Experimental Therapeutics, 344*(3), 624–636. https://doi.org/10.1124/jpet.112.200915.

Kalus, P., Müller, T. J., Zuschratter, W., & Senitz, D. (2000). The dendritic architecture of prefrontal pyramidal neurons in schizophrenic patients. *Neuroreport, 11*(16), 3621–3625. https://doi.org/10.1097/00001756-200011090-00044.

Kammermeier, P. J., & Worley, P. F. (2007). Homer 1a uncouples metabotropic glutamate receptor 5 from postsynaptic effectors. *Proceedings of the National Academy of Sciences of the United States of America, 104*(14), 6055–6060. https://doi.org/10.1073/pnas.0608991104.

Kang, J.-E., Lim, M. M., Bateman, R. J., Lee, J. J., Smyth, L. P., Cirrito, J. R., et al. (2009). Amyloid-beta dynamics are regulated by orexin and the sleep-wake cycle. *Science (New York, N.Y.), 326*(5955), 1005–1007. https://doi.org/10.1126/science.1180962.

Kantrowitz, J., & Javitt, D. C. (2012). Glutamatergic transmission in Schizophrenia. *Current Opinion in Psychiatry, 25*(2), 96–102. https://doi.org/10.1097/YCO.0b013e32835035b2.

Kaskie, R. E., Graziano, B., & Ferrarelli, F. (2017). Schizophrenia and sleep disorders: Links, risks, and management challenges. *Nature and Science of Sleep, 9*, 227–239. https://doi.org/10.2147/NSS.S121076.

Kato, T., Takata, M., Kitaichi, M., Kassai, M., Inoue, M., Ishikawa, C., et al. (2015). DSR-98776, a novel selective mGlu5 receptor negative allosteric modulator with potent antidepressant and antimanic activity. *European Journal of Pharmacology, 757*, 11–20. https://doi.org/10.1016/j.ejphar.2015.03.024.

Kawaura, K., Karasawa, J. I., & Hikichi, H. (2016). Stimulation of the metabotropic glutamate (mGlu) 2 receptor attenuates the MK-801-induced increase in the immobility time in the forced swimming test in rats. *Pharmacological Reports, 68*(1), 80–84. https://doi.org/10.1016/j.pharep.2015.05.027.

Kay, D. C. (1975). Human sleep during chronic morphine intoxication. *Psychopharmacologia, 44*(2), 117–124. https://doi.org/10.1007/BF00420997.

Keck, T. M., Yang, H.-J., Bi, G. H., Huang, Y., Zhang, H. Y., Srivastava, R., et al. (2013). Fenobam sulfate inhibits cocaine-taking and cocaine-seeking behavior in rats: Implications for addiction treatment in humans. *NIH Public Access, 229*(2), 253–265. https://doi.org/10.1007/s00213-013-3106-9.Fenobam.

Kékesi, K. A., Dobolyi, Á., Salfay, O., Nyitrai, G., & Juhász, G. (1997). Slow wave sleep is accompanied by release of certain amino acids in the thalamus of cats. *NeuroReport, 8*(5), 1183–1186. https://doi.org/10.1097/00001756-199703240-00025.

Keshavan, M. S., Miewald, J., Haas, G., Sweeney, J., Ganguli, R., & Reynolds, C. F. (1995). Slow-wave sleep and symptomatology in schizophrenia and related psychotic disorders. *Journal of Psychiatric Research.* https://doi.org/10.1016/0022-3956(95)00023-X.

Keshavan, M. S., Reynolds, C. F., Miewald, J. M., Montrose, D. M., Sweeney, J. A., Vasko, R. C., et al. (1998). Delta sleep deficits in Schizophrenia. *Archives of General Psychiatry, 55*(5), 443. https://doi.org/10.1001/archpsyc.55.5.443.

Kessler, R. C., Wai, T. C., Demler, O., & Walters, E. E. (2005). Prevalence, severity, and comorbidity of 12-month DSM-IV disorders in the National Comorbidity Survey replication. *Archives of General Psychiatry, 62*(6), 617–627. https://doi.org/10.1001/archpsyc.62.6.617.

Khazaie, H., & Masoudi, M. (2016). Sleep disturbances in veterans with chronic war-induced PTSD. *Journal of Injury and Violence Research, 8*(2), 99–107. https://doi.org/10.5249/jivr.v8i2.808.

Khazan, N., & Colasanti, B. (1972). Protracted rebound in rapid movement sleep time and electroencephalogram voltage output in morphine-dependent rats upon withdrawal. *Journal of Pharmacology and Experimental Therapeutics, 183*(1), 23–30.

Khazan, K., Weeks, J. R., & Schroeder, L. A. (1967). Electroencephalographic, electromyographic and behavioral correlates during a cycle of self-maintained morphine addiction in the rat. *The Journal of Pharmacology and Experimental Therapeutics, 155*(3), 521–531.

Killgore, W. D. S. (2013). Self-reported sleep correlates with prefrontal-amygdala functional connectivity and emotional functioning. *Sleep*, *36*(11), 1597–1608. https://doi.org/10.5665/sleep.3106.

Kilpatrick, D. G., Resnick, H. S., Milanak, M. E., Miller, M. W., Keyes, K. M., & Friedman, M. J. (2013). Prevalence Using DSM-IV and DSM-5 criteria. *J Trauma Stress*, *26*(5), 537–547. https://doi.org/10.1002/jts.21848.

Kim, J. H., Ahn, J. H., Min, C. Y., Yoo, D. M., & Choi, H. G. (2021). Association between sleep quality and subjective cognitive decline: Evidence from a community health survey. *Sleep Medicine*, *83*, 123–131. https://doi.org/10.1016/j.sleep.2021.04.031.

Kim, J., An, B., Kim, J., Park, S., Park, S., Hong, I., et al. (2015). MGluR2/3 in the lateral amygdala is required for fear extinction: Cortical input synapses onto the lateral amygdala as a target site of the mGluR2/3 action. *Neuropsychopharmacology*, *40*(13), 2916–2928. https://doi.org/10.1038/npp.2015.145.

Kim, S. H., Fraser, P. E., Westaway, D., St George-Hyslop, P. H., Ehrlich, M. E., & Gandy, S. (2010). Group II metabotropic glutamate receptor stimulation triggers production and release of Alzheimer's amyloid(beta)42 from isolated intact nerve terminals. *The Journal of Neuroscience*, *30*(11), 3870–3875. https://doi.org/10.1523/JNEUROSCI.4717-09.2010.

Kim, J. S., Schmid-Burgk, W., Claus, D., & Kornhuber, H. H. (1982). Increased serum glutamate in depressed patients. *Archiv Für Psychiatrie Und Nervenkrankheiten*, *232*(4), 299–304. https://doi.org/10.1007/BF00345492.

Kinney, G. G., Burno, M., Campbell, U. C., Hernandez, L. M., Rodriguez, D., Bristow, L. J., et al. (2003). Metabotropic glutamate subtype 5 receptors modulate locomotor activity and sensorimotor gating in rodents. *J Pharmacol Exp Ther*, *306*(1), 116–123. https://doi.org/10.1124/jpet.103.048702.

Klar, R., Walker, A. G., Ghose, D., Grueter, B. A., Engers, D. W., Hopkins, C. R., et al. (2015). Activation of metabotropic glutamate receptor 7 is required for induction of long-term potentiation at SC-CA1 synapses in the hippocampus. *The Journal of Neuroscience: The Official Journal of the Society for Neuroscience*, *35*(19), 7600–7615. https://doi.org/10.1523/JNEUROSCI.4543-14.2015.

Knackstedt, L. A., Trantham-Davidson, H. L., & Schwendt, M. (2014). The role of ventral and dorsal striatum mGluR5 in relapse to cocaine-seeking and extinction learning. *Addiction Biology*, *19*(1), 87–101. https://doi.org/10.1111/adb.12061.

Kobayashi, I., Boarts, J. M., & Delahanty, D. L. (2007). Polysomnographically measured sleep abnormalities in PTSD: A meta-analytic review. *Psychophysiology*, *44*(4), 660–669. https://doi.org/10.1111/j.1469-8986.2007.537.x.

Kobayashi, I., Huntley, E., Lavela, J., & Mellman, T. A. (2012). Subjectively and objectively measured sleep with and without posttraumatic stress disorder and trauma exposure. *Sleep*, *35*(7), 957–965. https://doi.org/10.5665/sleep.1960.

Koob, G. F., & Colrain, I. M. (2020). Alcohol use disorder and sleep disturbances: A feed-forward allostatic framework. *Neuropsychopharmacology*. https://doi.org/10.1038/s41386-019-0446-0.

Korenic, S. A., Klingaman, E. A., Wickwire, E. M., Gaston, F. E., Chen, H., Wijtenburg, S. A., et al. (2020). Sleep quality is related to brain glutamate and symptom severity in schizophrenia. *Journal of Psychiatric Research*, *120*(October 2019), 14–20. https://doi.org/10.1016/j.jpsychires.2019.10.006.

Korpi, E. R., den Hollander, B., Farooq, U., Vashchinkina, E., Rajkumar, R., Nutt, D. J., et al. (2015). Mechanisms of action and persistent neuroplasticity by drugs of abuse. *Pharmacological Reviews*, *67*(4), 872–1004. https://doi.org/10.1124/pr.115.010967.

Krakow, B. J., Ulibarri, V. A., Moore, B. A., & McIver, N. D. (2015). Posttraumatic stress disorder and sleep-disordered breathing: A review of comorbidity research. *Sleep Medicine Reviews*, *24*, 37–45. https://doi.org/10.1016/j.smrv.2014.11.001.

Krystal, J. H., Abi-Saab, W., Perry, E., D'Souza, D. C., Liu, N., Gueorguieva, R., et al. (2005). Preliminary evidence of attenuation of the disruptive effects of the NMDA glutamate receptor antagonist, ketamine, on working memory by pretreatment with the group II metabotropic glutamate receptor agonist, LY354740, in healthy human subjects. *Psychopharmacology*, *179*(1), 303–309. https://doi.org/10.1007/s00213-004-1982-8.

Krystal, J. H., Davis, L. L., Neylan, T. C., Raskind, M. A., Schnurr, P. P., Stein, M. B., et al. (2017). It is time to address the crisis in the pharmacotherapy of posttraumatic stress disorder: A consensus statement of the PTSD psychopharmacology working group. *Biological Psychiatry*, *82*(7), e51–e59. https://doi.org/10.1016/j.biopsych.2017.03.007.

Kumar, A., Dhull, D. K., & Mishra, P. S. (2015). Therapeutic potential of mGluR5 targeting in Alzheimer's disease. *Frontiers in Neuroscience*, *9*, 1–9. https://doi.org/10.3389/fnins.2015.00215.

Kumaresan, V., Yuan, M., Yee, J., Famous, K. R., Anderson, S. M., Schmidt, H. D., et al. (2009). Metabotropic glutamate receptor 5 (mGluR5) antagonists attenuate cocaine priming- and cue-induced reinstatement of cocaine seeking. *Behavioural Brain Research*. https://doi.org/10.1016/j.bbr.2009.03.039.

Kuo, C. Y., Hsiao, H. T., Lo, I. H., & Nikolai, T. (2021). Association between obstructive sleep apnea, its treatment, and Alzheimer's disease: Systematic mini-review. *Frontiers in Aging Neuroscience*, *12*, 1–6. https://doi.org/10.3389/fnagi.2020.591737.

Kupfer, D. J., Ehlers, C. L., Pollock, B. G., Swami Nathan, R., & Perel, J. M. (1989). Clomipramine and EEG sleep in depression. *Psychiatry Research*, *30*(2), 165–180. https://doi.org/10.1016/0165-1781(89)90158-3.

Kupfer, D. J., Reynolds, C. F., Ulrich, R. F., & Grochocinski, V. J. (1986). Comparison of automated REM and slow-wave sleep analysis in young and middle-aged depressed subjects. *Biological Psychiatry*, *21*(2), 189–200. https://doi.org/10.1016/0006-3223(86)90146-0.

Lan, M. J., McLoughlin, G. A., Griffin, J. L., Tsang, T. M., Huang, J. T. J., Yuan, P., et al. (2009). Metabonomic analysis identifies molecular changes associated with the pathophysiology and drug treatment of bipolar disorder. *Molecular Psychiatry*, *14*(3), 269–279. https://doi.org/10.1038/sj.mp.4002130.

Laniepce, A., Cabe, N., Andre, C., Bertran, F., Boudehent, C., Lahbairi, N., et al. (2020). The effect of alcohol withdrawal syndrome severity on sleep, brain and cognition. *Brain Communications*, *2*(2). https://doi.org/10.1093/braincomms/fcaa123.

Laurén, J., Gimbel, D. A., Nygaard, H. B., Gilbert, J. W., & Strittmatter, S. M. (2009). Cellular prion protein mediates impairment of synaptic plasticity by amyloid-B oligomers. *Nature*, *457*(7233), 1128–1132. https://doi.org/10.1038/nature07761.

Lauterbach, D., Behnke, C., & McSweeney, L. B. (2011). Sleep problems among persons with a lifetime history of posttraumatic stress disorder alone and in combination with a lifetime history of other psychiatric disorders: A replication and extension. *Comprehensive Psychiatry*, *52*(6), 580–586. https://doi.org/10.1016/j.comppsych.2011.01.007.

Lavreysen, H., Langlois, X., Ahnaou, A., Drinkenburg, W., te Riele, P., Biesmans, I., et al. (2013). Pharmacological characterization of JNJ-40068782, a new potent, selective, and systemically active positive allosteric modulator of the mGlu2 receptor and its radioligand [3H]JNJ-40068782. *The Journal of Pharmacology and Experimental Therapeutics*, *346*(3), 514–527. https://doi.org/10.1124/jpet.113.204990.

Lavreysen, H., Langlois, X., Donck, L. V., Nuñez, J. M., Pype, S., Lütjens, R., & Megens, A. (2015). Preclinical evaluation of the antipsychotic potential of the mGlu2-positive allosteric modulator JNJ-40411813. *Pharmacology Research & Perspectives*, *3*(2), e00097. https://doi.org/10.1002/prp2.97.

le Bon, O., Lanquart, J.-P., Hein, M., & Loas, G. (2019). Sleep ultradian cycling: Statistical distribution and links with other sleep variables, depression, insomnia and sleepiness—A retrospective study on 2,312 polysomnograms. *Psychiatry Research*, *279*, 140–147. https://doi.org/10.1016/j.psychres.2018.12.141.

le Bon, O., Verbanck, P., Hoffmann, G., Murphy, J. R., Staner, L., de Groote, D., et al. (1997). Sleep in detoxified alcoholics: Impairment of most standard sleep parameters and increased risk for sleep apnea, but not for myoclonias—A controlled study. *Journal of Studies on Alcohol, 58*(1), 30–36. https://doi.org/10.15288/jsa.1997. 58.30.

Lee, H., Ogawa, O., Zhu, X., O'Neill, M. J., Petersen, R. B., Castellani, R. J., et al. (2004). Aberrant expression of metabotropic glutamate receptor 2 in the vulnerable neurons of Alzheimer's disease. *Acta Neuropathologica, 107*(4), 365–371. https://doi.org/10.1007/s00401-004-0820-8.

Lee, B., Platt, D. M., Rowlett, J. K., Adewale, A. S., & Spealman, R. D. (2005). Attenuation of behavioral effects of cocaine by the metabotropic glutamate receptor 5 antagonist 2-methyl-6-(phenylethynyl)-pyridine in squirrel monkeys: Comparison with dizocilpine. *Journal of Pharmacology and Experimental Therapeutics, 312*(3), 1232–1240. https://doi.org/10.1124/jpet.104.078733.

Leeman-markowski, B. A., Meador, K. J., Moo, L. R., Cole, A. J., Hoch, D. B., Garcia, E., et al. (2020). *U.S. Department of Veterans Affairs, 40*(March 2005), 315–324. https://doi.org/10.1016/j.yebeh.2018.06.047.Does.

Levine, J., Panchalingam, K., Rapoport, A., Gershon, S., McClure, R. J., & Pettegrew, J. W. (2000). Increased cerebrospinal fluid glutamine levels in depressed patients. *Biological Psychiatry, 47*(7), 586–593. https://doi.org/10.1016/s0006-3223(99)00284-x.

Lewis, C., Lewis, K., Kitchiner, N., Isaac, S., Jones, I., & Bisson, J. I. (2020). Sleep disturbance in post-traumatic stress disorder (PTSD): A systematic review and meta-analysis of actigraphy studies. *European Journal of Psychotraumatology, 11*(1). https://doi.org/10.1080/20008198.2020.1767349.

Li, P., Gao, L., Yu, L., Zheng, X., Ulsa, M. C., Yang, H. W., et al. (2022). Daytime napping and Alzheimer's dementia: A potential bidirectional relationship. *Alzheimer's and Dementia, 2021*, 1–11. https://doi.org/10.1002/alz.12636.

Li, X., Xi, Z.-X., & Markou, A. (2013). Metabotropic glutamate 7 (mGlu7) receptor: A target for medication development for the treatment of cocaine dependence. *Neuropharmacology, 66*, 12–23. https://doi.org/10.1016/j.neuropharm.2012.04.010.

Lim, A. S. P., Kowgier, M., Yu, L., Buchman, A. S., & Bennett, D. A. (2013). Sleep fragmentation and the risk of incident Alzheimer's disease and cognitive decline in older persons. *Sleep, 36*(7), 1027–1032. https://doi.org/10.5665/sleep.2802.

Lin, X., Fisher, N. M., Dogra, S., Senter, R. K., Reed, C. W., Kalbfleisch, J. J., et al. (2022). Differential activity of mGlu7 allosteric modulators provides evidence for mGlu7/8 heterodimers at hippocampal Schaffer collateral-CA1 synapses. *The Journal of Biological Chemistry, 298*(10), 102458. https://doi.org/10.1016/j.jbc.2022.102458.

Lindemann, L., Porter, R. H., Scharf, S. H., Kuennecke, B., Bruns, A., von Kienlin, M., et al. (2015). Pharmacology of basimglurant (RO4917523, RG7090), a unique metabotropic glutamate receptor 5 negative allosteric modulator in clinical development for depression. *The Journal of Pharmacology and Experimental Therapeutics, 353*(1), 213–233. https://doi.org/10.1124/jpet.114.222463.

Linden, A. M., Johnson, B. G., Peters, S. C., Shannon, H. E., Tian, M., Wang, Y., et al. (2002). Increased anxiety-related behavior in mice deficient for metabotropic glutamate 8 (mGlu8) receptor. *Neuropharmacology, 43*(2), 251–259. https://doi.org/10.1016/S0028-3908(02)00079-5.

Linden, A. M., Shannon, H., Baez, M., Yu, J. L., Koester, A., & Schoepp, D. D. (2005). Anxiolytic-like activity of the mGLU2/3 receptor agonist LY354740 in the elevated plus maze test is disrupted in metabotropic glutamate receptor 2 and 3 knock-out mice. *Psychopharmacology, 179*(1), 284–291. https://doi.org/10.1007/s00213-004-2098-x.

Liriano, F., Hatten, C., & Schwartz, T. L. (2019). Ketamine as treatment for post-traumatic stress disorder: A review. *Drugs in Context, 8*, 1–7. https://doi.org/10.7573/dic.212305.

Littlewood, D. L., Gooding, P. A., Panagioti, M., & Kyle, S. D. (2016). Nightmares and suicide in posttraumatic stress disorder: The mediating role of defeat, entrapment, and hopelessness. *Journal of Clinical Sleep Medicine*, *12*(3), 393–399. https://doi.org/10.5664/jcsm.5592.

Livingstone, M. S., & Hubel, D. H. (1981). Effects of sleep and arousal on the processing of visual information in the cat. *Nature*, *291*(5816), 554–561. https://doi.org/10.1038/291554a0.

Llinás, R. R., & Steriade, M. (2006). Bursting of thalamic neurons and states of vigilance. *Journal of Neurophysiology*, *95*(6), 3297–3308. https://doi.org/10.1152/jn.00166.2006.

Logan, C. N., Bechard, A. R., Hamor, P. U., Wu, L., Schwendt, M., & Knackstedt, L. A. (2020). *Ceftriaxone and mGlu2/3 interactions in the nucleus accumbens core affect the reinstatement of cocaine-seeking in male and female rats*. https://doi.org/10.1007/s00213-020-05514-y/Published.

Logan, R. W., & McClung, C. A. (2019). Rhythms of life: Circadian disruption and brain disorders across the lifespan. *Nature Reviews. Neuroscience*, *20*(1), 49–65. https://doi.org/10.1038/s41583-018-0088-y.

Lopez, J., Hoffmann, R., Armitage, R., & Are, A. (2010). Reduced sleep spindle activity in early-onset and elevated risk for depression running head: Sleep spindles and adolescent MDD. *Journal of the American Academy of Child and Adolescent Psychiatry*, *49*(9), 934–943. https://doi.org/10.1016/j.jaac.2010.05.014.Reduced.

Lopez-Rodriguez, F., Medina-Ceja, L., Wilson, C. L., Jhung, D., & Morales-Villagran, A. (2007). Changes in extracellular glutamate levels in rat orbitofrontal cortex during sleep and wakefulness. *Archives of Medical Research*, *38*(1), 52–55. https://doi.org/10.1016/j.arcmed.2006.07.004.

Lorrain, D. S., Baccei, C. S., Bristow, L. J., Anderson, J. J., & Varney, M. A. (2003). Effects of ketamine and N-methyl-D-aspartate on glutamate and dopamine release in the rat prefrontal cortex: modulation by a group II selective metabotropic glutamate receptor agonist LY379268. *Neuroscience*, *117*(3), 697–706. https://doi.org/10.1016/s0306-4522(02)00652-8.

Lou, Z., Chen, L., Liu, H., Ruan, L., & Zhou, W. (2014). Blockade of mGluR5 in the nucleus accumbens shell but not core attenuates heroin seeking behavior in rats. *Acta Pharmacologica Sinica*, *35*(12), 1485–1492. https://doi.org/10.1038/aps.2014.93.

Lu, Y. M., Jia, Z., Janus, C., Henderson, J. T., Gerlai, R., Wojtowicz, J. M., et al. (1997). Mice lacking metabotropic glutamate receptor 5 show impaired learning and reduced CA1 long-term potentiation (LTP) but normal CA3 LTP. *Journal of Neuroscience*, *17*(13), 5196–5205. https://doi.org/10.1523/jneurosci.17-13-05196.1997.

Luby, E. D., Cohen, B. D., Rosenbaum, G., Gottlieb, J. S., & Kelley, R. (1959). Study of a new schizophrenomimetic drug; sernyl. *AMA Archives of Neurology and Psychiatry*, *81*(3), 363–369. https://doi.org/10.1001/archneurpsyc.1959.02340150095011.

Lucey, B. P., Mccullough, A., Landsness, E. C., Toedebusch, C. D., Mcleland, J. S., Zaza, A. M., et al. (2019). Reduced non-rapid eye movement sleep is associated with tau pathology in early Alzheimer's disease. *Science Translational Medicine*, *11*-(474). https://doi.org/10.1126/scitranslmed.aau6550.Reduced.

Luessen, D. J., & Conn, P. J. (2022). Allosteric modulators of metabotropic glutamate receptors as novel therapeutics for neuropsychiatric disease. *Pharmacological Reviews*, *74*(3), 630–661. https://doi.org/10.1124/pharmrev.121.000540.

Lüscher, C., & Huber, K. M. (2010). Group 1 mGluR-dependent synaptic long-term depression: Mechanisms and implications for circuitry and disease. *Neuron*, *65*(4), 445–459. https://doi.org/10.1016/j.neuron.2010.01.016.

Lutzu, S., & Castillo, P. E. (2021). Modulation of NMDA receptors by G-protein-coupled receptors: Role in synaptic transmission, plasticity and beyond. *Neuroscience*, *456*, 27–42. https://doi.org/10.1016/j.neuroscience.2020.02.019.

Luykx, J. J., Laban, K. G., van den Heuvel, M. P., Boks, M. P. M., Mandl, R. C. W., Kahn, R. S., et al. (2012). Region and state specific glutamate downregulation in major depressive disorder: A meta-analysis of 1H-MRS findings. *Neuroscience and Biobehavioral Reviews*, *36*(1), 198–205. https://doi.org/10.1016/j.neubiorev.2011.05.014.

Ly, S., Lee, D. A., Strus, E., Prober, D. A., & Naidoo, N. (2020). Evolutionarily conserved regulation of sleep by the protein translational regulator PERK. *Current Biology : CB*, *30*(9), 1639–1648.e3. https://doi.org/10.1016/j.cub.2020.02.030.

Lyashenko, E. A., Poluektov, M. G., Levin, O. S., & Pchelina, P. V. (2016). Age-related sleep changes and its implication in neurodegenerative diseases. *Current Aging Science*, *9*(1), 26–33. https://doi.org/10.2174/1874609809666151130220219.

Lydon-Staley, D. M., Harrington Cleveland, H., Huhn, A. S., Cleveland, M. J., Harris, J., Stankoski, D., et al. (2017). Daily sleep quality affects drug craving, partially through indirect associations with positive affect, in patients in treatment for nonmedical use of prescription drugs. *Paper Knowledge Toward a Media History of Documents*, *65*, 275–282. https://doi.org/10.1016/j.addbeh.2016.08.026.

Lyon, L., Burnet, P. W., Kew, J. N., Corti, C., Rawlins, J. N. P., Lane, T., et al. (2011). Fractionation of spatial memory in GRM2/3 (mGlu2/mGlu3) double knockout mice reveals a role for group II metabotropic glutamate receptors at the interface between arousal and cognition. *Neuropsychopharmacology*, *36*(13), 2616–2628. https://doi.org/10.1038/npp.2011.145.

Maglione, J. E., Ancoli-Israel, S., Peters, K. W., Paudel, M. L., Yaffe, K., Ensrud, K. E., et al. (2014). Subjective and objective sleep disturbance and longitudinal risk of depression in a cohort of older women. *Sleep*, *37*(7), 1179–1187. https://doi.org/10.5665/sleep.3834.

Maher, M. J., Rego, S. A., & Asnis, G. M. (2006). Sleep disturbances in patients with post-traumatic stress disorder: Epidemiology, impact and approaches to management. *CNS Drugs*, *20*(7), 567–590. https://doi.org/10.2165/00023210-200620070-00003.

Maksymetz, J., Moran, S. P., & Conn, P. J. (2017). Targeting metabotropic glutamate receptors for novel treatments of schizophrenia. *Molecular Brain*, *10*(1), 1–19. https://doi.org/10.1186/s13041-017-0293-z.

Manoach, D. S., Cain, M. S., Vangel, M. G., Khurana, A., Goff, D. C., & Stickgold, R. (2004). A failure of sleep-dependent procedural learning in chronic, medicated schizophrenia. *Biological psychiatry*, *56*(12), 951–956. https://doi.org/10.1016/j.biopsych.2004.09.012.

Manoach, D. S., & Stickgold, R. (2009). Does abnormal sleep impair memory consolidation in Schizophrenia? *Frontiers in Human Neuroscience*, *3*, 21. https://doi.org/10.3389/neuro.09.021.2009.

Manoach, D. S., & Stickgold, R. (2019). Abnormal sleep spindles, memory consolidation, and Schizophrenia. *Annual Review of Clinical Psychology*, *15*, 451–479. https://doi.org/10.1146/annurev-clinpsy-050718-095754.

Manoach, D. S., Thakkar, K. N., Stroynowski, E., Ely, A., McKinley, S. K., Wamsley, E., et al. (2010). Reduced overnight consolidation of procedural learning in chronic medicated schizophrenia is related to specific sleep stages. *Journal of Psychiatric Research*, *44*(2), 112–120. https://doi.org/10.1016/j.jpsychires.2009.06.011.

Marabese, I., Rossi, F., Palazzo, E., De Novellis, V., Starowicz, K., Cristino, L., et al. (2007). Periaqueductal gray metabotropic glutamate receptor subtype 7 and 8 mediate opposite effects on amino acid release, rostral ventromedial medulla cell activities, and thermal nociception. *Journal of Neurophysiology*, *98*(1), 43–53. https://doi.org/10.1152/jn.00356.2007.

Maret, S., Dorsaz, S., Gurcel, L., Pradervand, S., Petit, B., Pfister, C., et al. (2007). Homer1a is a core brain molecular correlate of sleep loss. *Proceedings of the National Academy of Sciences of the United States of America*, *104*(50), 20090–20095. https://doi.org/10.1073/pnas.0710131104.

Marrosu, F., Portas, C., Mascia, M. S., Casu, M. A., Fà, M., Giagheddu, M., et al. (1995). Microdialysis measurement of cortical and hippocampal acetylcholine release during sleep-wake cycle in freely moving cats. *Brain Research, 671*(2), 329–332. https://doi.org/10.1016/0006-8993(94)01399-3.

Marshall, R. D., Beebe, K. L., Oldham, M., & Zaninelli, R. (2001). Efficacy and safety of paroxetine treatment for chronic PTSD: A fixed-dose, placebo-controlled study. *American Journal of Psychiatry, 158*(12), 1982–1988. https://doi.org/10.1176/appi.ajp.158.12.1982.

Martin, S. C., Monroe, S. K., & Diering, G. H. (2019). Homer1a and mGluR1/5 signaling in homeostatic sleep drive and output. *Yale Journal of Biology and Medicine, 92*(1), 93–101.

Mateo, Y., Lack, C. M., Morgan, D., Roberts, D. C. S., & Jones, S. R. (2005). Reduced dopamine terminal function and insensitivity to cocaine following cocaine binge self-administration and deprivation. *Neuropsychopharmacology, 30*(8), 1455–1463. https://doi.org/10.1038/sj.npp.1300687.

Matosin, N., Fernandez-Enright, F., Lum, J. S., Andrews, J. L., Engel, M., Huang, X.-F., et al. (2015). Metabotropic glutamate receptor 5, and its trafficking molecules Norbin and Tamalin, are increased in the CA1 hippocampal region of subjects with schizophrenia. *Schizophrenia Research, 166*(1–3), 212–218. https://doi.org/10.1016/j.schres.2015.05.001.

Matosin, N., Newell, K. A., Quidé, Y., Andrews, J. L., Teroganova, N., Green, M. J., et al. (2018). Effects of common GRM5 genetic variants on cognition, hippocampal volume and mGluR5 protein levels in schizophrenia. *Brain Imaging and Behavior, 12*(2), 509–517. https://doi.org/10.1007/s11682-017-9712-0.

Matrisciano, F., Panaccione, I., Zusso, M., Giusti, P., Tatarelli, R., Iacovelli, L., et al. (2007). Group-II metabotropic glutamate receptor ligands as adjunctive drugs in the treatment of depression: A new strategy to shorten the latency of antidepressant medication? *Molecular Psychiatry, 12*(8), 704–706. https://doi.org/10.1038/sj.mp.4002005.

Matsunaga, S., Kishi, T., Nomura, I., Sakuma, K., Okuya, M., Ikuta, T., et al. (2018). The efficacy and safety of memantine for the treatment of Alzheimer's disease. *Expert Opinion on Drug Safety, 17*(10), 1053–1061. https://doi.org/10.1080/14740338.2018.1524870.

Matuskey, D., Pittman, B., Forselius, E., Malison, R. T., & Morgan, P. T. (2011). A multistudy analysis of the effects of early cocaine abstinence on sleep. *Drug and Alcohol Dependence, 115*(1–2), 62–66. https://doi.org/10.1016/j.drugalcdep.2010.10.015.

Mauri, M. C., Ferrara, A., Boscati, L., Bravin, S., Zamberlan, F., Alecci, M., et al. (1998). Plasma and platelet amino acid concentrations in patients affected by major depression and under fluvoxamine treatment. *Neuropsychobiology, 37*(3), 124–129. https://doi.org/10.1159/000026491.

Mazzitelli, M., Palazzo, E., Maione, S., & Neugebauer, V. (2018). Group II metabotropic glutamate receptors: Role in pain mechanisms and pain modulation. *Frontiers in Molecular Neuroscience, 11*, 1–11. https://doi.org/10.3389/fnmol.2018.00383.

Mccormick, D. A., & Bal, T. (1997). Sleep and arousal: Thalamocortical mechanisms. *Annual Review of Neuroscience, 20*, 185–215. https://doi.org/10.1146/annurev.neuro.20.1.185.

McCutcheon, J. E., Loweth, J. A., Ford, K. A., Marinelli, M., Wolf, M. E., & Tseng, K. Y. (2011). Group I mGlur activation reverses cocaine-induced accumulation of calcium-permeable AMPA receptors in nucleus accumbens synapses via a protein kinase C-dependent mechanism. *Journal of Neuroscience, 31*(41), 14536–14541. https://doi.org/10.1523/JNEUROSCI.3625-11.2011.

McGeehan, A. J., & Olive, M. F. (2003). The mGluR5 antagonist MPEP reduces the conditioned rewarding effects of cocaine but not other drugs of abuse. *Synapse, 47*(3), 240–242. https://doi.org/10.1002/syn.10166.

McGhee, L. L., Maani, C. V., Garza, T. H., Gaylord, K. M., & Black, I. H. (2008). The correlation between ketamine and posttraumatic stress disorder in burned service members. *The Journal of Trauma, 64*(2 Suppl). https://doi.org/10.1097/ta.0b013e318160ba1d.

Mcomish, C. E., Pavey, G., Gibbons, A., Hopper, S., Udawela, M., Scarr, E., et al. (2016). Lower [3H]LY341495 binding to mGlu2/3 receptors in the anterior cingulate of subjects with major depressive disorder but not bipolar disorder or Schizophrenia. *Journal of Affective Disorders*, *190*, 241–248. https://doi.org/10.1016/j.jad.2015.10.004.

Medrano, M. C., Mendiguren, A., & Pineda, J. (2015). Effect of ceftriaxone and topiramate treatments on naltrexone-precipitated morphine withdrawal and glutamate receptor desensitization in the rat locus coeruleus. *Psychopharmacology*, *232*(15), 2795–2809. https://doi.org/10.1007/s00213-015-3913-2.

Mellman, T. A., Bustamante, V., Fins, A. I., Pigeon, W. R., & Nolan, B. (2002). REM sleep and the early development of posttraumatic stress disorder. *American Journal of Psychiatry*, *159*(10), 1696–1701. https://doi.org/10.1176/appi.ajp.159.10.1696.

Mellman, T. A., Kobayashi, I., Lavela, J., Wilson, B., & Hall Brown, T. S. (2014). A relationship between REM sleep measures and the duration of posttraumatic stress disorder in a young adult urban minority population. *Sleep*, *37*(8), 1321–1326. https://doi.org/10.5665/sleep.3922.

Mellman, T. A., Pigeon, W. R., Nowell, P. D., & Nolan, B. (2007). Relationships between REM sleep findings and PTSD symptoms during the early aftermath of trauma. *Journal of Traumatic Stress*, *20*(5), 893–901. https://doi.org/10.1002/jts.20246.

Meyerhoff, D. J., Mon, A., Metzler, T., & Neylan, T. C. (2014). Cortical gamma-aminobutyric acid and glutamate in posttraumatic stress disorder and their relationships to self-reported sleep quality. *Sleep*, *37*(5), 893–900. https://doi.org/10.5665/sleep.3654.

Miguel-Hidalgo, J. J., Alvarez, X. A., Cacabelos, R., & Quack, G. (2002). Neuroprotection by memantine against neurodegeneration induced by β-amyloid(1-40). *Brain Research*, *958*(1), 210–221. https://doi.org/10.1016/S0006-8993(02)03731-9.

Miller, D. D. (2004). Atypical antipsychotics: Sleep, sedation, and efficacy. *Primary Care Companion to the Journal of Clinical Psychiatry*, *6*(Suppl. 2), 3–7.

Moghaddam, B., & Adams, B. W. (1998). Reversal of phencyclidine effects by a group II metabotropic glutamate receptor agonist in rats. *Science*, *281*(5381), 1349–1352. https://doi.org/10.1126/science.281.5381.1349.

Moghaddam, B., & Jackson, M. E. (2003). Glutamatergic animal models of Schizophrenia. *Annals Of The New York Academy of Sciences*, *1003*, 131–137. https://doi.org/10.1196/annals.1300.065.

Mondino, A., Cavelli, M., González, J., Murillo-Rodriguez, E., Torterolo, P., & Falconi, A. (2021). Effects of Cannabis consumption on sleep. *Advances in Experimental Medicine and Biology*, *1297*, 147–162. https://doi.org/10.1007/978-3-030-61663-2_11.

Montana, M. C., Cavallone, L. F., Stubbert, K. K., Stefanescu, A. D., Kharasch, E. D., & Gereau, R. W., IV. (2009). The metabotropic glutamate receptor subtype 5 antagonist fenobam is analgesic and has improved in vivo selectivity compared with the prototypical antagonist 2-methyl-6-(phenylethynyl)-pyridine. *Journal of Pharmacology and Experimental Therapeutics*, *330*(3), 834–843. https://doi.org/10.1124/jpet.109.154138.

Morgan, P. T., & Malison, R. T. (2007). Cocaine and sleep: Early abstinence. *The Scientific World Journal*, *7*(Suppl. 2), 223–230. https://doi.org/10.1100/tsw.2007.209.

Morgan, P. T., Pace-Schott, E. F., Sahul, Z. H., Coric, V., Stickgold, R., & Malison, R. T. (2006). Sleep, sleep-dependent procedural learning and vigilance in chronic cocaine users: Evidence for occult insomnia. *Drug and Alcohol Dependence*, *82*(3), 238–249. https://doi.org/10.1016/j.drugalcdep.2005.09.014.

Morgan, C. J. A., Perry, E. B., Cho, H. S., Krystal, J. H., & D'Souza, D. C. (2006). Greater vulnerability to the amnestic effects of ketamine in males. *Psychopharmacology*, *187*(4), 405–414. https://doi.org/10.1007/s00213-006-0409-0.

Morin, N., Grégoire, L., Gomez-Mancilla, B., Gasparini, F., & di Paolo, T. (2010). Effect of the metabotropic glutamate receptor type 5 antagonists MPEP and MTEP in

Parkinsonian monkeys. *Neuropharmacology*, *58*(7), 981–986. https://doi.org/10.1016/J.NEUROPHARM.2009.12.024.

Morphy, H., Dunn, K. M., Lewis, M., Boardman, H. F., & Croft, P. R. (2007). Epidemiology of insomnia: A longitudinal study in a UK population. *Sleep*, *30*(3), 274–280.

Moruzzi, G., & Magoun, H. W. (1949). Brain stem reticular formation and activation of the EEG. *Electroencephalography and Clinical Neurophysiology*, *1*(1–4), 455–473.

Moussawi, K., & Kalivas, P. W. (2010). Group II metabotropic glutamate receptors (mGlu2/3) in drug addiction. *European Journal of Pharmacology*, *639*(1–3), 115–122. https://doi.org/10.1016/j.ejphar.2010.01.030.

Moussawi, K., Zhou, W., Shen, H., Reichel, C. M., See, R. E., Carr, D. B., et al. (2011). Reversing cocaine-induced synaptic potentiation provides enduring protection from relapse. *Proceedings of the National Academy of Sciences of the United States of America*, *108*(1), 385–390. https://doi.org/10.1073/pnas.1011265108.

Nawreen, N., Baccei, M. L., & Herman, J. P. (2021). Single prolonged stress reduces intrinsic excitability and excitatory synaptic drive onto pyramidal neurons in the infralimbic prefrontal cortex of adult male rats. *Frontiers in Cellular Neuroscience*, *15*(July), 1–11. https://doi.org/10.3389/fncel.2021.705660.

Nedelcovych, M., Gould, R., Gong, X., Felts, A., Grannan, M., Thompson, A., et al. (2015). Selective antagonism of mGlu5 alters sleep-wake and spectral EEG and ameliorates behavioral abnormalities in a rodent model of traumatic stress. *The FASEB Journal*, *29*(S1). https://doi.org/10.1096/fasebj.29.1_supplement.615.8.

Neylan, T. C., Kessler, R. C., Ressler, K. J., Clifford, G., Beaudoin, F. L., An, X., et al. (2021). Prior sleep problems and adverse post-traumatic neuropsychiatric sequelae of motor vehicle collision in the AURORA study. *Sleep*, *44*(3). https://doi.org/10.1093/sleep/zsaa200.

Nicoletti, F., Bockaert, J., Collingridge, G. L., Conn, P. J., Ferraguti, F., Schoepp, D. D., et al. (2011). Metabotropic glutamate receptors: From the workbench to the bedside. *Neuropharmacology*, *60*(7–8), 1017–1041. https://doi.org/10.1016/j.neuropharm.2010.10.022.

Nicoletti, F., Orlando, R., Menna, L. D., Cannella, M., Notartomaso, S., Mascio, G., et al. (2019). Targeting mGlu receptors for optimization of antipsychotic activity and disease-modifying effect in Schizophrenia. *Frontiers in Psychiatry*, *10*, 1–10. https://doi.org/10.3389/fpsyt.2019.00049.

Niedzielska-Andres, E., Pomierny-Chamioło, L., Andres, M., Walczak, M., Knackstedt, L. A., Filip, M., et al. (2021). Cocaine use disorder: A look at metabotropic glutamate receptors and glutamate transporters. *Pharmacology and Therapeutics*, *221*. https://doi.org/10.1016/j.pharmthera.2020.107797.

Niswender, C. M., & Conn, P. J. (2010). Metabotropic glutamate receptors: physiology, pharmacology, and disease. *Annual Review of Pharmacology and Toxicology*, *50*, 295–322. https://doi.org/10.1146/annurev.pharmtox.011008.145533.

O'Connor, R. M., McCafferty, C. P., Bravo, J. A., Singewald, N., Holmes, A., & Cryan, J. F. (2019). Increased amygdalar metabotropic glutamate receptor 7 mRNA in a genetic mouse model of impaired fear extinction. *Psychopharmacology*, *236*(1), 265–272. https://doi.org/10.1007/s00213-018-5031-4.

O'Connor, R. M., Thakker, D. R., Schmutz, M., van der Putten, H., Hoyer, D., Flor, P. J., et al. (2013). Adult siRNA-induced knockdown of mGlu7 receptors reduces anxiety in the mouse. *Neuropharmacology*, *72*, 66–73. https://doi.org/10.1016/j.neuropharm.2013.03.028.

Obal, F., & Krueger, J. M. (2003). Biochemical regulation of non-rapid-eye-movement sleep. *Frontiers in Bioscience*, *8*, d520–d550. https://doi.org/10.2741/1033.

Ohayon, M. M., Carskadon, M. A., Guilleminault, C., & Vitiello, M. V. (2004). Meta-analysis of quantitative sleep parameters from childhood to old age in healthy individuals: Developing normative sleep values across the human lifespan. *Sleep, 27*(7), 1255–1273. https://doi.org/10.1093/sleep/27.7.1255.

Olive, M. F. (2010). Cognitive effects of Group I metabotropic glutamate receptor ligands in the context of drug addiction. *European Journal of Pharmacology, 639*(1–3), 47–58. https://doi.org/10.1016/j.ejphar.2010.01.029.

Pace-Schott, E. F., Germain, A., & Milad, M. R. (2015). Sleep and REM sleep disturbance in the pathophysiology of PTSD: The role of extinction memory. *Biology of Mood and Anxiety Disorders, 5*(1), 1–19. https://doi.org/10.1186/s13587-015-0018-9.

Pace-Schott, E. F., Stickgold, R., Muzur, A., Wigren, P. E., Ward, A. S., Hart, C. L., et al. (2005). Sleep quality deteriorates over a binge-abstinence cycle in chronic smoked cocaine users. *Psychopharmacology, 179*(4), 873–883. https://doi.org/10.1007/s00213-004-2088-z.

Page, G., Khidir, F. A. L., Pain, S., Barrier, L., Fauconneau, B., Guillard, O., et al. (2006). Group I metabotropic glutamate receptors activate the p70S6 kinase via both mammalian target of rapamycin (mTOR) and extracellular signal-regulated kinase (ERK 1/2) signaling pathways in rat striatal and hippocampal synaptoneurosomes. *Neurochemistry International, 49*(4), 413–421. https://doi.org/10.1016/j.neuint.2006.01.020.

Palazzo, E., Fu, Y., Ji, G., Maione, S., & Neugebauer, V. (2008). Group III mGluR7 and mGluR8 in the amygdala differentially modulate nocifensive and affective pain behaviors. *Neuropharmacology, 55*(4), 537–545. https://doi.org/10.1016/j.neuropharm.2008.05.007.

Palucha, A., & Pilc, A. (2007). Metabotropic glutamate receptor ligands as possible anxiolytic and antidepressant drugs. *Pharmacology & Therapeutics, 115*(1), 116–147. https://doi.org/10.1016/j.pharmthera.2007.04.007.

Pałucha-Poniewiera, A., & Pilc, A. (2013). A selective mGlu7 receptor antagonist MMPIP reversed antidepressant-like effects of AMN082 in rats. *Behavioural Brain Research, 238*, 109–112. https://doi.org/10.1016/j.bbr.2012.10.004.

Parmentier-Batteur, S., Obrien, J. A., Doran, S., Nguyen, S. J., Flick, R. B., Uslaner, J. M., et al. (2012). Differential effects of the mGluR5 positive allosteric modulator CDPPB in the cortex and striatum following repeated administration. *Neuropharmacology, 62*(3), 1453–1460. https://doi.org/10.1016/j.neuropharm.2010.11.013.

Pecknold, J. C., McClure, D. J., Appeltauer, L., Wrzesinski, L., & Allan, T. (1982). Treatment of anxiety using fenobam (a nonbenzodiazepine) in a double-blind standard (diazepam) placebo-controlled study. *Journal of Clinical Psychopharmacology, 2*(2), 129–133.

Pelkey, K. A., Lavezzari, G., Racca, C., Roche, K. W., & McBain, C. J. (2005). mGluR7 is a metaplastic switch controlling bidirectional plasticity of feedforward inhibition. *Neuron, 46*(1), 89–102. https://doi.org/10.1016/j.neuron.2005.02.011.

Pierri, J. N., Volk, C. L., Auh, S., Sampson, A., & Lewis, D. A. (2001). Decreased somal size of deep layer 3 pyramidal neurons in the prefrontal cortex of subjects with schizophrenia. *Archives of General Psychiatry, 58*(5), 466–473. https://doi.org/10.1001/archpsyc.58.5.466.

Pierri, J. N., Volk, C. L. E., Auh, S., Sampson, A., & Lewis, D. A. (2003). Somal size of prefrontal cortical pyramidal neurons in schizophrenia: Differential effects across neuronal subpopulations. *Biological Psychiatry, 54*(2), 111–120. https://doi.org/10.1016/s0006-3223(03)00294-4.

Pomierny-Chamiolo, L., Miszkiel, J., Frankowska, M., & Mizera, J. (2017). Neuroadaptive changes in metabotropic glutamate mGlu2/3R expression during different phases of cocaine addiction in rats. *Pharmacological Reports, 69*(5), 1073–1081.

Porkka-Heiskanen, T., Strecker, R. E., Thakkar, M., Bjorkum, A. A., Greene, R. W., & McCarley, R. W. (1997). Adenosine: A mediator of the sleep-inducing effects of prolonged wakefulness. *Science*, *276*(5316), 1265–1268. https://doi.org/10.1126/science.276.5316.1265.

Post, R. M., Gillin, J. C., Wyatt, R. J., & Goodwin, F. K. (1974). The effect of orally administered cocaine on sleep of depressed patients. *Psychopharmacologia*, *37*(1), 59–66. https://doi.org/10.1007/BF00426683.

Potter, L. E., Zanos, P., & Gould, T. D. (2020). Antidepressant effects and mechanisms of group II mGlu receptor-specific negative allosteric modulators. *Neuron*, *105*(1), 139–148. https://doi.org/10.1016/j.neuron.2019.12.011.

Potvin, O., Lorrain, D., Forget, H., Dubé, M., Grenier, S., Préville, M., et al. (2012). Sleep quality and 1-year incident cognitive impairment in community-dwelling older adults. *Sleep*, *35*(4), 491–499. https://doi.org/10.5665/sleep.1732.

Pritchett, D., Jagannath, A., Brown, L. A., Tam, S. K. E., Hasan, S., Gatti, S., et al. (2015). Deletion of metabotropic glutamate receptors 2 and 3 (mGlu2 & mGlu3) in mice disrupts sleep and wheel-running activity, and increases the sensitivity of the circadian system to light. *PLoS One*, *10*(5), 1–21. https://doi.org/10.1371/journal.pone.0125523.

Pritchett, D., Wulff, K., Oliver, P. L., Bannerman, D. M., Davies, K. E., Harrison, P. J., et al. (2012). Evaluating the links between schizophrenia and sleep and circadian rhythm disruption. *Journal of Neural Transmission*, *119*(10), 1061–1075. https://doi.org/10.1007/s00702-012-0817-8.

Quiroz, J. A., Tamburri, P., Deptula, D., Banken, L., Beyer, U., Rabbia, M., et al. (2016). Efficacy and safety of basimglurant as adjunctive therapy for major depression: A randomized clinical trial. *JAMA Psychiatry*, *73*(7), 675–684. https://doi.org/10.1001/jamapsychiatry.2016.0838.

Raab-Graham, K. F., Workman, E. R., Namjoshi, S., & Niere, F. (2016). Pushing the threshold: How NMDAR antagonists induce homeostasis through protein synthesis to remedy depression. *Brain Research*, *1647*, 94–104. https://doi.org/10.1016/j.brainres.2016.04.020.

Rady, A., Mekky, J., Moulokheya, T., & Elsheshai, A. (2020). Polysomnographic correlates for the risk of relapse in detoxified opiate-misuse patients. *Neuropsychiatric Disease and Treatment*, *16*, 3187–3196. https://doi.org/10.2147/NDT.S284337.

Rasch, B., & Born, J. (2013). About sleep's role in memory. *Physiological Reviews*, *93*(2), 681–766. https://doi.org/10.1152/physrev.00032.2012.

Rauch, S. L., Shin, L. M., & Phelps, E. A. (2006). Neurocircuitry models of posttraumatic stress disorder and extinction: Human neuroimaging research-past, present, and future. *Biological Psychiatry*, *60*(4), 376–382. https://doi.org/10.1016/j.biopsych.2006.06.004.

Reeve, S., Sheaves, B., & Freeman, D. (2015). The role of sleep dysfunction in the occurrence of delusions and hallucinations: A systematic review. *Clinical Psychology Review*, *42*, 96–115. https://doi.org/10.1016/j.cpr.2015.09.001.

Régio Brambilla, C., Veselinović, T., Rajkumar, R., Mauler, J., Orth, L., Ruch, A., et al. (2020). mGluR5 receptor availability is associated with lower levels of negative symptoms and better cognition in male patients with chronic schizophrenia. *Human Brain Mapping*, *41*(10), 2762–2781. https://doi.org/10.1002/hbm.24976.

Revett, T. J., Baker, G. B., Jhamandas, J., & Kar, S. (2013). Glutamate system, amyloid β peptides and tau protein: Functional interrelationships and relevance to Alzheimer disease pathology. *Journal of Psychiatry and Neuroscience*, *38*(1), 6–23. https://doi.org/10.1503/jpn.110190.

Reynolds, C. F., & Kupfer, D. J. (1987). Sleep research in affective illness: State of the art circa 1987. *Sleep*, *10*(3), 199–215. https://doi.org/10.1093/sleep/10.3.199.

Richards, G., Messer, J., Faull, R. L. M., Stadler, H., Wichmann, J., Huguenin, P., et al. (2010). Altered distribution of mGlu2 receptors in β-amyloid-affected brain regions of Alzheimer cases and aged PS2APP mice. *Brain Research, 1363,* 180–190. https://doi.org/10.1016/j.brainres.2010.09.072.

Riedel, G., Casabona, G., Platt, B., MacPhail, E. M., & Nicoletti, F. (2000). Fear conditioning-induced time- and subregion-specific increase in expression of mGlu5 receptor protein in rat hippocampus. *Neuropharmacology, 39*(11), 1943–1951. https://doi.org/10.1016/S0028-3908(00)00037-X.

Riemann, D., Berger, M., & Voderholzer, U. (2001). Sleep and depression—Results from psychobiological studies: An overview. *Biological Psychology, 57*(1–3), 67–103. https://doi.org/10.1016/S0301-0511(01)00090-4.

Riemann, D., Krone, L. B., Wulff, K., & Nissen, C. (2020). Sleep, insomnia, and depression. *Neuropsychopharmacology, 45*(1), 74–89. https://doi.org/10.1038/s41386-019-0411-y.

Robertson, J. A., Purple, R. J., Cole, P., Zaiwalla, Z., Wulff, K., & Pattinson, K. T. S. (2016). Sleep disturbance in patients taking opioid medication for chronic back pain. *Anaesthesia, 71*(11), 1296–1307. https://doi.org/10.1111/anae.13601.

Rodd, Z. A., McKinzie, D. L., Bell, R. L., McQueen, V. K., Murphy, J. M., Schoepp, D. D., et al. (2006). The metabotropic glutamate 2/3 receptor agonist LY404039 reduces alcohol-seeking but not alcohol self-administration in alcohol-preferring (P) rats. *Behavioural Brain Research, 171*(2), 207–215. https://doi.org/10.1016/j.bbr.2006.03.032.

Rodriguez, A. L., Nong, Y., Sekaran, N. K., Alagille, D., Tamagnan, G. D., & Conn, P. J. (2005). A close structural analog of 2-methyl-6-(phenylethynyl)-pyridine acts as a neutral allosteric site ligand on metabotropic glutamate receptor subtype 5 and blocks the effects of multiple allosteric modulators. *Molecular Pharmacology, 68*(6), 1793–1802. https://doi.org/10.1124/mol.105.016139.

Roehrs, T., Sibai, M., & Roth, T. (2021). Sleep and alertness disturbance and substance use disorders: A bi-directional relation. *Pharmacology Biochemistry and Behavior, 203,* 173153. https://doi.org/10.1016/j.pbb.2021.173153.

Rogawski, M. A., & Wenk, G. L. (2003). The neuropharmacological basis for the use of memantine in the treatment of Alzheimer's disease. *CNS Drug Reviews, 9*(3), 275–308. https://doi.org/10.1111/j.1527-3458.2003.tb00254.x.

Romanella, S. M., Roe, D., Tatti, E., Cappon, D., Paciorek, R., Testani, E., et al. (2021). The sleep side of aging and Alzheimer's disease. *Sleep Medicine, 77,* 209–225. https://doi.org/10.1016/j.sleep.2020.05.029.

Romano, C., Sesma, M. A., McDonald, C. T., O'Malley, K., van den Pol, A. N., & Olney, J. W. (1995). Distribution of metabotropic glutamate receptor mGluR5 immunoreactivity in rat brain. *The Journal of Comparative Neurology, 355*(3), 455–469. https://doi.org/10.1002/cne.903550310.

Rondina, R., 2nd, Olsen, R. K., McQuiggan, D. A., Fatima, Z., Li, L., Oziel, E., et al. (2016). Age-related changes to oscillatory dynamics in hippocampal and neocortical networks. *Neurobiology of Learning and Memory, 134,* 15–30. https://doi.org/10.1016/j.nlm.2015.11.017.

Ronesi, J. A., & Huber, K. M. (2008). Homer interactions are necessary for metabotropic glutamate receptor-induced long-term depression and translational activation. *The Journal of Neuroscience, 28*(2), 543–547. https://doi.org/10.1523/JNEUROSCI.5019-07.2008.

Rosen, I. M., Aurora, R. N., Kirsch, D. B., Carden, K. A., Malhotra, R. K., Ramar, K., et al. (2019). Chronic opioid therapy and sleep: An American academy of sleep medicine position statement. *Journal of Clinical Sleep Medicine, 15*(11), 1671–1673. https://doi.org/10.5664/jcsm.8062.

Rosenwasser, A. M., & Turek, F. W. (2015). Neurobiology of circadian rhythm regulation. *Sleep Medicine Clinics, 10*(4), 403–412. https://doi.org/10.1016/j.jsmc.2015.08.003.

Rousseau, P. F., Vallat, R., Coste, O., Cadis, H., Nicolas, F., Trousselard, M., et al. (2021). Sleep parameters improvement in PTSD soldiers after symptoms remission. *Scientific Reports*, *11*(1), 1–13. https://doi.org/10.1038/s41598-021-88337-x.

Salling, M. C., Grassetti, A., Ferrera, V. P., Martinez, D., & Foltin, R. W. (2021). Negative allosteric modulation of metabotropic glutamate receptor 5 attenuates alcohol self-administration in baboons. *Pharmacology, Biochemistry, and Behavior*, *208*, 173227. https://doi.org/10.1016/j.pbb.2021.173227.

Sanacora, G., Treccani, G., & Popoli, M. (2012). Towards a glutamate hypothesis of depression: An emerging frontier of neuropsychopharmacology for mood disorders. *Neuropharmacology*, *62*(1), 63–77. https://doi.org/10.1016/j.neuropharm.2011.07.036.

Sansig, G., Bushell, T. J., Clarke, V. R., Rozov, A., Burnashev, N., Portet, C., et al. (2001). Increased seizure susceptibility in mice lacking metabotropic glutamate receptor 7. *The Journal of Neuroscience: The Official Journal of the Society for Neuroscience*, *21*(22), 8734–8745. https://doi.org/10.1523/JNEUROSCI.21-22-08734.2001.

Saper, C. B., & Fuller, P. M. (2017). Wake-sleep circuitry: An overview. *Current Opinion in Neurobiology*, *44*, 186–192. https://doi.org/10.1016/j.conb.2017.03.021.

Saper, C. B., Fuller, P. M., Pedersen, N. P., Lu, J., & Scammell, T. E. (2010). Sleep state switching. *Neuron*, *68*(6), 1023–1042. https://doi.org/10.1016/j.neuron.2010.11.032.

Saper, C. B., Scammell, T. E., & Lu, J. (2005). Hypothalamic regulation of sleep and circadian rhythms. *Nature*, *437*(7063), 1257–1263. https://doi.org/10.1038/nature04284.

Scammell, T. E., Arrigoni, E., & Lipton, J. O. (2017). Neural circuitry of wakefulness and sleep. *Neuron*, *93*(4), 747–765. https://doi.org/10.1016/j.neuron.2017.01.014.

Schierenbeck, T., Riemann, D., Berger, M., & Hornyak, M. (2008). Effect of illicit recreational drugs upon sleep: Cocaine, ecstasy and marijuana. *Sleep Medicine Reviews*, *12*(5), 381–389. https://doi.org/10.1016/j.smrv.2007.12.004.

Scullin, M. K., & Bliwise, D. L. (2015a). Is cognitive aging associated with levels of REM sleep or slow wave sleep? *Sleep*, *38*(3), 335–336. https://doi.org/10.5665/sleep.4482.

Scullin, M. K., & Bliwise, D. L. (2015b). Sleep, cognition, and normal aging: Integrating a half century of multidisciplinary research. *Perspectives on Psychological Science: A Journal of the Association for Psychological Science*, *10*(1), 97–137. https://doi.org/10.1177/1745691614556680.

Sekiya, Y., Nakagawa, T., Ozawa, T., Minami, M., & Satoh, M. (2004). Facilitation of morphine withdrawal symptoms and morphine-induced conditioned place preference by a glutamate transporter inhibitor DL-threo-β-benzyloxyaspartate in rats. *European Journal of Pharmacology*, *485*(1–3), 201–210. https://doi.org/10.1016/j.ejphar.2003.11.062.

Sengmany, K., & Gregory, K. J. (2016). Metabotropic glutamate receptor subtype 5: Molecular pharmacology, allosteric modulation and stimulus bias. *British Journal of Pharmacology*, *173*(20), 3001–3017. https://doi.org/10.1111/bph.13281.

Shankar, G. M., Bloodgood, B. L., Townsend, M., Walsh, D. M., Selkoe, D. J., & Sabatini, B. L. (2007). Natural oligomers of the Alzheimer amyloid-β protein induce reversible synapse loss by modulating an NMDA-type glutamate receptor-dependent signaling pathway. *Journal of Neuroscience*, *27*(11), 2866–2875. https://doi.org/10.1523/JNEUROSCI.4970-06.2007.

Sharkey, K. M., Kurth, M. E., Anderson, B. J., Corso, R. P., Millman, R. P., & Stein, M. D. (2011). Assessing sleep in opioid dependence: A comparison of subjective ratings, sleep diaries, and home polysomnography in methadone maintenance patients. *Drug and Alcohol Dependence*, *113*(2–3), 245–248. https://doi.org/10.1016/j.drugalcdep.2010.08.007.Assessing.

Sharma, P. (2016). Excessive daytime sleepiness in Schizophrenia: A naturalistic clinical study. *Journal of Clinical and Diagnostic Research*, *10*–*12*. https://doi.org/10.7860/jcdr/2016/21272.8627.

Shaw, I. R., Lavigne, G., & Mayer, P. (2006). Acute intravenous administration of morphine perturbs sleep architecture in healthy painfree young adults: A preliminary study. *Sleep*, *28*(6), 677–682. https://doi.org/10.1093/sleep/28.6.677.

Shepherd, J. D. (2012). Memory, plasticity and sleep—A role for calcium permeable AMPA receptors? *Frontiers in Molecular Neuroscience*, *5*, 49. https://doi.org/10.3389/fnmol.2012.00049.

Sherin, J. E., Elmquist, J. K., Torrealba, F., & Saper, C. B. (1998). Innervation of histaminergic tuberomammillary neurons by GABAergic and galaninergic neurons in the ventrolateral preoptic nucleus of the rat. *The Journal of Neuroscience*, *18*(12), 4705–4721. https://doi.org/10.1523/JNEUROSCI.18-12-04705.1998.

Shi, G., Yin, C., Fan, Z., Xing, L., Mostovoy, Y., Kwok, P. Y., et al. (2021). Mutations in metabotropic glutamate receptor 1 contribute to natural short sleep trait. *Current Biology*, *31*(1), 13–24.e4. https://doi.org/10.1016/j.cub.2020.09.071.

Shigemoto, R., Nakanishi, S., & Mizuno, N. (1992). Distribution of the mRNA for a metabotropic glutamate receptor (mGluR1) in the central nervous system: An in situ hybridization study in adult and developing rat. *The Journal of Comparative Neurology*, *322*(1), 121–135. https://doi.org/10.1002/cne.903220110.

Shigemoto, R., Nomura, S., Ohishi, H., Sugihara, H., Nakanishi, S., & Mizuno, N. (1993). Immunohistochemical localization of a metabotropic glutamate receptor, mGluR5, in the rat brain. *Neuroscience Letters*, *163*(1), 53–57. https://doi.org/10.1016/0304-3940(93)90227-c.

Shin, H.-Y., Gadzhanova, S., Roughead, E. E., Ward, M. B., & Pont, L. G. (2016). The use of antipsychotics among people treated with medications for dementia in residential aged care facilities. *International Psychogeriatrics*, *28*(6), 977–982. https://doi.org/10.1017/S1041610215002434.

Shiraishi-Yamaguchi, Y., & Furuichi, T. (2007). The Homer family proteins. *Genome Biology*, *8*(2), 1–12. https://doi.org/10.1186/gb-2007-8-2-206.

Siciliano, C. A., Ferris, M. J., & Jones, S. R. (2016). Cocaine self-administration disrupts mesolimbic dopamine circuit function and attenuates dopaminergic responsiveness to cocaine. *Behavioural Neuroscience*, *42*(4), 2091–2096. https://doi.org/10.1111/ejn.12970.

Sokolenko, E., Hudson, M. R., Nithianantharajah, J., & Jones, N. C. (2019). The mGluR2/3 agonist LY379268 reverses NMDA receptor antagonist effects on cortical gamma oscillations and phase coherence, but not working memory impairments, in mice. *Journal of Psychopharmacology*, *33*(12), 1588–1599. https://doi.org/10.1177/0269881119875976.

Song, Y., Blackwell, T., Yaffe, K., Ancoli-Israel, S., Redline, S., & Stone, K. L. (2015). Relationships between sleep stages and changes in cognitive function in older men: The MrOS Sleep Study. *Sleep*, *38*(3), 411–421. https://doi.org/10.5665/sleep.4500.

Sprecher, K. E., Ferrarelli, F., & Benca, R. M. (2015). Sleep and plasticity in schizophrenia. *Current Topics in Behavioral Neurosciences*, *25*, 433–458. https://doi.org/10.1007/7854_2014_366.

Srinivasan, V., Pandi-Perumal, S. R., Cardinali, D. P., Poeggeler, B., & Hardeland, R. (2006). Melatonin in Alzheimer's disease and other neurodegenerative disorders. *Behavioral and Brain Functions*, *2*, 1–23. https://doi.org/10.1186/1744-9081-2-15.

Steiger, A., & Kimura, M. (2010). Wake and sleep EEG provide biomarkers in depression. *Journal of Psychiatric Research*, *44*(4), 242–252. https://doi.org/10.1016/j.jpsychires.2009.08.013.

Steiger, A., & Pawlowski, M. (2019). Depression and sleep. *International Journal of Molecular Sciences*, *20*(3), 1–14. https://doi.org/10.3390/ijms20030607.

Steiger, A., Pawlowski, M., & Kimura, M. (2015). Sleep electroencephalography as a biomarker in depression. *ChronoPhysiology and Therapy*, *15*. https://doi.org/10.2147/cpt.s41760.

Stein, M. D., Herman, D. S., Bishop, S., Lassor, J. A., Weinstock, M., Anthony, J., et al. (2004). Sleep disturbances among methadone maintained patients. *Journal of Substance Abuse Treatment*, *26*(3), 175–180. https://doi.org/10.1016/S0740-5472(03)00191-0.

Steriade, M., Domich, L., Oakson, G., & Deschfines, M. (1987). The deafferented reticular thalamic nucleus generates spindle rhythmicity. *Journal of Neurophysiology*, *57*(1), 260–273. https://doi.org/10.1152/jn.1987.57.1.260.

Steriade, M., Iosif, G., & Apostol, V. (1968). Responsiveness of thalamic and cortical motor relays during arousal and various stages of sleep. *Journal of Neurophysiology*, *32*(2), 251–265. https://doi.org/10.1152/jn.1969.32.2.251.

Stopa, E. G., Volicer, L., Kuo-Leblanc, V., Harper, D., Lathi, D., Tate, B., et al. (1999). Pathologic evaluation of the human suprachiasmatic nucleus in severe dementia. *Neuropathology and Experimental Neurology*, *58*(1), 29–39. https://doi.org/10.1097/00005072-199901000-00004.

Straus, L. D., Acheson, D. T., Risbrough, V. B., Sean, P. A., Diego, S., Diego, S., et al. (2018). Sleep deprivation disrupts recall of conditioned fear extinction. Biological psychiatry. *Cognitive Neuroscience and Neuroimaging*, *2*(2), 123–129. https://doi.org/10.1016/j.bpsc.2016.05.004.

Sun, Y. G., Rupprecht, V., Zhou, L., Dasgupta, R., Seibt, F., & Beierlein, M. (2016). MGluR1 and mGluR5 synergistically control cholinergic synaptic transmission in the thalamic reticular nucleus. *Journal of Neuroscience*, *36*(30), 7886–7896. https://doi.org/10.1523/JNEUROSCI.0409-16.2016.

Suntsova, N., Szymusiak, R., Alam, M. N., Guzman-Marin, R., & McGinty, D. (2002). Sleep-waking discharge patterns of median preoptic nucleus neurons in rats. *The Journal of Physiology*, *543*, 665–677. https://doi.org/10.1113/jphysiol.2002.023085.

Swanson, C. J., Baker, D. A., Carson, D., Worley, P. F., & Kalivas, P. W. (2001). Repeated cocaine administration attenuates group I metabotropic glutamate receptor-mediated glutamate release and behavioral activation: A potential role for homer. *The Journal of Neuroscience: The Official Journal of the Society for Neuroscience*, *21*(22), 9043–9052. https://doi.org/10.1523/JNEUROSCI.21-22-09043.2001.

Sweeten, B. L. W., Adkins, A. M., Wellman, L. L., & Sanford, L. D. (2021). Group II metabotropic glutamate receptor activation in the basolateral amygdala mediates individual differences in stress-induced changes in rapid eye movement sleep. *Progress in Neuro-Psychopharmacology & Biological Psychiatry*, *104*, 110014. https://doi.org/10.1016/j.pnpbp.2020.110014.

Swift, K. M. (2020). Sleep and PTSD: Delving deeper to understand a complicated relationship. *Sleep*, *43*(9), 1–3. https://doi.org/10.1093/sleep/zsaa074.

Tadavarty, R., Rajput, P. S., Wong, J. M., Kumar, U., & Sastry, B. R. (2011). Sleep-deprivation induces changes in GABAb and mglu receptor expression and has consequences for synaptic long-term depression. *PLoS One*, *6*(9). https://doi.org/10.1371/journal.pone.0024933.

Taylor, D. L., Diemel, L. T., Cuzner, M. L., & Pocock, J. M. (2002). Activation of group II metabotropic glutamate receptors underlies microglial reactivity and neurotoxicity following stimulation with chromogranin A, a peptide up-regulated in Alzheimer's disease. *Journal of Neurochemistry*, *82*(5), 1179–1191. https://doi.org/10.1046/j.1471-4159.2002.01062.x.

Taylor, D. J., Pruiksma, K. E., Hale, W., McLean, C. P., Zandberg, L. J., Brown, L., et al. (2020). Sleep problems in active duty military personnel seeking treatment for post-traumatic stress disorder: Presence, change, and impact on outcomes. *Sleep*, *43*(10), 1–13. https://doi.org/10.1093/sleep/zsaa065.

Tesler, N., Gerstenberg, M., Franscini, M., Jenni, O. G., Walitza, S., & Huber, R. (2016). Increased frontal sleep slow wave activity in adolescents with major depression. *NeuroImage: Clinical*, *10*, 250–256. https://doi.org/10.1016/j.nicl.2015.10.014.

Tok, S., Ahnaou, A., & Drinkenburg, W. (2021). Functional neurophysiological biomarkers of early-stage Alzheimer's Disease: A perspective of network hyperexcitability in disease progression. *Journal of Alzheimer's Disease, 1–28*. https://doi.org/10.3233/jad-210397.

Tokuyama, S., Wakabayashi, H., & Ho, I. K. (1995). Direct evidence for a role of glutamate in the expression of the opioid withdrawal syndrome. *European Journal of Pharmacology, 295*, 123–129. https://doi.org/10.1007/978-1-4020-5614-7_3754.

Tononi, G., & Cirelli, C. (2014). Sleep and the price of plasticity: From synaptic and cellular homeostasis to memory consolidation and integration. *Neuron, 81*(1), 12–34. https://doi.org/10.1016/j.neuron.2013.12.025.

Tripathi, R., Rao, R., Dhawan, A., Jain, R., & Sinha, S. (2020). Opioids and sleep—A review of literature. *Sleep Medicine, 67*, 269–275. https://doi.org/10.1016/j.sleep.2019.06.012.

Troxel, W. M., Rodriguez, A., Seelam, R., Tucker, J. S., Shih, R. A., Dong, L., et al. (2021). Longitudinal associations of sleep problems with alcohol and cannabis use from adolescence to emerging adulthood. *Sleep, 44*(10). https://doi.org/10.1093/sleep/zsab102.

Trullas, R., & Skolnick, P. (1990). Functional antagonists at the NMDA receptor complex exhibit antidepressant actions. *European Journal of Pharmacology, 185*(1), 1–10. https://doi.org/10.1016/0014-2999(90)90204-J.

Um, J. W., Kaufman, A. C., Kostylev, M., Heiss, J. K., Stagi, M., Takahashi, H., et al. (2013). Metabotropic glutamate receptor 5 is a coreceptor for Alzheimer Aβ oligomer bound to cellular prion protein. *Neuron, 79*(5), 887–902. https://doi.org/10.1016/j.neuron.2013.06.036.

Umbricht, D., Niggli, M., Sanwald-Ducray, P., Deptula, D., Moore, R., Grünbauer, W., et al. (2020). Randomized, double-blind, placebo-controlled trial of the mGlu2/3 negative allosteric modulator decoglurant in partially refractory major depressive disorder. *Journal of Clinical Psychiatry, 81*(4). https://doi.org/10.4088/JCP.18m12470.

Valentino, R. J., & Volkow, N. D. (2020). Drugs, sleep, and the addicted brain. *Neuropsychopharmacology, 45*(1), 3–5. https://doi.org/10.1038/s41386-019-0465-x.

Valladares, E. M., & Irwin, M. R. (2007). Polysomnographic sleep dysregulation in cocaine dependence. *The Scientific World Journal, 7*(Suppl. 2), 213–216. https://doi.org/10.1100/tsw.2007.264.

van der Kam, E. L., de Vry, J., & Tzschentke, T. M. (2007). Effect of 2-methyl-6-(phenylethynyl) pyridine on intravenous self-administration of ketamine and heroin in the rat. *Behavioural Pharmacology, 18*(8), 717–724. https://doi.org/10.1097/FBP.0b013e3282f18d58.

van Huijstee, A. N., & Mansvelder, H. D. (2014). Glutamatergic synaptic plasticity in the mesocorticolimbic system in addiction. *Frontiers in Cellular Neuroscience, 8*, 466. https://doi.org/10.3389/fncel.2014.00466.

van Liempt, S., van Zuiden, M., Westenberg, H., Super, A., & Vermetten, E. (2013). Impact of impaired sleep on the development of PTSD symptoms in combat veterans: A prospective longitudinal cohort study. *Depression and Anxiety, 30*(5), 469–474. https://doi.org/10.1002/da.22054.

Veatch, O. J., Sutcliffe, J. S., Warren, Z. E., Keenan, B. T., Potter, M. H., & Malow, B. A. (2017). Shorter sleep duration is associated with social impairment and comorbidities in ASD. *Autism Research: Official Journal of the International Society for Autism Research, 10*(7), 1221–1238. https://doi.org/10.1002/aur.1765.

Veeneman, M. M. J., Boleij, H., Broekhoven, M. H., Snoeren, E. M. S., Guitart Masip, M., Cousijn, J., et al. (2011). Dissociable roles of mGlu5 and dopamine receptors in the rewarding and sensitizing properties of morphine and cocaine. *Psychopharmacology, 214*(4), 863–876. https://doi.org/10.1007/s00213-010-2095-1.

Verret, L., Goutagny, R., Fort, P., Cagnon, L., Salvert, D., Léger, L., et al. (2003). A role of melanin-concentrating hormone producing neurons in the central regulation of paradoxical sleep. *BMC Neuroscience, 4*, 19. https://doi.org/10.1186/1471-2202-4-19.

Vyazovskiy, V., Cirelli, C., Pfister-Genskow, M., Faraguna, U., & Tononi, G. (2008). Molecular and electrophysiological evidence for net synaptic potentiation in wake and depression in sleep. *Nature Neuroscience*, *11*(2), 200–208. https://doi.org/10.1038/nn2035.

Walker, A. G., Wenthur, C. J., Xiang, Z., Rook, J. M., Emmitte, K. A., Niswender, C. M., et al. (2015). Metabotropic glutamate receptor 3 activation is required for long-term depression in medial prefrontal cortex and fear extinction. *Proceedings of the National Academy of Sciences of the United States of America*, *112*(4), 1196–1201. https://doi.org/10.1073/pnas.1416196112.

Wamsley, E. J., Tucker, M. A., Shinn, A. K., Ono, K. E., McKinley, S. K., Ely, A. V., et al. (2012). Reduced sleep spindles and spindle coherence in schizophrenia: Mechanisms of impaired memory consolidation? *Biological Psychiatry*, *71*(2), 154–161. https://doi.org/10.1016/j.biopsych.2011.08.008.

Wang, H. E., Campbell-Sills, L., Kessler, R. C., Sun, X., Heeringa, S. G., Nock, M. K., et al. (2019). Pre-deployment insomnia is associated with post-deployment post-traumatic stress disorder and suicidal ideation in US Army soldiers. *Sleep*, *42*(2), 1–9. https://doi.org/10.1093/sleep/zsy229.

Wang, Y. Q., Liu, W. Y., Li, L., Qu, W. M., & Huang, Z. L. (2021). Neural circuitry underlying REM sleep: A review of the literature and current concepts. *Progress in Neurobiology*, *204*, 102106. https://doi.org/10.1016/j.pneurobio.2021.102106.

Wang, D., & Teichtahl, H. (2007). Opioids, sleep architecture and sleep-disordered breathing. *Sleep Medicine Reviews*, *11*(1), 35–46. https://doi.org/10.1016/j.smrv.2006.03.006.

Wang, H., & Zhuo, M. (2012). Group I metabotropic glutamate receptor-mediated gene transcription and implications for synaptic plasticity and diseases. *Frontiers in Pharmacology*, *3*, 1–8. https://doi.org/10.3389/fphar.2012.00189.

Watanabe, M., Nakamura, M., Sato, K., Kano, M., Simon, M. I., & Inoue, Y. (1998). Patterns of expression for the mRNA corresponding to the four isoforms of phospholipase Cβ in mouse brain. *European Journal of Neuroscience*, *10*(6), 2016–2025. https://doi.org/10.1046/j.1460-9568.1998.00213.x.

Watson, C. J., Lydic, R., & Baghdoyan, H. A. (2011). Sleep duration varies as a function of glutamate and GABA in rat pontine reticular formation. *Journal of Neurochemistry*, *118*(4), 571–580. https://doi.org/10.1111/j.1471-4159.2011.07350.x.

Weber, F., & Dan, Y. (2016). Circuit-based interrogation of sleep control. *Nature*, *538*(7623), 51–59. https://doi.org/10.1038/nature19773.

Weigend, S., Holst, S. C., Treyer, V., O'Gorman Tuura, R. L., Meier, J., Ametamey, S. M., et al. (2019). Dynamic changes in cerebral and peripheral markers of glutamatergic signaling across the human sleep-wake cycle. *Sleep*, *42*(11), 1–11. https://doi.org/10.1093/sleep/zsz161.

Welsch, L., Bailly, J., Darcq, E., & Kieffer, B. L. (2020). The negative affect of protracted opioid abstinence: Progress and perspectives from rodent models. *Biological Psychiatry*, *87*(1), 54–63. https://doi.org/10.1016/j.biopsych.2019.07.027.

Weng, Y. Y., Lei, X., & Yu, J. (2020). Sleep spindle abnormalities related to Alzheimer's disease: A systematic mini-review. *Sleep Medicine*, *75*, 37–44. https://doi.org/10.1016/j.sleep.2020.07.044.

Westermeyer, J., Khawaja, I. S., Freerks, M., Sutherland, R. J., Engle, K., Johnson, D., et al. (2010). Quality of sleep in patients with posttraumatic stress disorder. *Psychiatry (Edgemont)*, *7*(9), 21–27.

Wichniak, A., Antczak, J., Wierzbicka, A., & Jernajczyk, W. (2002). Alterations in pattern of rapid eye movement activity during REM sleep in depression. *Acta Neurobiologiae Experimentalis*, *62*(4), 243–250.

Wichniak, A., Wierzbicka, A., & Jernajczyk, W. (2013). Sleep as a biomarker for depression. *International Review of Psychiatry*, *25*(5), 632–645. https://doi.org/10.3109/09540261.2013.812067.

Wichniak, A., Wierzbicka, A., Walęcka, M., & Jernajczyk, W. (2017). Effects of antidepressants on sleep. *Current Psychiatry Reports, 19*(9), 1–7. https://doi.org/10.1007/s11920-017-0816-4.

Williams, J. A., Comisarow, J., Day, J., Fibiger, H. C., & Reiner, P. B. (1994). State-dependent release of acetylcholine in rat thalamus measured by in vivo microdialysis. *The Journal of Neuroscience: The Official Journal of the Society for Neuroscience, 14*(9), 5236–5242. https://doi.org/10.1523/JNEUROSCI.14-09-05236.1994.

Winkelman, J. W., & de Lecea, L. (2020). Sleep and neuropsychiatric illness. *Neuropsychopharmacology, 45*(1). https://doi.org/10.1038/s41386-019-0514-5.

Winship, I. R., Dursun, S. M., Baker, G. B., Balista, P. A., Kandratavicius, L., Maia-de-Oliveira, J. P., et al. (2019). An overview of animal models related to Schizophrenia. *Canadian Journal of Psychiatry, 64*(1), 5–17. https://doi.org/10.1177/0706743718773728.

Witkin, J. M., Monn, J. A., Schoepp, D. D., Li, X., Overshiner, C., Mitchell, S. N., et al. (2016). The rapidly acting antidepressant ketamine and the mGlu2/3 receptor antagonist LY341495 rapidly engage dopaminergic mood circuits. *Journal of Pharmacology and Experimental Therapeutics, 358*(1), 71–82. https://doi.org/10.1124/jpet.116.233627.

Wolf, M. E. (2016). Synaptic mechanisms underlying persistent cocaine craving. *Nature Reviews Neuroscience, 17*(6), 351–365. Nature Publishing Group https://doi.org/10.1038/nrn.2016.39.

Wong, B., Zimbelman, A. R., Milovanovic, M., Wolf, M. E., & Stefanik, M. T. (2022). GluA2-lacking AMPA receptors in the nucleus accumbens core and shell contribute to the incubation of oxycodone craving in male rats. *Addiction Biology, 27*(6), e13237. https://doi.org/10.1111/adb.13237.

Wood, C. M., Wafford, K. A., McCarthy, A. P., Hewes, N., Shanks, E., Lodge, D., et al. (2018). Investigating the role of mGluR2 versus mGluR3 in antipsychotic-like effects, sleep-wake architecture and network oscillatory activity using novel Han Wistar rats lacking mGluR2 expression. *Neuropharmacology, 140*, 246–259. https://doi.org/10.1016/j.neuropharm.2018.07.013.

Woodward, S. H., Leskin, G. A., & Sheikh, J. I. (2002). Movement during sleep: Associations with posttraumatic stress disorder, nightmares, and comorbid panic disorder. *Sleep, 25*(6), 681–688. https://doi.org/10.1093/sleep/25.6.669.

Wright, R. A., Arnold, M. B., Wheeler, W. J., Ornstein, P. L., & Schoepp, D. D. (2001). [3H]LY341495 binding to group II metabotropic glutamate receptors in rat brain. *The Journal of Pharmacology and Experimental Therapeutics, 298*(2), 453–460.

Wulff, K., Gatti, S., Wettstein, J. G., & Foster, R. G. (2010). Sleep and circadian rhythm disruption in psychiatric and neurodegenerative disease. *Nature Reviews Neuroscience, 11*(8), 589–599. https://doi.org/10.1038/nrn2868.

Xie, L., Kang, H., Xu, Q., Chen, M., Liao, Y., Thiyagarajan, M., et al. (2013). Sleep drives metabolite clearance from the adult brain. *NIH, 342*(6156). https://doi.org/10.1126/science.1241224.

Yeh, M. S. L., Poyares, D., Coimbra, B. M., Mello, A. F., Tufik, S., & Mello, M. F. (2021). Subjective and objective sleep quality in young women with posttraumatic stress disorder following sexual assault: A prospective study. *European Journal of Psychotraumatology, 12*-(1). https://doi.org/10.1080/20008198.2021.1934788.

Yokoi, M., Kobayashi, K., Manabe, T., Takahashi, T., Sakaguchi, I., Katsuura, G., et al. (1996). Impairment of hippocampal mossy fiber LTD in mice lacking mGluR2. *Science (New York, N.Y.), 273*(5275), 645–647. https://doi.org/10.1126/science.273.5275.645.

Yoo, S.-S., Gujar, N., Hu, P., Jolesz, F. A., & Walker, M. P. (2007). The human emotional brain without sleep—A prefrontal amygdala disconnect. *Current Biology: CB, 17*(20), R877–R878. https://doi.org/10.1016/j.cub.2007.08.007.

Yu, X., Li, W., Ma, Y., Tossell, K., Harris, J. J., Harding, E. C., et al. (2019). GABA and glutamate neurons in the VTA regulate sleep and wakefulness. *Nature Neuroscience, 22*(1), 106–119. https://doi.org/10.1038/s41593-018-0288-9.

Zarate, C. A., Singh, J. B., Carlson, P. J., Brutsche, N. E., Ameli, R., Luckenbaugh, D. A., et al. (2006). A randomized trial of an N-methyl-D-aspartate antagonist in treatment-resistant major depression. *North, 63*(8), 793–802.

Zhai, J., Tian, M. T., Wang, Y., Yu, J. L., Köster, A., Baez, M., et al. (2002). Modulation of lateral perforant path excitatory responses by metabotropic glutamate 8 (mGlu8) receptors. *Neuropharmacology, 43*(2), 223–230. https://doi.org/10.1016/S0028-3908(02)00087-4.

Zhang, Y., Ren, R., Yang, L., Zhang, H., Shi, Y., Okhravi, H. R., et al. (2022). Sleep in Alzheimer's disease: A systematic review and meta-analysis of polysomnographic findings. *Translational Psychiatry, 12*(1). https://doi.org/10.1038/s41398-022-01897-y.

Zhao, W., van Someren, E. J. W., Li, C., Chen, X., Gui, W., Tian, Y., et al. (2021). EEG spectral analysis in insomnia disorder: A systematic review and meta-analysis. In *Vol. 59. Sleep Medicine Reviews* W.B. Saunders Ltd. https://doi.org/10.1016/j.smrv.2021.101457.

Zhu, H., Rockhold, R. W., & Ho, I. K. (1998). The role of glutamate in physical dependence on opioids. *Japanese Journal of Pharmacology, 76*(1), 1–14. https://doi.org/10.1254/jjp.76.1.

CHAPTER FOUR

The role of metabotropic glutamate receptors in neurobehavioral effects associated with methamphetamine use

Peter U. Hámor[a,b,c], Lori A. Knackstedt[a,b], and Marek Schwendt[a,b,*]

[a]Department of Psychology, University of Florida, Gainesville, FL, United States
[b]Center for Addiction Research and Education, University of Florida, Gainesville, FL, United States
[c]Department of Pharmacology, Weill Cornell Medicine, Cornell University, New York, NY, United States
*Corresponding author: e-mail address: schwendt@ufl.edu

Contents

1. Methamphetamine use disorder (MUD)—Characteristics	178
2. Metabotropic glutamate receptors (mGlu)—An overview	182
3. mGlu receptors and methamphetamine-associated neural changes	184
3.1 Methamphetamine-induced neurotoxicity	184
3.2 Methamphetamine-induced changes mGlu receptor expression and function	186
4. mGlu receptors and meth-induced behavioral effects	188
4.1 Psychomotor effects of methamphetamine	188
4.2 Methamphetamine-associated learning, reward, and reinforcement	189
4.3 Reinstatement of methamphetamine-seeking	191
4.4 Methamphetamine-associated extinction learning	194
4.5 Methamphetamine-associated changes in non-drug related memory processes	196
5. mGlu receptor interaction with other neurotransmitter receptors in the context of methamphetamine-associated neural and behavioral effects	198
5.1 mGlu—Dopamine receptor interactions	198
5.2 mGlu-adenosine receptor interactions	200
5.3 mGlu-serotonin receptor interactions	202
6. Conclusions	203
Support	204
Conflict of interest	204
References	204

International Review of Neurobiology, Volume 168
ISSN 0074-7742
https://doi.org/10.1016/bs.irn.2022.10.005

Copyright © 2023 Elsevier Inc.
All rights reserved.

Abstract

Metabotropic glutamate (mGlu) receptors are expressed throughout the central nervous system and act as important regulators of drug-induced neuroplasticity and behavior. Preclinical research suggests that mGlu receptors play a critical role in a spectrum of neural and behavioral consequences arising from methamphetamine (meth) exposure. However, an overview of mGlu-dependent mechanisms linked to neurochemical, synaptic, and behavioral changes produced by meth has been lacking. This chapter provides a comprehensive review of the role of mGlu receptor subtypes (mGlu1-8) in meth-induced neural effects, such as neurotoxicity, as well as meth-associated behaviors, such as psychomotor activation, reward, reinforcement, and meth-seeking. Additionally, evidence linking altered mGlu receptor function to post-meth learning and cognitive deficits is critically evaluated. The chapter also considers the role of receptor-receptor interactions involving mGlu receptors and other neurotransmitter receptors in meth-induced neural and behavioral changes. Taken together, the literature indicates that mGlu5 regulates the neurotoxic effects of meth by attenuating hyperthermia and possibly through altering meth-induced phosphorylation of the dopamine transporter. A cohesive body of work also shows that mGlu5 antagonism (and mGlu2/3 agonism) reduce meth-seeking, though some mGlu5-blocking drugs also attenuate food-seeking. Further, evidence suggests that mGlu5 plays an important role in extinction of meth-seeking behavior. In the context of a history of meth intake, mGlu5 also co-regulates aspects of episodic memory, with mGlu5 stimulation restoring impaired memory. Based on these findings, we propose several avenues for the development of novel pharmacotherapies for Methamphetamine Use Disorder based on the selective modulation mGlu receptor subtype activity.

1. Methamphetamine use disorder (MUD)—Characteristics

Methamphetamine (meth) is a widely abused synthetic psychostimulant drug belonging to the amphetamines class. Recent estimates suggest that meth, together with other amphetamine-type stimulants, represents the second most commonly used illicit drug class worldwide (United Nations Office on Drugs and Crime (UNDOC), 2020). Recent surveys suggest that in the United States alone, approximately 2–2.5 million people reported past-year methamphetamine use, a sharp increase when compared to the 2010 statistic (Global Burden of Disease (GBD) Collaborative Network, 2021; Substance Abuse and Mental Health Services Administration (SAMHSA), 2021; with 52.9% of individuals reporting problematic meth use that met the criterion for methamphetamine use disorder (MUD; Jones, Compton, & Mustaquim, 2020). Although in the U.S., meth is also available as a prescription medication (Desoxyn®), the vast majority of recreationally used meth is produced and distributed illicitly rather than diverted from the prescription pool (U.S. Drug Enforcement Administration (DEA), 2020). The factors driving illicit meth use are both the relative ease and the low cost of synthesis, as well as the potent and long-lasting behavioral effects produced by meth. Meth is known to produce intense euphoria, alertness, and enhances energy levels (Kish, 2008). However, binge-like use of meth and/or development of MUD is accompanied by undesirable or aversive effects, such as increased incidence of aggression, dysphoria, anxiety, cognitive dysfunction, and recurrent psychosis. Excessive amounts of meth or extended periods of high-frequency meth use can also produce neurotoxicity and neuroinflammation, and outside the brain, adversely affecting cardiovascular, dermatological, and oral health (Kevil et al., 2019; Shetty et al., 2010; Yu, Zhu, Shen, Bai, & Di, 2015). The risk of overdose represents a significant concern, as annual meth-related deaths have recently increased to 3.9 per 100,000, with a significant share of drug overdose deaths now including combinations of opioids and meth (Al-Tayyib, Koester, Langegger, & Raville, 2017; Gladden, O'Donnell, Mattson, & Seth, 2019). The aforementioned complications, coupled with the strong addictive potential of meth, have largely compromised the utility of meth in the clinic. An oral formulation of meth (Desoxyn[®]) is currently only recommended for the treatment of attention deficit hyperactivity disorder (ADHD) and short-term treatment of obesity (U.S. Food and Drug Administration (FDA). Medication Guide, 2017). On the contrary, there is no FDA-approved pharmacotherapy for MUD.

It is well established that acute neurobehavioral effects of meth are primarily mediated by the meth-induced reversal of monoamine transporter function and dramatic and prolonged rise in synaptic monoamine levels (Kogan, Nichols, & Gibb, 1976; O'Dell, Weihmuller, & Marshall, 1991; Robinson, Yew, Paulson, & Camp, 1990; Sulzer et al., 1995). Meth also increases monoamine levels via the inhibition/reversal of vesicular monoamine transporter-2 (VMAT2; Sulzer et al., 1995) and by blocking the activity of monoamine oxidases (MAOs). Meth interacts with a host of other targets, including trace amine-associated receptor 1 (TAAR1), alpha-2 adrenergic receptors, and sigma receptors, though the consequences of meth binding to these receptors are not well-understood (Cisneros & Ghorpade, 2014; Kaushal & Matsumoto, 2011; Kikuchi-Utsumi et al., 2013). With regards to anatomical sites through which meth exerts its effects during acute and prolonged use, the midbrain dopamine system and its projections to the striatum and the frontal cortex have been the most thoroughly characterized. A large body of preclinical and clinical literature provided collective evidence that acute meth produces a dramatic and sustained increase of dopamine from nigrostriatal and mesostriatal terminals, with moderate increases detected at the terminal sites of mesocortical projections (Kogan et al., 1976; O'Dell et al., 1991; Robinson et al., 1990). In contrast, prolonged and binge-like use of meth produces dopamine depletion and downregulation of dopamine receptors and transporters, which has been interpreted either as a sign of neurotoxicity, or as a neuroadaptive mechanism related to meth tolerance (Kitamura, Tokunaga, Gotohda, & Kubo, 2007; Moszczynska et al., 2004; Woolverton, Ricaurte, Forno, & Seiden, 1989; Yu et al., 2015). Acute or binge-like administration of meth also increases glutamate levels across several forebrain regions, including the prefrontal cortex (PFC; Crocker et al., 2014; Stephans & Yamamoto, 1994, 1995), striatum and nucleus accumbens (Str, NAc, (Mark, Soghomonian, & Yamamoto, 2004; Miyazaki et al., 2013) as well as the hippocampus (Raudensky & Yamamoto, 2007; Zhang et al., 2014). Meth-induced increase in local glutamate levels is thought to be mediated via distant/circuitry-based mechanisms, such as activation of nigrostriatal, striatopallidal (Mark et al., 2004; Shen, Chen, Marino, McDevitt, & Xi, 2021) or thalamocortical projections (Mark et al., 2004) to produce elevated glutamate levels in the striatum and cortex, respectively. Alternatively, meth can directly affect presynaptic glutamate release (Andres et al., 2015; Shen et al., 2021), or neural activity of local glutamate neurons (Pu, Broening, & Vorhees, 1996; Stephans & Yamamoto, 1995). Though, it should be noted that both the mechanisms and magnitude of meth-induced glutamate levels likely depend both on the duration of meth exposure and the dose of meth used/administered (Earle & Davies, 1991; Jones et al., 2021; Pereira et al., 2012; Stephans & Yamamoto, 1995; Szumlinski et al., 2017).

Preclinical research has also demonstrated that contingent and noncontingent delivery of meth over prolonged periods of time leads to a spectrum of glutamatergic aberrations. A signature glutamatergic neuro-adaptation across several different chronic meth paradigms is that of "sensitized" glutamate increase in response to meth injection or after the re-exposure to meth-associated cues (Ito, Abekawa, & Koyama, 2006; Lominac et al., 2016; Stephans & Yamamoto, 1995). It has been documented that this sudden rise in glutamatergic tone is behaviorally significant for cue-induced meth craving and seeking (Abulseoud, Miller, Wu, Choi, & Holschneider, 2012; Kufahl et al., 2013; Parsegian & See, 2014; Watterson et al., 2013); see the section on Reinstatement of meth-seeking below).

Specifically, an increase in glutamate levels above the post-meth baseline has been detected in the ventral striatum and the prefrontal cortex (PFC), while inhibition of cortico-striatal glutamate input reduced reinstatement of meth-seeking (Caprioli et al., 2015). It should be noted that chronic meth self-administration also alters basal synaptic glutamate. In rats that underwent 10 days of extinction training (following 10 d FR1 2h/d self-administration; Parsegian & See, 2014) observed, decreased basal glutamate levels in both the PFC and NAc, while (Lominac, Sacramento, Szumlinski, & Kippin, 2012) using an analogous self-administration paradigm, but omitting extinction, reported an increase in basal glutamate levels in the NAc after 1- and 21-day withdrawal period. Thus, it is possible that the dysregulation of basal glutamate after meth depends on neural processes recruited during extinction training, analogously to what has been shown for glutamate receptors after extinction of meth and cocaine-seeking (Knackstedt et al., 2010; Schwendt, Reichel, & See, 2012).

Interestingly in humans, meth dependence is associated with increased glutamate levels (at least in some brain regions; Yang et al., 2018), while abstinent meth users show a decrease in glutamate metabolism across several cortical and subcortical regions (Ernst & Chang, 2008). This supports the idea that dysregulation of glutamate homeostasis is produced during active use and persists during abstinence from chronic meth. Several neurobiological mechanisms that may underlie wide-spread dysregulation of forebrain glutamatergic systems have been identified in preclinical studies. This includes meth-induced changes in neurophysiological characteristics of glutamatergic neurons, such as altered firing rate and activity (Mishra, Pena-Bravo, Leong, Lavin, & Reichel, 2017; Parsegian & See, 2014), dysregulation of vesicular glutamate transporters (vGluT; Earle & Davies, 1991; He et al., 2014; Jones et al., 2021; Mark, Quinton, Russek, & Yamamoto, 2007) and glutamate receptors, including mGlu2/3 and mGlu7 (Schwendt et al., 2012).

This chapter focuses on the role of the glutamatergic system, specifically metabotropic glutamate (mGlu) receptors, in neural, physiological, and behavioral changes produced by meth. However, since several neurotransmitter systems are known to be involved in regulating addictive behaviors, we will also discuss the possible role of receptor-receptor interactions involving mGlu receptors and other neurotransmitter receptors in meth-induced neural and behavioral changes. Finally, based on the review of the available evidence, this review will propose several avenues for the development of MUD pharmacotherapies based on modulating the activity of mGlu receptors.

2. Metabotropic glutamate receptors (mGlu)—An overview

Receptors for the major excitatory neurotransmitter glutamate can be found throughout the central nervous system and mediate a variety of cellular responses depending on their type (ionotropic versus metabotropic) or their synaptic location (presynaptic, postsynaptic, glia, endothelial cells); see: (Karakas, Regan, & Furukawa, 2015; Markou, 2007). Besides the fast ligand-gated ion channels mediating excitatory neurotransmission (ionotropic glutamate receptors or iGlus; Karakas et al., 2015; Traynelis et al., 2010), there are also eight different metabotropic glutamate receptors (or mGlus). These are part of the G protein-coupled receptor (GPCR) superfamily and can be categorized into three groups: Group I (mGlu1, mGlu5), Group II (mGlu2, mGlu3), and Group III (mGlu4, mGlu6, mGlu7, mGlu8), for review see: Niswender and Conn (2010). Unlike iGlus, activation of these receptors does not result in an opening of an ion-permeable channel but initiates complex protein signaling cascades that can influence other proteins and receptors (Gabriel et al., 2012). mGlus are coupled with and activate G-proteins. Group I proteins interact with $G\alpha_{q/11}$ G-proteins, which are further involved in the $PKC/IP_3/DAG$ signal transduction. On the other hand, Groups II and III are coupled predominantly to $G\alpha_{i/o}$ proteins, which classically inhibit the function of adenylyl cyclase and directly regulate ion channels via $G_{\beta\gamma}$ subunits of the G-protein. Additionally, mGlus can activate other downstream effectors, including phospholipase D and protein kinase pathways, such as CK1, CDK5, Jun, MAPK/ERK, and mTOR/p70 S6 pathways (Conn & Pin, 1997; Niswender & Conn, 2010). Besides the signaling diversity, individual mGlu receptor subtypes are located at a variety of synaptic and extrasynaptic sites, and this topographical distribution can further vary between brain regions, as reviewed in Niswender and Conn (2010). In short, mGlu1 and 5 are predominately present at postsynaptic sites, and less frequently located on glia, with only sparse presynaptic expression. On the other hand, mGlu2, 3 and 7 are almost exclusively expressed on presynaptic sites, with mGlu3 showing moderate expression in glia and postsynaptic sites. Though mGlu3 postsynaptic expression might be more frequent in the cortex in comparison to other subcortical sites (such as the striatum, amygdala, or midbrain; (Neki et al., 1996; Petralia, Wang, Niedzielski, & Wenthold, 1996;

Tamaru, Nomura, Mizuno, & Shigemoto, 2001)). mGlu4 and mGlu8 expression is less understood. In general, mGlu4 functions as an inhibitory autoreceptor and is primarily localized presynaptically, mainly in the cerebellar cortex, basal ganglia, thalamus, and hippocampus (Volpi, Fallarino, Mondanelli, Macchiarulo, & Grohmann, 2018). Even less is known about mGlu8 expression patterns, but it is found near active presynaptic sites in the hippocampus (Shigemoto et al., 1997). mGlu7 is the most expressed mGlu in the brain, and is found mainly in the amygdala, hippocampus, and hypothalamus (Kinoshita, Shigemoto, Ohishi, van der Putten, & Mizuno, 1998). mGlu6, the last from the Group III mGlus, is almost exclusively expressed postsynaptically on bipolar cells of the mammalian retina (Testa, Catania, & Young, 1994; Tian & Kammermeier, 2006), and thus it will not be the focus of the current review.

Presynaptic mGlu receptors are predominately $G\alpha_{i/o}$ – coupled and are involved in inhibitory control over neurotransmitter release, often acting as autoreceptors for glutamate. For example, activation of mGlu2, 3 or 7 receptors on cortico-striatal terminals significantly dampens down glutamate release during reinstatement, and reduced drug- (including meth-) seeking (Kufahl, Watterson, et al., 2013; Li, Li, Gardner, & Xi, 2010). Postsynaptic mGlu receptors that are typically $G\alpha_q/G_{11}$-coupled often participate in long-term changes in synaptic plasticity, such as mGlu1,5-induced long-term depression (LTD) detected in the striatum, hippocampus and ventral tegmental area (VTA; Bellone & Lüscher, 2005; O'Riordan, Hu, & Rowan, 2018; Scheyer, Christian, Wolf, & Tseng, 2018), while the onset of long-term potentiation after activation of mGlu5 has been observed in the amygdala (Fendt & Schmid, 2002). Psychostimulants, such as cocaine, are known to disrupt mGlu5-dependent LTD in several brain regions, though this has not yet been described for meth. Restoring mGlu5-dependent LTD diminishes cocaine seeking, suggesting a causal role of drug-induced changes in mGlu5-dependent plasticity. For a more in-depth review of synaptic plasticity controlled by individual mGlu subtypes, see: (Bellone & Mameli, 2012; Chiamulera, Piva, & Abraham, 2021).

The following sections will discuss the role of individual mGlu receptors in meth-associated changes in neural function and behavior. Thought it should be noted that mGlu receptors are involved in regulating a wide variety of behaviors and behavioral pathologies, including anxiety, depression, pain, epilepsy, and others that are outside of the scope of the current review.

3. mGlu receptors and methamphetamine-associated neural changes

3.1 Methamphetamine-induced neurotoxicity

It is well accepted that meth can have neurotoxic effects, though the extent, the reversibility, and the mechanisms of meth-induced toxicity are still being debated (Jayanthi, Daiwile, & Cadet, 2021; Yang et al., 2018). For the sake of this review (which does not specifically focus on neurotoxicity), the years of research in this area can be briefly summarized below. Acute administration of high doses of meth, or frequent exposure to moderate-to-high meth doses, produces scalable damage to the dopaminergic system that is reversible under certain conditions. The mechanisms of meth-induced neuronal damage are multiple and include neuroinflammation, oxidative stress, hyperthermia, and mitochondrial dysfunction (Kim, Yun, & Park, 2020; Matsumoto et al., 2014; Shaerzadeh, Streit, Heysieattalab, & Khoshbouei, 2018; Shin et al., 2018). Excessive glutamate release and supranormal extracellular glutamate concentrations are thought to contribute to mechanisms of meth-induced neurotoxicity. Elevated glutamate levels have been observed both in the striatum of animals after administration of high doses of meth, as well as in the brain of active meth users diagnosed with MUD (Ernst & Chang, 2008; Yang, Yang, et al., 2018). While inhibition of NMDA glutamate receptors reliably reduces meth-induced neurotoxicity, mGlu5 (and possibly also mGlu2/3) inhibitors also display neuroprotective properties (Battaglia et al., 2002; Gołembiowska, Konieczny, Ossowska, & Wolfarth, 2002; Gołembiowska, Konieczny, Wolfarth, & Ossowska, 2003). This suggests that mGlu5 activation is a critical event 'upstream' from meth-induced dopaminergic neurotoxicity (Battaglia et al., 2002). Though it cannot be excluded that neuroprotective properties of mGlu5 co-depend on its intracellular interaction with NMDA receptors (Tu et al., 1999), as supported by observations that neurotoxicity induced by direct administration of NMDA agonists is diminished in cells expressing mGlu5 (Bruno et al., 2000). In addition, the frequently used mGlu5 antagonist/inverse agonist MPEP (2-methyl-6-(phenylethynyl)pyridine), can also inhibit NMDA receptors (or have other off-target effects) at higher concentrations (Movsesyan et al., 2001) therefore its neuroprotective effects can be partly attributed to the blockade of non-mGlu5 targets (Lea IV, Movsesyan, & Faden, 2005).

Hyperthermia, or elevated body temperature due to failed thermoregulation, can cause severe damage to the CNS (Burke & Hanani, 2012). High temperatures can induce neuronal death, and trigger apoptosis-like molecular signaling cascades that further damage the brain (White et al., 2007). The use of meth (and other amphetamine-type stimulants) may result in hyperthermia, which not only directly damages neuronal cells, but also results in the breakdown of the blood-brain barrier (BBB), which further exacerbates brain injury (Bowyer & Hanig, 2014; Kiyatkin & Sharma, 2009). The role of excitatory amino acids in thermoregulatory and hyperthermia has long been recognized (Cremades & Peñafiel, 1982; Monda, Viggiano, Sullo, & de Luca, 1998; Sengupta, Jaryal, Kumar, & Mallick, 2014), and several studies have reported that pharmacological inhibition or genetic ablation of mGlu2/3 and/or mGlu5 receptors reduce stress-induced hyperthermia (Brodkin et al., 2002; Iijima, Shimazaki, Ito, & Chaki, 2007; Spooren et al., 2000). Analogously, Gołembiowska et al. (2003) found that systemic administration of MPEP (5 mg/kg) reduced meth-induced neurotoxicity and hyperthermia in rats after a binge-like meth regimen (five doses of 10 mg/kg meth administered every 2 h; Gołembiowska et al., 2003). Though MPEP might provide only transient relief from meth-induced hyperthermia (Battaglia et al., 2002), this initial MPEP effect contributes to the reduction of long-lasting meth-induced neurotoxicity (Gołembiowska et al., 2003).

The ability of MPEP to reduce meth-induced neurotoxicity (as well as meth-induced behaviors; see below), has been attributed to the drug's ability to counteract meth-induced dopamine efflux in the striatum. Several potential mechanisms underlying the effect of MPEP on dopamine release have been put forward, including polysynaptic circuitry mechanisms involving cortical mGlu5, group I mGlu expressed on midbrain dopamine neurons, or local mechanisms involving mGlu5 expressed on striatal dopaminergic terminals (Hubert, Paquet, & Smith, 2001; Page, Peeters, Najimi, Maloteaux, & Hermans, 2001). With regards to "local" effects of MPEP, the work by Page et al. has shown that mGlu5 expressed in rat striatal synaptosomes can modulate the activity of DAT via a phosphorylation-dependent process (Page et al., 2001). It is well established that DAT contains multiple functionally relevant phosphorylation sites that modulate DA transport, in part, via promoting the internalization of the transporter (Foster & Vaughan, 2017). Meth and other amphetamine-type stimulants induce DAT internalization and loss of function through recruiting PKC and other relevant kinases (Challasivakanaka et al., 2017; Sandoval, Riddle, Ugarte, Hanson, & Fleckenstein, 2001; Saunders et al., 2000). Therefore, it is possible that MPEP-dependent neuroprotection is through inhibition of DAT phosphorylation and internalization, which in turn aids with moderating synaptic dopamine levels.

There are additional mechanisms through which mGlu receptors can regulate meth-induced neurotoxicity and thus can be exploited to develop mGlu-based interventions. This includes the ability of mGlu5 antagonists to suppress meth-elicited production of reactive oxygen species (Battaglia et al., 2002), or ameliorate meth-associated neuroinflammation and glial activation (Spampinato, Copani, Nicoletti, Sortino, & Caraci, 2018). It should be noted that the reason why only mGlu5 blockers have been tested for their effects on meth-induced neurotoxicity is that orthosteric agonists of mGlu5 could be neurotoxic on their own (Wíniewski & Car, 2002). Therefore, the behavioral and neural effects of mGlu5 activation have been predominately explored using mGlu5 positive allosteric modulators (PAMs); see below.

3.2 Methamphetamine-induced changes mGlu receptor expression and function

Meth exposure produces a spectrum of short-term and long-term neurochemical and neurobiological changes in the brain. Emerging evidence suggests that mGlu receptors themselves are subject to meth-induced neuroadaptations. Here we discuss the studies published to date that evaluated changes in function, synaptic/cellular location and/or expression produced by various regimens of meth exposure in animals. Though cocaine, alcohol, or opioid use have all been shown to dynamically alter mGlu5 availably in the human brain, dysregulation of human mGlu receptors after meth has not yet been assessed, and therefore, it will not be discussed here.

Early evidence that mGlu receptor function could be altered after amphetamine-type stimulants originates from studies using d-amphetamine. Mao and Wang (2001) reported that repeated administration of amphetamine downregulates the expression of mGlu5, but not mGlu1, mRNA in the striatum for up to 28 days after the last dose of the stimulant (Mao & Wang, 2001). Using a similar experimental design, our laboratory found that repeated amphetamine administration does not alter mGlu5 receptor protein in the striatum after 21 days of withdrawal, but downregulates several proteins within the mGlu5 signaling complex, suggesting lasting receptor dysfunction after amphetamine (Schwendt & McGinty, 2007). Repeated amphetamine also produces an early and sustained increase in mGlu8 mRNA levels in the striatum and cingulate cortex (Parelkar & Wang, 2008). Another early report provided indirect evidence that mGlu receptor function is altered after meth. Hashimoto et al. (2007) reported that a single high dose of meth (40 mg/kg) increased the mRNA and protein levels of a mGlu1/5-specific scaffolding protein Homer 1a across the rat striatum (Hashimoto et al., 2007). Another study found that repeated administration of low meth doses (2 mg/kg/day for 10 days) produced robust behavioral sensitization in mice and increased mGlu5 protein levels in the hippocampus, as detected after 10 days of withdrawal (Liao et al., 2018). Herrold, Voigt, and Napier (2011) explored changes in mGlu5 cell surface expression using a BS^3 crosslinking assay 24 h after the expression of conditioned meth seeking utilizing a CPP paradigm that included three days of daily meth administration (1 mg/kg; Herrold et al., 2011). The study analyzed mGlu5 expression across the three limbic brain regions relevant for drug-seeking—mPFC, NAc, and ventral pallidum (VP), but found that a change in mGlu5 distribution (a decrease in the surface-to-intracellular receptor ratio) was limited to the mPFC (Herrold et al., 2011). However, as the rat tissue was analyzed after a meth-seeking test, it remains undetermined whether the redistribution (internalization) of mGlu5 was driven by prior meth-conditioning or, more acutely, by re-exposure to the meth context. Further, when analyzed 14 days after the last meth injection with no conditioned placement test, Herrold, Persons, and Napier (2013) found an increase in surface-to-intracellular mGlu5 ratio exclusively in the VP, and not in the other two brain areas (Herrold, Persons, & Napier, 2013). In contingent meth exposure paradigms, our group has shown that extended access to meth self-administration (6 h/day for 14 days) reduced mGlu5, but not mGlu2/3, protein levels selectively in the perirhinal cortex (PrH), as measured 14 days after the discontinuation of meth (Reichel, Schwendt, McGinty, Olive, & See, 2011). Interestingly the expression of mGlu5 in the PFC and PrH negatively correlated with prior meth intake (Reichel et al., 2011). In a separate study, utilizing the identical self-administration paradigm, we have reported that meth self-administration followed by 14 days of abstinence reduced cell surface and total levels of mGlu2/3 in the striatum, while in the PFC, only a loss of surface mGlu2/3 and mGlu7 receptors was detected (Schwendt et al., 2012). Interestingly, 14 days of extinction training reversed all identified meth-induced changes, except for the reduced cell surface mGlu2/3 levels in the PFC (Schwendt et al., 2012). As neurotransmitter receptors located on the cell surface are thought to represent the functional receptor pool, the analysis of mGlu receptor cell surface expression provides an indirect measure of their function. Though it should be noted that other mechanisms, such as changes within the receptor signalosome, can alter mGlu receptor function. Accordingly, Murray et al. (2021) reported that disrupted mGlu5-dependent synaptic plasticity (long-term depression) found in the NAc core during early withdrawal from chronic meth self-administration occurs without a concomitant change in the mGlu5 surface expression (Murray et al., 2021). Curiously, the authors have failed to identify the exact cellular or trans-synaptic mechanism responsible for the apparent mGlu5 dysfunction. In summary, when interpreting meth-induced changes in mGlu receptors, the collective evidence highlights the necessity to consider a host of variables such as: the amount of drug taken (Reichel et al., 2011), the length of withdrawal (Murray et al., 2021), and the character of post-meth experience such as extinction training (Schwendt et al., 2012). Outside meth, there are indications that mGlu5 receptor availability in the brain of subjects diagnosed with SUD changes depending on such variables (Ceccarini et al., 2020). And further, mGlu5 availability is predictive of several SUD-associated behaviors, such as drug craving or novelty-seeking (Ceccarini et al., 2020; Leurquin-Sterk et al., 2016). Therefore, future research into this area needs to be based on well-controlled studies that incorporate these variables into a relevant meth self-administration paradigm (such as an extended drug access paradigm) to collect data that will be both comprehensive and useful for the development of novel therapies for MUD.

4. mGlu receptors and meth-induced behavioral effects
4.1 Psychomotor effects of methamphetamine

One of the early pivotal studies that established the role of a mGlu receptor in the psychomotor activating effects of stimulants was the study in which Chiamulera et al. (2001) demonstrated a lack of cocaine-induced hyperactivity in mGlu5 (*GRM5* null) knockout mice (Chiamulera et al., 2001). A later study found that systemic inhibition of mGlu5 (with the receptor antagonist MPEP), dose-dependently inhibited the acute locomotor effects of d-amphetamine (Mcgeehan, Janak, & Olive, 2004). The first direct evidence that mGlu receptors regulate acute meth-induced hyperactivity was presented by Satow et al. (2008). Here the authors evaluated the effects of the mGlu1 allosteric antagonist FTIDC (4-[1-(2-fluoropyridine-3-yl)-5-methyl-1*H*-1,2,3-triazol-4-yl]-*N*-isopropyl-*N*-methyl-3,6-dihydropyridine-1(2*H*)-carboxamide), the mGlu2/3 agonist LY379268 and the mGlu5 antagonist MPEP on meth-induced hyperlocomotion in mice. All three tested compounds dose-dependently reduced the acute locomotor effects of meth, although MPEP and LY379268 were effective only at doses (30 mg/kg and 3 mg/kg, respectively) that also caused hypolocomotion in drug-naïve animals (Satow et al., 2008). Repeated administration of meth, akin to other stimulants, reliably produces behavioral sensitization to a subsequent (typically lower) dose of meth (Fujiwara, Kazahaya, Nakashima, Sato, & Otsuki, 1987; Pierce & Kalivas, 1997). Using this paradigm, Kufahl, Nemirovsky, Watterson, Zautra, and Olive (2013) found that systemic mGlu5 inhibition with fenobam (an anxiolytic-like drug and mGlu5 NAM/inverse agonist (Porter et al., 2005), reduced the expression of locomotor sensitization to meth (Kufahl, Nemirovsky, et al., 2013). In addition, fenobam was partially effective at attenuating locomotion in saline-treated rats that received a single, low dose of meth (0.5 mg/kg; (Kufahl, Nemirovsky, et al., 2013). The same study also showed that a mGlu5 PAM CDPPB (3-Cyano-N-(1,3-diphenyl-1H-pyrazol-5-yl) benzamide) failed to attenuate meth-induced behavioral sensitization. While the neural mechanisms through which group I and group II mGlu receptors can modulate acute and conditioned psychomotor effects of meth are not fully understood, limited evidence suggests that it likely involves the regulation of striatal dopamine release (Arai et al., 1997; Shimazoe et al., 2002), and/or maintenance of glutamate homeostasis in the PFC (Lominac et al., 2016).

4.2 Methamphetamine-associated learning, reward, and reinforcement

Several studies explored the role of mGlu receptors in mediating meth-associated reward, learning, and reinforcement using a gene deletion approach. In this regard, Chesworth, Brown, Kim, and Lawrence (2013) found that mGlu5 KO mice self-administered meth under a fixed-ratio or progressive-ratio condition at a similar rate as their wild-type counterparts (Chesworth et al., 2013). However, mGlu5 deletion slowed the extinction of meth cue–associated responding, which is in agreement with other research on the critical role of mGlu5 in extinction learning (see below). In the same study, mGlu5 deletion increased cue-primed meth reinstatement, as well-conditioned hyperactivity in the CPP test (Chesworth et al., 2013). It is possible that overall hyperactivity and generalized learning deficits observed in mGlu5 KO mice contributed to these effects (Hamilton, Esseltine, Devries, Cregan, & Ferguson, 2014; Jew et al., 2013; Xu et al., 2021). Along the same lines, recent evidence suggests that data collected using a constitutive mGlu5 deletion should be interpreted with caution, as altered developmental plasticity might lead to divergent behavioral manifestations and thus inaccurate interpretation of the role of mGlu5 in psychostimulant-associated behaviors (e.g., see: Chiamulera et al., 2001 vs. Fowler, Varnell, & Cooper, 2011). In support, adaptive changes following the constitutive deletion of other mGlu receptors have been observed. For example, mGlu2 and mGlu3 KO mice display dopaminergic aberrations, such as elevated D2 receptors in the striatum and marked agonist–induced dopamine supersensitivity (Seeman, Battaglia, Corti, Corsi, & Bruno, 2009). mGlu2 KO mice also exhibit hyperactivity in a novel environment, augmented dopamine and glutamate release in response to cocaine, and elevated locomotor sensitization and conditioned responses to cocaine (Morishima et al., 2005). And finally, mGlu2-deficient rats (due to a non-sense *GRM2* mutation) displayed decreased reward efficacy of cocaine, an effect driven by adaptations in basal glutamate regulation (Yang et al., 2017). Akin to the effects of mGlu2 and 3 deletion, genetic ablation of another (predominately presynaptic) mGlu receptor—mGlu7, increased alcohol consumption (Gyetvai et al., 2011). Unfortunately, none of these studies were conducted using meth, so the off-target effects of constitutive mGlu gene deletion are not known.

Thanks to advances in the field of mGlu receptor pharmacology, a large array of experimental mGlu ligands exist, with several of them evaluated in animal models of meth use. With regards to mGlu5, Osborne and Olive (2008) reported that systemic administration of mGlu5 NAM 3-[(2-methyl-1,3-thiazol-4-yl)ethynyl]pyridine (MTEP), dose-dependently (1–3 mg/kg) reduced meth self-administration, under the fixed ratio 1 (FR1) schedule of reinforcement (Osborne & Olive, 2008). In a follow-up study, Gass, Osborne, Watson, Brown, and Olive (2009) reported the ability of MTEP (3 mg/kg) to also reduce responding under a progressive ratio (PR) schedule (Gass et al., 2009). Specifically, MTEP dose-dependently reduced the total number of reinforcers earned, independently of intravenous meth concentration available. Temporal analysis of the data revealed that MTEP (3 mg/kg) produced a delay in escalated PR responding and reduced the overall number of responses, but the time to reach the breakpoint was not altered. This suggests that MTEP may reduce the reinforcement efficacy of meth, without causing a premature cessation of PR responding. On the other hand, the study found that MTEP (1–3 mg/kg) did not alter food reinforcement (Gass et al., 2009). Another study evaluated the effects of mGlu1 and mGlu5 NAMs for their ability to alter maintenance of meth-induced CPP

(Herrold, Voigt, & Napier, 2013). Each NAM was administered daily over the course of two days, once the CPP had already been established. Administration of mGluR1 NAM, JNJ16259685 (0.3 mg/kg) did not alter the conditioned effects of meth, while mGluR5 NAMs MTEP (3 mg/kg) and MPEP (30 mg/kg) inhibited the expression of established CPP behavior (Herrold, Voigt, & Napier, 2013). Finally, the effects of the mGlu1 NAM JNJ16259685 on meth reinforcement were also evaluated in non-human primates (Achat-Mendes, Platt, & Spealman, 2012). Systemic mGlu1 NAM administration dose-dependently inhibited meth reinforcement under a second-order schedule (FI5-FR10), with no significant effects on food-reinforced responding (Achat-Mendes et al., 2012). The effects of other mGlu ligands on meth-associated reward and reinforcement have yet to be explored, though studies show that mGlu2/3 agonists or mGlu2 PAMs decrease self-administration of other stimulants (e.g., nicotine (Justinova et al., 2015, Justinova, le Foll, Redhi, Markou, & Goldberg, 2016).

4.3 Reinstatement of methamphetamine-seeking

Reinstatement of the drug-seeking response is a key behavioral variable in preclinical research and is thought to relate to relapse in human subjects. Reducing relapse rates is a major indicator of treatment success in SUD (Venniro, Caprioli, & Shaham, 2016). In animal studies, drug-seeking behavior can be observed and tested either by between-session (self-administration is followed by extinction training, during which the animal is learning that the conditioned behavior does not deliver reinforcer anymore, followed by reinstatement test), within-session (test for reinstatement is performed right after the extinction of the drug-reinforced behavior), or the between-within session (test is conducted on the same day after different days of drug withdrawal). Reinstatement can also be observed after a period of abstinence, although the term "relapse" is used for this experimental design. The test itself can be either cue-primed (environmental cues connected to the reinforcer administration, such as visual, auditory, olfactory, or tactile cues), or drug-primed (administration of a dose of the drug used in self-administration); see: Shaham, Shalev, Lu, de Wit, and Stewart (2003).

Here a review of systemic mGlu manipulations is provided first, followed by the discussion of intracranial mGlu manipulations, with both topics focused on meth-seeking. Watterson et al. (2013) explored the effect of fenobam (5, 10 and 15 mg/kg) on reinstatement of meth-seeking in male rats with a history of meth self-administration (FR1: 2 h daily access to meth for a minimum of 8 days) and extinction (Watterson et al., 2013). The two higher doses of fenobam reduced drug- and cue-induced reinstatement of meth-seeking. However, fenobam also attenuated cue-induced reinstatement of sucrose- and food-seeking at all doses tested, indicating potential side-effects of this drug on motivation to seek natural reinforcers. Gass et al. (2009) analyzed the effect of systemic administration of mGlu5 NAM MTEP on cue- and drug-primed reinstatement of meth-seeking (Gass et al., 2009). While the two lower doses of MTEP used (0.3, 1 mg/kg) had no effect on reinstatement, the highest dose (3 mg/kg) significantly attenuated meth-seeking behavior in both cue- and drug-primed tests (Gass et al., 2009). None of the MTEP doses used in this study affected cue-induced reinstatement of food-seeking behavior. This is likely due to the higher potency and selectivity of MTEP when compared to fenobam. MTEP also shows fast brain penetration and rapid onset of receptor occupation (reduced meth-seeking already within the first 30 min), in contrast to the unfavorable pharmacokinetic profile of fenobam in rodents and humans (Busse et al., 2004; Cavallone et al., 2020; Gass et al., 2009). With regards to mGlu receptors other than mGlu5, Kufahl, Nemirovsky, et al. (2013), Kufahl, Watterson, et al. (2013) investigated the effects of systemic mGlu2/3 activation (using a mGlu2/3 agonist LY379268) on cue- and drug-primed reinstatement of meth seeking in male rats (Kufahl, Watterson, et al., 2013). This study also compared rats with a history of limited (FR1: 90 min/day for 16 days) and extended (8 days of limited access, followed by FR1: 6 h/day for 8 days) access to meth self-administration. While all rats displayed similar responding during cue- and drug-primed tests, extended-access rats demonstrated higher "sensitivity" to mGlu2/3 activation (at 0.3 mg/kg) to reduce cue-primed reinstatement. Interestingly, the dose–dependent reduction of drug-primed reinstatement by LY379268 was similar in both groups of rats regardless of past meth access (Kufahl, Watterson, et al., 2013). Again, as was the case with the discussed mGlu5 ligands, LY379268 might have a limited therapeutic window as sucrose seeking in drug-naïve rats was significantly suppressed by the two higher doses of the LY379268 drug (1 and 3 mg/kg). Several studies evaluated the effects of mGlu5 PAMs on extinction (see the section below) and reinstatement. Widholm, Gass, Cleva, and Olive (2012) reported that systemic administration of CDPPB (60 mg/kg) failed to reduce context-primed reinstatement of meth-seeking after extinction. The authors utilized male rats that underwent standard meth self-administration (FR1: 2 h/daily for 12 days) followed by extinction training in a different context (Widholm et al., 2012). The lack of the CDPPB effect is somewhat surprising, as in our recent study, this compound (CDPPB, 30 mg/kg) significantly reduced context-primed relapse to cocaine seeking after abstinence (Gobin & Schwendt, 2020). Differences in psychostimulant type/mechanism of action and incorporation of extinction training possibly account for this discrepancy. It should also be noted that CDPPB displays a non-linear dose-effect profile with regard to its neural and behavioral effects, with lower/mid-range doses having higher efficacy (Uslaner et al., 2009).

Two studies published to date have evaluated the effects of local, intracranial administration of mGlu5 PAMs or NAMs on reinstatement of meth seeking. Peters, Scofield, Ghee, Heinsbroek, and Reichel (2016) used a novel mGlu5 PAM DPFE (1-(4-(2,4-difluorophenyl) piperazin-1-yl)-2-((4-fluorobenzyl)oxy)-ethanone) administered directly into the PrH and evaluated meth-seeking in rats with a history of limited (FR1: 1 h/day for 21 days) and extended (FR1: 1 h/day for 7 days, followed by 6 h/day for 14 days) meth self-administration. In addition, this study evaluated the possible effect of novel cue (novel stimulus light and additional lever) presentation on meth-seeking driven by "standard" conditioned cues (light and tone). The rationale of this study was to investigate the role of PrH mGlu5s under the conditions in which novelty competes with the conditioned rewarding effects of the drug, as shown for cocaine (Reichel & Bevins, 2008). The study found that animals with a history of extended access to meth (higher number of cue-drug pairings during self-administration) preferentially responded to meth-associated cues, while limited-access animals equally divided time to responding to novel and meth-related cues. Local administration of DPFE did not alter meth-seeking under standard conditions but redistributed responding away from a meth-paired cue to a novelty cue in rats with a history of extended access to meth. This suggests that mGlu5 receptors (in the PrH and elsewhere) can play a more complex role in regulating meth-seeking, especially under "real-life" relapse conditions during which a mix of conditioned, neutral, and novel cues is often present. These brain-region and cue-specific specific effects of DPFE might also explain the inability of systemic mGlu5 PAM to reduce meth-seeking elicited by the drug-conditioned context (Widholm et al., 2012). Additionally, local intra-NAc inhibition of mGlu5 receptors (using mGlu5 NAM MTEP) does not alter cue-primed meth seeking under the "standard" (no novel cues are present) conditions (Murray et al., 2021). Therefore, given the currently incomplete understanding regarding the location or the involvement of mGlu5 receptors in meth-seeking more research into this topic is needed. At the same time, findings by Peter et al. indicate that pharmacotherapies that rescue mGlu5 function in the PrH disrupted by chronic meth, might restore novelty salience and reduce relapse in MUD.

4.4 Methamphetamine-associated extinction learning

Extinction is learning that the stimulus or response previously paired with reinforcer delivery is no longer paired and leads to the reduction of stimuli-induced responses (for an introduction to this topic, see: Bouton, 2004). Within the preclinical research on motivated behavior and drug addiction, this process is typically studied in animals that acquired stable self-administration behavior and are then re-exposed to the drug-associated context and/or cues in which the reinforcer contingency (drug delivery) has been removed. Extinction learning is typically studied in a within-session design (in which operant responding during a single session is analyzed over individual time bins), or between-session design (in which operant responding is measured per session over the course of several days). Extinction training is considered completed once animals reach a predetermined criterion, typically corresponding to low levels (15–30%) of the initial response rate.

Among the mGlu receptors, mGlu5 in particular, has been found to be necessary for various forms of extinction learning (fear cue extinction, drug cue extinction etc.), although the following discussion will focus solely on findings with meth. In this regard, Chesworth et al. (2013) found that constitutive deletion of mGlu5 (in *GRM5* KO mice) significantly impeded between-session extinction learning following the discontinuation of meth self-administration (Chesworth et al., 2013). The self-administration regimen used in this study included up to 3 weeks of FR1 2h access and two independent PR sessions, which is considered an extensive self-administration experience for mice. The *GRM5* KO mice required a significantly longer time (7 days vs 3 days) to extinguish meth seeking when compared to their wild-type counterparts; 30% of averaged stable FR1 active lever responding was used as extinction criterion (Chesworth et al., 2013).

While the effects of pharmacological mGlu5 blockade on extinction of meth seeking have not been investigated to date, several studies evaluated the hypothesis that mGlu5 activation would accelerate extinction learning. Widholm et al. (2012) investigated the effect of systemic high-dose CDPPB (60 mg/kg) on extinction of meth-seeking in an alternative context. In their experimental design, after the completion of meth self-administration

(FR1: 2 h/day for 12 days), rats underwent daily extinction training in a context that was distinct from the self-administration context, while the presentation of drug-associated cues mirrored the meth self-administration conditions. Surprisingly, the authors failed to observe the effect of daily pre-session CDPPB administration on the rate of extinction when compared to vehicle-treated rats. This finding is in contrast with studies using other stimulants, such as cocaine, in which CDPPB accelerated instrumental and CPP-based extinction (Cleva et al., 2011; Gass & Olive, 2009). A separate study by the same research group investigated the effects of systemic CDPPB administration on instrumental extinction of meth-seeking in the same context (Kufahl et al., 2012). Here rats underwent either limited (1 h) or extended (6 h) access meth self-administration (FR1: 1–6 h/day for 16 days) followed by a "standard" daily extinction training in the self-administration context, with no cues present. Daily pretreatment with a moderate dose of CDPPB (30 mg/kg) significantly enhanced the daily extinction rate, mainly in the initial phase of extinction (days 1–7), regardless of the self-administration access history (Kufahl et al., 2012). Interestingly, CDPPB-facilitated extinction had no lasting moderating effect on subsequent cue-induced reinstatement of meth seeking, or on the additional post-test extinction training (Kufahl et al., 2012). Though, it should be noted that the rates of instrumental responding remained low after the cue test. The observed discrepancies between the two studies can be attributed to a different circuitry and salience of discrete vs contextual cues, or to the recently described non-linear dose-dependent effect of CDPPB on learning (Uslaner et al., 2009). Huang, Yu, Chang, and Gean (2016), investigated the role and the brain location of mGlu5 receptors involved in extinction of conditioned meth responses, using a conditioned place preference (CPP) task in adult male mice (Huang et al., 2016). The authors focused on the basolateral amygdala (BLA), as prior research suggested that encoding of drug-context associations (using an amphetamine-induced CPP) is accompanied by an increase in the number and the efficacy of excitatory/glutamatergic synapses within this brain region (Rademacher, Rosenkranz, Morshedi, Sullivan, & Meredith, 2010). The study showed that a 3-day meth conditioning paradigm (2 mg/kg of meth) produced a robust conditioned place preference and increased excitatory transmission in the BLA, as evidenced by increased AMPA/NMDA ratios and GluA2 surface expression in acute BLA slices (Huang et al., 2016). Consequently, the prolonged (10-day), but not standard (6-day) extinction procedure was able to suppress meth-primed renewal of conditioned memory and reverse AMPA and

NMDA receptor changes. In relevance to mGlu5, repeated daily administration of CDPPB (10 mg/kg) accelerated extinction-related outcomes, and now rats that underwent only the standard extinction procedure were protected against meth-primed relapse and did not display meth-induced changes in AMPA/NMDA availability and function (Huang et al., 2016). On the contrary, administration of mGlu5 NAM (MTEP: 20 mg/kg) prior to daily extinction session, blocked the beneficial effects of the prolonged extinction procedure (Huang et al., 2016). It should be noted that chronic meth exposure can reduce the expression/availability of mGlu receptors (including mGlu2/3, 5 and 7) in several brain regions (see above), potentially compromising the efficacy of mGlu PAMs to enhance extinction and/or provide protection against meth relapse. On the other hand, we have shown that extinction training itself can recruit a specific type of glutamatergic plasticity (including the rescue of mGlu function and availability) that may have a protective, anti-relapse effect (Knackstedt et al., 2010; Schwendt et al., 2012; Schwendt & Knackstedt, 2021). Additional research is needed to characterize and understand the role of individual mGlu receptors in extinction of meth-seeking behavior.

4.5 Methamphetamine-associated changes in non-drug related memory processes

Problematic meth use in humans has been associated with a spectrum of cognitive impairments, affecting executive function, attention, social cognition, flexibility, working memory, information processing speed, and language (Scott et al., 2007). These impairments not only reduce the quality of life, but together with altered inhibitory control and impulsivity, may also contribute to the high risk of relapse, even after prolonged periods of successful abstinence (Paulus, Tapert, & Schuckit, 2005; Simon, Dacey, Glynn, Rawson, & Ling, 2004; Wang et al., 2013). Due to the significance of this problem, many preclinical studies (300+ PubMed-indexed publications to date) investigated the characteristics and neural basis of meth-induced memory and cognitive deficits. For the most recent overview of the preclinical research on this topic, see: (Bernheim, See, & Reichel, 2016; Mizoguchi & Yamada, 2019; Shukla & Vincent, 2021). Given the focus of this review, here we will primarily discuss the available evidence on the role of mGlu receptors in meth-induced disruption in cognitive functioning.

In this regard, the limited evidence centers on the role of mGlu receptors and meth-induced disruption of episodic memory. Episodic memory disruptions have been well documented both in meth users and experimental animals (for review see: Reichel, Ramsey, Schwendt, McGinty, & See, 2012). In experimental animals, a particular aspect of episodic memory, a recognition memory, is typically investigated using variations of an object recognition task. In these tasks, it is common that animals are first introduced and are able to freely explore a set of objects, with unique locations or tactile or visual characteristics. After the familiarization phase, innate novelty-triggered object exploration/interaction is assessed, with the changed location, changed object identity, or order representing the novelty stimulus. It is well established that episodic memory requires intact hippocampal function. Though, depending on whether all or some of the "what/where/when" episodic memory aspects are being tested, additional cortical areas (such as the PFC and the PrH) are recruited (Barker et al., 2017). Both non-contingent administration and self-administration of meth impair recognition memory in male rats (Bernheim et al., 2016; Reichel et al., 2011, 2012), a disruption that coincided with altered excitatory/glutamatergic neurophysiology in the PFC (Mishra et al., 2017, 2021) and disrupted long-term depression in the PrL (Scofield et al., 2015). This is significant as glutamate receptors (including several mGlu receptors) within this PFC ↔ hippocampus (HPC) ↔ PrH anatomical circuit are important for acquisition, consolidation, and retrieval of recognition memory (Barker et al., 2006; Barker, Bashir, Brown, & Warburton, 2006; Winters & Bussey, 2005), though relatively limited information exists regarding the role of individual mGlu receptors in meth-induced deficits of recognition memory. Two recent publications investigated the role of glutamate receptors (including mGlu2, 3 and 5) in the PrH in meth-induced deficits in object recognition memory and cue-induced meth-seeking (Peters et al., 2016; Reichel et al., 2011). Both implemented self-administration paradigm of 7 days of limited (1 h), followed by either 14 days of limited or extended (6 h) access. The original investigation by Reichel et al. (2011) reported that extended access to self-administered meth reduced mGlu5, but not mGlu2/3, protein levels in the PFC and PrH in adult male rats (Reichel et al., 2011). Interestingly, there was a negative correlation between the meth intake and both mGlu5 expression and recognition memory performance, indicating that elevated meth intake can simultaneously lower the expression of PrH mGlu5 and PrH-dependent memory function. Utilizing the mGlu5 PAM CDPPB (30 mg/kg) to augment the function of the existing mGlu5 receptors (brain-wide), reversed the meth-induced recognition memory deficit (Reichel et al., 2011). The PAM was administered immediately after the familiarization phase in order to improve memory consolidation.

Though it should be noted that CDPPB only improved the short-term (90 min), but not the long-term (24 h) memory retention/recall. This suggests that CDPPB did not produce a long-term reversal of neurobiology related to impaired memory; rather, its effect was limited to the drug's own pharmacokinetic window of activity. In a follow-up study, Peters et al. (2016), showed that local, intra-PrH administration of a different mGlu5 PAM (DPFE) also rescued novel object recognition memory in adult male rats with a history of extended access to self-administered meth (Peters et al., 2016). See the section below for additional information regarding the effects of DPFE meth seeking also investigated in this study. Collectively, these results indicate that PrH mGlu5 receptors play a dual role in regulating both drug-related (meth-conditioned) and drug-unrelated (episodic) memory. Thus, selective targeting of these receptors might alleviate certain aspects of cognitive dysfunction in MUD, as well as reduce drug craving, particularly in situations where novel stimuli compete with conditioned stimuli for control over meth seeking (Peters et al., 2016).

5. mGlu receptor interaction with other neurotransmitter receptors in the context of methamphetamine-associated neural and behavioral effects

5.1 mGlu—Dopamine receptor interactions

Neural and behavioral consequences of meth use have been attributed, in part, to its ability to simultaneously increase dopamine and glutamate levels in the striatum. The combined effects of these two neurotransmitters on meth-induced cellular signaling, neurotoxicity, activity of striatal neural outputs and on various meth-associated behaviors are well-studied (Mark et al., 2004; Marshall, O'Dell, & Weihmuller, 1993; Miyazaki et al., 2013; Parsegian & See, 2014; Stephans & Yamamoto, 1994). The integration of striatal dopamine and glutamate signal can occur indirectly, via the activation of glutamatergic and dopaminergic inputs (Brown, 2021), or via overlapping striatal microcircuits (Nolan et al., 2020), through interaction between intracellular pathways (Voulalas, Holtzclaw, Wolstenholme, Russell, & Hyman, 2005), or activation of recently discovered dopamine-glutamate receptor heterocomplexes, including the D1-mGlu5 and D2-mGlu5 heteromers (Beggiato et al., 2016; Cabello et al., 2009; Sebastianutto et al., 2020). Due to the prevalence of such interactions, glutamate-dopamine interactions have been evaluated as a possible target for the pharmacotherapy of various psychopathologies, including SUD.

Specifically, mGlu–dopamine interactions after meth have been investigated with respect to meth-induced neurotoxicity. Battaglia et al. (2002) examined the amount of dopamine/dopamine metabolites in the dorsal striatum, their relationship to meth-induced neurotoxicity, and their regulation by mGlu5 receptors in adult male rats (Battaglia et al., 2002). A binge-like regimen of meth administration (3 x 5 mg/kg meth in 2-h intervals) produced a dramatic reduction of basal extracellular dopamine in the striatum that was attenuated by the repeated pretreatment with mGlu5 NAMs, MPEP, and SIB-1893 (both at 10 mg/kg, (Battaglia et al., 2002). Although this data supported the role of mGlu5 in meth-induced dopaminergic insult, it did not directly prove the ability of MPEP to block meth-induced striatal dopamine release. Indeed, systemic (10 mg/kg) or intra-striatal (100 µM) administration of MPEP failed to attenuate the acute meth-induced rise in striatal dopamine (Battaglia et al., 2002). A subsequent study by Gołembiowska et al. (2003), reported a similar restorative effect of prolonged mGlu5 blockade on meth-induced dopamine depletion. In this study, MPEP was co-injected with each meth administration (5 x 10 mg/kg meth every 2 h; Gołembiowska et al., 2003). Systemic MPEP (5 mg/kg) significantly reduced the rise in striatal dopamine levels after a single injection of meth (10 mg/kg). This effect was not replicated when MPEP was infused directly into the striatum across a range of MPEP doses tested, although the highest dose of MPEP (500 µM) paradoxically increased dopamine levels (Gołembiowska et al., 2003). This suggests that extra-striatal populations of mGlu5 receptors (such mGlu located in the VTA) might be responsible for the dopamine release-controlling effect. Indeed, mGlu5 receptors are expressed on both dopamine and GABA VTA neurons (Merrill, Friend, Newton, Hopkins, & Edwards, 2015). It is possible that prolonged exposure to meth dysregulates mGlu5 receptors in the mesostriatal dopamine pathways, which leads to the inability of mGlu5 blockers (such as MPEP) to have an effect on the meth-induced rise in striatal dopamine levels.

Unlike mGlu5, mGlu2/3 receptors in the striatum are almost exclusively located on presynaptic terminals that have been identified as glutamatergic afferents. And while there is a lack of direct evidence that mGlu2/3 are expressed directly on dopamine terminals, several studies reproduced the finding regarding the ability of the systemic or intra-striatal administration of mGlu2/3 (and group III mGlu) agonists to reduce basal or (meth) amphetamine-induced dopamine levels (Chaki, Yoshikawa, & Okuyama, 2006; Mao, Lau, & Wang, 2000; Pehrson & Moghaddam, 2010). Genetic ablation of mGlu3 or both mGlu2 and mGlu3 reduced the ability of acute meth or amphetamine to elevate dopamine levels (Fujioka et al., 2014; Lane et al., 2013). It has been proposed that this mGlu2/3-mediated effect on dopamine is mediated via local striatal microcircuitry that relies on cholinergic interneurons (Johnson, Mateo, & Lovinger, 2017). Perhaps it is not surprising that the detailed mechanism of mGlu-dopamine interactions is not yet fully understood, as meth triggers local dopamine release via a mix of non-vesicular and standard vesicular mechanisms and also activates multiple brain circuits that in turn, hyperactivate midbrain dopamine neurons.

5.2 mGlu-adenosine receptor interactions

A wealth of evidence supports the concept that glutamate–adenosine interactions play a critical role in regulating normal and abnormal brain function, including in Major Depression, Parkinson's disease, and SUD. Specifically, a number of studies have investigated the interaction between adenosine (A1, A2a) and mGlu receptors (for review see: León-Navarro, Albasanz, & Martín, 2018). Though the interaction between mGlu5 and A2a receptors in the striatum is the best characterized, an interaction between group III mGlu and A2a receptors in the striatum and between mGlu1 and A1 receptors in the cortex has also been detected (Ciruela et al., 2001; Lopez et al., 2008). It has been established that mGlu5 and A2a are co-expressed on D2-positive medium spiny neurons in the striatum, where they form a heteromeric complex (together with D2 receptors), or otherwise interact through their respective downstream intracellular signaling (Cabello et al., 2009; Ferré et al., 2002; Hámor, Gobin, & Schwendt, 2020; Nishi et al., 2003). The interaction between mGlu5 and A2a receptors is synergistic and serves to augment striatopallidal GABAergic output, and more broadly, regulates the function of the basal ganglia as a whole (Beggiato et al., 2016). While most studies investigated the mGlu5-A2a interaction as a potential target for the development of novel anti-Parkinson's agents (e.g., Coccurello, Breysse, and Amalric (2004) and others), a recent study by Wright et al. (2016) examined the functional significance of mGlu5-A2a interaction in the context of meth-associated behaviors using wild type and A2a KO adult male mice (Wright et al., 2016). In this study, a simultaneous inhibition of mGlu5 and A2a receptors (using MTEP and SCH 58261, respectively), dose-dependently and synergistically reduced meth-induced hyperactivity. Specifically, administration of combined subthreshold doses of A2a and mGlu5 antagonists prevented the meth-induced increase in locomotion and stereotypic rearing behavior

(Wright et al., 2016), similar to what has been observed for cocaine (Brown, Duncan, Stagnitti, Ledent, & Lawrence, 2012). Analogously, simultaneous inhibition of mGlu5 and A2a (using subthreshold doses of each antagonist) prevented expression of meth-induced CPP. This study also indicated that cooperative regulation of meth-associated behaviors and reward by A2a and mGlu5 might be altered by meth itself, as repeated meth administration (1 mg/kg/day for 10 days) increased mGlu5 binding in the NAc core and shell in wild-type (but not in A2a KO) mice. Although this study did not evaluate expression of A2a receptors, other evidence suggests that A2a and mGlu5 receptors in the striatum might undergo dynamic regulation after discontinuation of meth treatment. During early (24 h) abstinence from chronic meth self-administration, an increase in A2a protein levels in the NAc shell has been detected (Kavanagh, Schreiner, Levis, O'Neill, & Bachtell, 2015), while our data suggest that A2a (but not mGlu5) is upregulated after extended (30 days) abstinence, selectively in the NAc core (Fig. 1). It should be noted that our findings (and other published data), do not provide detailed evidence regarding how (up/down-regulation) and where (synaptic location) meth dysregulates mGlu5-A2a interactions. Therefore, additional research is

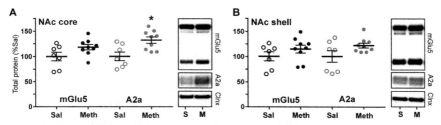

Fig. 1 A history of extended-access meth self-administration upregulates A2a protein levels in the NAc core (A), but not shell (B). Rats were trained to self-administer meth for 21 days (FR1: 7 days or 1 h/day access, followed by 14 days of 6 h/day access). A control group of rats had access to intravenous saline under identical conditions. The mean cumulative meth intake over the course of self-administration was 61 ± 5 mg/kg. Brain tissue was collected after 30 days of abstinence in the home cage and the NAc core and shell were dissected according to the coordinates based on the rat brain atlas (Paxinos & Watson, 2007). Protein levels of mGlu5 and A2a receptors in the total tissue lysate were assessed by immunoblotting (see Hámor et al. (2020) for more details) using validated antibodies for mGlu5 (AB5675, MilliporeSigma), A2a (sc-32261, Santa Cruz Biotechnology) and housekeeping control calnexin (Clnx, ADI-SPA-860, Enzo Life Sciences). Note that both mGlu5 dimer (upper band) and monomer (lower band) were co-detected in the NAc tissue samples. Data represent a normalized immunoblotting signal for individual samples and the mean $\pm$ SEM ($n=7$–9/group). *$P<0.05$, Meth vs Saline.

needed to understand the role of mGlu5-A2a receptor cooperation in meth addiction. For example, some striatal A2a and mGlu5, are co-expressed on the presynaptic terminals and on glia, and activation of these receptor populations can have divergent effects on synaptic transmission and behavior (Rodrigues, Alfaro, Rebola, Oliveira, & Cunha, 2005). This could explain why either A2a antagonists (SCH 58261; Wright et al., 2016) or agonists (CGS 21680; Acosta-García, Jiménez, & Miranda, 2017) both show functional synergism with mGlu5 antagonists in reducing meth-associated behaviors. Still, this initial research, together with data on other drugs of abuse (e.g., alcohol; Adams, Cowen, Short, & Lawrence, 2008), suggests that mGlu5-A2a interaction represents a promising clinical target for SUD therapy. As side effects of currently evaluated pre-clinical interventions are common, utilizing additive effects of subthreshold mGlu5+A2a drug combinations holds promise for the development of safer SUD pharmacotherapies.

5.3 mGlu-serotonin receptor interactions

A reciprocal interplay between serotoninergic and glutamatergic systems takes place primarily in the frontal cortex, and dysregulation of this relationship has been implicated in cognitive impairment, as well as in the pathophysiology of schizophrenia and (meth-induced) psychosis (for review see: Shah & González-Maeso, 2019). An overlapping cortical distribution of mGlu2/3 and 5-HT2a receptors, mostly investigated in the PFC, has led to a hypothesis of functional, physiological, and even physical interaction between the two receptors (Gewirtz & Marek, 2000; Marek, Wright, Schoepp, Monn, & Aghajanian, 2000). In this regard, Marek et al. (2000) showed that selective mGlu2/3 agonism suppressed glutamate release induced by 5-HT2a receptor activation in the mPFC, while mGlu2/3 antagonism increased the amplitude and frequency of serotonin-induced EPSCs, suggesting that receptor crosstalk mediate the action of endogenous glutamate on autoreceptors (Marek et al., 2000). Further, an increase in impulsive choice behavior after intra-PFC administration of direct 5-HT2a agonist DOI [(+/-)-1-(2,5-dimethoxy-4-iodophenyl)-2-aminopropan hydrochloride] was prevented by a co-administration of a mGlu2/3 agonist (Wischhof, Hollensteiner, & Koch, 2011). In relevance to psychosis and schizophrenia, studies show that the ability of clozapine (an atypical anti-psychotic and 5-HT2a antagonist) to reduce the impairing effects of dissociative drugs such phencyclidine depends on mGlu2/3 receptor activation (Hideshima et al., 2018; Horiguchi, Huang, & Meltzer, 2011).

Therefore, combined administration of 5-HT2a antagonists (e.g., atypical antipsychotic medications) with mGlu2/3 agonists might provide an additive benefit for the treatment of psychosis (Wischhof & Koch, 2016). It has been proposed that functional antagonism of 5-HT2a by mGlu2/3 receptors could be directly attributed to a formation of a 5-HT2a-mGlu2/3 receptor heterodimer (González-Maeso et al., 2008), although the role of receptor signaling crosstalk (independent of heteromer formation) cannot be discounted (Delille, Mezler, & Marek, 2013). We have been able to co-immunoprecipitate mGlu2 with 5-HT2a in the PrH, but not PFC (Hámor et al., 2018), suggesting that depending on the brain region, either mechanism could contribute to the observed functional interaction between the two receptors. Our study also revealed that a history of extended access meth self-administration (7 days of 1 h access, followed by 14 days of 6 h access to meth) increased the 5-HT2a-to-mGlu2 expression ratio in the PFC and PrH in adult male rats (Hámor et al., 2018). This finding is in agreement with other reports that detected augmented behavioral sensitivity to 5-HT2a agonists and increased 5-HT2a-to-mGlu2 expression ratio in the PFC after a binge-like meth administration in adult male mice (Chiu et al., 2014). This suggests that chronic methamphetamine produces a pattern of 5-HT2a/mGlu2 dysregulation that may contribute to the emergence of post-meth psychosis and cognitive deficits.

6. Conclusions

While acute or prolonged exposure to meth directly regulates the levels of dopamine (and other monoamines), accumulated evidence suggests that glutamate plays a role in a plethora of neural and behavioral effects of meth. This book chapter specifically focused on the role of mGlu receptors in meth-induced effects. Although, our understanding of the role of mGlu within the context of meth-induced effects is currently incomplete, the studies discussed in this chapter offer a compelling argument for further investigation of these receptors. The supporting evidence encouraging future research into the role of mGlu in meth-induced effects include: the observed ability of mGlu5 inhibitors to attenuate meth-induced toxicity, meth-associated reward, and meth-seeking, as well as pro-cognitive effects of mGlu5 activators. Further, emerging research into group II and III mGlu receptors is motivated by the findings of dysregulated mGlu2/3 and mGlu7 after chronic meth that might be related to the aberrant activity of cortical and striatal neurons. Taken together, future research into mGlu receptors might shed light on the synaptic mechanism and/or the identity of mGlu-expressing neuronal populations involved in meth effects and identify novel, mGlu-based interventions to ameliorate or reverse meth-induced neurobehavioral deficits.

Support

MS is supported by DA050118 and DA049212. These funding sources had no role in the writing or publication of this manuscript.

Conflict of interest

The authors have no conflict of interest to report.

References

Abulseoud, O. A., Miller, J. D., Wu, J., Choi, D. S., & Holschneider, D. P. (2012). Ceftriaxone upregulates the glutamate transporter in medial prefrontal cortex and blocks reinstatement of methamphetamine seeking in a condition place preference paradigm. *Brain Research, 1456*, 14–21. https://doi.org/10.1016/J.BRAINRES.2012.03.045.

Achat-Mendes, C., Platt, D. M., & Spealman, R. D. (2012). Antagonism of metabotropic glutamate 1 receptors attenuates behavioral effects of cocaine and methamphetamine in squirrel monkeys. *The Journal of Pharmacology and Experimental Therapeutics, 343*(1), 214–224. https://doi.org/10.1124/JPET.112.196295.

Acosta-García, J., Jiménez, J. C., & Miranda, F. (2017). Additive effects of coadministration of A2A receptor agonist CGS-21680 and mGluR5 antagonist MPEP on the development and expression of methamphetamine-induced locomotor sensitization in rats. *Journal of Drug and Alcohol Research, 6*(1), 1–11. https://doi.org/10.4303/JDAR/236038.

Adams, C. L., Cowen, M. S., Short, J. L., & Lawrence, A. J. (2008). Combined antagonism of glutamate mGlu5 and adenosine A2A receptors interact to regulate alcohol-seeking in rats. *International Journal of Neuropsychopharmacology, 11*(2), 229–241. https://doi.org/10.1017/S1461145707007845.

Al-Tayyib, A., Koester, S., Langegger, S., & Raville, L. (2017). Heroin and methamphetamine injection: An emerging drug use pattern. *Substance Use & Misuse, 52*(8), 1051–1058. https://doi.org/10.1080/10826084.2016.1271432.

Andres, M. A., Cooke, I. M., Bellinger, F. P., Berry, M. J., Zaporteza, M. M., Rueli, R. H., et al. (2015). Methamphetamine acutely inhibits voltage-gated calcium channels but chronically upregulates L-type channels. *Journal of Neurochemistry, 134*(1), 56. https://doi.org/10.1111/JNC.13104.

Arai, I., Shimazoe, T., Shibata, S., Inoue, H., Yoshimatsu, A., & Watanabe, S. (1997). Methamphetamine-induced sensitization of dopamine release via a metabotropic glutamate receptor mediated pathway in rat striatal slices. *Japanese Journal of Pharmacology, 73*(3), 243–246. https://doi.org/10.1254/jjp.73.243.

Barker, G. R. I., Banks, P. J., Scott, H., Ralph, G. S., Mitrophanous, K. A., Wong, L. F., et al. (2017). Separate elements of episodic memory subserved by distinct hippocampal-prefrontal connections. *Nature Neuroscience, 20*(2), 242–250. https://doi.org/10.1038/NN.4472.

Barker, G. R. I., Bashir, Z. I., Brown, M. W., & Warburton, E. C. (2006). A temporally distinct role for group I and group II metabotropic glutamate receptors in object recognition memory. *Learning & Memory (Cold Spring Harbor, N.Y.), 13*(2), 178–186. https://doi.org/10.1101/LM.77806.

Barker, G. R. I., Warburton, E. C., Koder, T., Dolman, N. P., More, J. C. A., Aggleton, J. P., et al. (2006). The different effects on recognition memory of perirhinal kainate and NMDA glutamate receptor antagonism: implications for underlying plasticity mechanisms. *The Journal of Neuroscience: The Official Journal of the Society for Neuroscience, 26*(13), 3561–3566. https://doi.org/10.1523/JNEUROSCI.3154-05.2006.

Battaglia, G., Fornai, F., Busceti, C. L., Aloisi, G., Cerrito, F., de Blasi, A., et al. (2002). Selective blockade of mGlu5 metabotropic glutamate receptors is protective against methamphetamine neurotoxicity. *The Journal of Neuroscience: The Official Journal of the Society for Neuroscience, 22*(6), 2135–2141. https://doi.org/10.1523/JNEUROSCI.22-06-02135.2002.

Beggiato, S., Tomasini, M. C., Borelli, A. C., Borroto-Escuela, D. O., Fuxe, K., Antonelli, T., et al. (2016). Functional role of striatal A2A, D2, and mGlu5 receptor interactions in regulating striatopallidal GABA neuronal transmission. *Journal of Neurochemistry, 138*(2), 254–264. https://doi.org/10.1111/jnc.13652.

Bellone, C., & Lüscher, C. (2005). mGluRs induce a long-term depression in the ventral tegmental area that involves a switch of the subunit composition of AMPA receptors. *The European Journal of Neuroscience, 21*(5), 1280–1288. https://doi.org/10.1111/J.1460-9568.2005.03979.X.

Bellone, C., & Mameli, M. (2012). MGluR-dependent synaptic plasticity in drug-seeking. *Frontiers in Pharmacology, 3*, 159. https://doi.org/10.3389/FPHAR.2012.00159/BIBTEX.

Bernheim, A., See, R. E., & Reichel, C. M. (2016). Chronic methamphetamine self-administration disrupts cortical control of cognition. *Neuroscience and Biobehavioral Reviews, 69*, 36–48. https://doi.org/10.1016/J.NEUBIOREV.2016.07.020.

Bouton, M. E. (2004). Context and behavioral processes in extinction. *Learning & Memory, 11*(5), 485–494. https://doi.org/10.1101/LM.78804.

Bowyer, J. F., & Hanig, J. P. (2014). Amphetamine- and methamphetamine-induced hyperthermia: Implications of the effects produced in brain vasculature and peripheral organs to forebrain neurotoxicity. *Temperature: Multidisciplinary Biomedical Journal, 1*(3), 172. https://doi.org/10.4161/23328940.2014.982049.

Brodkin, J., Bradbury, M., Busse, C., Warren, N., Bristow, L. J., & Varney, M. A. (2002). Reduced stress-induced hyperthermia in mGluR5 knockout mice. *The European Journal of Neuroscience, 16*(11), 2241–2244. https://doi.org/10.1046/J.1460-9568.2002.02294.X.

Brown, C. (2021). *Investigating the role of corticostriatal glutamate signaling in methamphetamine reward and reinforcement.* [Merritt ID: ark:/13030/m5f261n9, UC Santa Barbara]. https://escholarship.org/uc/item/24w0336m.

Brown, R. M., Duncan, J. R., Stagnitti, M. R., Ledent, C., & Lawrence, A. J. (2012). mGlu5 and adenosine A2A receptor interactions regulate the conditioned effects of cocaine. *The International Journal of Neuropsychopharmacology, 15*(07), 995–1001. https://doi.org/10.1017/S146114571100126X.

Bruno, V., Ksiazek, I., Battaglia, G., Lukic, S., Leonhardt, T., Sauer, D., et al. (2000). Selective blockade of metabotropic glutamate receptor subtype 5 is neuroprotective. *Neuropharmacology, 39*(12), 2223–2230. https://doi.org/10.1016/S0028-3908(00)00079-4.

Burke, S., & Hanani, M. (2012). The actions of hyperthermia on the autonomic nervous system: Central and peripheral mechanisms and clinical implications. *Autonomic Neuroscience: Basic & Clinical, 168*(1–2), 4–13. https://doi.org/10.1016/j.autneu.2012.02.003.

Busse, C. S., Brodkin, J., Tattersall, D., Andersen, J. J., Warren, N., Tehrani, L., et al. (2004). The behavioral profile of the potent and selective mGlu5 receptor antagonist 3-[(2-methyl-1,3-thiazol-4-yl)ethynyl] pyridine (MTEP) in rodent models of anxiety. *Neuropsychopharmacology, 29*(11), 1971–1979. https://doi.org/10.1038/sj.npp.1300540.

Cabello, N., Gandía, J., Bertarelli, D. C. G., Watanabe, M., Lluís, C., Franco, R., et al. (2009). Metabotropic glutamate type 5, dopamine D2 and adenosine A2a receptors form higher-order oligomers in living cells. *Journal of Neurochemistry, 109*(5), 1497–1507. https://doi.org/10.1111/J.1471-4159.2009.06078.X.

Caprioli, D., Venniro, M., Zeric, T., Li, X., Adhikary, S., Madangopal, R., et al. (2015). Effect of the novel positive allosteric modulator of mGluR2 AZD8529 on incubation of methamphetamine craving after prolonged voluntary abstinence in a rat model. *Biological Psychiatry, 78*(7), 463. https://doi.org/10.1016/J.BIOPSYCH.2015.02.018.

Cavallone, L. F., Montana, M. C., Frey, K., Kallogjeri, D., Wages, J. M., Rodebaugh, T. L., et al. (2020). The metabotropic glutamate receptor 5 negative allosteric modulator fenobam: Pharmacokinetics, side effects, and analgesic effects in healthy human subjects. *Pain, 161*(1), 135. https://doi.org/10.1097/J.PAIN.0000000000001695.

Ceccarini, J., Leurquin-Sterk, G., Crunelle, C. L., de Laat, B., Bormans, G., Peuskens, H., et al. (2020). Recovery of decreased metabotropic glutamate receptor 5 availability in abstinent alcohol-dependent patients. *Journal of Nuclear Medicine: Official Publication, Society of Nuclear Medicine, 61*(2), 256–262. https://doi.org/10.2967/JNUMED.119.228825.

Chaki, S., Yoshikawa, R., & Okuyama, S. (2006). Group II metabotropic glutamate receptor-mediated regulation of dopamine release from slices of rat nucleus accumbens. *Neuroscience Letters, 404*(1–2), 182–186. https://doi.org/10.1016/j.neulet.2006.05.043.

Challasivakanaka, S., Zhen, J., Smith, M. E., Reith, M. E. A., Foster, J. D., & Vaughan, R. A. (2017). Dopamine transporter phosphorylation site threonine 53 is stimulated by amphetamines and regulates dopamine transport, efflux, and cocaine analog binding. *Journal of Biological Chemistry, 292*(46), 19066–19075. https://doi.org/10.1074/jbc.M117.787002.

Chesworth, R., Brown, R. M., Kim, J. H., & Lawrence, A. J. (2013). The metabotropic glutamate 5 receptor modulates extinction and reinstatement of methamphetamine-seeking in mice. *PLoS One, 8*(7), e68371. https://doi.org/10.1371/journal.pone.0068371.

Chiamulera, C., Epping-Jordan, M. P., Zocchi, A., Marcon, C., Cottiny, C., Tacconi, S., et al. (2001). Reinforcing and locomotor stimulant effects of cocaine are absent in mGluR5 null mutant mice. *Nature Neuroscience, 4*(9), 873–874. https://doi.org/10.1038/nn0901-873.

Chiamulera, C., Piva, A., & Abraham, W. C. (2021). Glutamate receptors and metaplasticity in addiction. *Current Opinion in Pharmacology, 56*, 39–45. https://doi.org/10.1016/J.COPH.2020.09.005.

Chiu, H. Y., Chan, M. H., Lee, M. Y., Chen, S. T., Zhan, Z. Y., & Chen, H. H. (2014). Long-lasting alterations in 5-HT2A receptor after a binge regimen of methamphetamine in mice. *The International Journal of Neuropsychopharmacology, 17*(10), 1647–1658. https://doi.org/10.1017/S1461145714000455.

Ciruela, F., Escriche, M., Burgueño, J., Angulo, E., Casadó, V., Soloviev, M. M., et al. (2001). Metabotropic glutamate 1α and adenosine A1 receptors assemble into functionally interacting complexes. *Journal of Biological Chemistry, 276*(21), 18345–18351. https://doi.org/10.1074/jbc.M006960200.

Cisneros, I. E., & Ghorpade, A. (2014). Methamphetamine and HIV-1-induced neurotoxicity: Role of trace amine associated receptor 1 cAMP signaling in astrocytes. *Neuropharmacology, 85*, 499. https://doi.org/10.1016/J.NEUROPHARM.2014.06.011.

Cleva, R. M., Hicks, M. P., Gass, J. T., Wischerath, K. C., Plasters, E. T., Widholm, J. J., et al. (2011). mGluR5 positive allosteric modulation enhances extinction learning following cocaine self-administration. *Behavioral Neuroscience, 125*(1), 10–19. https://doi.org/10.1037/a0022339.

Coccurello, R., Breysse, N., & Amalric, M. (2004). Simultaneous blockade of adenosine A2A and metabotropic glutamate mGlu5 receptors increase their efficacy in reversing Parkinsonian deficits in rats. *Neuropsychopharmacology, 29*(8), 1451–1461. https://doi.org/10.1038/sj.npp.1300444.

Conn, P. J., & Pin, J.-P. (1997). Pharmacology and functions of metabotropic glutamate receptors. *Annual Review of Pharmacology and Toxicology, 37*(1), 205–237. https://doi.org/10.1146/annurev.pharmtox.37.1.205.

Cremades, A., & Peñafiel, R. (1982). Hyperthermia and brain neurotransmitter amino acid levels in infant rats. *General Pharmacology, 13*(4), 347–350. https://doi.org/10.1016/0306-3623(82)90056-8.

Crocker, C. E., Bernier, D. C., Hanstock, C. C., Lakusta, B., Purdon, S. E., Seres, P., et al. (2014). Prefrontal glutamate levels differentiate early phase schizophrenia and methamphetamine addiction: A1H MRS study at 3 Tesla. *Schizophrenia Research, 157*(1–3), 231–237. https://doi.org/10.1016/j.schres.2014.05.004.

Delille, H. K., Mezler, M., & Marek, G. J. (2013). The two faces of the pharmacological interaction of mGlu2 and 5-HT2A—Relevance of receptor heterocomplexes and interaction through functional brain pathways. *Neuropharmacology, 70*, 296–305. https://doi.org/10.1016/J.NEUROPHARM.2013.02.005.

Earle, M. L., & Davies, J. A. (1991). The effect of methamphetamine on the release of glutamate from striatal slices. *Journal of Neural Transmission General Section, 86*(3), 217–222. https://doi.org/10.1007/BF01250707.

Ernst, T., & Chang, L. (2008). Adaptation of brain glutamate plus glutamine during abstinence from chronic methamphetamine use. *Journal of Neuroimmune Pharmacology, 3*(3), 165–172. https://doi.org/10.1007/S11481-008-9108-4/FIGURES/3.

Fendt, M., & Schmid, S. (2002). Metabotropic glutamate receptors are involved in amygdaloid plasticity. *The European Journal of Neuroscience, 15*(9), 1535–1541. https://doi.org/10.1046/J.1460-9568.2002.01988.X.

Ferré, S., Karcz-Kubicha, M., Hope, B. T., Popoli, P., Burgueño, J., Gutiérrez, M. A., et al. (2002). Synergistic interaction between adenosine A2A and glutamate mGlu5 receptors: Implications for striatal neuronal function. *Proceedings of the National Academy of Sciences of the United States of America, 99*(18), 11940–11945. https://doi.org/10.1073/pnas.172393799.

Foster, J. D., & Vaughan, R. A. (2017). Phosphorylation mechanisms in dopamine transporter regulation. *Journal of Chemical Neuroanatomy, 83–84*, 10. https://doi.org/10.1016/J.JCHEMNEU.2016.10.004.

Fowler, M. A., Varnell, A. L., & Cooper, D. C. (2011). mGluR5 knockout mice exhibit normal conditioned place-preference to cocaine. *Nature Precedings, 2011*, 1. https://doi.org/10.1038/npre.2011.6180.1.

Fujioka, R., Nii, T., Iwaki, A., Shibata, A., Ito, I., Kitaichi, K., et al. (2014). Comprehensive behavioral study of mGluR3 knockout mice: Implication in schizophrenia related endophenotypes. *Molecular Brain, 7*(1), 1–18. https://doi.org/10.1186/1756-6606-7-31.

Fujiwara, Y., Kazahaya, Y., Nakashima, M., Sato, M., & Otsuki, S. (1987). Behavioral sensitization to methamphetamine in the rat: An ontogenic study. *Psychopharmacology, 91*(3), 316–319. https://doi.org/10.1007/BF00518183.

Gabriel, L., Lvov, A., Orthodoxou, D., Rittenhouse, A. R., Kobertz, W. R., & Melikian, H. E. (2012). The acid-sensitive, anesthetic-activated potassium leak channel, KCNK3, is regulated by 14-3-3β-dependent, protein kinase C (PKC)-mediated endocytic trafficking. *The Journal of Biological Chemistry, 287*(39), 32354–32366. https://doi.org/10.1074/jbc.M112.391458.

Gass, J. T., & Olive, M. F. (2009). Positive allosteric modulation of mGluR5 receptors facilitates extinction of a cocaine contextual memory. *Biological Psychiatry, 65*(8), 717–720. https://doi.org/10.1016/j.biopsych.2008.11.001.

Gass, J. T., Osborne, M. P. H., Watson, N. L., Brown, J. L., & Olive, M. F. (2009). mGluR5 antagonism attenuates methamphetamine reinforcement and prevents reinstatement of methamphetamine-seeking behavior in rats. *Neuropsychopharmacology: Official Publication of the American College of Neuropsychopharmacology, 34*(4), 820–833. https://doi.org/10.1038/NPP.2008.140.

Gewirtz, J. C., & Marek, G. J. (2000). Behavioral evidence for interactions between a hallucinogenic drug and group II metabotropic glutamate receptors. *Neuropsychopharmacology: Official Publication of the American College of Neuropsychopharmacology, 23*(5), 569–576. https://doi.org/10.1016/S0893-133X(00)00136-6.

Gladden, R. M., O'Donnell, J., Mattson, C. L., & Seth, P. (2019). Changes in opioid-involved overdose deaths by opioid type and presence of benzodiazepines, cocaine, and methamphetamine—25 States, July–December 2017 to January–June 2018. *Morbidity and Mortality Weekly Report, 68*(34), 737–744. https://doi.org/10.15585/MMWR.MM6834A2.

Global Burden of Disease (GBD) Collaborative Network. (2021). *Global Burden of Disease Study*. https://doi.org/10.6069/1D4Y-YQ37.

Gobin, C., & Schwendt, M. (2020). The cognitive cost of reducing relapse to cocaine-seeking with mGlu5 allosteric modulators. *Psychopharmacology, 237*(1), 115–125. https://doi.org/10.1007/s00213-019-05351-8.

Gołembiowska, K., Konieczny, J., Ossowska, K., & Wolfarth, S. (2002). The role of striatal metabotropic glutamate receptors in degeneration of dopamine neurons: Review article. *Amino Acids, 23*(1), 199–205. https://doi.org/10.1007/S00726-001-0129-Z.

Gołembiowska, K., Konieczny, J., Wolfarth, S., & Ossowska, K. (2003). Neuroprotective action of MPEP, a selective mGluR5 antagonist, in methamphetamine-induced dopaminergic neurotoxicity is associated with a decrease in dopamine outflow and inhibition of hyperthermia in rats. *Neuropharmacology, 45*(4), 484–492. https://doi.org/10.1016/S0028-3908(03)00209-0.

González-Maeso, J., Ang, R. L., Yuen, T., Chan, P., Weisstaub, N.v., López-Giménez, J. F., et al. (2008). Identification of a serotonin/glutamate receptor complex implicated in psychosis. *Nature, 452*(7183), 93–97. https://doi.org/10.1038/NATURE06612.

Gyetvai, B., Simonyi, A., Oros, M., Saito, M., Smiley, J., & Vadász, C. (2011). mGluR7 genetics and alcohol: Intersection yields clues for addiction. *Neurochemical Research, 36*(6), 1087. https://doi.org/10.1007/S11064-011-0452-Z.

Hamilton, A., Esseltine, J. L., Devries, R. A., Cregan, S. P., & Ferguson, S. S. G. (2014). Metabotropic glutamate receptor 5 knockout reduces cognitive impairment and pathogenesis in a mouse model of Alzheimer's disease. *Molecular Brain, 7*(1), 1–12. https://doi.org/10.1186/1756-6606-7-40.

Hámor, P. U., Gobin, C. M., & Schwendt, M. (2020). The role of glutamate mGlu5 and adenosine A2a receptor interactions in regulating working memory performance and persistent cocaine seeking in rats. *Progress in Neuro-Psychopharmacology and Biological Psychiatry, 103*, 109979. https://doi.org/10.1016/j.pnpbp.2020.109979.

Hámor, P. U., Šírová, J., Páleníček, T., Zaniewska, M., Bubeníková-Valešová, V., & Schwendt, M. (2018). Chronic methamphetamine self-administration dysregulates 5-HT2A and mGlu2 receptor expression in the rat prefrontal and perirhinal cortex: Comparison to chronic phencyclidine and MK-801. *Pharmacology, Biochemistry, and Behavior, 175*, 89. https://doi.org/10.1016/J.PBB.2018.09.007.

Hashimoto, K., Nakahara, T., Yamada, H., Hirano, M., Kuroki, T., & Kanba, S. (2007). A neurotoxic dose of methamphetamine induces gene expression of Homer 1a, but not Homer 1b or 1c, in the striatum and nucleus accumbens. *Neurochemistry International, 51*(2–4), 227–232. https://doi.org/10.1016/J.NEUINT.2007.05.017.

He, Z., Chen, Y., Dong, H., Su, R., Gong, Z., & Yan, L. (2014). Inhibition of vesicular glutamate transporters contributes to attenuate methamphetamine-induced conditioned place preference in rats. *Behavioural Brain Research, 267*, 1–5. https://doi.org/10.1016/J.BBR.2014.02.047.

Herrold, A. A., Persons, A. L., & Napier, T. C. (2013). Cellular distribution of AMPA receptor subunits and mGlu5 following acute and repeated administration of morphine or methamphetamine. *Journal of Neurochemistry, 126*(4), 503–517. https://doi.org/10.1111/jnc.12323.

Herrold, A. A., Voigt, R. M., & Napier, T. C. (2011). Brain region-selective cellular redistribution of mGlu5 but not GABA B receptors following methamphetamine-induced associative learning. *Synapse, 65*(12), 1333–1343. https://doi.org/10.1002/syn.20968.

Herrold, A. A., Voigt, R. M., & Napier, T. C. (2013). mGluR5 is necessary for maintenance of methamphetamine-induced associative learning. *European Neuropsychopharmacology: The Journal of the European College of Neuropsychopharmacology, 23*(7), 691–696. https://doi.org/10.1016/J.EURONEURO.2012.05.014.

Hideshima, K. S., Hojati, A., Saunders, J. M., On, D. M., de la Fuente Revenga, M., Shin, J. M., et al. (2018). Role of mGlu2 in the 5-HT2A receptor-dependent antipsychotic activity of clozapine in mice. *Psychopharmacology, 235*(11), 3149. https://doi.org/10.1007/S00213-018-5015-4.

Horiguchi, M., Huang, M., & Meltzer, H. Y. (2011). Interaction of mGlu2/3 agonism with clozapine and lurasidone to restore novel object recognition in subchronic phencyclidine-treated rats. *Psychopharmacology, 217*(1), 13–24. https://doi.org/10.1007/S00213-011-2251-2.

Huang, C. H., Yu, Y. J., Chang, C. H., & Gean, P. W. (2016). Involvement of metabotropic glutamate receptor 5 in the inhibition of methamphetamine-associated contextual memory after prolonged extinction training. *Journal of Neurochemistry, 137*(2), 216–225. https://doi.org/10.1111/jnc.13525.

Hubert, G. W., Paquet, M., & Smith, Y. (2001). Differential subcellular localization of mGluR1a and mGluR5 in the rat and monkey Substantia nigra. *The Journal of Neuroscience: The Official Journal of the Society for Neuroscience, 21*(6), 1838–1847. https://doi.org/10.1523/JNEUROSCI.21-06-01838.2001.

Iijima, M., Shimazaki, T., Ito, A., & Chaki, S. (2007). Effects of metabotropic glutamate 2/3 receptor antagonists in the stress-induced hyperthermia test in singly housed mice. *Psychopharmacology, 190*(2), 233–239. https://doi.org/10.1007/S00213-006-0618-6.

Ito, K., Abekawa, T., & Koyama, T. (2006). Relationship between development of cross-sensitization to MK–801 and delayed increases in glutamate levels in the nucleus accumbens induced by a high dose of methamphetamine. *Psychopharmacology, 187*(3), 293–302. https://doi.org/10.1007/S00213-006-0423-2/FIGURES/6.

Jayanthi, S., Daiwile, A. P., & Cadet, J. L. (2021). Neurotoxicity of methamphetamine: Main effects and mechanisms. *Experimental Neurology, 344.* https://doi.org/10.1016/j.expneurol.2021.113795.

Jew, C. P., Wu, C. S., Sun, H., Zhu, J., Huang, J. Y., Yu, D., et al. (2013). mGluR5 ablation in cortical glutamatergic neurons increases novelty-induced locomotion. *PLoS One, 8*(8), 70415. https://doi.org/10.1371/JOURNAL.PONE.0070415.

Johnson, K. A., Mateo, Y., & Lovinger, D. M. (2017). Metabotropic glutamate receptor 2 inhibits thalamically-driven glutamate and dopamine release in the dorsal striatum. *Neuropharmacology, 117,* 114–123. https://doi.org/10.1016/J.NEUROPHARM.2017.01.038.

Jones, B., Balasubramaniam, M., Lebowitz, J. J., Taylor, A., Villalta, F., Khoshbouei, H., et al. (2021). Activation of proline biosynthesis is critical to maintain glutamate homeostasis during acute methamphetamine exposure. *Scientific Reports, 11*(1), 1–16. https://doi.org/10.1038/s41598-020-80917-7.

Jones, C. M., Compton, W. M., & Mustaquim, D. (2020). Patterns and characteristics of methamphetamine use among adults—United States, 2015–2018. *Morbidity and Mortality Weekly Report, 69*(12), 317–323. https://doi.org/10.15585/MMWR.MM6912A1.

Justinova, Z., le Foll, B., Redhi, G. H., Markou, A., & Goldberg, S. R. (2016). Differential effects of the metabotropic glutamate 2/3 receptor agonist LY379268 on nicotine versus cocaine self-administration and relapse in squirrel monkeys. *Psychopharmacology*, *233*(10), 1791–1800. https://doi.org/10.1007/S00213-015-3994-Y.

Justinova, Z., Panlilio, L.v., Secci, M. E., Redhi, G. H., Schindler, C. W., Cross, A. J., et al. (2015). The novel metabotropic glutamate receptor 2 positive allosteric modulator, AZD8529, decreases nicotine self-administration and relapse in squirrel monkeys. *Biological Psychiatry*, *78*(7), 452–462. https://doi.org/10.1016/J.BIOPSYCH.2015.01.014.

Karakas, E., Regan, M. C., & Furukawa, H. (2015). Emerging structural insights into the function of ionotropic glutamate receptors. *Trends in Biochemical Sciences*, *40*(6), 328–337. https://doi.org/10.1016/j.tibs.2015.04.002.

Kaushal, N., & Matsumoto, R. R. (2011). Role of sigma receptors in methamphetamine-induced neurotoxicity. *Current Neuropharmacology*, *9*(1), 54–57. https://doi.org/10.2174/157015911795016930.

Kavanagh, K. A., Schreiner, D. C., Levis, S. C., O'Neill, C. E., & Bachtell, R. K. (2015). Role of adenosine receptor subtypes in methamphetamine reward and reinforcement. *Neuropharmacology*, *89*, 265–273. https://doi.org/10.1016/J.NEUROPHARM.2014.09.030.

Kevil, C. G., Goeders, N. E., Woolard, M. D., Bhuiyan, M. S., Dominic, P., Kolluru, G. K., et al. (2019). Methamphetamine use and cardiovascular disease. *Arteriosclerosis, Thrombosis, and Vascular Biology*, *39*(9), 1739–1746. https://doi.org/10.1161/ATVBAHA.119.312461.

Kikuchi-Utsumi, K., Ishizaka, M., Matsumura, N., Watabe, M., Aoyama, K., Sasakawa, N., et al. (2013). Involvement of the α1D-adrenergic receptor in methamphetamine-induced hyperthermia and neurotoxicity in rats. *Neurotoxicity Research*, *24*(2), 130–138. https://doi.org/10.1007/S12640-012-9369-9/FIGURES/7.

Kim, B., Yun, J., & Park, B. (2020). Methamphetamine-induced neuronal damage: Neurotoxicity and neuroinflammation. *Biomolecules & Therapeutics*, *28*(5), 381–388. https://doi.org/10.4062/BIOMOLTHER.2020.044.

Kinoshita, A., Shigemoto, R., Ohishi, H., van der Putten, H., & Mizuno, N. (1998). Immunohistochemical localization of metabotropic glutamate receptors, mGluR7a and mGluR7b, in the central nervous system of the adult rat and mouse: A light and electron microscopic study. *J. Comp. Neurol*, *393*, 332–352. https://doi.org/10.1002/(SICI)1096-9861(19980413)393:3.

Kish, S. J. (2008). Pharmacologic mechanisms of crystal meth. *CMAJ*, *178*(13), 1679–1682. Canadian Medical Association https://doi.org/10.1503/cmaj.071675.

Kitamura, O., Tokunaga, I., Gotohda, T., & Kubo, S. I. (2007). Immunohistochemical investigation of dopaminergic terminal markers and caspase-3 activation in the striatum of human methamphetamine users. *International Journal of Legal Medicine*, *121*(3), 163–168. https://doi.org/10.1007/S00414-006-0087-9/FIGURES/2.

Kiyatkin, E. A., & Sharma, H. S. (2009). Acute methamphetamine intoxication: Brain hyperthermia, blood–brain barrier, brain edema, and morphological cell abnormalities. *International Review of Neurobiology*, *88*(C), 65–100. https://doi.org/10.1016/S0074-7742(09)88004-5.

Knackstedt, L. A., Moussawi, K., Lalumiere, R., Schwendt, M., Klugmann, M., & Kalivas, P. W. (2010). Extinction training after cocaine self-administration induces glutamatergic plasticity to inhibit cocaine seeking. *Journal of Neuroscience*, *30*(23), 7984–7992. https://doi.org/10.1523/JNEUROSCI.1244-10.2010.

Kogan, F. J., Nichols, W. K., & Gibb, J. W. (1976). Influence of methamphetamine on nigral and striatal tyrosine hydroxylase activity and on striatal dopamine levels. *European Journal of Pharmacology*, *36*(2), 363–371. https://doi.org/10.1016/0014-2999(76)90090-X.

Kufahl, P. R., Hood, L. E., Nemirovsky, N. E., Barabas, P., Halstengard, C., Villa, A., et al. (2012). Positive allosteric modulation of mGluR5 accelerates extinction learning but not relearning following methamphetamine self-administration. *Frontiers in Pharmacology*, *3*, 194. https://doi.org/10.3389/fphar.2012.00194.

Kufahl, P. R., Nemirovsky, N. E., Watterson, L. R., Zautra, N., & Olive, M. F. (2013). Positive or negative allosteric modulation of metabotropic glutamate receptor 5 (mGluR5) does not alter expression of behavioral sensitization to methamphetamine. *F1000Research, 2*, 84. https://doi.org/10.12688/F1000RESEARCH.2-84.V1.

Kufahl, P. R., Watterson, L. R., Nemirovsky, N. E., Hood, L. E., Villa, A., Halstengard, C., et al. (2013). Attenuation of methamphetamine seeking by the mGluR2/3 agonist LY379268 in rats with histories of restricted and escalated self-administration. *Neuropharmacology, 66*, 290–301. https://doi.org/10.1016/j.neuropharm.2012.05.037.

Lane, T. A., Boerner, T., Bannerman, D. M., Kew, J. N. C., Tunbridge, E. M., Sharp, T., et al. (2013). Decreased striatal dopamine in group II metabotropic glutamate receptor (mGlu2/mGlu3) double knockout mice. *BMC Neuroscience, 14*(1), 1–7. https://doi.org/10.1186/1471-2202-14-102/FIGURES/2.

Lea, P. M., IV, Movsesyan, V. A., & Faden, A. I. (2005). Neuroprotective activity of the mGluR5 antagonists MPEP and MTEP against acute excitotoxicity differs and does not reflect actions at mGluR5 receptors. *British Journal of Pharmacology, 145*(4), 527–534. https://doi.org/10.1038/sj.bjp.0706219.

León-Navarro, D. A., Albasanz, J. L., & Martín, M. (2018). Functional cross-talk between adenosine and metabotropic glutamate receptors. *Current Neuropharmacology, 17*(5), 422–437. https://doi.org/10.2174/1570159X16666180416093717.

Leurquin-Sterk, G., van den Stock, J., Crunelle, C. L., de Laat, B., Weerasekera, A., Himmelreich, U., et al. (2016). Positive association between limbic metabotropic glutamate receptor 5 availability and novelty-seeking temperament in humans: An 18F-FPEB PET study. *Journal of Nuclear Medicine, 57*(11), 1746–1752. https://doi.org/10.2967/jnumed.116.176032.

Li, X., Li, J., Gardner, E. L., & Xi, Z. X. (2010). Activation of mGluR7s inhibits cocaine-induced reinstatement of drug-seeking behavior by a nucleus accumbens glutamate-mGluR2/3 mechanism in rats. *Journal of Neurochemistry, 114*(5), 1368–1380. https://doi.org/10.1111/J.1471-4159.2010.06851.X.

Liao, Y. H., Wang, Y. H., Sun, L. H., Deng, W. T., te Lee, H., & Yu, L. (2018). mGluR5 upregulation and the effects of repeated methamphetamine administration and withdrawal on the rewarding efficacy of ketamine and social interaction. *Toxicology and Applied Pharmacology, 360*, 58–68. https://doi.org/10.1016/J.TAAP.2018.09.035.

Lominac, K. D., Quadir, S. G., Barrett, H. M., Mckenna, C. L., Schwartz, L. M., Ruiz, P. N., et al. (2016). Prefrontal glutamate correlates of methamphetamine sensitization and preference. *European Journal of Neuroscience, 43*(5), 689–702. https://doi.org/10.1111/EJN.13159.

Lominac, K. D., Sacramento, A. D., Szumlinski, K. K., & Kippin, T. E. (2012). Distinct neurochemical adaptations within the nucleus accumbens produced by a history of self-administered vs non-contingently administered intravenous methamphetamine. *Neuropsychopharmacology, 37*(3), 707–722. https://doi.org/10.1038/npp.2011.248.

Lopez, S., Turle-Lorenzo, N., Johnston, T. H., Brotchie, J. M., Schann, S., Neuville, P., et al. (2008). Functional interaction between adenosine A2A and group III metabotropic glutamate receptors to reduce parkinsonian symptoms in rats. *Neuropharmacology, 55*(4), 483–490. https://doi.org/10.1016/J.NEUROPHARM.2008.06.038.

Mao, L., Lau, Y. S., & Wang, J. Q. (2000). Activation of group III metabotropic glutamate receptors inhibits basal and amphetamine-stimulated dopamine release in rat dorsal striatum: An in vivo microdialysis study. *European Journal of Pharmacology, 404*(3), 289–297. https://doi.org/10.1016/S0014-2999(00)00633-6.

Mao, L., & Wang, J. Q. (2001). Differentially altered mGluR1 and mGluR5 mRNA expression in rat caudate nucleus and nucleus accumbens in the development and expression of behavioral sensitization to repeated amphetamine administration. *Synapse (New York, N.Y.), 41*(3), 230–240. https://doi.org/10.1002/SYN.1080.

Marek, G. J., Wright, R. A., Schoepp, D. D., Monn, J. A., & Aghajanian, G. K. (2000). Physiological antagonism between 5-hydroxytryptamine(2A) and group II metabotropic glutamate receptors in prefrontal cortex. *The Journal of Pharmacology and Experimental Therapeutics*, *292*(1), 76–87. http://www.ncbi.nlm.nih.gov/pubmed/10604933.

Mark, K. A., Quinton, M. S., Russek, S. J., & Yamamoto, B. K. (2007). Dynamic changes in vesicular glutamate transporter 1 function and expression related to methamphetamine-induced glutamate release. *Journal of Neuroscience*, *27*(25), 6823–6831. https://doi.org/10.1523/JNEUROSCI.0013-07.2007.

Mark, K. A., Soghomonian, J. J., & Yamamoto, B. K. (2004). High-dose methamphetamine acutely activates the striatonigral pathway to increase striatal glutamate and mediate long-term dopamine toxicity. *Journal of Neuroscience*, *24*(50), 11449–11456. https://doi.org/10.1523/JNEUROSCI.3597-04.2004.

Markou, A. (2007). The role of metabotropic glutamate receptors in drug reward, motivation and dependence. *Drug News & Perspectives*, *20*(2), 103. https://doi.org/10.1358/dnp.2007.20.2.1083435.

Marshall, J. F., O'Dell, S. J., & Weihmuller, F. B. (1993). Dopamine-glutamate interactions in methamphetamine-induced neurotoxicity. *Journal of Neural Transmission*, *91*(2–3), 241–254. https://doi.org/10.1007/BF01245234.

Matsumoto, R. R., Seminerio, M. J., Turner, R. C., Robson, M. J., Nguyen, L., Miller, D. B., et al. (2014). Methamphetamine-induced toxicity: an updated review on issues related to hyperthermia. *Pharmacology & Therapeutics*, *144*(1), 28–40. https://doi.org/10.1016/J.PHARMTHERA.2014.05.001.

Mcgeehan, A. J., Janak, P. H., & Olive, M. F. (2004). Effect of the mGluR5 antagonist 6-methyl-2-(phenylethynyl)pyridine (MPEP) on the acute locomotor stimulant properties of cocaine, D-amphetamine, and the dopamine reuptake inhibitor GBR12909 in mice. *Psychopharmacology*, *174*(2), 266–273. https://doi.org/10.1007/s00213-003-1733-2.

Merrill, C. B., Friend, L. N., Newton, S. T., Hopkins, Z. H., & Edwards, J. G. (2015). Ventral tegmental area dopamine and GABA neurons: Physiological properties and expression of mRNA for endocannabinoid biosynthetic elements. *Scientific Reports*, *5*(1), 1–16. https://doi.org/10.1038/srep16176.

Mishra, D., Pena-Bravo, J. I., Ghee, S. M., Berini, C., Reichel, C. M., & Lavin, A. (2021). Effects of high dosage methamphetamine on glutamatergic neurotransmission in the nucleus accumbens and prefrontal cortex. *BioRxiv*, *2021*(4), 440987. https://doi.org/10.1101/2021.04.22.440987.

Mishra, D., Pena-Bravo, J. I., Leong, K. C., Lavin, A., & Reichel, C. M. (2017). Methamphetamine self-administration modulates glutamate neurophysiology. *Brain Structure and Function*, *222*(5), 2031–2039. https://doi.org/10.1007/s00429-016-1322-x.

Miyazaki, M., Noda, Y., Mouri, A., Kobayashi, K., Mishina, M., Nabeshima, T., et al. (2013). Role of convergent activation of glutamatergic and dopaminergic systems in the nucleus accumbens in the development of methamphetamine psychosis and dependence. *International Journal of Neuropsychopharmacology*, *16*(6), 1341–1350. https://doi.org/10.1017/S1461145712001356.

Mizoguchi, H., & Yamada, K. (2019). Methamphetamine use causes cognitive impairment and altered decision-making. *Neurochemistry International*, *124*, 106–113. https://doi.org/10.1016/J.NEUINT.2018.12.019.

Monda, M., Viggiano, A., Sullo, A., & de Luca, V. (1998). Aspartic and glutamic acids increase in the frontal cortex during prostaglandin E1 hyperthermia. *Neuroscience*, *83*(4), 1239–1243. https://doi.org/10.1016/S0306-4522(97)00448-X.

Morishima, Y., Miyakawa, T., Furuyashiki, T., Tanaka, Y., Mizuma, H., & Nakanishi, S. (2005). Enhanced cocaine responsiveness and impared motor coordination in metabotropic glutamate receptor subtype 2 knockout mice. *Proceedings of the National Academy of Sciences of the United States of America*, *102*(11), 4170–4175. https://doi.org/10.1073/PNAS.0500914102/SUPPL_FILE/00914TABLE1.PDF.

Moszczynska, A., Fitzmaurice, P., Ang, L., Kalasinsky, K. S., Schmunk, G. A., Peretti, F. J., et al. (2004). Why is parkinsonism not a feature of human methamphetamine users? *Brain: A Journal of Neurology, 127*(Pt. 2), 363–370. https://doi.org/10.1093/BRAIN/AWH046.

Movsesyan, V. A., O'Leary, D. M., Fan, L., Bao, W., Mullins, P. G., Knoblach, S. M., et al. (2001). mGluR5 antagonists 2-methyl-6-(phenylethynyl)-pyridine and (E)-2-methyl-6-(2-phenylethenyl)-pyridine reduce traumatic neuronal injury in vitro and in vivo by antagonizing N-methyl-D-aspartate receptors. *The Journal of Pharmacology and Experimental Therapeutics, 296*(1), 41–47.

Murray, C. H., Christian, D. T., Milovanovic, M., Loweth, J. A., Hwang, E.-K., Caccamise, A. J., et al. (2021). mGlu5 function in the nucleus accumbens core during the incubation of methamphetamine craving. *Neuropharmacology, 186*, 108452. https://doi.org/10.1016/j.neuropharm.2021.108452.

Neki, A., Ohishi, H., Kaneko, T., Shigemoto, R., Nakanishi, S., & Mizuno, N. (1996). Pre- and postsynaptic localization of a metabotropic glutamate receptor, mGluR2, in the rat brain: An immunohistochemical study with a monoclonal antibody. *Neuroscience Letters, 202*(3), 197–200. https://doi.org/10.1016/0304-3940(95)12248-6.

Nishi, A., Liu, F., Matsuyama, S., Hamada, M., Higashi, H., Nairn, A. C., et al. (2003). Metabotropic mGlu5 receptors regulate adenosine A2A receptor signaling. *Proceedings of the National Academy of Sciences of the United States of America, 100*(3), 1322–1327. https://doi.org/10.1073/PNAS.0237126100.

Niswender, C. M., & Conn, P. J. (2010). Metabotropic glutamate receptors: Physiology, pharmacology, and disease. *Annual Review of Pharmacology and Toxicology, 50*, 295. https://doi.org/10.1146/ANNUREV.PHARMTOX.011008.145533.

Nolan, S. O., Zachry, J. E., Johnson, A. R., Brady, L. J., Siciliano, C. A., & Calipari, E. S. (2020). Direct dopamine terminal regulation by local striatal microcircuitry. *Journal of Neurochemistry, 155*(5), 475–493. https://doi.org/10.1111/JNC.15034.

O'Dell, S. J., Weihmuller, F. B., & Marshall, J. F. (1991). Multiple methamphetamine injections induce marked increases in extracellular striatal dopamine which correlate with subsequent neurotoxicity. *Brain Research, 564*(2), 256–260. https://doi.org/10.1016/0006-8993(91)91461-9.

O'Riordan, K. J., Hu, N. W., & Rowan, M. J. (2018). Physiological activation of mGlu5 receptors supports the ion channel function of NMDA receptors in hippocampal LTD induction in vivo. *Scientific Reports, 8*(1). https://doi.org/10.1038/S41598-018-22768-X.

Osborne, M. P. H., & Olive, M. F. (2008). A role for mGluR5 receptors in intravenous methamphetamine self-administration. *Annals of the New York Academy of Sciences, 1139*(1), 206–211. https://doi.org/10.1196/annals.1432.034.

Page, G., Peeters, M., Najimi, M., Maloteaux, J.-M., & Hermans, E. (2001). Modulation of the neuronal dopamine transporter activity by the metabotropic glutamate receptor mGluR5 in rat striatal synaptosomes through phosphorylation mediated processes. *Journal of Neurochemistry, 76*(5), 1282–1290. https://doi.org/10.1046/j.1471-4159.2001.00179.x.

Parelkar, N. K., & Wang, J. Q. (2008). Upregulation of metabotropic glutamate receptor 8 mRNA expression in the rat forebrain after repeated amphetamine administration. *Neuroscience Letters, 433*(3), 250–254. https://doi.org/10.1016/J.NEULET.2008.01.015.

Parsegian, A., & See, R. E. (2014). Dysregulation of dopamine and glutamate release in the prefrontal cortex and nucleus accumbens following methamphetamine self-administration and during reinstatement in rats. *Neuropsychopharmacology: Official Publication of the American College of Neuropsychopharmacology, 39*(4), 811–822. https://doi.org/10.1038/npp.2013.231.

Paulus, M. P., Tapert, S. F., & Schuckit, M. A. (2005). Neural activation patterns of methamphetamine-dependent subjects during decision making predict relapse. *Archives of General Psychiatry, 62*(7), 761–768. https://doi.org/10.1001/ARCHPSYC.62.7.761.

Paxinos, G., & Watson, C. (2007). *The rat brain in stereotaxic coordinates* (6th ed.). Academic Press/Elsevier.

Pehrson, A. L., & Moghaddam, B. (2010). Impact of metabotropic glutamate 2/3 receptor stimulation on activated dopamine release and locomotion. *Psychopharmacology, 211*(4), 443. https://doi.org/10.1007/S00213-010-1914-8.

Pereira, F. C., Cunha-Oliveira, T., Viana, S. D., Travassos, A. S., Nunes, S., Silva, C., et al. (2012). Disruption of striatal glutamatergic/GABAergic homeostasis following acute methamphetamine in mice. *Neurotoxicology and Teratology, 34*(5), 522–529. https://doi.org/10.1016/J.NTT.2012.07.005.

Peters, J., Scofield, M. D., Ghee, S. M., Heinsbroek, J. A., & Reichel, C. M. (2016). Perirhinal cortex mGlu5 receptor activation reduces relapse to methamphetamine seeking by restoring novelty salience. *Neuropsychopharmacology, 41*(6), 1477–1485. https://doi.org/10.1038/npp.2015.283.

Petralia, R. S., Wang, Y. X., Niedzielski, A. S., & Wenthold, R. J. (1996). The metabotropic glutamate receptors, mGluR2 and mGluR3, show unique postsynaptic, presynaptic and glial localizations. *Neuroscience, 71*(4), 949–976. https://doi.org/10.1016/0306-4522(95)00533-1.

Pierce, R. C., & Kalivas, P. W. (1997). A circuitry model of the expression of behavioral sensitization to amphetamine-like psychostimulants. *Brain Research Reviews, 25*(2), 192–216. https://doi.org/10.1016/S0165-0173(97)00021-0.

Porter, R. H. P., Jaeschke, G., Spooren, W., Ballard, T. M., Büttelmann, B., Kolczewski, S., et al. (2005). Fenobam: A clinically validated nonbenzodiazepine anxiolytic is a potent, selective, and noncompetitive mGlu5 receptor antagonist with inverse agonist activity. *Journal of Pharmacology and Experimental Therapeutics, 315*(2), 711–721. https://doi.org/10.1124/JPET.105.089839.

Pu, C., Broening, H. W., & Vorhees, C.v. (1996). Effect of methamphetamine on glutamate-positive neurons in the adult and developing rat somatosensory cortex. *Synapse, 231328–231334.* https://doi.org/10.1002/(SICI)1098-2396(199608)23:4.

Rademacher, D. J., Rosenkranz, J. A., Morshedi, M. M., Sullivan, E. M., & Meredith, G. E. (2010). Amphetamine-associated contextual learning is accompanied by structural and functional plasticity in the basolateral amygdala. *Journal of Neuroscience, 30*(13), 4676–4686. https://doi.org/10.1523/JNEUROSCI.6165-09.2010.

Raudensky, J., & Yamamoto, B. K. (2007). Effects of chronic unpredictable stress and methamphetamine on hippocampal glutamate function. *Brain Research, 1135*(1), 129–135. https://doi.org/10.1016/j.brainres.2006.12.002.

Reichel, C. M., & Bevins, R. A. (2008). Competition between the conditioned rewarding effects of cocaine and novelty. *Behavioral Neuroscience, 122*(1), 140–150. https://doi.org/10.1037/0735-7044.122.1.140.

Reichel, C. M., Ramsey, L. A., Schwendt, M., McGinty, J. F., & See, R. E. (2012). Methamphetamine-induced changes in the object recognition memory circuit. *Neuropharmacology, 62*(2), 1119–1126. https://doi.org/10.1016/J.NEUROPHARM.2011.11.003.

Reichel, C. M., Schwendt, M., McGinty, J. F., Olive, M. F., & See, R. E. (2011). Loss of object recognition memory produced by extended access to methamphetamine self-administration is reversed by positive allosteric modulation of metabotropic glutamate receptor 5. *Neuropsychopharmacology: Official Publication of the American College of Neuropsychopharmacology, 36*(4), 782–792. https://doi.org/10.1038/NPP.2010.212.

Robinson, T. E., Yew, J., Paulson, P. E., & Camp, D. M. (1990). The long-term effects of neurotoxic doses of methamphetamine on the extracellular concentration of dopamine measured with microdialysis in striatum. *Neuroscience Letters, 110*(1–2), 193–198. https://doi.org/10.1016/0304-3940(90)90810-V.

Rodrigues, R. J., Alfaro, T. M., Rebola, N., Oliveira, C. R., & Cunha, R. A. (2005). Co-localization and functional interaction between adenosine A2A and metabotropic group 5 receptors in glutamatergic nerve terminals of the rat striatum. *Journal of Neurochemistry, 92*(3), 433–441. https://doi.org/10.1111/j.1471-4159.2004.02887.x.

Sandoval, V., Riddle, E. L., Ugarte, Y. V., Hanson, G. R., & Fleckenstein, A. E. (2001). Methamphetamine-induced rapid and reversible changes in dopamine transporter function: An in vitro model. *The Journal of Neuroscience: The Official Journal of the Society for Neuroscience, 21*(4), 1413–1419. https://doi.org/10.1523/JNEUROSCI.21-04-01413.2001.

Satow, A., Maehara, S., Ise, S., Hikichi, H., Fukushima, M., Suzuki, G., et al. (2008). Pharmacological effects of the metabotropic glutamate receptor 1 antagonist compared with those of the metabotropic glutamate receptor 5 antagonist and metabotropic glutamate receptor 2/3 agonist in rodents: Detailed investigations with a selective allosteric metabotropic glutamate receptor 1 antagonist, FTIDC [4-[1-(2-fluoropyridine-3-yl)-5-methyl-1H-1,2,3-triazol-4-yl]-N-isopropyl-N-methyl-3,6-dihydropyridine-1(2H)-carboxamide]. *The Journal of Pharmacology and Experimental Therapeutics, 326*(2), 577–586. https://doi.org/10.1124/JPET.108.138107.

Saunders, C., Ferrer, J.v., Shi, L., Chen, J., Merrill, G., Lamb, M. E., et al. (2000). Amphetamine-induced loss of human dopamine transporter activity: An internalization-dependent and cocaine-sensitive mechanism. *Proceedings of the National Academy of Sciences of the United States of America, 97*(12), 6850–6855. https://doi.org/10.1073/pnas.110035297.

Scheyer, A. F., Christian, D. T., Wolf, M. E., & Tseng, K. Y. (2018). Emergence of endocytosis-dependent mGlu1 LTD at nucleus accumbens synapses after withdrawal from cocaine self-administration. *Frontiers in Synaptic Neuroscience, 10.* https://doi.org/10.3389/FNSYN.2018.00036.

Schwendt, M., & Knackstedt, L. A. (2021). Extinction vs. abstinence: A review of the molecular and circuit consequences of different post-cocaine experiences. *International Journal of Molecular Sciences, 22*(11). https://doi.org/10.3390/IJMS22116113.

Schwendt, M., & McGinty, J. F. (2007). Regulator of G-protein signaling 4 interacts with metabotropic glutamate receptor subtype 5 in rat striatum: Relevance to amphetamine behavioral sensitization. *The Journal of Pharmacology and Experimental Therapeutics, 323*(2), 650–657. https://doi.org/10.1124/JPET.107.128561.

Schwendt, M., Reichel, C. M., & See, R. E. (2012). Extinction-dependent alterations in corticostriatal mGluR2/3 and mGluR7 receptors following chronic methamphetamine self-administration in rats. *PLoS One, 7*(3). https://doi.org/10.1371/JOURNAL.PONE.0034299.

Scofield, M. D., Trantham-Davidson, H., Schwendt, M., Leong, K. C., Peters, J., See, R. E., et al. (2015). Failure to recognize novelty after extended methamphetamine self-administration results from loss of long-term depression in the perirhinal cortex. *Neuropsychopharmacology: Official Publication of the American College of Neuropsychopharmacology, 40*(11), 2526–2535. https://doi.org/10.1038/NPP.2015.99.

Scott, J. C., Woods, S. P., Matt, G. E., Meyer, R. A., Heaton, R. K., Atkinson, J. H., et al. (2007). Neurocognitive effects of methamphetamine: A critical review and meta-analysis. *Neuropsychology Review, 17*(3), 275–297. https://doi.org/10.1007/s11065-007-9031-0.

Sebastianutto, I., Goyet, E., Andreoli, L., Font-Ingles, J., Moreno-Delgado, D., Bouquier, N., et al. (2020). D1-mGlu5 heteromers mediate noncanonical dopamine signaling in Parkinson's disease. *The Journal of Clinical Investigation, 130*(3), 1168. https://doi.org/10.1172/JCI126361.

Seeman, P., Battaglia, G., Corti, C., Corsi, M., & Bruno, V. (2009). Glutamate receptor mGlu2 and mGlu3 knockout striata are dopamine supersensitive, with elevated D2(High) receptors and marked supersensitivity to the dopamine agonist (+)PHNO. *Synapse (New York, N.Y.), 63*(3), 247–251. https://doi.org/10.1002/SYN.20607.

Sengupta, T., Jaryal, A. K., Kumar, V. M., & Mallick, H. N. (2014). L-glutamate microinjection in the preoptic area increases brain and body temperature in freely moving rats. *Neuroreport*, *25*(1), 28–33. https://doi.org/10.1097/WNR.0000000000000035.

Shaerzadeh, F., Streit, W. J., Heysieattalab, S., & Khoshbouei, H. (2018). Methamphetamine neurotoxicity, microglia, and neuroinflammation. *Journal of Neuroinflammation*, *15*(1), 1–6. https://doi.org/10.1186/S12974-018-1385-0/FIGURES/2.

Shah, U. H., & González-Maeso, J. (2019). Serotonin and glutamate interactions in preclinical schizophrenia models. *ACS Chemical Neuroscience*, *10*(7), 3068–3077. https://doi.org/10.1021/ACSCHEMNEURO.9B00044/ASSET/IMAGES/LARGE/CN-2019-00044M_0003.JPEG.

Shaham, Y., Shalev, U., Lu, L., de Wit, H., & Stewart, J. (2003). The reinstatement model of drug relapse: history, methodology and major findings. *Psychopharmacology*, *168*(1–2), 3–20. https://doi.org/10.1007/s00213-002-1224-x.

Shen, H., Chen, K., Marino, R. A. M., McDevitt, R. A., & Xi, Z. X. (2021). Deletion of VGLUT2 in midbrain dopamine neurons attenuates dopamine and glutamate responses to methamphetamine in mice. *Pharmacology Biochemistry and Behavior*, *202*, 173104. https://doi.org/10.1016/J.PBB.2021.173104.

Shetty, V., Mooney, L. J., Zigler, C. M., Belin, T. R., Murphy, D., & Rawson, R. (2010). The relationship between methamphetamine use and increased dental disease. *The Journal of the American Dental Association*, *141*(3), 307–318. https://doi.org/10.14219/JADA.ARCHIVE.2010.0165.

Shigemoto, R., Kinoshita, A., Wada, E., Nomura, S., Ohishi, H., Takada, M., et al. (1997). Differential presynaptic localization of metabotropic glutamate receptor subtypes in the rat hippocampus. *Journal of Neuroscience*, *17*(19), 7503–7522. https://doi.org/10.1523/JNEUROSCI.17-19-07503.1997.

Shimazoe, T., Doi, Y., Arai, I., Yoshimatsu, A., Fukumoto, T., & Watanabe, S. (2002). Both metabotropic glutamate I and II receptors mediate augmentation of dopamine release from the striatum in methamphetamine-sensitized rats. *Japanese Journal of Pharmacology*, *89*(1), 85–88. https://doi.org/10.1254/JJP.89.85.

Shin, E. J., Tran, H. Q., Nguyen, P. T., Jeong, J. H., Nah, S. Y., Jang, C. G., et al. (2018). Role of mitochondria in methamphetamine-induced dopaminergic neurotoxicity: Involvement in oxidative stress, neuroinflammation, and pro-apoptosis—A review. *Neurochemical Research*, *43*(1), 57–69. https://doi.org/10.1007/S11064-017-2318-5.

Shukla, M., & Vincent, B. (2021). Methamphetamine abuse disturbs the dopaminergic system to impair hippocampal-based learning and memory: An overview of animal and human investigations. *Neuroscience and Biobehavioral Reviews*, *131*, 541–559. https://doi.org/10.1016/J.NEUBIOREV.2021.09.016.

Simon, S. L., Dacey, J., Glynn, S., Rawson, R., & Ling, W. (2004). The effect of relapse on cognition in abstinent methamphetamine abusers. *Journal of Substance Abuse Treatment*, *27*(1), 59–66. https://doi.org/10.1016/J.JSAT.2004.03.011.

Spampinato, S. F., Copani, A., Nicoletti, F., Sortino, M. A., & Caraci, F. (2018). Metabotropic glutamate receptors in glial cells: A new potential target for neuroprotection? *Frontiers in Molecular Neuroscience*, *11*, 414. https://doi.org/10.3389/FNMOL.2018.00414/BIBTEX.

Spooren, W. P. J. M., Vassout, A., Neijt, H. C., Kuhn, R., Gasparini, F., Roux, S., et al. (2000). Anxiolytic-like effects of the prototypical metabotropic glutamate receptor 5 antagonist 2-methyl-6-(phenylethynyl)pyridine in rodents. *The Journal of Pharmacology and Experimental Therapeutics*, *295*(3), 1267–1275. https://pubmed.ncbi.nlm.nih.gov/11082464/.

Stephans, S. E., & Yamamoto, B. K. (1994). Methamphetamine-induced neurotoxicity: Roles for glutamate and dopamine efflux. *Synapse (New York, N.Y.)*, *17*(3), 203–209. https://doi.org/10.1002/SYN.890170310.

Stephans, S. E., & Yamamoto, B. K. (1995). Effect of repeated methamphetamine administrations on dopamine and glutamate efflux in rat prefrontal cortex. *Brain Research*, *700*(1–2), 99–106. https://doi.org/10.1016/0006-8993(95)00938-M.

Substance Abuse and Mental Health Services Administration (SAMHSA). (2021). *2020 National Survey on Drug Use and Health (NSDUH)*. https://www.samhsa.gov/data/data-we-collect/nsduh-national-survey-drug-use-and-health.

Sulzer, D., Chen, T. K., Lau, Y. Y., Kristensen, H., Rayport, S., & Ewing, A. (1995). Amphetamine redistributes dopamine from synaptic vesicles to the cytosol and promotes reverse transport. *The Journal of Neuroscience: The Official Journal of the Society for Neuroscience*, *15*(5 Pt. 2), 4102–4108.

Szumlinski, K. K., Lominac, K. D., Campbell, R. R., Cohen, M., Fultz, E. K., Brown, C. N., et al. (2017). Methamphetamine addiction vulnerability: The glutamate, the bad and the ugly. *Biological Psychiatry*, *81*(11), 959. https://doi.org/10.1016/J.BIOPSYCH.2016.10.005.

Tamaru, Y., Nomura, S., Mizuno, N., & Shigemoto, R. (2001). Distribution of metabotropic glutamate receptor mGluR3 in the mouse CNS: Differential location relative to pre- and postsynaptic sites. *Neuroscience*, *106*(3), 481–503. https://doi.org/10.1016/S0306-4522(01)00305-0.

Testa, C. M., Catania, M. V., & Young, A. B. (1994). Anatomical distribution of metabotropic glutamate receptors in mammalian brain. *The Metabotropic Glutamate Receptors*, 99–123. https://doi.org/10.1007/978-1-4757-2298-7_4.

Tian, L., & Kammermeier, P. J. (2006). G protein coupling profile of mGluR6 and expression of G alpha proteins in retinal ON bipolar cells. *Visual Neuroscience*, *23*(6), 909–916. https://doi.org/10.1017/S0952523806230268.

Traynelis, S. F., Wollmuth, L. P., McBain, C. J., Menniti, F. S., Vance, K. M., Ogden, K. K., et al. (2010). Glutamate receptor ion channels: Structure, regulation, and function. *Pharmacological Reviews*, *62*(3), 405–496. https://doi.org/10.1124/pr.109.002451.

Tu, J. C., Xiao, B., Naisbitt, S., Yuan, J. P., Petralia, R. S., Brakeman, P., et al. (1999). Coupling of mGluR/Homer and PSD-95 complexes by the Shank family of postsynaptic density proteins. *Neuron*, *23*(3), 583–592. https://doi.org/10.1016/S0896-6273(00)80810-7.

U.S. Drug Enforcement Administration (DEA). (2020). *Methamphetamine—Drug fact sheet*. https://www.deadiversion.usdoj.gov/drug_chem_info/meth.pdf.

U.S. Food and Drug Administration (FDA). Medication Guide. (2017). *Desoxyn*. https://www.accessdata.fda.gov/drugsatfda_docs/label/2017/005378s034lbl.pdf.

United Nations Office on Drugs and Crime (UNDOC). (2020). World Drug Report. In *United Nations publication, Sales No. E.20.XI.6*. https://wdr.unodc.org/wdr2020/index2020.html.

Uslaner, J. M., Parmentier-Batteur, S., Flick, R. B., Surles, N. O., Lam, J. S. H., McNaughton, C. H., et al. (2009). Dose-dependent effect of CDPPB, the mGluR5 positive allosteric modulator, on recognition memory is associated with GluR1 and CREB phosphorylation in the prefrontal cortex and hippocampus. *Neuropharmacology*, *57*(5–6), 531–538. https://doi.org/10.1016/j.neuropharm.2009.07.022.

Venniro, M., Caprioli, D., & Shaham, Y. (2016). Animal models of drug relapse and craving: From drug priming-induced reinstatement to incubation of craving after voluntary abstinence. *Progress in Brain Research*, *224*, 25–52. https://doi.org/10.1016/BS.PBR.2015.08.004.

Volpi, C., Fallarino, F., Mondanelli, G., Macchiarulo, A., & Grohmann, U. (2018). Opportunities and challenges in drug discovery targeting metabotropic glutamate receptor 4. *Expert Opinion on Drug Discovery*, *13*(5), 411–423. https://doi.org/10.1080/17460441.2018.1443076.

Voulalas, P. J., Holtzclaw, L., Wolstenholme, J., Russell, J. T., & Hyman, S. E. (2005). Metabotropic glutamate receptors and dopamine receptors cooperate to enhance extracellular signal-regulated kinase phosphorylation in striatal neurons. *Journal of Neuroscience, 25*(15), 3763–3773. https://doi.org/10.1523/JNEUROSCI.4574-04.2005.

Wang, G., Shi, J., Chen, N., Xu, L., Li, J., Li, P., et al. (2013). Effects of length of abstinence on decision-making and craving in methamphetamine abusers. *PLoS One, 8*(7), e68791. https://doi.org/10.1371/JOURNAL.PONE.0068791.

Watterson, L. R., Kufahl, P. R., Nemirovsky, N. E., Sewalia, K., Hood, L. E., & Olive, M. F. (2013). Attenuation of reinstatement of methamphetamine-, sucrose-, and food-seeking behavior in rats by fenobam, a metabotropic glutamate receptor 5 negative allosteric modulator. *Psychopharmacology, 225*(1), 151–159. https://doi.org/10.1007/s00213-012-2804-z.

White, M. G., Luca, L. E., Nonner, D., Saleh, O., Hu, B., Barrett, E. F., et al. (2007). Cellular mechanisms of neuronal damage from hyperthermia. *Progress in Brain Research, 162*, 347–371. https://doi.org/10.1016/S0079-6123(06)62017-7.

Widholm, J. J., Gass, J. T., Cleva, R. M., & Olive, M. F. (2012). The Mglur5 positive allosteric modulator CDPPB does not alter extinction or contextual reinstatement of methamphetamine-seeking behavior in rats. *Journal of Addiction Research & Therapy, 01*(S1), 1–12. https://doi.org/10.4172/2155-6105.S1-004.

Winiewski, K., & Car, H. (2002). (S)-3,5-DHPG: A review. *CNS Drug Reviews, 8*(1), 101–116. https://doi.org/10.1111/J.1527-3458.2002.TB00218.X.

Winters, B. D., & Bussey, T. J. (2005). Glutamate receptors in perirhinal cortex mediate encoding, retrieval, and consolidation of object recognition memory. *Journal of Neuroscience, 25*(17), 4243–4251. https://doi.org/10.1523/JNEUROSCI.0480-05.2005.

Wischhof, L., Hollensteiner, K. J., & Koch, M. (2011). Impulsive behaviour in rats induced by intracortical DOI infusions is antagonized by co-administration of an mGlu2/3 receptor agonist. *Behavioural Pharmacology, 22*(8), 805–813. https://doi.org/10.1097/FBP.0B013E32834D6279.

Wischhof, L., & Koch, M. (2016). 5-HT2A and mGlu2/3 receptor interactions: On their relevance to cognitive function and psychosis. *Behavioural Pharmacology, 27*(1), 1–11. https://doi.org/10.1097/FBP.0000000000000183.

Woolverton, W. L., Ricaurte, G. A., Forno, L. S., & Seiden, L. S. (1989). Long-term effects of chronic methamphetamine administration in rhesus monkeys. *Brain Research, 486*(1), 73–78. https://doi.org/10.1016/0006-8993(89)91279-1.

Wright, S. R., Zanos, P., Georgiou, P., Yoo, J. H., Ledent, C., Hourani, S. M., et al. (2016). A critical role of striatal A2AR–mGlu5R interactions in modulating the psychomotor and drug-seeking effects of methamphetamine. *Addiction Biology, 21*(4), 811–825. https://doi.org/10.1111/ADB.12259.

Xu, J., Marshall, J. J., Kraniotis, S., Nomura, T., Zhu, Y., & Contractor, A. (2021). Genetic disruption of Grm5 causes complex alterations in motor activity, anxiety and social behaviors. *Behavioural Brain Research, 411*. https://doi.org/10.1016/J.BBR.2021.113378.

Yang, X., Wang, Y., Li, Q., Zhong, Y., Chen, L., Du, Y., et al. (2018). The main molecular mechanisms underlying methamphetamine- induced neurotoxicity and implications for pharmacological treatment. *Frontiers in Molecular Neuroscience, 11*. https://doi.org/10.3389/FNMOL.2018.00186.

Yang, W., Yang, R., Luo, J., He, L., Liu, J., & Zhang, J. (2018). Increased absolute glutamate concentrations and Glutamate-to-creatine ratios in patients with methamphetamine use disorders. *Frontiers in Psychiatry, 9*(Aug), 368. https://doi.org/10.3389/FPSYT.2018.00368/BIBTEX.

Yang, H. J., Zhang, H. Y., Bi, G. H., He, Y., Gao, J. T., & Xi, Z. X. (2017). Deletion of type 2 metabotropic glutamate receptor decreases sensitivity to cocaine reward in rats. *Cell Reports, 20*(2), 319–332. https://doi.org/10.1016/J.CELREP.2017.06.046.

Yu, S., Zhu, L., Shen, Q., Bai, X., & Di, X. (2015). Recent advances in methamphetamine neurotoxicity mechanisms and its molecular pathophysiology. *Behavioural Neurology, 2015*. https://doi.org/10.1155/2015/103969.

Zhang, S., Jin, Y., Liu, X., Yang, L., Juan, G. Z., Wang, H., et al. (2014). Methamphetamine modulates glutamatergic synaptic transmission in rat primary cultured hippocampal neurons. *Brain Research, 1582*, 1–11. https://doi.org/10.1016/j.brainres.2014.07.040.

CHAPTER FIVE

The role of mGlu receptors in susceptibility to stress-induced anhedonia, fear, and anxiety-like behavior

Cassandra G. Modrak[a,b,c], Courtney S. Wilkinson[a,b,c], Harrison L. Blount[a,b,c], Marek Schwendt[a,b,c], and Lori A. Knackstedt[a,b,c],*

[a]Department of Psychology, University of Florida, Gainesville, FL, United States
[b]Center for Addiction Research and Education, University of Florida, Gainesville, FL, United States
[c]Center for OCD, Anxiety, and Related Disorders, University of Florida, Gainesville, FL, United States
*Corresponding author: e-mail address: knack@ufl.edu

Contents

1. Introduction	222
2. Animal models of stress induction	224
2.1 Predator stress	224
2.2 Social defeat stress (SDS)	225
2.3 Foot shock Stress	226
2.4 Chronic variable stress (CVS)	227
2.5 Perinatal and early life stress	227
2.6 Chronic corticosterone	228
3. Behavioral tests for anxiety-like behavior, anhedonia, and fear	228
3.1 Anxiety-like behavior assessment	228
3.2 Tools for the assessment of depression-like behavior	231
3.3 Assessment of fear	233
4. Brain-wide mGlu5 and stress-induced anhedonia, fear, and anxiety-like behavior	233
4.1 mGlu5 and anhedonia	233
4.2 mGlu5 and anxiety-like behavior	235
4.3 mGlu5 and fear conditioning	236
5. The mGlu5-containing circuitry underlying stress-induced anhedonia, fear, and anxiety-like behavior	237
5.1 mGlu5: Amygdala and extended amygdala	237
5.2 mGlu5: Bed nucleus of the stria terminalis (BNST)	239
5.3 mGlu5: Medial prefrontal cortex (mPFC)	240
5.4 mGlu5: Hippocampus (HPC)	241
5.5 mGlu5: Nucleus accumbens (NA)	243

International Review of Neurobiology, Volume 168
ISSN 0074-7742
https://doi.org/10.1016/bs.irn.2022.10.006

6. Summary of mGlu5 role in anhedonia, fear, and anxiety-like behavior 244
7. mGlu2 and mGlu3 receptors and stress-induced anhedonia, fear, and anxiety-like behavior 247
   7.1 mGlu2 and mGlu3 receptors and resilience to anhedonia 247
   7.2 mGlu2 and mGlu3 receptors and resilience to anxiety-like behavior 250
   7.3 mGlu2 and mGlu3 receptors and resilience to fear conditioning 250
8. The mGlu2/3-containing circuitry underlying stress-induced anhedonia, fear, and anxiety-like behavior 250
   8.1 mGlu2/3: Amygdala 250
   8.2 mGlu2/3: Bed nucleus of the stria terminalis (BNST) 251
   8.3 mGlu2/3: Medial prefrontal cortex (mPFC) 251
   8.4 mGlu2/3: Hippocampus (HPC) 252
   8.5 Lateral septal nuclei 254
9. mGlu2/3 conclusions 254
10. General conclusions 255
References 256

Abstract

Stress and trauma exposure contribute to the development of psychiatric disorders such as post-traumatic stress disorder (PTSD) and major depressive disorder (MDD) in a subset of people. A large body of preclinical work has found that the metabotropic glutamate (mGlu) family of G protein-coupled receptors regulate several behaviors that are part of the symptom clusters for both PTSD and MDD, including anhedonia, anxiety, and fear. Here, we review this literature, beginning with a summary of the wide variety of preclinical models used to assess these behaviors. We then summarize the involvement of Group I and II mGlu receptors in these behaviors. Bringing together this extensive literature reveals that mGlu5 signaling plays distinct roles in anhedonia, fear, and anxiety-like behavior. mGlu5 promotes susceptibility to stress-induced anhedonia and resilience to stress-induced anxiety-like behavior, while serving a fundamental role in the learning underlying fear conditioning. The medial prefrontal cortex, basolateral amygdala, nucleus accumbens, and ventral hippocampus are key regions where mGlu5, mGlu2, and mGlu3 regulate these behaviors. There is strong support that stress-induced anhedonia arises from decreased glutamate release and post-synaptic mGlu5 signaling. Conversely, decreasing mGlu5 signaling increases *resilience* to stress-induced anxiety-like behavior. Consistent with opposing roles for mGlu5 and mGlu2/3 in anhedonia, evidence suggests that increased glutamate transmission may be therapeutic for the extinction of fear learning. Thus, a large body of literature supports the targeting of pre- and post-synaptic glutamate signaling to ameliorate post-stress anhedonia, fear, and anxiety-like behavior.

1. Introduction

Exposure to stressors is nearly inevitable over the course of the human lifespan. Exposure to a severe stressor, or trauma, is also prevalent, with estimates of lifetime exposure to trauma ranging from 54% to 89% depending on the geographical region assessed (Kessler, Sonnega, Bromet, Hughes, & Nelson, 1995; Mylle & Maes, 2004). A subpopulation of trauma-exposed individuals later develops Post-traumatic Stress Disorder (PTSD). PTSD is a multidimensional anxiety disorder. DSM-5 criteria for PTSD diagnosis include the presence of symptoms in each three categories (intrusion of trauma-related images and memories, avoidance of trauma-related stimuli, and negative alterations in cognition and/or mood) that last for more than a month after the trauma and have a negative impact on the quality of life (American Psychological Association, 2013). Symptoms of increased arousal (e.g., heightened startle response, hypervigilance) may also be present but are not required for diagnosis. Symptoms in the "negative alterations in cognition and/or mood" category of PTSD diagnostic criteria overlap with those for Major Depressive Disorder, which is also exacerbated by stress, including: loss of interest or pleasure in daily activities, anhedonia, negative affect, insomnia, and guilt (American Psychological Association, 2013).

The neurobiology underlying susceptibility to stress-induced disorders is complex and involves many hormone and neurotransmitter systems. The metabotropic glutamate receptors (mGlu's) have been implicated in fear, anxiety, anhedonia and susceptibility to the development of PTSD and as such, are the focus of the present review. mGlu receptors are G protein-coupled receptors (GPCRs). There are three families, or "groups" of mGlu receptors: Group I, II, and III. These groups are defined based on their G-protein coupling and signaling as well as amino acid sequences (4). Group I mGlu receptors include mGlu1 and mGlu5 receptors, group II includes mGlu2 and mGlu3 receptors, and group III includes mGlu4, mGlu6, mGlu7, and mGlu8 receptors. Group I mGlu receptors are coupled to $G_{q/11}$ proteins and group II and III receptors are coupled to $G_{i/o}$ proteins (Niswender & Conn, 2010). With the exception of mGlu6, which is expressed in the retina, mGlu receptors are found at moderate-to-high levels of expression throughout the CNS, primarily in neurons. mGlu5 is also expressed in astrocytes. However, the synaptic localization differs by group, with Group I mGlu's predominantly expressed post-synaptically. Group II and III mGlu's are expressed post-synaptically as well, but are primarily expressed pre-synaptically where they can serve as autoreceptors to reduce glutamate release. See (Niswender & Conn, 2010) for a thorough review of mGlu expression and pharmacology.

Studies using genetic or pharmacological manipulation of mGlu function have found roles for these receptors in fear, anxiety, and anhedonia. The majority of this research focuses on three mGlu receptors: the Group II receptors (mGlu2 and mGlu3) and the Group I receptor (mGlu5). Here, we will review the preclinical literature investigating the role of these receptors in mediating fear, anxiety and anhedonia arising from exposure to a stressor, with a focus on their role in susceptibility and resilience to the long-term consequences of stress. First, we will summarize the rodent models used to induce stress and to assess anhedonia, fear, and anxiety-like behavior in studies that assess the involvement of mGlu receptors. This will be followed by a summary of the results of studies assessing the role of mGlu2, mGlu3, and mGlu5 in such behaviors. For the most part, we will not discuss studies which assessed these behaviors in the absence of a stressor, unless necessary to clarify results. Unless noted, all studies were done using male rodents.

2. Animal models of stress induction
2.1 Predator stress

Predator scent stress (PSS) and predator threat stress are ethologically relevant stressors that induce differential neural, behavioral and physiological responses. Generally, predator stress exposure occurs in chambers with no route of escape for the rodent. This denies rodents the ability to engage in most defensive behaviors (except freezing), decreases non-defensive behaviors, and results in persistent fear, anhedonia, and anxiety-like behavior (Apfelbach, Blanchard, Blanchard, Hayes, & McGregor, 2005; Tyler, Bluitt, Engers, Lindsley, & Besheer, 2022). Predator stress causes acute and long-lasting changes in the hypothalamic-pituitary-adrenal axis and monoaminergic and glutamatergic systems (Adamec, Kent, Anisman, Shallow, & Merali, 1998; Apfelbach et al., 2005; Belzung, El Hage, Moindrot, & Griebel, 2001; Hayley, Borowski, Merali, & Anisman, 2001; Shallcross et al., 2019; Shallcross, Wu, Knackstedt, & Schwendt, 2020; Tyler et al., 2022). Induction of predator scent stress and duration of exposure differs by group. Common methods include exposure to bobcat urine, soiled cat litter or sand, previously worn cat collars, and the fox pheromone, 2,5-dihydro-2,4,5-trimethylthiazoline (TMT; (Albrechet-Souza & Gilpin, 2019; Cohen et al., 2012; Dremencov et al., 2019; Mackenzie, Nalivaiko, Beig, Day, & Walker, 2010; Schwendt et al., 2018; Shallcross et al., 2019, 2020; Zohar, Matar, Ifergane, Kaplan, & Cohen, 2008). Behaviors measured during predator scent stress include time spent freezing, avoidance of the area where the scent is introduced, digging (if bedding is used) and fecal boli production. Predator threat stress involves exposure to a live predator, such as a cat. The predator is placed in the testing arena first, and the rodent is introduced shortly after. Predator exposure lasts 5–10 min (Adamec, Head, Blundell, Burton, & Berton, 2006; Adamec & Shallow, 1993; Adamec, Walling, & Burton, 2004). During predator threat exposure, active escape (pushing or biting the predator, sideways or upright postures) and passive escape (freezing, attempts to leave the exposure arena) behaviors are recorded. Susceptibility to the long-term effects of predator stress can be assessed through testing for anxiety-like behavior and/or the ability of the stress-paired context to induced conditioned fear (e.g., freezing).

2.2 Social defeat stress (SDS)

Rodents are a social species and thus are vulnerable to psychosocial stress. The social defeat stress paradigm has strong predictive, discriminative, and face validity capable of inducing anxiety, anhedonia, and fear (Berton et al., 2006; Krishnan et al., 2007; Tsankova et al., 2006). To provoke social defeat, an aggressive rodent is placed in a test chamber as the "resident" of the test chamber and permitted to habituate to the chamber overnight, prior to testing. Rodents experiencing social defeat stress are placed into the resident cage as "intruders" for 5–10 min where residents and intruders freely interact. Intruders can then be transferred to a compartment with a perforated or glass divider separating them from the resident, but with visible contact, for a longer duration (<24 h) to induce further psychological stress (Golden, Covington, Berton, & Russo, 2011; Jing et al., 2021; Yashiro & Seki, 2017). This process is repeated for several days with novel residents introduced periodically. The pattern of exploratory behavior of the intruder, as well as rearing, tail rattling, wrestling, and biting, are quantified. Of note, aggression is variable within and across strains; appropriately aggressive rodents must be identified using screening processes prior to test start. The resident is often from a different strain than the intruders or is a socially isolated conspecific.

Shortly after chronic social defeat stress, intruders are subjected to a social interaction test. This test uses an open field arena with a mesh enclosure and surrounding "interaction zone" on one wall, with the rest of the arena as the "surrounding zone." The test rodent is placed in the arena and allowed to freely roam for 2–3 min in two separate sessions. In the first session, the mesh cage is empty. In the second session, a novel rodent to the intruder is placed in the mesh enclosure. Time spent in the interaction zone with and without the novel rodent is recorded. From these parameters, a social interaction ratio is calculated (time spent in the interaction zone of the social stimulus divided by the time spent in the interaction zone of without a social stimulus) and commonly used to delineate stress-susceptible vs stress-resilient rodents (Golden et al., 2011).

Another stressor that relies on the social structure of rodents is the chronic subordinate colony housing model. In this model, mice are first housed individually. Then, a group of mice is housed with a dominant male for several consecutive days–weeks to induce chronic psychosocial stress. Evaluation of subordination and dominance is monitored periodically throughout this procedure according to the number of defensive behaviors (darting, submissive upright postures) or offensive behaviors (attacks, mounts, chasing), respectively. Dominant males are replaced by a novel dominant male weekly to avoid habituation. In this model, the control is typically one of single-housed control mice (Langgartner, Füchsl, Uschold-Schmidt, Slattery, & Reber, 2015; Peterlik et al., 2017; Reber et al., 2016).

2.3 Foot shock Stress

Foot shock stress is a painful, aversive stimulus that induces both emotional and physical distress in rodents. Unlike other aversive stimuli (e.g., loud noises), rodents generally do not habituate to foot shock. First gaining recognition in 1908, the foot shock stress paradigm is a classic method for investigating underlying stress neurobiology and neuropsychiatric disorders, such as anxiety, panic disorders, PTSD, and depression (Bali & Jaggi, 2015; Yerkes & Dodson, 1908). The number of shocks, intensity, and duration of shock are varied across experiments, as are the conditions under which shock is administered. Foot shock stress can be presented as an inescapable stressor (e.g., Inescapable Shock; IES) in which electric shock is delivered regardless of any rodent action, and escape attempt number is viewed as a measure of learned helplessness or stress-susceptibility (Bali & Jaggi, 2015). Escapable foot shock models allow rodents to stop or avoid the shocking floor via "safe" areas or a lever that discontinues shock and can be used to condition avoidance (Rorick-Kehn et al., 2007; Shors, Seib, Levine, & Thompson, 1989). In operant conflict paradigms, a reward is paired with a foot shock. Operant conflict tasks, fear-potentiated startle tasks, and shock-induced ultrasonic vocalization for the assessment of anxiety-like behavior have been pharmacologically validated for anxiolytic drugs (De Vry, Benz, Schreiber, & Traber, 1993; Sánchez, 1993). Similarly,

The role of mGlu receptors in susceptibility   227

foot shock–induced conditioned fear and models of learned helplessness are pharmacologically validated models of depression (Overmier & Seligman, 1967; Seligman & Maier, 1967).

2.4 Chronic variable stress (CVS)

Chronic variable stress models involve the repeated exposure of rodents to a series of stressors that are presented in pseudo-random order. Typical stressors used include restraint stress, cage tilt (45° for 4–10 h), foot shock, soiled cage, introduction of a novel object into the home cage, bedding removal, home cage crowding, social defeat, social instability, social isolation, intermittent illumination during the dark phase of the light cycle, reversal of the light cycle, stroboscopic light, food/water deprivation, cage vibration, exposure to loud noises, and cold or forced swim. Experimenters choose 2–7 of these mild stressors to administer at a single time point or multiple times a day over a period of days or weeks. After chronic variable stress exposure, sucrose preference is commonly used to identify stress–susceptible and –resilient subpopulations (Musazzi, Tornese, Sala, & Popoli, 2018; Palmfeldt, Henningsen, Eriksen, Müller, & Wiborg, 2016; Sun et al., 2017). Exposure to chronic variable stress increases adrenal weights, reduces thymus weights, and alterations of the HPA axis and induces long-term anxiety- and depression-like behavior (Cullinan & Wolfe, 2000; Herman, Adams, & Prewitt, 1995; Palmfeldt et al., 2016; Peterlik et al., 2017). This model is also referred to as Chronic Unpredictable Stress (CUS).

2.5 Perinatal and early life stress

To induce perinatal stress (PNS), pregnant female rodents, or dams, are exposed to restraint stress, unpredictable variable stress, social stress, foot shock, loud noise, or bright light during gestation. Dams are exposed to stress sessions daily for the last few weeks of gestation. To capture the extent to which perinatal stress is induced, ultrasonic vocalizations of newly born pups can be recorded during short-term isolation periods (Laloux et al., 2012; Maccari, Krugers, Morley-Fletcher, Szyf, & Brunton, 2014). Early life, or postnatal, stress is produced by maternal deprivation or isolation, both of which increase HPA axis activity, corticosterone secretion, and anxiety-like behaviors in adulthood (Kalinichev, Easterling, Plotsky, & Holtzman, 2002; Ogawa et al., 1994). Both perinatal and early adolescence are periods in which the developing brain is vulnerable to stress-induced remodeling that can increase HPA axis response and duration in adulthood and decrease stress-resilience in adulthood following further stress-exposure, with males displaying increased susceptibility in adulthood after perinatal stress in comparison to females (Brunton & Russell, 2010; Darnaudéry & Maccari, 2008; Maccari et al., 2003; Meaney, Szyf, & Seckl, 2007; Seckl & Holmes, 2007; Weinstock, 2008; Zuena et al., 2008).

2.6 Chronic corticosterone

Chronic exogenous corticosterone (CORT) treatment induces persistent depressive-like behavior and is implemented by interchanging the home cage water bottle with a corticosterone hemisuccinate solution (Gourley & Taylor, 2009). Chronic oral corticosterone exposure is permitted for a total of 20 days, with the first 2 weeks at higher corticosterone concentrations (5–6 mg/kg/day). The concentration is then lowered every 3 days to reduce acute withdrawal and facilitate recovery of endogenous corticosterone secretion. Rats return to normal home cage water for a washout period prior to subsequent behavioral testing. Chronic corticosterone exposure increases immobility in the forced swim task and inhibits reward sensitivity that is rescued by antidepressant treatment (Gourley, Kiraly, Howell, Olausson, & Taylor, 2008; Gourley & Taylor, 2009; Gourley et al., 2008; Joffe et al., 2020). This paradigm shows high face, predictive, and construct validity for depression (Gourley & Taylor, 2009).

3. Behavioral tests for anxiety-like behavior, anhedonia, and fear

3.1 Anxiety-like behavior assessment

Behavioral assays of anxiety-like behavior exploit rodents' innate exploratory and foraging nature and preference for dark, closed environments to avoid predation. The elevated plus, elevated zero, and elevated Y maze all produce a conflict for rodents: the propensity to explore and a competing drive for safety. Elevated maze tasks are short in duration (~5 min) and occur on elevated platforms approximately 50 cm from the floor with "closed" (lined with walls) and "open" (no walls) portions of the maze. Rodents often spend more time in the enclosed portions of these mazes, but drugs can alter the ratio of time spent in the open and closed areas of these mazes. Clinically effective anxiolytic drugs increase exploration in the open areas, and anxiogenic drugs decrease exploration, pharmacologically validating these mazes as effective models of anxiety (Kulkarni, Singh, & Bishnoi, 2008; Montgomery, 1955; Pellow, Chopin, File, & Briley, 1985; Shepherd, Grewal, Fletcher, Bill, & Dourish, 1994).

For all elevated mazes, typical dependent variables include time spent in the open area, time spent in the closed area, number of open area entries and number of closed area entries, ratio of time spent in open and closed areas, locomotion, latency to the open or closed areas, head dips over the edge of the platform, and stretched attend postures. However, not all elevated mazes are created equal. Given the cross shape of the elevated plus maze (EPM), the correct classification of time spent in the center zone formed by the opposing arms is commonly critiqued. The annular shape of the elevated zero maze (EM) allows for continuous exploration and eliminates any ambiguity in interpreting center zone engagements of the EPM.

The light–dark (LD) box is used to assess anxiety-like behavior in rodents. This apparatus is composed of a chamber fitted with two compartments. Rodents are placed into the apparatus for approximately 5 min without prior habituation. One portion of the box is darkened, and the remainder of the box is open to light with an opening in the dividing wall for rodents to move between compartments. At task start, rats are placed into one compartment allowing measurement of the following parameters: latency to light/dark side, number of transitions between sides, locomotor activity in light/dark side, time in light/dark side, and number of rears (Bourin & Hascoët, 2003; Buonaguro et al., 2020; Peterlik et al., 2017). Innately, rodents show a significant preference for the dark side of the box (Misslin, Belzung, & Vogel, 1989), however anxiolytic drugs increase exploration and time spent in the light side of the box (Crawley, 1981; Hascoët & Bourin, 1998; López-Rubalcava, Saldívar, & Fernández-Guasti, 1992; Vicente & Zangrossi, 2014).

There are different approaches for using the light–dark box, with differences observed in task duration, compartment size, and the ratio of light-to-dark compartments. The light–dark box is a pharmacologically validated model of anxiety, with time spent in the light compartments identified as the most reliable parameter for assessing anxiogenic and anxiolytic effects (Hascoët & Bourin, 1998). Evaluation of anxiety-like behavior may be confounded by sedation effects or change in exploration, and false positives may occur if manipulations increase locomotion. Considerations and assessment of locomotor changes are encouraged for accurate interpretation of the task.

The open field test (OFT) can be used to assess anxiety-like behavior. Rodents are placed in the center (or along the wall) of an enclosed square platform. The test arena is divided into a center zone and peripheral zones adjacent to the walls. Locomotion during the task (2–20 min) is video recorded and analyzed for time spent in the center vs at the walls, grooming, rearing, and fecal boli are recorded. Increased center exploration (relative to thigmotaxis) without an increase in locomotion indicates lower anxiety. Platforms can vary in shape (circular, square, rectangle), but measure the same parameters (Kraeuter, Guest, & Sarnyai, 2019; Prut & Belzung, 2003; Sturman, Germain, & Bohacek, 2018). Open field test as a model of anxiety is strong in construct and face validity, but a thorough review of pharmacological validity is less convincing (Prut & Belzung, 2003). Therefore, prior to drawing conclusions on anxiety-like behavior, experimenters should consider using multiple tasks to form accurate interpretations.

The acoustic startle response (ASR) task is one of the most translationally relevant and consistent tasks across species due to the evolutionarily conserved unconditional startle reflex (Landis & Hunt, 1939; Swerdlow, Braff, & Geyer, 1999). Rodents are enclosed in stabilimeter chamber enclosed in a sound attenuating chamber box connected to a startle detector. The task begins with habituation to a low decibel (68 dB) white noise for a period of 3–5 min followed by 30 startle trials of 110 dB white noise at 30–45 s inter-trial intervals. Startle amplitude is assessed. Under normal conditions, rodents will reduce startle amplitude over trials as they habituate to the loud stimulus. A lack of habituation to the stimulus or increase in startle response throughout the task indicates greater anxiety-like behavior (Shallcross et al., 2019; Shallcross, Wu, Wilkinson, Knackstedt, & Schwendt, 2021; Tyler et al., 2022).

Stress-induced hyperthermia (SIH) is an observable phenomenon conserved across species and used as a rodent model of anxiety (Marazziti, Di Muro, & Castrogiovanni, 1992). First observed in a group of three rodents housed together, when removed one-by-one, the last rodent showed increased body temperature (Borsini, Lecci, Volterra, & Meli, 1989; Rodgers, Cole, & Harrison-Phillips, 1994). Studies since confirm rectal temperature increases after, or in anticipation of, stress within 10 min, and recovery to basal temperatures takes 1–2 h (Van der Heyden, Zethof, & Olivier, 1997). In single-housed mice, the act of measuring rectal temperature is the stressor that induces hyperthermia; to do so a lubricated thermometer is inserted 2 cm into the rectum until the temperature reads stably for 20 s. This highly reproducible measurement has strong pharmacological, face, and construct validity, as classic anxiolytics, but not antidepressants, reduce hyperthermia (Borsini et al., 1989; Lecci, Borsini, Volterra, & Meli, 1990; Marazziti et al., 1992; Van der Heyden et al., 1997; Zethof, Van der Heyden, Tolboom, & Olivier, 1994). SIH is accompanied by increased plasma ACTH, corticosterone, and glucose levels that return to baseline in a manner parallel to changes in temperature (Groenink, van der Gugten, Zethof, van der Heyden, & Olivier, 1994).

3.2 Tools for the assessment of depression-like behavior

The forced swim test (FST) was first established to identify antidepressant therapeutics (Porsolt, Anton, Blavet, & Jalfre, 1978). In this task, rodents are placed in an inescapable cylinder filled with water to a depth that does not allow the rodent to rest at the bottom of the chamber (temperature of 23 °C) for 4–6 min. Parameters assessed include time spent immobile, latency to immobility, and time spent in struggle (active coping). Immobility is defined by a ceasing of active coping or motionless floating except for motions to keep the head above water. Lower latency to immobility and greater time spent immobile indicate greater depressive-like behavior. Forced swim is pharmacologically validated for depression, however, a continued argument persists in the translation of decreased immobility as depressive-like behavior. It is possible that rodents reduce mobility to conserve energy; an adaptive response rather than an expression of learned helplessness (Arai, Tsuyuki, Shiomoto, Satoh, & Otomo, 2000; Holmes, 2003). While forced swim shows reproducible and reliable findings (Petit-Demouliere, Chenu, & Bourin, 2005), outcomes are sensitive to strain and sex (Alonso, Castellano, Afonso, & Rodriguez, 1991; David, Nic Dhonnchadha, Jolliet, Hascoët, & Bourin, 2001; Lucki, Dalvi, & Mayorga, 2001; Võikar, Kõks, Vasar, & Rauvala, 2001). The studies summarized throughout this chapter largely use the forced swim test to evaluate stress–induced depression, however, it can be used to *induce* stress as well.

Inspired by the forced swim test, Steru, Chermat, Thierry, and Simon (1985) established the tail suspension test (TST) which evaluates similar behavior but does not require a rodent to be able to swim, an advantage over forced swim. This task exploits rodents' inherent preference to correct themselves from being inverted. In this task, rodents are suspended upside down by their tails for ~6 min. Agitation (escape behavior) and immobility are measured throughout the task. Agitation is defined as periods of intense movement toward the goal of escape and this behavior is interspersed with energy-saving periods where escape behavior ceases (immobility).

This method is pharmacologically validated for depression; mice show reduced immobility after acute or chronic antidepressants administration and the tail suspension test shows stronger predictive validity of antidepressant activity than forced swim (Cryan, Mombereau, & Vassout, 2005; Rupniak, 2003; Steru et al., 1985). There are important considerations for tail suspension. First, this paradigm should only be used in mice, as the weight of an adult rat would cause pain and stress. Additionally, some mouse strains climb their own tails to escape (i.e., C57Bl/6 mice) and mice displaying this behavior should be excluded from analyses (Cryan et al., 2003; Mayorga & Lucki, 2001). Like forced swim, tail suspension can be used to evaluated depression–like behavior after stress exposure or as a stressor itself.

Anhedonia is defined as the inability to experience pleasure from reward and is present in depressive disorders. The sucrose preference task permits rodents 48 h of continuous 2-bottle choice access to water and a sucrose solution (generally between 1% and 2% w/v but higher concentrations are also used) in the home cage. Rodents prefer sweet solutions to water, however, rats displaying anhedonia show decreased sucrose consumption and preference (Highland, Zanos, Georgiou, & Gould, 2019; Joffe et al., 2020; Liu et al., 2018; Sobrian, Marr, & Ressman, 2003; Yashiro & Seki, 2017; Zhou et al., 2011). Sucrose preference is calculated by dividing the amount of sucrose consumed by the total fluid intake (water + sucrose). Clinically effective antidepressants restore sucrose preference (Liu et al., 2015). Individual variability in rodent weight and thirst affects sucrose consumption. Therefore, sucrose preference is considered a more stable comparison parameter. In addition to individual differences in third, circadian rhythm impacts rodent drinking behavior, and time of test must be considered when comparing results across subjects and studies.

Rodents are inherently cautious to consume new types of food, likely due to their inability to vomit. Novelty-suppressed feeding, or novelty-induced hypophagia, can be used to assess anxiety-like behavior. Twenty-four hours prior to task start, rats are food deprived with ad libitum access to water. At the start of the task, rodents are placed along the periphery of an open, square arena subdivided into a center zone and periphery. Standard rodent chow (or highly palatable food if not food deprived) is placed in the center of the arena, and latency to food approach and latency to eat is recorded. Rats are returned to their home cage, where they are given a predetermined amount of food. The amount of food consumed in a familiar environment (home cage) is recorded to assess motivation (Jing et al., 2021; Toma et al., 2017). Chronic treatment with SSRIs decreases novelty-suppressed feeding (Bodnoff, Suranyi-Cadotte, Quirion, & Meaney, 1989; Dulawa & Hen, 2005; Oh, Zupan, Gross, & Toth, 2009; Santarelli et al., 2003). Further, this task has high predictive validity for drug efficacy and onset of anxiolytic and antidepressant drug activity (Dulawa & Hen, 2005).

3.3 Assessment of fear

Pavlovian fear conditioning relies on the associations formed between a fear-inducing stimulus and the cues and contexts surrounding this exposure (Pavlov, 1927). While many stressors can induce fear conditioning, the "conditioned foot shock" or "classical fear conditioning" (CFC) model is most commonly used. The CFC model administers shock in a unique context, often paired with discrete cues such as lights and tones, and assesses time spent freezing as a measure of fear during the exposure and during a test in which no shock is administered. The ability to extinguish the conditioned freezing response in the context is typically assessed and manipulated as a screen to develop treatments for stress-related disorders. During conditioning, the stress stimulus is presented simultaneously with a cue (tone or light) or in an environment (context) specific to the stress stimulus exposure until the stimuli and the cue are paired. On test day, the stress stimulus is removed, and the rodent is exposed to the previously paired cue or context. Time spent freezing (ceasing of movement except for respiration) is recorded as an indication of fear. Other displays of fear are also recorded, including context aversion, ultrasonic vocalizations, blood pressure, and changes in respiration (Myers & Davis, 2007).

There are other assays for anhedonia and anxiety not reviewed here either because their utility is controversial (e.g., marble burying task), or because they were not used in any studies on the role of mGlu receptors and stress-induced behaviors. Many studies, particularly those with a focus on anhedonia and anxiety-like behavior, use a battery of tests in a single study in order to strengthen the ability to draw conclusions.

4. Brain-wide mGlu5 and stress-induced anhedonia, fear, and anxiety-like behavior

4.1 mGlu5 and anhedonia

mGlu5 KO mice are a useful model for exploring the role of this receptor in fear, anxiety, and anhedonia. mGlu5 KO mice are viable, and exhibit a modest decrease in body weight but no changes in locomotion relative to WT (Shin et al., 2015). In a foot shock-based model of learned helplessness, mGlu5 KO mice and mGlu5 heterozygote mice displayed increased susceptibility to learned helplessness relative to control mice subjected to the same procedures. There were no genotypic differences in pain sensitivity to explain this effect. Prior to SDS, mGlu5 KO mice and controls showed no difference in social interaction. However, after 3 days of SDS, KO's show reduced social interaction but no changes in total locomotion. To further understand the role of mGlu5, KO and WT mice were exposed to 3 days of restraint stress, with a baseline and post-stress sucrose preference test used to assess stress-induced anhedonia. There were no effects of mGlu5 KO on baseline anhedonia, but following restraint stress, mGlu5 KO mice displayed reduced sucrose preference. There was no effect of mGlu5 KO on acute, single-session FST and TST, but following repeated testing, mGlu5 KO mice displayed greater immobility, a marker of anhedonia (Shin et al., 2015). Thus, mGlu5 KO alone does not produce anhedonia, but following exposure to a variety of types of stressors, a greater susceptibility to develop anhedonia is unmasked.

In support of the idea that mGlu5 signaling confers resilience to stress-induced anhedonia, the mGlu5 positive allosteric modulator (PAM) 3-Cyano-N-(1,3-diphenyl-1H-pyrazol-5-yl)benzamide (CDPPB) was found to reduce depressive-like symptoms following shock-induced learned helplessness procedure (Shin et al., 2015). In this model, mGlu5 heterozygote mice display more escape failures than controls. In mGlu5 heterozygotes, CDPPB (10 mg/kg but not 1 mg/kg) administered three times across the 24 h following the fourth and final foot shock exposure yielded lower mean escape latency than vehicle-treated mice. This effect was not observed in unshocked mice, negating the possibility for non-specific effects on the ability to escape. Conversely, the mGlu5 NAM 3-((2-Methyl-4-thiazolyl)ethynyl)pyridine (MTEP) at a dose of 10 mg/kg but not 1 mg/kg, administered to WT mice in the 24 h following social defeat stress decreases social interaction (Shin et al., 2015).

In a model where mice were exposed to two 4-day cycles of 16 h/day alcohol vapor exposure and acute restraint stress, MTEP was able to block the expression of anhedonia in the novelty-induced hypophagia test following restraint stress in both alcohol-exposed and alcohol-naïve female adult mice (Kasten et al., 2020). This indicates that alcohol exposure influences the neurobiology underlying stress-induced anhedonia.

4.2 mGlu5 and anxiety-like behavior

In contrast to the role for mGlu5 in susceptibility to stress-induced anhedonia, mGlu5 KO mice were found to display resilience to the effects of chronic subordinate colony housing (CSC) on anxiety-like behavior. When male mGlu5 KO mice and their male WT littermates were exposed to either single-housed control (SHC) or CSC conditions for 14–20 days, mGlu5 KO prevented stress-induced hyperthermia (SIH) and increased adrenal weight (Peterlik et al., 2017). In the absence of exposure to stress prior to testing for SIH, mGlu5 KO continues to prevent the expression of SIH (Brodkin et al., 2002). Thus, while mGlu5 is necessary for resilience to the effects of chronic stress on measures of depressive-like behavior/anhedonia (Shin et al., 2015), it is necessary for susceptibility to SIH. However, chronic administration of the selective mGlu5 NAM 2-chloro-4-((2,5-dimethyl-1-(4-(trifluoromethoxy)phenyl)-1H-imidazol-4-yl)ethynyl)pyridine (CTEP; 2 mg/kg/day) via osmotic minipumps during the CSC protocol was unable to prevent stress-induced hyperthermia as did mGlu5 knock out (Peterlik et al., 2017), indicating a potential developmental role for mGlu5 in regulating this behavior. Interestingly, in the absence of chronic stress, acute administration of the mGlu5 allosteric antagonist MTEP (16 mg/kg) immediately prior to a SIH protocol attenuates SIH (Brodkin et al., 2002).

When a pharmacological approach was used to probe the role of CNS mGlu5 in other tests for anxiety-like behavior, mGlu5 was found to be necessary for susceptibility to the ability of CSC to induce anxiety-like behavior in the light-dark box. The mGlu5 inverse agonist CTEP (2 mg/kg) was administered via chronic osmotic minipumps for the duration of CSC, preventing the ability of CSC to reduce time spent in the light side of the light-dark box (Peterlik et al., 2017). Interestingly, the same study found that CTEP did not alter CSC-induced anxiety-like behavior in the EPM (Peterlik et al., 2017). In this study, CTEP was administered during the course of CSC and during anxiety-testing. Similarly, when CTEP (2 mg/kg p.o.) was administered immediately prior to CSDS, CSDS-induced deficits in social interaction and FST are unaffected (Wagner et al., 2015). CTEP treatment had no effects on basal or circulating corticosterone nor adrenal gland mass in this study. Thus, mGlu5 antagonism was protective against CSC-induced anxiety-like behavior in the light dark box only.

Akin to the ability of systemic mGlu5 antagonism to be unable to alter stress-induced anxiety-like behavior, systemic mGlu5 positive allosteric modulation does not alter PSS-induced anxiety in stress-susceptible rats (Shallcross et al., 2019). In this study, male Sprague-Dawley rats underwent a single brief exposure to inescapable predator scent stress (TMT). One week later, they were tested for anxiety-like behavior in the EPM and ASR, and classified as stress-susceptible and -resilient phenotypes. Two weeks after EPM and ASR testing, the stress-susceptible rats showed increased anxiety in the light-dark box test, which was unaffected by administration of the mGlu5 PAM CDPPB (30 mg/kg). However, if mGlu5 signaling mediates increased (susceptibility to) anxiety-like behavior, the inability of CDPPB to alter anxiety-like behavior may be due to a "ceiling effect" of maximal anxiety exhibited by stress-susceptible rats in this model.

Taken together, the results of these studies suggest that while mGlu5 expression throughout the CNS is necessary for resilience to physiological responses to CSC-induced stress (i.e., SIH), not all stress-induced anxiety behaviors are influenced by chronic mGlu5 downregulation. Additionally, the role of mGlu5 in CSC-induced SIH may be developmental in nature, as only mGlu5 KO mice show changes in SIH while chronic antagonism does not influence this behavior. However, mGlu5 signaling also regulates SIH in the absence of chronic stress. It is unclear why mGlu5 KO mice would be found to be susceptible to stress-induced anhedonia but resilient to stress-induced SIH and adrenal gland adaptations. One explanation is that SIH is more of a physiological measure, dependent on different CNS or PNS mechanisms than are depressive-like behaviors.

4.3 mGlu5 and fear conditioning

mGlu5 signaling is consistently found to play a role in conditioned fear responses. Several mGlu5 PAMs have been found to enhance extinction of conditioned fear. The PAM CDPPB (20 mg/kg IP) enhances within-session extinction of conditioned foot shock and reduces freezing during the retention test in an un-drugged condition (Sethna & Wang, 2014). The same has been found to be true in a PSS model, where CDPPB (30 mg/kg) facilitates extinction of contextual fear in PSS-susceptible rats (Shallcross et al., 2019). While (Ganella, Thangaraju, Lawrence, & Kim, 2016) failed to find an acute effect of systemic CDPPB on fear extinction, the dose employed in this study was 10-fold lower than the other studies (3 mg/kg IP). While acute administration of another mGluR5 PAM, ADX 47273 (30 mg/kg IP) had no effect on fear extinction, when administered during extinction trials, it reduced freezing during a retention trial that was conducted without ADX on board (Xu et al., 2013). The mGlu5 allosteric antagonist MTEP (2mg/kg) inhibits extinction of conditioned foot shock (Ganella et al., 2016). The mGlu5 KO mouse has impaired acquisition and extinction of conditioned foot shock, which may not necessarily be indicative of mGlu5's role in fear, but rather its importance for conditioning in general (Jacob et al., 2009; Xu, Zhu, Contractor, & Heinemann, 2009). Interestingly, chronic delivery of CTEP (2mg/kg) did not alter the acquisition and expression of conditioned foot shock (Peterlik et al., 2017).

5. The mGlu5-containing circuitry underlying stress-induced anhedonia, fear, and anxiety-like behavior

5.1 mGlu5: Amygdala and extended amygdala

There is evidence from PSS models that mGlu5 expression in the amygdala underlies resilience. In one study, male Sprague-Dawley rats were exposed to TMT once for 10min and later assessed for anxiety-like behavior in the EPM and ASR. In this model, rats are characterized as stress-resilient, -susceptible, or of an intermediate phenotype based on a double median split conducted on time spent in the open arms of the EPM and on the amount of habituation of the acoustic startle response. When rats were re-exposed to the TMT context with no TMT present, rats that had been classified as stress-susceptible, displayed greater freezing than both unstressed controls and resilient rats. Rats were killed immediately after this re-exposure for the assessment of genes of interest via qPCR. Resilient rats displayed greater mGlu5 mRNA expression in the amygdala relative to both unstressed controls and susceptible rats (Schwendt et al., 2018). In this study, tissue punches contained both the central nucleus of the amygdala (CeA) and basolateral amygdala (BLA). In a follow-up study by the same group (Shallcross et al., 2021), fluorescent in situ hybridization in ex vivo slices revealed that mGlu5 was only upregulated in the BLA and not the CeA of resilient rats. Such upregulation was only present in vGluT-positive cells and not GAD-positive cells, indicating that this upregulation occurs in glutamatergic neurons. This study also found such upregulation was present regardless of whether rats were re-exposed to the TMT context or not. In this model, intra-BLA mGlu5 antagonism with MTEP increased contextual fear (freezing upon re-exposure to the TMT context with no TMT present) in stress-resilient but not in control rats. Such freezing rapidly extinguished over subsequent exposure to the TMT chamber when MTEP was not administered. Conversely, in another study by the same group, the mGlu5 positive allosteric modulator, CDPPB, administered systemically, facilitated fear extinction learning in susceptible rats (Shallcross et al., 2019). However, CDPPB did not attenuate anxiety in the light-dark box in the same susceptible rats. Thus, BLA mGlu5 seems to mediate conditioned fear in the PSS model, and not unconditioned anxiety.

These findings are in contrast to a role for BLA mGlu5 in anxiety-like behavior in the absence of a history of stress: intra-BLA MPEP, an mGlu5 antagonist, is anxiolytic in the light-dark box and EPM tests in the absence of a stressor (Pérez de la Mora et al., 2006). In addition, in vivo administration of mGlu5 agonist, DHPB, into the BLA promotes conditioned fear in adolescent Sprague-Dawley rats (Rahman, Kedia, Fernandes, & Chattarji, 2017). After repeated exposures to a tone (CS−) or a tone paired with a low-intensity footshock (CS+), rats were injected with bilateral intra-BLA infusions of DHPG, which resulted in freezing to both the CS− and the CS+ as well as the context itself (prior to tone) compared to saline controls. Freezing between the CS− or CS+ was equivocal. One day after injections, rats froze only in response to the conditioned stimulus, despite equal freezing to both stimuli and the context the day before, suggesting DHPG amplified conditioning of the weak footshock. Considering DHPG treatment initially evoked indiscriminate anxiogenic behaviors that did not impact subsequent fear conditioning, it is possible that mGlu5 in the BLA exerts its effects on fear learning and anxiety in ways that are distinct from one another. In a follow-up in vitro analysis, DHPG promoted intrinsic and postsynaptic excitation of LA neurons in vitro while facilitating LTP activity when paired with the weak CS+/US association. Thus, intra-BLA mGlu5 antagonism increases contextual fear conditioning in a PSS model, while intra-BLA mGlu5 agonism enhances cue-conditioned fear to footshock. Besides the stress paradigm, these studies differed in that Shallcross et al. (2021) studied fear extinction following prior PSS exposure, while the latter study focused on mGlu5 during fear encoding and recall. Therefore, it is also possible that the contrasting findings between these two studies relies on the cognitive capacity required of the models used and their relationship with prior stress.

Decreased BLA and CeA mGlu5 mRNA expression is also observed in Sprague-Dawley offspring whose mothers underwent maternal stress procedures (Buonaguro et al., 2020). Pregnant dams were restrained three times a day until delivery, beginning on day 11 of gestation. Pups were weaned

21 days after birth and allowed to develop regularly until 4 months of age. Anxiety-like behaviors were quantified via light-dark box testing. Compared to controls, prenatally-stressed (PRS) 4-month-old pups spent greater time in the dark box and exhibited lower latency to initially enter the light box which was accompanied by greater basolateral and medial amygdala Homer1a mRNA. In contrast, PRS rats expressed reduced Homer1b and mGluR5 mRNA expression in these regions. This is significant because Homer1a disrupts mGlu5-Homer1b coupling, which is necessary for canonical mGlu5 signaling. However, PRS results in increased expression of mGlu5 receptors in the amygdala when assessed earlier in life, at PND22 (Laloux et al., 2012).

Thus, glutamate signaling in the BLA has consistently been found to play a role in stress-related anxiety and fear. To determine the downstream mediators of these effects, pathway-specific bidirectional optogenetic strategies were used to investigate the BLA-vHPC and BLA-mPFC pathways in the long-term effects of CSDS. CSDS was found to reduce excitatory transmission along the BLA-vHPC and BLA-mPFC pathways (Kim, Kang, Choi, Chang, & Koo, 2022). Inhibiting these pathways immediately after a social defeat episode led to increased depressive-like behaviors and reductions in mGlu5 mRNA expression in the target regions. Conversely, stimulating each projection separately or simultaneous stimulation of both of these projections immediately after a social defeat episode decreases anhedonia (sucrose preference) and social avoidance (Kim et al., 2022). Thus, stimulating glutamate release along the BLA-vHPC and BLA-mPFC pathways reduces the pro-depressive effects of CSDS. The importance of mGlu5 in mediating these effects was determined by overexpressing mGlu5 with a lentivirus in the BLA, finding that this overexpression reduced the ability of CSDS to produce anhedonia. Conversely, using shRNA to knockdown mGlu5 in the mPFC or vHPC produces decreased sucrose preference and social interaction following CSDS and prevents the ability of optogenetic stimulation of BLA-vHPC and BLA-mPFC pathways to reduce CSDS-induced anhedonia (Kim et al., 2022).

5.2 mGlu5: Bed nucleus of the stria terminalis (BNST)

Following two 4-day cycles of 16 h/day alcohol vapor exposure during adolescence, in adulthood, acute restraint stress unmasked persistent changes in BNST LTD (Kasten et al., 2020). The same group followed up on this finding, and instead of using acute mGlu5 NAMs, used a genetic strategy to reduce mGlu5 signaling (Kasten, Holmgren, Lerner, & Wills, 2021). Site-selective knockdown of mGlu5 was accomplished via the use of Grm5^{loxp} mice and injection of AAVs expressing either Cre or control GFP into the BNST between PND22 and 29. The mice were then exposed to either alcohol vapor or clean air for two 4-day cycles of exposure. Mice were then left undisturbed until adulthood (PND70), when they were tested for novelty-induced hypophagia following restraint stress, basal anxiety-like activity in the open field, and foot shock–induced freezing. mGlu5 knockdown in the BNST did not alter novelty-induced hypophagia. However, it did reduce time spent in the center of the open field in male, but not female alcohol-naïve mice, indicating that mGlu5 in the BNST is protective against anxiety. BNST mGlu5 knockdown also prevented extinction of the tone–shock relationship in female and male alcohol-naïve mice. Thus, in the absence of a history of alcohol, mGlu5 BNST is necessary for the extinction of conditioned fear and, in male mice only, anxiety-like behavior.

5.3 mGlu5: Medial prefrontal cortex (mPFC)

Rats resilient to the anxiety- and fear-provoking capability of a single PSS exposure show increased mGlu5 mRNA expression in the mPFC relative to both stress-susceptible and unstress control male rats (Schwendt et al., 2018). In the same model, this was later determined to be present in both the infralimbic (IL) and prelimbic (PL) regions of the mPFC (Shallcross et al., 2021). In agreement with these data, following CSDS, reduced mGluR5 protein expression was found in the mPFC of susceptible, but not resilient, mice (Kim et al., 2022). Following SDS, mGluR5 knockdown in the mPFC significantly reduced social interaction and sucrose preference (Kim et al., 2022). There are several signaling pathways initiated by mGlu5 binding, including the canonical Gq-dependent mechanism which forms inositol-1, 4, 5-trisphosphate (ITP) and diacylglycerol (DAG) to activate protein kinase C (PKC). mGlu5 signaling can be Gq-independent, through Homer1b/c (Mao et al., 2005) to form PI3K/Akt/mTOR or through β-arrestin2 to activate ERK (Stoppel et al., 2017). CSDS was found to downregulate Homer1b/c expression and phosphorylation of Akt in the mPFC of susceptible, but not resilient mice, with no changes in pPKC and pERK levels (Kim et al., 2022). To determine whether these changes in expression were associated with susceptibility to stress, a selective PI3K-Akt inhibitor, LY294002, or a selective MAPK/ERK kinase inhibitor, U0126, was infused directly into the mPFC immediately prior to a single SDS session (SSDS). Infusion of LY294002, but not of U0126, into the mPFC reduced social interaction and sucrose preference. Thus, the mGluR5-Homer1b-AKT signaling pathway in the mPFC is important for mediating the pro-depressive effects of social defeat stress (Kim et al., 2022).

Activation of mGlu5 receptors facilitates long-term potentiation (LTP) of excitatory transmission onto PFC somatostatin-expressing interneurons (SST-Ins) after restraint stress, biasing information processing within PFC toward BLA-driven feedforward inhibition (Joffe et al., 2022). SST-mGlu5 −/− mice showed impaired tone-shock fear conditioning relative to WT. Thus, mGlu5 expression in SST-expressing cells of the mPFC is necessary for fear conditioning.

5.4 mGlu5: Hippocampus (HPC)

Six days of restraint stress yields heterogenous changes in ventral hippocampal (vHPC) mGlu5 expression, with some rats displaying lower mGlu5 protein expression (low expression or LE group) relative to controls and some rats displaying higher mGlu5 protein expression (high expression or HE group) relative to the control group (Yim, Han, Seo, Kim, & Kim, 2018). In both groups, protein expression changes were mirrored by changes in mGlu5 mRNA expression, indicating that in this brain region, there is a good correlation between mGlu5 mRNA and protein expression. Additionally, the HE group showed reduced methylation of CpG sites and reduced basal theta spectral power on stress day 6. LE rats showed increased methylation of the mGlu5 CpG island and unaltered basal EEG theta spectral power. Knockdown of mGlu5, specifically in the hippocampus via lentiviruses expressing mGlu5 shRNA rescued the reduction in baseline theta power in HE rats. In general, increases in theta band power in the hippocampus have been associated with "normal" memory encoding, suggesting that local mGlu5 knockdown, restores hippocampal neural activity in HE rats. Thus, repetitive restraint stress produces heterogenous changes in hippocampal mGlu5 expression that underlie changes in neural activity. However, it is not clear if such individual differences in hippocampal mGlu5 expression are accompanied by differential susceptibility to long-term behavioral (e.g., anxiety or anhedonia) consequences of restraint stress as has been shown for amygdala and PFC mGlu5 following predator scent stress.

In the vHPC of mice susceptible to CSDS based on the social interaction and sucrose preference tests, mGluR5 protein expression was decreased relative to unstressed control mice (Kim et al., 2022). This was accompanied by decreased phosphorylation of downstream signaling molecules pPKC and pERK. No changes in Homer1b/c or pAKT levels were observed in the vHPC. Following SDS, mGluR5 knockdown in the vHPC significantly reduced social interaction and sucrose preference (Kim et al., 2022). The local infusion of U0126 (MAPK/ERK kinase inhibitor), but not LY294002 (PI3K-Akt inhibitor), into the vHPC decreased social interaction and sucrose preference following CSDS. Together with the findings from the same model presented above, these findings suggest that different mGlu5 downstream signaling in the mPFC and vHPC mediate susceptibility to depressive-like symptoms after CSDS. Unlike in the mPFC, mGlu5 mediates these effects in a Gq-PKC-ERK-dependent manner in the vHPC (Kim et al., 2022). PRS also results in reduced expression of mGlu5 in the hippocampus when assessed at PND14 and PND22 (Laloux et al., 2012). Thus, there is evidence that vHPC mGlu5 signaling mediates resilience to anhedonia following stress exposure.

Contrary to these results, in a model of CVS, hippocampal mGlu5 was found to be associated with susceptibility to stress (Sun et al., 2017). Male Sprague-Dawley rats underwent 8 weeks of 14 h/day CVS. Resilient and susceptible phenotypes were determined with the sucrose preference test. Resilient rats responded to 8 weeks of CVS with less than a 10% in reduction of sucrose preference from control consumption, and susceptible rats exhibited a >30% reduction in baseline sucrose preference. Hippocampal mGlu5 mRNA and protein levels were increased in the susceptible rats. Intrahippocampal administration of a glucocorticoid receptors (GR) antagonist restored mGlu5 expression in susceptible rats, as did a cannabinoid receptor 2 (CB2) receptor antagonist. Such treatment also restored sucrose preference in susceptible rats. Thus, hippocampal glucocorticoid and cannabinoid receptors modulate the expression of hippocampal mGlu5, with increased GR and CB2 signaling promoting increased mGlu5 expression and stress-susceptibility. These results are in opposition to the pro-resilient role of vHPC mGlu uncovered by Kim et al. (2022), potentially due to targeting a more dorsal portion of the hippocampus in this study. The stereotactic coordinates provided by Sun et al. (2017) were more consistent with the dorsal hippocampus, potentially explaining the discrepancy in findings. However, as neither Kim et al. (2022) or Sun et al. (2017) provided schematics of the specific hippocampal region (dorsal vs ventral) targeted this is difficult to state for certain.

Supporting a role for decreased vHPC glutamate signaling through mGlu5 in underlying stress-susceptibility, reduced glutamate release has been observed in the vHPC of stress-susceptible rats. After 5 weeks of CMS, rats that were found to be stress-susceptible with the sucrose preference test (exhibiting <55% sucrose preference after CMS but not before) displayed reduced basal and K^+-evoked glutamate release in the hippocampus (Tornese et al., 2019). Interestingly, in susceptible rats, acute treatment with ketamine normalized sucrose preference and basal glutamate levels but not evoked glutamate release, suggesting a change in postsynaptic and/or glial vs presynaptic glutamate mechanisms.

5.5 mGlu5: Nucleus accumbens (NA)

Following 10–12 weeks of Social Isolation Stress (SIS), male Wistar rats consumed less sucrose and spent less time in EPM open arms compared to controls (Mao & Wang, 2018). This prolonged social isolation-induced anhedonia and anxiety-like behavior was accompanied by increased NA intracellular and membrane-bound mGlu5 protein expression. The ratio of surface/internalized mGlu5 expression was higher in SIS-exposed subjects than in controls. However, this study did not separate rats into stress-Susceptible and resilient phenotypes, which may have influenced the results. In a study that did separate PSS-exposed *male* Sprague-Dawley rats into stress-susceptible based on anxiety-like behavior in the EPM and ASR, there were no observed changes in NA mGlu5 mRNA expression relative to stress-resilient or unstressed control rats (Schwendt et al., 2018). In female rats that are found to be resilient to a single exposure to PSS based on sucrose preference and time spent in the open arms of the EPM, NA core and shell mGlu5 expression is increased relative to stress-susceptible and unstressed control rats (Blount, Dee, Wu, Schwendt, & Knackstedt, 2023). When wild-type mice receiving SDS or electrical foot shock were segregated into susceptible and resilient phenotypes based on social avoidance, susceptible mice had less mGlu5 expression in the NA than susceptible and control mice (Shin et al., 2015). Thus, the type of stressor and sex of the rodent influences NA mGlu5 expression changes, with increases accompanying SIS, SDS, and footshock-induced anhedonia and anxiety-like behavior in males, no change in susceptible male rats after PSS and increases in expression accompanying resilience to anhedonia and anxiety induced by PSS in females.

Additional support for the protective role of NA mGlu5 is that intra-NA infusion of the mGlu5 agonist CHPG promotes stress resilience after CSDS (Xu et al., 2021). Overexpression of mGlu5 in the NAc shell prevents

CSDS-induced anhedonia, while NA core overexpression produces resilience to pain (Xu et al., 2021). Lentiviral rescue of mGluR5 in the NAc shell but not the NA core rescues the stress-induced depression-like phenotype of mGlu5 KO mice (Shin et al., 2015). Activation of mGlu5stimulates serum response factor (SRF) phosphorylation at Ser-103, which in turn induces deltaFosB expression. Consequently, NA shell mGlu5-dependent deltaFosB signaling was found to be critical for resilience to stress-induced anhedonia (Shin et al., 2015). These findings suggest that mGlu5 signaling in the NA core and shell is necessary for resilience to stress-induced pain and anhedonia.

Following 10 sessions of CSDS, optogenetic bidirectional manipulation of mPFC, BLA, and vHPC glutamatergic (CaMKII1a-expressing) neurons was conducted to investigate the role of such NA afferents in the expression of stress-induced anhedonia (Bagot et al., 2015). Resilient and susceptible phenotypes were determined by their peer engagement in a social interaction test 24h following 10-day CSDS where mice with ratios <1 were deemed susceptible and >1 resilient. LTD at vHPC-NA synapses induced a resilient phenotype in a social interaction test following CSDS. Acute opto-stimulation of BLA-NA and mPFC-NA pathways was pro-resilient in a social interaction test; in contrast, acute stimulation of vHPC terminals in the NA promoted susceptible-like behavior in a social interaction test following CSDS. Thus, glutamatergic signaling from the ventral hippocampus to the nucleus accumbens inhibits social interaction, but BLA-NA and PFC-NA glutamatergic projections promote social interaction. These results are consistent with a role for glutamate signaling through NA mGlu5 to mediate resilience to stress.

6. Summary of mGlu5 role in anhedonia, fear, and anxiety-like behavior

Systemic knockout (mGlu5 KO), knockdown (mGlu5 heterozygote), and pharmacological stimulation/inhibition of mGlu5 produce distinct effects on post-stress behaviors dependent on the type of assay used (see Table 1). mGlu5 stimulation (e.g., via PAMs) consistently facilitates the extinction of conditioned fear. Reducing mGlu5 expression and signaling increases susceptibility to stress-induced anhedonia in many rodent assays, while mGlu5 stimulation increases resilience (decreases anhedonia). Conversely, decreasing mGlu5 signaling increases *resilience* to stress-induced anxiety-like behavior.

Table 1 Summary of results from systemic mGlu5 knockout (KO) and pharmacological stimulation/inhibition.

Manipulation	Behavior	Pro-resilience	No effect	Pro-susceptibility
mGlu5 KO	Anxiety-like behavior	↓ CSC-induced SIH (Peterlik et al., 2017); ↓ SIH (Brodkin et al., 2002)		
	Anhedonia/ depressive like symptoms			↑ Learned helplessness, SDS-induced social interaction, restraint stress-induced sucrose preference, immobility after repeated TST and FST (Shin et al., 2015)
	Fear	↓ Impaired CFC (Xu et al., 2009) **reflects a learning deficit, not necessarily susceptibility		↓ Impaired extinction of CFC (Xu et al., 2009) **reflects a learning deficit, not necessarily resilience
mGlu5 antagonist/ NAM	Anxiety-like behavior	↓ SIH; MTEP 16 mg/kg (Brodkin et al., 2002); ↓ CSC-induced anxiety in light-dark box; CTEP 2 mg/kg (Peterlik et al., 2017)	CSDS+EPM; CTEP 2 mg/kg (Peterlik et al., 2017)	
	Anhedonia/ depressive-like symptoms		CSDS-induced deficits in social interaction and FST; CTEP 2 mg/kg (Wagner et al., 2015)	↑ Anhedonia after SDS; MTEP 10 mg/kg (Shin et al., 2015)
	Fear			↑ Extinction impairments in CFC model; MTEP 2 mg/kg; (Ganella et al., 2016)

Continued

Table 1 Summary of results from systemic mGlu5 knockout (KO) and pharmacological stimulation/inhibition.—cont'd

Manipulation	Behavior	Pro-resilience	No effect	Pro-susceptibility
mGlu5 PAM	Anxiety-like behavior		PSS-induced anxiety-like behavior in the light-dark box; CDPPB 30 mg/kg (Shallcross et al., 2019)	
	Anhedonia/ depressive-like symptoms	↓ Learned helplessness; CDPPB 10 mg/kg (Shin et al., 2015)		
	Fear	↑ Extinction of conditioned fear; CDPPB 20 and 30 mg/kg; ADX 47273 30 g/kg; (Sethna & Wang, 2014, Shallcross et al., 2019, Xu et al., 2013)	Extinction of conditioned fear; CDPPB 3 mg/kg (Ganella et al., 2016)	

CFC, Classical fear conditioning; CSC, Chronic subordinate colony housing; CSDS, Chronic social defeat stress; EPM, Elevated plus maze; FST, Forced swim test; PSS,: Predator scent stress; SIH, Stress-induced hyperthermia; SDS, Social defeat stress; TST, Tail suspension test.

Examining the role of mGlu5 in discrete brain regions and signaling pathways sheds provides potential explanations for the discrepant role of mGlu5 in anhedonia and anxiety-like behavior. mGlu5 expression has been found to be decreased in the BLA, mPFC, vHPC, and NA of stress-susceptible rodents. Manipulation of mGlu5 in these circuits finds that BLA mGlu5 mediates resilience to the expression of conditioned fear in the PSS model, and not unconditioned anxiety. mPFC and vHPC mGlu5 mediate resilience to anxiety and anhedonia produced by stress exposure. mPFC mGlu5 receptors, particularly in SST-interneurons, mediate fear conditioning. NA shell, but not core, mGlu5 mediates resilience to stress-induced anhedonia. Thus, distinct regions are responsible for stress-induced anhedonia, fear, and anxiety-like behavior following exposure to a stressor.

7. mGlu2 and mGlu3 receptors and stress-induced anhedonia, fear, and anxiety-like behavior

The Group II family of metabotropic glutamate receptors (mGluRs) include mGlu2 and mGlu3. These members of the Group II mGlu family are Gi-coupled receptors, inhibiting adenylyl cyclase, allowing them to negatively regulate synaptic glutamate release. Historically it has been difficult to study these receptors separately as their protein sequences are highly conserved, and antibodies have only recently been developed to distinguish the two receptors. Additionally, orthosteric agonists and antagonists bind both mGlu2 and 3. In recent years, more selective tools for stimulating/inhibiting mGlu2 and mGlu3 separately, have been developed, including positive and negative allosteric modulators (PAMs and NAMs, respectively). Here, we present research indicating the potential role of both receptors in mediating anhedonia, fear conditioning, and anxiety.

7.1 mGlu2 and mGlu3 receptors and resilience to anhedonia

A substantial body of literature finds that mGlu2/3 antagonists such as LY341495 have anti-depressive effects in rodent models of stress-induced anhedonia. Chronic LY341495 treatment during 4 weeks of CUS normalized CUS-induced immobility in the FST (Seo et al., 2022). However, when LY341495 was not administered during CUS, but acutely 24-h prior to testing in the FST or for 3 days between the conclusion of CUS and testing in the FST, no effect on immobility was observed (Pałucha-Poniewiera, Podkowa, & Rafało-Ulińska, 2021). However, this acute and subchronic treatment with LY341495 was able to reduce immobility in the TST in the same animals (Pałucha-Poniewiera et al., 2021). Following 10 days of CSDS, the mGlu2/3 receptor antagonist TP0178894 administered prior to testing attenuated immobility in the TST and FST and increased sucrose preference in a dose-dependent manner (Dong et al., 2022). The same group used this model to find that another mGlu2/3 antagonist, MGS0039, had similar effects on TST and FST (Dong et al., 2017). Male CD-1 mice treated with the mGlu2/3 receptor antagonist LY341495 immediately prior to IES, with and without exogenous CORT administration, show decreased escape failures (Highland et al., 2019). LY341495 was also effective when administered for 1 or 3 days prior to exposure to IES. Conversely, pretreatment with the mGlu2/3 receptor agonist LY379268 increased escape failures in the IES model. Thus, mGlu2/3 antagonists are consistently found to decrease stress-induced anhedonia across multiple animal models.

In recent years, pharmacological and genetic tools have been developed to distinguish the role of mGlu2 in anhedonia from that of mGlu3. Evaluation of a highly selective mGlu3 NAM (VU0650786) finds a role for mGlu3 in stress-induced anhedonia (Engers et al., 2015). VU0650786 administered before an FST session to male Sprague-Dawley rats reduced time spent immobile in the FST. This was only observed at the highest dose and was potentially due to a basement effect; these rats were not stressed prior to the FST, indicating that mGlu3 plays a role in anhedonic effects in the absence of a history of stress. A later study investigated the same compound for its ability to attenuate CVS-induced depressive-like behavior. Joffe et al. (2020) subjected C57BL/6J mice to 4 weeks of CVS or exogenous CORT administration, followed by a battery of behavioral assessments 1 week later, including the TST, FST, and sucrose preference. Though the present study did not undergo stress-susceptible/-resilient characterizations, mice having undergone either chronic CORT treatment or CVS exposure showed similar anxiogenic and depressive-like effects. Administration of the mGlu3 NAM 45 min prior to FST and TST and 24 h prior to the sucrose preference test normalized the CVS- and CORT-induced depressive-like symptoms. However, with the exception of immobility in the FST following CVS, an mGlu2 NAM was found to have the same anti-depressive effects in this study. Thus, pharmacological tools reveal that both mGlu2 and mGlu3 receptors mediate anhedonia and depressive-like behaviors following exposure to stress or exogenous cortisol.

However, there is evidence suggesting that the genetic deletion (knock out) of mGlu2, but not mGlu3, promotes resilience to stress in rodents. Both mGlu2 and mGlu3 KO mice are viable and exhibit no changes in locomotor activity compared to WT mice. Male and female mGlu2 and mGlu3 knockout mice (on a CD-1 background) and CD-1 wild type mice were investigated for their susceptibility to stress-induced anhedonia (Highland et al., 2019). mGlu2 KO mice displayed decreased immobility during the FST and fewer escape failures during IES, while mGlu3 KO mice exhibited similar immobility and escape behaviors as WT mice. In the same study, mGlu2 KO mice were resilient to the IES escape deficits induced by CORT treatment in WT mice. When tested for sucrose preference following 10 days of CSDS, mGlu2 KO mice did not display the same reduction in sucrose preference (relative to unstressed controls) that WT mice did. Thus, while acute antagonism of mGlu3 prevents the ability of stress and exogenous CORT to produce anhedonia, the genetic deletion of this gene does not have the same effect.

In contrast to the findings of Highland et al. (2019), another study using mGlu2 KO found that the absence of this gene conferred susceptibility, and not resilience, to the effects of CUS on stress-induced immobility in the FST (Nasca, Bigio, Zelli, Nicoletti, & McEwen, 2015). Nasca et al. (2015) exposed male C57bl WT and mGlu2 KO mice to chronic unpredictable stress (CUS) and evaluated CUS-induced vulnerability by testing mice on FST prior to and following CUS exposure. mGlu2 KO mice displayed no difference in immobility at baseline. After 4 weeks of CUS, both WT and mGlu2 KO displayed greater immobility in the FST, with KO's exhibiting greater immobility than WT. No other tests for anhedonia were done in this study. While both studies utilized an mGlu2 KO model to characterize stress-resilient and -susceptible mice, Highland et al. (2019) used male and female mGlu2 KO mice bred on a CD-1 background, while Nasca et al. (2015) used male mGlu2 KO mice bred on a C57bl background. In addition to the potential role of sex differences, CD-1 mice exhibit less anxiety than C57bl mice in an open field test which can also contribute to these contradictory findings (Michalikova, van Rensburg, Chazot, & Ennaceur, 2010). Thus, it is possible that the resilient behaviors of mGlu2 KO mice bred on a CD1 background but not C57bl were due to sex or strain differences. However, this is an unlikely explanation as a large body of evidence implicates mGlu2/3 signaling in pro-depressive effects. Additionally, while the stress models and measures of depressive-like behaviors were different between Highland et al. (2019) and Nasca et al. (2015), both Nasca et al. (2015) and Joffe et al. (2020) exposed mice to CVS for 4 weeks, following by testing in the FST. Thus, it is not clear at this time why there are discrepant findings between these studies, but those of Nasca et al. (2015) seem to be at odds with the majority of the literature on the protective effects of decreasing mGlu2 signaling on depressive-like behavior.

7.2 mGlu2 and mGlu3 receptors and resilience to anxiety-like behavior

The mGlu3-specific NAM, VU6017587, has anxiogenic effects in the EPM. Additional evidence from a PSS model indicates that mGlu3 is implicated in vulnerability to develop anxiety-like behavior following a stressor (Tyler et al., 2022). In this study, male Long-Evans rats were treated with the mGlu3 NAM VU6010572 45 min prior to TMT exposure. Two weeks later, rats were tested on the elevated zero maze (EZM) and ASR to evaluate TMT-induced anxiety behaviors in an undrugged condition. Rats that had received VU6010572 weeks earlier displayed greater habituation of the ASR and time spent in the open arms of the EZM. However, in this study, TMT did not increase either of these anxiety-like behaviors relative to the control condition. Thus, mGlu3 blockade is effective at reducing anxiety, but whether it can do so in a condition of heightened anxiety is unclear at this time.

7.3 mGlu2 and mGlu3 receptors and resilience to fear conditioning

Negative modulation of mGlu3 specifically reduces fear-related behavior in the PSS model (Tyler et al., 2022). Male Long-Evans rats were treated with the mGlu3 NAM VU6010572 prior to TMT exposure. Rats underwent an additional context re-exposure session 2 weeks later, in which no TMT was present, and no drugs were administered. TMT-exposed rats exhibited several stress-reactive behaviors in the form of elevated freezing, digging, and decreased locomotion, all of which were unaffected by the VU6010572 treatment. However, the VU6010572 treatment prior to TMT exposure later protected rats from exhibiting conditioned fear behaviors during re-exposure to the TMT context. Thus, mGlu3 receptor signaling may mediate the lasting effects of TMT-induced fear.

8. The mGlu2/3-containing circuitry underlying stress-induced anhedonia, fear, and anxiety-like behavior

8.1 mGlu2/3: Amygdala

BLA mGlu2/3 is necessary for stress-induced disruptions to REM sleep following fear conditioning (Sweeten, Adkins, Wellman, & Sanford, 2021).

Male Wistar rats underwent a sleep recording to evaluate baseline REM via EEG, followed by conditioned foot shock. Rats were returned to their home cage for an additional sleep assessment and subsequently characterized as resilient if showing a <50% decrease in REM during the first 4 h of sleep relative to baseline, and susceptible if this decrease was >50%. One week following this initial shock test, rats were bilaterally infused with either the mGlu2/3 agonist LY379268 or vehicle into the BLA 30 min prior to an extinction session followed by EEG in the home cage. LY379268 normalized sleep in susceptible rats in contrast to vehicle-treated rats of the same classification and had no effect on sleep in the resilient group. However, vulnerability defined as shock-induced changes in REM sleep did not confer vulnerability in terms of time spent freezing during the shock training or during the shock context re-exposure. Finally, all rats, independent of their drug treatment, showed stress-induced hypothermia immediately following both the initial shock test and context re-exposure (Sweeten et al., 2021). Thus, while mGlu2/3 in the BLA can regulate stress-induced sleep disturbances but not SIH, the relationship of such sleep disturbances to other behavioral effects of stress exposure are unclear at this point. In a PRS model, PN22 offspring displayed anxiety-like behavior in the EPM and OFT. However, this was not accompanied by changes in amygdala mGlu2/3 expression (Laloux et al., 2012).

8.2 mGlu2/3: Bed nucleus of the stria terminalis (BNST)

One region of the extended amygdala, the BNST, demonstrates mGlu3-regulated plasticity in response to stress (Tyler et al., 2022). Rats exposed to PSS (TMT) showed an increase in the NMDA receptor subunit transcripts GriN2B and GriN3B, whereas stressed rats that were treated with an mGlu3 NAM VU6010572 showed decreased GluN2A, GluN3B, and mGlu2. Thus, NMDA and mGlu2 receptor expression and possibly function is modulated by mGlu3 activity in the BNST, a mechanism that may contribute to the modulation of fear memories.

8.3 mGlu2/3: Medial prefrontal cortex (mPFC)

There are various regions within the cortex that have shown promise as a target treatment for stress-related disorders, but those with a more predominant focus are the prelimbic cortex (PrL), medial PFC (mPFC), and insular cortex. Jing et al. (2021) exposed male C57 mice to the CSDS paradigm and assessed anxiety- and anhedonia-like behaviors through the use of the OFT and SI. Mice were considered susceptible if their mean SI ratio on the initial SI test was decreased to below two standard deviations from SI ratio mean of the unstressed group. Susceptible animals displayed greater mGlu2/3 expression in the PrL, but not in other cortical regions (Jing et al., 2021). On the other hand, Nasca et al. (2015) found a drastic downregulation in mGlu2 protein expression in the mPFC in both stress-susceptible and stress-resilient mice. This may be due to differences in the timing of the assessment of these proteins relative to stress exposure. Jing et al. assessed mGlu2/3 protein expression 5 days after the last exposure, while Nasca et al. assessed mGlu2 expression 24 h following the last CUS exposure. Of course, differences in findings may also be explained by the use of an antibody that did not distinguish between mGlu2 and mGlu3 by Jing et al. (2021).

Pretreatment with the mGlu2/3 antagonist LY341495 into the PL prior to a SI test dose-dependently reduced the social avoidance impairments found in susceptible mice, but had no effect on unstressed or resilient mice, nor did it alter immobility on the OFT (Jing et al., 2021). LY341495 was effective at decreasing escape failures when administered into the PL on the day prior to, but not on the day of, exposure to IES (Highland et al., 2019). Thus, mGlu2/3 signaling in the PL is necessary for susceptibility to the depressive-like effects of stress. In addition, administration of either an mGlu2 or mGlu3 NAM (VU6001966 and VU0650786, respectively) increases neuronal activity in the mPFC, and potentiates PFC → MDT excitatory transmission. The MDT is one region downstream of the PFC that may be mediating the ability of mGlu2 and 3 NAMs to reduce anhedonia, as optogenetic inhibition of the PFC → MDT pathway reverses the anxiolytic and antidepressant effects of these substances (Joffe et al., 2020).

TMT exposure among rats upregulates mGlu3 mRNA in the insular cortex, and treatment with mGlu3 NAM VU6010572 blocked TMT-induced upregulation of NMDA receptor GluN3B subunit transcript (Tyler et al., 2022). Given the same study found that VU6010572 blocked stress-induced fear and anxiety responses, it could be argued that mGlu3 modulation of GluN3B in the insular cortex may modulate stress vulnerability in rodents. Overall, this study indicates that within the cortex specifically, elevated mGlu2 and 3 activity appears to promote stress-susceptible behaviors, and pharmacologically blocking/reversing this effect may be protective against maladaptive behaviors in response to stress.

8.4 mGlu2/3: Hippocampus (HPC)

The hippocampus is a cortical area of the brain that plays a crucial role in the formation and retrieval of memory. In the context of PTSD and other stress-related behaviors, it seems unsurprising that this region may be involved, considering the role the development of fear memories plays a role in stress-induced behaviors. Evidence from prenatal restraint stress (PRS) models in adult rodents suggests that decreased mGlu2/3 expression in the ventral hippocampus may underlie anxiety-like behaviors. Laloux et al. (2012) exposed female Sprague-Dawley dams to PRS 3 times daily from experimental day 11 until delivery. Ultrasonic vocalizations (USV) of the male pups were measured on postnatal D6, D10, and D14, during maternal potentiation (MP) as well as exposure to an unfamiliar male odor (UMO). Following weaning at PND21, rats demonstrated PRS-induced anxiety behaviors using OFT and EPM on PND22. Immunoblotting revealed downregulation of mGlu2/3 in the hippocampus of PRS pups at PND22 (Laloux et al., 2012). Given reduced mGlu2/3 coincided with heightened anxiety at PND22, it could be posited that mGlu2/3 mediates stress-induced anxiety in the hippocampus. However, this should be followed up with mechanistic studies that test the necessity of mGlu2 and 3 for PRS-induced anxiety.

Scholler et al. (2017) exposed male C57Bl6/J mice to conditioned fear, followed by an infusion of either the mGlu2 PAM DN13, mGlu2/3 agonist DCG-IV, or both into the CA3 region of the hippocampus. Mice were then tested on contextual fear 1 day following treatment, then a cued fear test 24 h later. Pretreatment with the higher dose of DCG-IV (0.25 nmol) alone and DCG-IV + DN13 attenuated freezing during the contextual fear test. While there was no effect of DN13 on contextual fear conditioning, it was able to enhance the efficacy of a dose of DCG-IV that alone was unable to alter conditioning (Scholler et al., 2017). Thus, the present study provides support for a possible role of hippocampal mGlu2 in fear conditioning.

In addition, Nasca et al. (2015) demonstrated CUS-induced downregulation of mGlu2 transcription in the HPC of susceptible mice, an effect that was blocked following both treatment of MR antagonists spironolactone and combined treatment of spironolactone and GR antagonist RU486. Spironolactone attenuated anxiety during LDT in the form of increased time spent in the light chamber. In this context, it appears that stress-induced mGlu2 downregulation in the hippocampus promotes stress-susceptible behaviors, an effect mediated by elevated MR activity. Such findings provide more clarity on why such studies contradict those emphasizing the cortex, as it appears that in contrast to the cortex, mGlu2/3 expression appears downregulated within the hippocampus, and interventions designed to positively modulate these receptors may encourage anxiolytic and stress-resilient behaviors.

8.5 Lateral septal nuclei

To examine the role of the lateral septal nucleus in resilience to CSDS, male C57BL/6J mice were exposed to CSDS for 10 days (Wang et al., 2022). Mice then underwent a battery of behavioral assessments for anxiety- and depressive-like behaviors: the open field test (OFT), elevated plus maze (EPM), social interaction test (SI), and tail suspension test (TST). Classification as stress-resilient or stress-susceptible in this study was determined by performance on the SI test, in which a ratio score of <1 regarding time spent in the zone in the presence of a naive Kunming mouse vs time spent in zone without was considered susceptible, whereas a ratio of ≥ 1 was regarded as resilient to social defeat stress. Time spent in the central arena of the OFT following CSDS exposure was also decreased in susceptible mice. Stress-susceptible mice displayed reduced mGlu2/3 expression in the ventral, but not dorsal lateral septal nucleus. The mGlu2/3 agonist LY379268 ($1 \mu g/\mu L$) infused into the ventral lateral septal nucleus (LSv) reversed CSDS-induced social withdrawal impairments on the SI test among susceptible mice, in a dose-dependent manner.

9. mGlu2/3 conclusions

The literature provides ample evidence for the involvement of Group II mGlu activity in mediating vulnerability to stress. Systemic blockade of both mGlu2/3 is consistently found ameliorate depressive-like symptoms such as immobility and sucrose preference. A smaller body of evidence indicates that such blockade is also able to reduce anxiety and fear, particularly when mGlu3 is targeted alone. Evidence suggests an upregulation in mGlu2/3 within the PFC and BNST, while mGlu2/3 is downregulated in the hippocampus, BLA, and LSv in susceptible rodents. In support of these findings, pharmacologically blocking and/or negatively modulating both mGlu2/3 within the PFC and BNST promotes stress resilience, attenuating stress-induced anhedonia and anxiety as well as reinstatement across a wide variety of stress models. However, given mGlu2 deletion and intra-mPFC treatment of mGlu2/3 antagonist LY379268 enhanced stress resilience (Highland et al., 2019), it is likely that mGlu2 confers stress resilience in the PFC. mGlu3 NAM treatment in the BNST attenuated TMT-induced fear and anxiety and posits that mGlu3 specifically may be driving persistent stress-induced behaviors within this region (Tyler et al., 2022). Taken together, mGlu2/3 plays a role in stress vulnerability, though the neuroanatomy driving certain behaviors needs to be considered when pharmacologically treating stress susceptible behaviors.

10. General conclusions

The results presented here provide compelling evidence for the role of mGlu receptors in stress-susceptibility and resilience, however, there are several limitations to consider. The findings in this chapter are only as useful as the models are valid. Beyond model limitations, researchers are met with a challenge of accurate classification of differential stress responses. Experimenters commonly employ nonparametric analyses to subdivide rats into stress-susceptible and stress-resilient phenotypes. While advantageous for identifying stress-response extremes, this method can cloud continuous relationships. Further, groups differ in the measures they use to subdivide groups and the time point at which subdivision occurs. For example, some groups segregate rats into stress groups based on behavior (1) during stress exposure (e.g., freezing, helplessness), (2) 24h after stress exposure, or (3) days to weeks after stress exposure. To phenotype, some groups use double median split analyses of acoustic startle response and EPM (Schwendt et al., 2018; Shallcross et al., 2019, 2021), while others classify based on social interaction ratio (Bagot et al., 2015; Kim et al., 2022; Shin et al., 2015; Tse et al., 2019; Wagner et al., 2015; Wang et al., 2022), the number of standard deviations away from the mean of stress exposed (Jing et al., 2021) or controls (Nasca et al., 2015), a cutoff of greater than the 95% confidence interval of the comparison group (Shin et al., 2015), or degree of change from baseline (Palmfeldt et al., 2016; Sun et al., 2017; Sweeten et al., 2021). Lastly, our interpretation of the findings highlighted in this chapter are limited, as the majority of publications on this topic used only male rodents, and marked sex differences in stress responsivity have been reported elsewhere (Adamec et al., 2006; Cohen & Yehuda, 2011; Weinstock, Razin, Schorer-Apelbaum, Men, & McCarty, 1998).

Increased mGlu5 expression and signaling is associated with resilience to stress-induced anhedonia, but not anxiety. Decreased expression of inhibitory, Gi-coupled mGlu2/3 expression and signaling mediates resilience to stress-induced anhedonia. Together with the observation that at least in some circuits (i.e., BLA-vHPC and BLA-mPFC), stress exposure reduces glutamate transmission, and optogenetic stimulation of these pathways reduces stress-induced anhedonic behaviors, there is strong support that anhedonia arises from decreased glutamate release and post-synaptic mGlu5 signaling. Conversely, decreasing mGlu5 signaling increases *resilience* to stress-induced anxiety-like behavior. The involvement of mGlu5 in fear conditioning and extinction is difficult to disentangle from its role in basic learning processes, however, stimulating mGlu5 signaling facilitates the extinction of conditioned fear, and mGlu5 KO prevents fear conditioning and extinction. Little work has been done on the role of mGlu2/3 receptors in conditioned fear, but consistent with opposing roles for mGlu5 and mGlu2/3 in anhedonia, an mGlu3 NAM attenuates conditioned fear, indicating that increased glutamate transmission may be therapeutic for fear conditioning.

References

Adamec, R., Head, D., Blundell, J., Burton, P., & Berton, O. (2006). Lasting anxiogenic effects of feline predator stress in mice: Sex differences in vulnerability to stress and predicting severity of anxiogenic response from the stress experience. *Physiology & Behavior, 88*(1–2), 12–29.

Adamec, R., Kent, P., Anisman, H., Shallow, T., & Merali, Z. (1998). Neural plasticity, neuropeptides and anxiety in animals—Implications for understanding and treating affective disorder following traumatic stress in humans. *Neuroscience and Biobehavioral Reviews, 23*(2), 301–318.

Adamec, R. E., & Shallow, T. (1993). Lasting effects on rodent anxiety of a single exposure to a cat. *Physiology & Behavior, 54*(1), 101–109.

Adamec, R., Walling, S., & Burton, P. (2004). Long-lasting, selective, anxiogenic effects of feline predator stress in mice. *Physiology & Behavior, 83*(3), 401–410.

Albrechet-Souza, L., & Gilpin, N. W. (2019). The predator odor avoidance model of post-traumatic stress disorder in rats. *Behavioural Pharmacology, 30*(2 and 3-Spec Issue), 105–114.

Alonso, S. J., Castellano, M. A., Afonso, D., & Rodriguez, M. (1991). Sex differences in behavioral despair: Relationships between behavioral despair and open field activity. *Physiology & Behavior, 49*(1), 69–72.

American Psychological Association (Ed.). (2013). *Diagnostic and statistical manual of mental disorders (5th Edition).* Washington, DC: American Psychiatric Association Publishing. Retrieved June 7, 2022, from https://doi.org/10.1176/appi.books.9780890425596.

Apfelbach, R., Blanchard, C. D., Blanchard, R. J., Hayes, R. A., & McGregor, I. S. (2005). The effects of predator odors in mammalian prey species: A review of field and laboratory studies. *Neuroscience and Biobehavioral Reviews, 29*(8), 1123–1144.

Arai, I., Tsuyuki, Y., Shiomoto, H., Satoh, M., & Otomo, S. (2000). Decreased body temperature dependent appearance of behavioral despair in the forced swimming test in mice. *Pharmacological Research, 42*(2), 171–176.

Bagot, R. C., Parise, E. M., Peña, C. J., Zhang, H.-X., Maze, I., Chaudhury, D., et al. (2015). Ventral hippocampal afferents to the nucleus accumbens regulate susceptibility to depression. *Nature Communications, 6*, 7062.

Bali, A., & Jaggi, A. S. (2015). Electric foot shock stress: A useful tool in neuropsychiatric studies. *Reviews in the Neurosciences, 26*(6), 655–677.

Belzung, C., El Hage, W., Moindrot, N., & Griebel, G. (2001). Behavioral and neurochemical changes following predatory stress in mice. *Neuropharmacology, 41*(3), 400–408.

Berton, O., McClung, C. A., Dileone, R. J., Krishnan, V., Renthal, W., Russo, S. J., et al. (2006). Essential role of BDNF in the mesolimbic dopamine pathway in social defeat stress. *Science, 311*(5762), 864–868.

Blount, H. L., Dee, J., Wu, L., Schwendt, M., & Knackstedt, L. A. (2023). Resilience to anxiety and anhedonia after predator scent stress is accompanied by increased nucleus accumbens mGlu5 in female rats. *Behavioural Brain Research, 436*, 114090.

Bodnoff, S. R., Suranyi-Cadotte, B., Quirion, R., & Meaney, M. J. (1989). A comparison of the effects of diazepam versus several typical and atypical anti-depressant drugs in an animal model of anxiety. *Psychopharmacology, 97*(2), 277–279.

Borsini, F., Lecci, A., Volterra, G., & Meli, A. (1989). A model to measure anticipatory anxiety in mice? *Psychopharmacology, 98*(2), 207–211.

Bourin, M., & Hascoët, M. (2003). The mouse light/dark box test. *European Journal of Pharmacology, 463*(1–3), 55–65.

Brodkin, J., Bradbury, M., Busse, C., Warren, N., Bristow, L. J., & Varney, M. A. (2002). Reduced stress-induced hyperthermia in mGluR5 knockout mice. *The European Journal of Neuroscience, 16*(11), 2241–2244.

Brunton, P. J., & Russell, J. A. (2010). Prenatal social stress in the rat programmes neuroendocrine and behavioural responses to stress in the adult offspring: Sex-specific effects. *Journal of Neuroendocrinology, 22*(4), 258–271.

Buonaguro, E. F., Morley-Fletcher, S., Avagliano, C., Vellucci, L., Iasevoli, F., Bouwalerh, H., et al. (2020). Glutamatergic postsynaptic density in early life stress programming: Topographic gene expression of mGlu5 receptors and Homer proteins. *Progress in Neuro-Psychopharmacology & Biological Psychiatry, 96*, 109725.

Cohen, H., Liu, T., Kozlovsky, N., Kaplan, Z., Zohar, J., & Mathé, A. A. (2012). The neuropeptide Y (NPY)-ergic system is associated with behavioral resilience to stress exposure in an animal model of post-traumatic stress disorder. *Neuropsychopharmacology, 37*(2), 350–363.

Cohen, H., & Yehuda, R. (2011). Gender differences in animal models of posttraumatic stress disorder. *Disease Markers, 30*(2–3), 141–150.

Crawley, J. N. (1981). Neuropharmacologic specificity of a simple animal model for the behavioral actions of benzodiazepines. *Pharmacology, Biochemistry, and Behavior, 15*(5), 695–699.

Cryan, J. F., Kelly, P. H., Neijt, H. C., Sansig, G., Flor, P. J., & van Der Putten, H. (2003). Antidepressant and anxiolytic-like effects in mice lacking the group III metabotropic glutamate receptor mGluR7. *The European Journal of Neuroscience, 17*(11), 2409–2417.

Cryan, J. F., Mombereau, C., & Vassout, A. (2005). The tail suspension test as a model for assessing antidepressant activity: Review of pharmacological and genetic studies in mice. *Neuroscience and Biobehavioral Reviews, 29*(4–5), 571–625.

Cullinan, W. E., & Wolfe, T. J. (2000). Chronic stress regulates levels of mRNA transcripts encoding beta subunits of the GABA(A) receptor in the rat stress axis. *Brain Research, 887*(1), 118–124.

Darnaudéry, M., & Maccari, S. (2008). Epigenetic programming of the stress response in male and female rats by prenatal restraint stress. *Brain Research Reviews, 57*(2), 571–585.

David, D. J., Nic Dhonnchadha, B. A., Jolliet, P., Hascoët, M., & Bourin, M. (2001). Are there gender differences in the temperature profile of mice after acute antidepressant administration and exposure to two animal models of depression? *Behavioural Brain Research, 119*(2), 203–211.

De Vry, J., Benz, U., Schreiber, R., & Traber, J. (1993). Shock-induced ultrasonic vocalization in young adult rats: A model for testing putative anti-anxiety drugs. *European Journal of Pharmacology, 249*(3), 331–339.

Dong, C., Tian, Z., Fujita, Y., Fujita, A., Hino, N., Iijima, M., et al. (2022). Antidepressant-like actions of the mGlu2/3 receptor antagonist TP0178894 in the chronic social defeat stress model: Comparison with escitalopram. *Pharmacology, Biochemistry, and Behavior, 212*, 173316.

Dong, C., Zhang, J.-C., Yao, W., Ren, Q., Ma, M., Yang, C., et al. (2017). Rapid and sustained antidepressant action of the mGlu2/3 receptor antagonist MGS0039 in the social defeat stress model: Comparison with ketamine. *The International Journal of Neuropsychopharmacology, 20*(3), 228–236.

Dremencov, E., Lapshin, M., Komelkova, M., Alliluev, A., Tseilikman, O., Karpenko, M., et al. (2019). Chronic predator scent stress alters serotonin and dopamine levels in the rat thalamus and hypothalamus, respectively. *General Physiology and Biophysics, 38*(2), 187–190.

Dulawa, S. C., & Hen, R. (2005). Recent advances in animal models of chronic antidepressant effects: The novelty-induced hypophagia test. *Neuroscience and Biobehavioral Reviews, 29*(4–5), 771–783.

Engers, J. L., Rodriguez, A. L., Konkol, L. C., Morrison, R. D., Thompson, A. D., Byers, F. W., et al. (2015). Discovery of a selective and CNS penetrant negative allosteric modulator of metabotropic glutamate receptor subtype 3 with antidepressant and anxiolytic activity in rodents. *Journal of Medicinal Chemistry, 58*(18), 7485–7500.

Ganella, D. E., Thangaraju, P., Lawrence, A. J., & Kim, J. H. (2016). Fear extinction in 17 day old rats is dependent on metabotropic glutamate receptor 5 signaling. *Behavioural Brain Research, 298*(Pt A), 32–36.

Golden, S. A., Covington, H. E., Berton, O., & Russo, S. J. (2011). A standardized protocol for repeated social defeat stress in mice. *Nature Protocols, 6*(8), 1183–1191.

Gourley, S. L., Kiraly, D. D., Howell, J. L., Olausson, P., & Taylor, J. R. (2008). Acute hippocampal brain-derived neurotrophic factor restores motivational and forced swim performance after corticosterone. *Biological Psychiatry, 64*(10), 884–890.

Gourley, S. L., & Taylor, J. R. (2009). Recapitulation and reversal of a persistent depression-like syndrome in rodents. *Current Protocols in Neuroscience. Chapter 9*, Unit 9.32.

Gourley, S. L., Wu, F. J., Kiraly, D. D., Ploski, J. E., Kedves, A. T., Duman, R. S., et al. (2008). Regionally specific regulation of ERK MAP kinase in a model of antidepressant-sensitive chronic depression. *Biological Psychiatry, 63*(4), 353–359.

Groenink, L., van der Gugten, J., Zethof, T., van der Heyden, J., & Olivier, B. (1994). Stress-induced hyperthermia in mice: Hormonal correlates. *Physiology & Behavior, 56*(4), 747–749.

Hascoët, M., & Bourin, M. (1998). A new approach to the light/dark test procedure in mice. *Pharmacology, Biochemistry, and Behavior, 60*(3), 645–653.

Hayley, S., Borowski, T., Merali, Z., & Anisman, H. (2001). Central monoamine activity in genetically distinct strains of mice following a psychogenic stressor: Effects of predator exposure. *Brain Research, 892*(2), 293–300.

Herman, J. P., Adams, D., & Prewitt, C. (1995). Regulatory changes in neuroendocrine stress-integrative circuitry produced by a variable stress paradigm. *Neuroendocrinology, 61*(2), 180–190.

Highland, J. N., Zanos, P., Georgiou, P., & Gould, T. D. (2019). Group II metabotropic glutamate receptor blockade promotes stress resilience in mice. *Neuropsychopharmacology, 44*(10), 1788–1796.

Holmes, P. V. (2003). Rodent models of depression: Reexamining validity without anthropomorphic inference. *Critical Reviews in Neurobiology, 15*(2), 143–174.

Jacob, W., Gravius, A., Pietraszek, M., Nagel, J., Belozertseva, I., Shekunova, E., et al. (2009). The anxiolytic and analgesic properties of fenobam, a potent mGlu5 receptor antagonist, in relation to the impairment of learning. *Neuropharmacology, 57*(2), 97–108.

Jing, X.-Y., Wang, Y., Zou, H.-W., Li, Z.-L., Liu, Y.-J., & Li, L.-F. (2021). mGlu2/3 receptor in the prelimbic cortex is implicated in stress resilience and vulnerability in mice. *European Journal of Pharmacology, 906*, 174231.

Joffe, M. E., Maksymetz, J., Luschinger, J. R., Dogra, S., Ferranti, A. S., Luessen, D. J., et al. (2022). Acute restraint stress redirects prefrontal cortex circuit function through mGlu5 receptor plasticity on somatostatin-expressing interneurons. *Neuron, 110*(6), 1068–1083.

Joffe, M. E., Santiago, C. I., Oliver, K. H., Maksymetz, J., Harris, N. A., Engers, J. L., et al. (2020). mGlu2 and mGlu3 negative allosteric modulators divergently enhance thalamo-cortical transmission and exert rapid antidepressant-like effects. *Neuron, 105*(1), 46–59.

Kalinichev, M., Easterling, K. W., Plotsky, P. M., & Holtzman, S. G. (2002). Long-lasting changes in stress-induced corticosterone response and anxiety-like behaviors as a consequence of neonatal maternal separation in Long-Evans rats. *Pharmacology, Biochemistry, and Behavior, 73*(1), 131–140.

Kasten, C. R., Carzoli, K. L., Sharfman, N. M., Henderson, T., Holmgren, E. B., Lerner, M. R., et al. (2020). Adolescent alcohol exposure produces sex differences in negative affect-like behavior and group I mGluR BNST plasticity. *Neuropsychopharmacology, 45*(8), 1306–1315.

Kasten, C. R., Holmgren, E. B., Lerner, M. R., & Wills, T. A. (2021). BNST specific mGlu5 receptor knockdown regulates sex-dependent expression of negative affect produced by adolescent ethanol exposure and adult stress. *Translational Psychiatry, 11*(1), 178.

Kessler, R. C., Sonnega, A., Bromet, E., Hughes, M., & Nelson, C. B. (1995). Posttraumatic stress disorder in the National Comorbidity Survey. *Archives of General Psychiatry, 52*(12), 1048–1060.

Kim, J., Kang, S., Choi, T.-Y., Chang, K.-A., & Koo, J. W. (2022). Metabotropic glutamate receptor 5 in amygdala target neurons regulates susceptibility to chronic social stress. *Biological Psychiatry*.

Kraeuter, A.-K., Guest, P. C., & Sarnyai, Z. (2019). The open field test for measuring locomotor activity and anxiety-like behavior. *Methods in Molecular Biology, 1916*, 99–103.

Krishnan, V., Han, M.-H., Graham, D. L., Berton, O., Renthal, W., Russo, S. J., et al. (2007). Molecular adaptations underlying susceptibility and resistance to social defeat in brain reward regions. *Cell, 131*(2), 391–404.

Kulkarni, S. K., Singh, K., & Bishnoi, M. (2008). Comparative behavioural profile of newer antianxiety drugs on different mazes. *Indian Journal of Experimental Biology, 46*(9), 633–638.

Laloux, C., Mairesse, J., Van Camp, G., Giovine, A., Branchi, I., Bouret, S., et al. (2012). Anxiety-like behaviour and associated neurochemical and endocrinological alterations in male pups exposed to prenatal stress. *Psychoneuroendocrinology, 37*(10), 1646–1658.

Landis, C., & Hunt, W. A. (1939). *The startle pattern*. Farrar & Rinehart.

Langgartner, D., Füchsl, A. M., Uschold-Schmidt, N., Slattery, D. A., & Reber, S. O. (2015). Chronic subordinate colony housing paradigm: A mouse model to characterize the consequences of insufficient glucocorticoid signaling. *Frontiers in Psychiatry, 6*, 18.

Lecci, A., Borsini, F., Volterra, G., & Meli, A. (1990). Pharmacological validation of a novel animal model of anticipatory anxiety in mice. *Psychopharmacology, 101*(2), 255–261.

Liu, X.-L., Luo, L., Mu, R.-H., Liu, B.-B., Geng, D., Liu, Q., et al. (2015). Fluoxetine regulates mTOR signalling in a region-dependent manner in depression-like mice. *Scientific Reports, 5*, 16024.

Liu, M.-Y., Yin, C.-Y., Zhu, L.-J., Zhu, X.-H., Xu, C., Luo, C.-X., et al. (2018). Sucrose preference test for measurement of stress-induced anhedonia in mice. *Nature Protocols, 13*(7), 1686–1698.

López-Rubalcava, C., Saldívar, A., & Fernández-Guasti, A. (1992). Interaction of GABA and serotonin in the anxiolytic action of diazepam and serotonergic anxiolytics. *Pharmacology Biochemistry and Behavior, 43*(2), 433–440.

Lucki, I., Dalvi, A., & Mayorga, A. J. (2001). Sensitivity to the effects of pharmacologically selective antidepressants in different strains of mice. *Psychopharmacology, 155*(3), 315–322.

Maccari, S., Darnaudery, M., Morley-Fletcher, S., Zuena, A. R., Cinque, C., & Van Reeth, O. (2003). Prenatal stress and long-term consequences: Implications of glucocorticoid hormones. *Neuroscience and Biobehavioral Reviews, 27*(1–2), 119–127.

Maccari, S., Krugers, H. J., Morley-Fletcher, S., Szyf, M., & Brunton, P. J. (2014). The consequences of early-life adversity: Neurobiological, behavioural and epigenetic adaptations. *Journal of Neuroendocrinology, 26*(10), 707–723.

Mackenzie, L., Nalivaiko, E., Beig, M. I., Day, T. A., & Walker, F. R. (2010). Ability of predator odour exposure to elicit conditioned versus sensitised post traumatic stress disorder-like behaviours, and forebrain deltaFosB expression, in rats. *Neuroscience, 169*(2), 733–742.

Mao, L.-M., & Wang, J. Q. (2018). Alterations in mGlu5 receptor expression and function in the striatum in a rat depression model. *Journal of Neurochemistry, 145*(4), 287–298.

Mao, L., Yang, L., Tang, Q., Samdani, S., Zhang, G., & Wang, J. Q. (2005). The scaffold protein Homer1b/c links metabotropic glutamate receptor 5 to extracellular signal-regulated protein kinase cascades in neurons. *The Journal of Neuroscience, 25*(10), 2741–2752.

Marazziti, D., Di Muro, A., & Castrogiovanni, P. (1992). Psychological stress and body temperature changes in humans. *Physiology & Behavior, 52*(2), 393–395.

Mayorga, A. J., & Lucki, I. (2001). Limitations on the use of the C57BL/6 mouse in the tail suspension test. *Psychopharmacology, 155*(1), 110–112.

Meaney, M. J., Szyf, M., & Seckl, J. R. (2007). Epigenetic mechanisms of perinatal programming of hypothalamic-pituitary-adrenal function and health. *Trends in Molecular Medicine, 13*(7), 269–277.

Michalikova, S., van Rensburg, R., Chazot, P. L., & Ennaceur, A. (2010). Anxiety responses in Balb/c, c57 and CD-1 mice exposed to a novel open space test. *Behavioural Brain Research, 207*(2), 402–417.

Misslin, R., Belzung, C., & Vogel, E. (1989). Behavioural validation of a light/dark choice procedure for testing anti-anxiety agents. *Behavioural Processes, 18*(1–3), 119–132.

Montgomery, K. C. (1955). The relation between fear induced by novel stimulation and exploratory drive. *Journal of Comparative and Physiological Psychology, 48*(4), 254–260.

Musazzi, L., Tornese, P., Sala, N., & Popoli, M. (2018). What acute stress protocols can tell us about PTSD and stress-related neuropsychiatric disorders. *Frontiers in Pharmacology, 9*, 758.

Myers, K. M., & Davis, M. (2007). Mechanisms of fear extinction. *Molecular Psychiatry, 12*(2), 120–150.

Mylle, J., & Maes, M. (2004). Partial posttraumatic stress disorder revisited. *Journal of Affective Disorders, 78*(1), 37–48.

Nasca, C., Bigio, B., Zelli, D., Nicoletti, F., & McEwen, B. S. (2015). Mind the gap: Glucocorticoids modulate hippocampal glutamate tone underlying individual differences in stress susceptibility. *Molecular Psychiatry, 20*(6), 755–763.

Niswender, C. M., & Conn, P. J. (2010). Metabotropic glutamate receptors: Physiology, pharmacology, and disease. *Annual Review of Pharmacology and Toxicology, 50*, 295–322.

Ogawa, T., Mikuni, M., Kuroda, Y., Muneoka, K., Mori, K. J., & Takahashi, K. (1994). Periodic maternal deprivation alters stress response in adult offspring: Potentiates the negative feedback regulation of restraint stress-induced adrenocortical response and reduces the frequencies of open field-induced behaviors. *Pharmacology, Biochemistry, and Behavior, 49*(4), 961–967.

Oh, J., Zupan, B., Gross, S., & Toth, M. (2009). Paradoxical anxiogenic response of juvenile mice to fluoxetine. *Neuropsychopharmacology, 34*(10), 2197–2207.

Overmier, J. B., & Seligman, M. E. (1967). Effects of inescapable shock upon subsequent escape and avoidance responding. *Journal of Comparative and Physiological Psychology, 63*(1), 28–33.

Palmfeldt, J., Henningsen, K., Eriksen, S. A., Müller, H. K., & Wiborg, O. (2016). Protein biomarkers of susceptibility and resilience to stress in a rat model of depression. *Molecular and Cellular Neurosciences, 74*, 87–95.

Pałucha-Poniewiera, A., Podkowa, K., & Rafało-Ulińska, A. (2021). The group II mGlu receptor antagonist LY341495 induces a rapid antidepressant-like effect and enhances the effect of ketamine in the chronic unpredictable mild stress model of depression in C57BL/6J mice. *Progress in Neuro-Psychopharmacology & Biological Psychiatry, 109*, 110239.

Pavlov, P. I. (1927). Conditioned reflexes: An investigation of the physiological activity of the cerebral cortex. *Annals of Neurosciences, 17*(3), 136–141.

Pellow, S., Chopin, P., File, S. E., & Briley, M. (1985). Validation of open:Closed arm entries in an elevated plus-maze as a measure of anxiety in the rat. *Journal of Neuroscience Methods, 14*(3), 149–167.

Pérez de la Mora, M., Lara-García, D., Jacobsen, K. X., Vázquez-García, M., Crespo-Ramírez, M., Flores-Gracia, C., et al. (2006). Anxiolytic-like effects of the selective metabotropic glutamate receptor 5 antagonist MPEP after its intra-amygdaloid microinjection in three different non-conditioned rat models of anxiety. *The European Journal of Neuroscience, 23*(10), 2749–2759.

Peterlik, D., Stangl, C., Bauer, A., Bludau, A., Keller, J., Grabski, D., et al. (2017). Blocking metabotropic glutamate receptor subtype 5 relieves maladaptive chronic stress consequences. *Brain, Behavior, and Immunity, 59*, 79–92.

Petit-Demouliere, B., Chenu, F., & Bourin, M. (2005). Forced swimming test in mice: A review of antidepressant activity. *Psychopharmacology, 177*(3), 245–255.

Porsolt, R. D., Anton, G., Blavet, N., & Jalfre, M. (1978). Behavioural despair in rats: A new model sensitive to antidepressant treatments. *European Journal of Pharmacology, 47*(4), 379–391.

Prut, L., & Belzung, C. (2003). The open field as a paradigm to measure the effects of drugs on anxiety-like behaviors: A review. *European Journal of Pharmacology, 463*(1–3), 3–33.

Rajbhandari, A. K., Gonzalez, S. T., Fanselow, M. S. Gordon, J. A., Moghaddam, B., Giocomo, L., Nirenberg, S., et al. (2018).

Ravinder, S., Burghardt, N. S., Brodsky, R., Bauer, E. P., & Chattarji, S. (2017). Activation of the same amygdala causes divergent effects on specific versus indiscriminate fear. *Elife, 6*, e25665.

Reber, S. O., Langgartner, D., Foertsch, S., Postolache, T. T., Brenner, L. A., Guendel, H., et al. (2016). Chronic subordinate colony housing paradigm: A mouse model for mechanisms of PTSD vulnerability, targeted prevention, and treatment-2016 Curt Richter Award Paper. *Psychoneuroendocrinology, 74*, 221–230.

Rodgers, R. J., Cole, J. C., & Harrison-Phillips, D. J. (1994). "Cohort removal" induces hyperthermia but fails to influence plus-maze behaviour in male mice. *Physiology & Behavior, 55*(1), 189–192.

Rorick-Kehn, L. M., Johnson, B. G., Knitowski, K. M., Salhoff, C. R., Witkin, J. M., Perry, K. W., et al. (2007). In vivo pharmacological characterization of the structurally novel, potent, selective mGlu2/3 receptor agonist LY404039 in animal models of psychiatric disorders. *Psychopharmacology, 193*(1), 121–136.

Rupniak, N. M. J. (2003). Animal models of depression: Challenges from a drug development perspective. *Behavioural Pharmacology, 14*(5–6), 385–390.

Sánchez, C. (1993). Effect of serotonergic drugs on footshock-induced ultrasonic vocalization in adult male rats. *Behavioural Pharmacology, 4*(3), 269–277.

Santarelli, L., Saxe, M., Gross, C., Surget, A., Battaglia, F., Dulawa, S., et al. (2003). Requirement of hippocampal neurogenesis for the behavioral effects of antidepressants. *Science, 301*(5634), 805–809.

Scholler, P., Nevoltris, D., de Bundel, D., Bossi, S., Moreno-Delgado, D., Rovira, X., et al. (2017). Allosteric nanobodies uncover a role of hippocampal mGlu2 receptor homodimers in contextual fear consolidation. *Nature Communications, 8*(1), 1967.

Schwendt, M., Shallcross, J., Hadad, N. A., Namba, M. D., Hiller, H., Wu, L., et al. (2018). A novel rat model of comorbid PTSD and addiction reveals intersections between stress susceptibility and enhanced cocaine seeking with a role for mGlu5 receptors. *Translational Psychiatry, 8*(1), 209.

Seckl, J. R., & Holmes, M. C. (2007). Mechanisms of disease: Glucocorticoids, fetal and placental metabolism and fetal "programming" of adult pathophysiology. *Nature Clinical Practice. Endocrinology & Metabolism, 3*(6), 479–488.

Seligman, M. E., & Maier, S. F. (1967). Failure to escape traumatic shock. *Journal of Experimental Psychology, 74*(1), 1–9.

Seo, M. K., Lee, J. A., Jeong, S., Seog, D.-H., Lee, J. G., & Park, S. W. (2022). Effects of chronic LY341495 on hippocampal mTORC1 signaling in mice with chronic unpredictable stress-induced depression. *International Journal of Molecular Sciences, 23*(12).

Sethna, F., & Wang, H. (2014). Pharmacological enhancement of mGluR5 facilitates contextual fear memory extinction. *Learning & Memory, 21*(12), 647–650.

Shallcross, J., Hámor, P., Bechard, A. R., Romano, M., Knackstedt, L., & Schwendt, M. (2019). The divergent effects of CDPPB and cannabidiol on fear extinction and anxiety in a predator scent stress model of PTSD in rats. *Frontiers in Behavioral Neuroscience, 13*, 91.

Shallcross, J., Wu, L., Knackstedt, L. A., & Schwendt, M. (2020). Increased mGlu5 mRNA expression in BLA glutamate neurons facilitates resilience to the long-term effects of a single predator scent stress exposure. *BioRxiv.*

Shallcross, J., Wu, L., Wilkinson, C. S., Knackstedt, L. A., & Schwendt, M. (2021). Increased mGlu5 mRNA expression in BLA glutamate neurons facilitates resilience to the long-term effects of a single predator scent stress exposure. *Brain Structure & Function, 226*(7), 2279–2293.

Shepherd, J. K., Grewal, S. S., Fletcher, A., Bill, D. J., & Dourish, C. T. (1994). Behavioural and pharmacological characterisation of the elevated "zero-maze" as an animal model of anxiety. *Psychopharmacology, 116*(1), 56–64.

Shin, S., Kwon, O., Kang, J. I., Kwon, S., Oh, S., Choi, J., et al. (2016). mGluR5 in the nucleus accumbens is critical for promoting resilience to chronic stress. *Nature Neuroscience, 18*(7), 1017–1024.

Shors, T. J., Seib, T. B., Levine, S., & Thompson, R. F. (1989). Inescapable versus escapable shock modulates long-term potentiation in the rat hippocampus. *Science, 244*(4901), 224–226.

Sobrian, S. K., Marr, L., & Ressman, K. (2003). Prenatal cocaine and/or nicotine exposure produces depression and anxiety in aging rats. *Progress in Neuro-Psychopharmacology & Biological Psychiatry, 27*(3), 501–518.

Steru, L., Chermat, R., Thierry, B., & Simon, P. (1985). The tail suspension test: A new method for screening antidepressants in mice. *Psychopharmacology, 85*(3), 367–370.

Stoppel, L. J., Auerbach, B. D., Senter, R. K., Preza, A. R., Lefkowitz, R. J., & Bear, M. F. (2017). β-Arrestin2 couples metabotropic glutamate receptor 5 to neuronal protein synthesis and is a potential target to treat fragile X. *Cell Reports, 18*(12), 2807–2814.

Sturman, O., Germain, P.-L., & Bohacek, J. (2018). Exploratory rearing: A context- and stress-sensitive behavior recorded in the open-field test. *Stress (Amsterdam, Netherlands), 21*(5), 443–452.

Sun, H., Su, R., Zhang, X., Wen, J., Yao, D., Gao, X., et al. (2017). Hippocampal GR- and CB1-mediated mGluR5 differentially produces susceptibility and resilience to acute and chronic mild stress in rats. *Neuroscience, 357*, 295–302.

Sweeten, B. L. W., Adkins, A. M., Wellman, L. L., & Sanford, L. D. (2021). Group II metabotropic glutamate receptor activation in the basolateral amygdala mediates individual differences in stress-induced changes in rapid eye movement sleep. *Progress in Neuro-Psychopharmacology & Biological Psychiatry, 104*, 110014.

Swerdlow, N. R., Braff, D. L., & Geyer, M. A. (1999). Cross-species studies of sensorimotor gating of the startle reflex. *Annals of the New York Academy of Sciences, 877*, 202–216.

Toma, W., Kyte, S. L., Bagdas, D., Alkhlaif, Y., Alsharari, S. D., Lichtman, A. H., et al. (2017). Effects of paclitaxel on the development of neuropathy and affective behaviors in the mouse. *Neuropharmacology, 117*, 305–315.

Tornese, P., Sala, N., Bonini, D., Bonifacino, T., La Via, L., Milanese, M., et al. (2019). Chronic mild stress induces anhedonic behavior and changes in glutamate release, BDNF trafficking and dendrite morphology only in stress vulnerable rats. The rapid restorative action of ketamine. *Neurobiology of Stress, 10*, 100160.

Tsankova, N. M., Berton, O., Renthal, W., Kumar, A., Neve, R. L., & Nestler, E. J. (2006). Sustained hippocampal chromatin regulation in a mouse model of depression and antidepressant action. *Nature Neuroscience, 9*(4), 519–525.

Tse, Y. C., Lopez, J., Moquin, A., Wong, S.-M. A., Maysinger, D., & Wong, T. P. (2019). The susceptibility to chronic social defeat stress is related to low hippocampal extrasynaptic NMDA receptor function. *Neuropsychopharmacology, 44*(7), 1310–1318.

Tyler, R. E., Bluitt, M. N., Engers, J. L., Lindsley, C. W., & Besheer, J. (2022). The effects of predator odor (TMT) exposure and mGlu3 NAM pretreatment on behavioral and NMDA receptor adaptations in the brain. *Neuropharmacology, 207*, 108943.

Van der Heyden, J. A., Zethof, T. J., & Olivier, B. (1997). Stress-induced hyperthermia in singly housed mice. *Physiology & Behavior, 62*(3), 463–470.

Vicente, M. A., & Zangrossi, H. (2014). Involvement of 5-HT2C and 5-HT1A receptors of the basolateral nucleus of the amygdala in the anxiolytic effect of chronic antidepressant treatment. *Neuropharmacology, 79*, 127–135.

Võikar, V., Kõks, S., Vasar, E., & Rauvala, H. (2001). Strain and gender differences in the behavior of mouse lines commonly used in transgenic studies. *Physiology & Behavior, 72*(1–2), 271–281.

Wagner, K. V., Hartmann, J., Labermaier, C., Häusl, A. S., Zhao, G., Harbich, D., et al. (2015). Homer1/mGluR5 activity moderates vulnerability to chronic social stress. *Neuropsychopharmacology, 40*(5), 1222–1233.

Wang, Y., Jiang, Y., Song, B.-L., Zou, H.-W., Li, Z.-L., Li, L.-F., et al. (2022). mGlu2/3 receptors within the ventral part of the lateral septal nuclei modulate stress resilience and vulnerability in mice. *Brain Research, 1779*, 147783.

Weinstock, M. (2008). The long-term behavioural consequences of prenatal stress. *Neuroscience and Biobehavioral Reviews, 32*(6), 1073–1086.

Weinstock, M., Razin, M., Schorer-Apelbaum, D., Men, D., & McCarty, R. (1998). Gender differences in sympathoadrenal activity in rats at rest and in response to footshock stress. *International Journal of Developmental Neuroscience, 16*(3–4), 289–295.

Xu, X., Wu, K., Ma, X., Wang, W., Wang, H., Huang, M., et al. (2021). mGluR5-mediated eCB signaling in the nucleus accumbens controls vulnerability to depressive-like behaviors and pain after chronic social defeat stress. *Molecular Neurobiology, 58*(10), 4944–4958.

Xu, J., Zhu, Y., Contractor, A., & Heinemann, S. F. (2009). mGluR5 has a critical role in inhibitory learning. *The Journal of Neuroscience, 29*(12), 3676–3684.

Xu, J., Zhu, Y., Kraniotis, S., He, Q., Marshall, J. J., Nomura, T., et al. (2013). Potentiating mGluR5 function with a positive allosteric modulator enhances adaptive learning. *Learning & Memory, 20*(8), 438–445.

Yashiro, S., & Seki, K. (2017). Association of social defeat stress-induced anhedonia-like symptoms with mGluR1-dependent decrease in membrane-bound AMPA-GluR1 in the mouse ventral midbrain. *Stress (Amsterdam, Netherlands), 20*(4), 404–418.

Yerkes, R. M., & Dodson, J. D. (1908). The relation of strength of stimulus to rapidity of habit-formation. *Journal of Comparative Neurology and Psychology, 18*(5), 459–482.

Yim, Y. S., Han, W., Seo, J., Kim, C. H., & Kim, D. G. (2018). Differential mGluR5 expression in response to the same stress causes individually adapted hippocampal network activity. *Biochemical and Biophysical Research Communications, 495*(1), 1305–1311.

Zethof, T. J., Van der Heyden, J. A., Tolboom, J. T., & Olivier, B. (1994). Stress-induced hyperthermia in mice: A methodological study. *Physiology & Behavior, 55*(1), 109–115.

Zhou, Q.-G., Hu, Y., Wu, D.-L., Zhu, L.-J., Chen, C., Jin, X., et al. (2011). Hippocampal telomerase is involved in the modulation of depressive behaviors. *The Journal of Neuroscience, 31*(34), 12258–12269.

Zohar, J., Matar, M. A., Ifergane, G., Kaplan, Z., & Cohen, H. (2008). Brief post-stressor treatment with pregabalin in an animal model for PTSD: Short-term anxiolytic effects without long-term anxiogenic effect. *European Neuropsychopharmacology, 18*(9), 653–666.

Zuena, A. R., Mairesse, J., Casolini, P., Cinque, C., Alemà, G. S., Morley-Fletcher, S., et al. (2008). Prenatal restraint stress generates two distinct behavioral and neurochemical profiles in male and female rats. *PLoS One, 3*(5), e2170.

CHAPTER SIX

The metabotropic glutamate receptor 5 as a biomarker for psychiatric disorders

Ruth H. Asch[a,*], Ansel T. Hillmer[a,b], Stephen R. Baldassarri[c,d], and Irina Esterlis[a,e,f]

[a]Department of Psychiatry, Yale University, New Haven, CT, United States
[b]Department of Radiology and Biomedical Imaging, New Haven, CT, United States
[c]Yale Program in Addiction Medicine, Yale University, New Haven, CT, United States
[d]Department of Internal Medicine, Yale University, New Haven, CT, United States
[e]Department of Psychology, Yale University, New Haven, CT, United States
[f]Clinical Neurosciences Division, U.S. Department of Veterans Affairs National Center for Posttraumatic Stress Disorder, Veterans Affairs Connecticut Healthcare System, West Haven, CT, United States
*Corresponding author: e-mail address: ruth.asch@yale.edu

Contents

1. Introduction	266
2. Quantifying mGlu5 with PET	267
3. Mood disorders	268
3.1 Major depressive disorder	268
3.2 Bipolar disorder	272
3.3 Treatment trials in mood disorders	273
4. Anxiety and trauma disorders	277
4.1 Posttraumatic stress disorder (PTSD)	278
4.2 Obsessive compulsive disorder (OCD)	281
4.3 Treatment trials in anxiety and trauma disorders	281
5. Substance use: Nicotine, cannabis, and alcohol	284
5.1 Nicotine use disorder	284
5.2 Cannabis use disorder (CUD)	286
5.3 Alcohol use disorder (AUD)	287
5.4 Treatment trials in substance use disorders	289
6. Summary and commentary	291
References	292

Abstract

The role of glutamate system in the etiology and pathophysiology of psychiatric disorders has gained considerable attention in the past two decades, including dysregulation of the metabotropic glutamatergic receptor subtype 5 (mGlu5). Thus, mGlu5 may represent a promising therapeutic target for psychiatric conditions, particularly stress-related disorders. Here, we describe mGlu5 findings in mood disorders,

International Review of Neurobiology, Volume 168
ISSN 0074-7742
https://doi.org/10.1016/bs.irn.2022.10.007

Copyright © 2023 Elsevier Inc.
All rights reserved.

anxiety, and trauma disorders, as well as substance use (specifically nicotine, cannabis, and alcohol use). We highlight insights gained from positron emission tomography (PET) studies, where possible, and discuss findings from treatment trials, when available, to explore the role of mGlu5 in these psychiatric disorders. Through the research evidence reviewed in this chapter, we make the argument that, not only is dysregulation of mGlu5 evident in numerous psychiatric disorders, potentially functioning as a disease "biomarker," the normalization of glutamate neurotransmission via changes in mGlu5 expression and/or modulation of mGlu5 signaling may be a needed component in treating some psychiatric disorders or symptoms. Finally, we hope to demonstrate the utility of PET as an important tool for investigating mGlu5 in disease mechanisms and treatment response.

1. Introduction

Glutamate is the predominant excitatory neurotransmitter in the mammalian central nervous system, and glutamatergic neurotransmission is regulated by ionotropic and metabotropic glutamate receptors (mGluRs). The mGluRs are divided into three families based on sequence homology, their interaction with intracellular signaling cascades, and pharmacological properties: group I (mGluR1 and 5), group II (mGluR2 and 3) and group III (mGluR4, 6, 7, 8) (Conn & Pin, 1997; Coutinho & Knöpfel, 2002). Specifically, the group I mGlu5, is expressed mostly in post-synaptic neurons (Shigemoto et al., 1997; Sistiaga, Herrero, Conquet, & Sánchez-Prieto, 1998; Takumi, Matsubara, Rinvik, & Ottersen, 1999), and is located on the cellular and intracellular membranes (Hubert, Paquet, & Smith, 2001; López-Bendito, Shigemoto, Fairén, & Luján, 2002; O'Malley, Jong, Gonchar, Burkhalter, & Romano, 2003) of glutamatergic (excitatory) neurons, gamma-aminobutyric acid-ergic (GABA-ergic, inhibitory) neurons, and on glial cells in the perisynaptic space (Lavialle et al., 2011; Lujan, Nusser, Roberts, Shigemoto, & Somogyi, 1996). Regions with the highest mGlu5 density in humans are the hippocampus and putamen, followed by the caudate and cerebral cortex ($\sim$10–15% lower than highest density regions), and thalamus ($\sim$45–50% lower than highest density regions), and lowest expression in the cerebellum ($\sim$65% lower than highest density regions) (Daggett et al., 1995).

Signaling through mGlu5s can be potentiated by positive allosteric modulators (PAMs) and inhibited by negative allosteric modulators (NAMs) (Purgert et al., 2014). PAMs enhance the activation of receptors or prevent agonist-related desensitization, while NAMs can be either non-competitive receptor antagonists or inverse agonists. In recent years, allosteric modulation of mGlu5 has gained attention as a potential therapeutic target in the treatment of psychiatric disorders (Ritzén, Mathiesen, & Thomsen, 2005; Rocher et al., 2011). Drugs that target mGlu5 may bring about a therapeutic effect through secondary N-methyl-D-aspartate receptor (NMDAR) modulation, with scaffolding proteins such as Homer, Shank, and PSD-95 playing an important role in the physical proximity of these receptors at the postsynaptic density, as well as impacting signaling mechanisms of both mGlu5 and NMDAR (Piers et al., 2012; Tu et al., 1999). Moreover, activation of mGlu5 potentiates NMDAR activity while mGlu5 NAMs potentiate the effects of NMDAR antagonists (Attucci, Carlà, Mannaioni, & Moroni, 2001; Awad, Hubert, Smith, Levey, & Conn, 2000; Jin, Xue, Mao, & Wang, 2015; Pisani et al., 2001).

Here, we discuss the role of mGlu5 in psychiatric disorders, including mood disorders, trauma and anxiety disorders, as well as substance use disorders (nicotine, cannabis, and alcohol will specifically be discussed) providing information from preclinical animal models, postmortem, and clinical studies. Clinical investigations include magnetic resonance spectroscopy (MRS), but where possible, we specifically highlight positron emission tomography (PET) studies and results from drug trials that have targeted mGlu5. Further, we provide results from our recent investigations and, when conflicting, equivocal, or limited findings are presented, we present potential hypotheses that might unify results.

2. Quantifying mGlu5 with PET

PET can be used to quantify mGlu5 availability in vivo. (However, it is important to note that receptor function may not be directly related to availability). Current mGlu5 radioligands bind to a site in the transmembrane domain of mGlu5 and not to the N-terminal orthosteric site where glutamate binds (Abdallah et al., 2017; DeLorenzo et al., 2017). Predominantly used radioligands are [^{11}C]ABP688: (E)-3-((6-methylpyridin-2-yl)ethynyl) cyclohex-2-en-1-one-O-[^{11}C]methyloxime(Ametamey et al., 2006); and [^{18}F]FPEB: 3-[^{18}F]fluoro-5-[(pyridin-3-yl)ethynyl]benzonitrile (Patel et al., 2007). These two radioligands offer different advantages/advantages related to quantification techniques and half-life of the isotope (Abdallah, Mason, et al., 2017; DeLorenzo et al., 2017). Both appear to target cell surface receptors only and are not able to penetrate the cell membrane to bind to internalized receptors (Lin et al., 2015). As such, it is expected that mGlu5 levels measured with PET in vivo reflect the population of receptors at the cell surface available for neurotransmitter (or drug) binding (i.e., mGlu5 receptor *availability*), and does not account for perisynaptic, extrasynaptic, or internalized mGlu5 receptor pools (Lavialle et al., 2011; Lujan et al., 1996; Luján, Roberts, Shigemoto, Ohishi, & Somogyi, 1997).

In PET studies, volume of distribution (V_T, a ratio of radiotracer in the regions of interest to that in the plasma at equilibrium) is a commonly used outcome measure. However, V_T represents the combination of radiotracer specifically bound to target as well as non-specific (that is, "off-target") uptake. There is no evidence that non-specific uptake (which is the sum of non-displaceable binding and free/unbound radiotracer) differs in disease conditions from that of healthy controls, however, this has not been fully explored. Notably, calculating V_T necessitates blood sampling in order to measure metabolite-corrected plasma radiotracer concentrations. When blood sampling is not performed, the outcome measures commonly used are either the distribution volume ratio (DVR; the ratio of binding in the region(s) of interest relative to a reference region) or the binding potential (BP_{ND}; equal to DVR-1). The DVR and BP_{ND} calculations assume the reference region to be devoid of target, thus representing non-displaceable binding and free/unbound radiotracer. In mGlu5 imaging, a region devoid of mGlu5 does not exist in the human brain, and caution must be taken when evaluating DVR or BP_{ND} as the outcome measure.

3. Mood disorders
3.1 Major depressive disorder

There is mounting preclinical and clinical evidence of the role of glutamate in mood disorders, including major depressive disorder (MDD) (Chowdhury et al., 2016; Duman, Aghajanian, Sanacora, & Krystal, 2016; Gerhard, Wohleb, & Duman, 2016; Sanacora, Rothman, Mason, & Krystal, 2003; Sanacora, Treccani, & Popoli, 2012). In a number of animal models of depressive behavior, including chronic unpredictable mild stress(Mishra et al., 2020), chronic social defeat stress (Mishra, Kumar, Behar, & Patel, 2018; Veeraiah et al., 2014), and chronic forced swim stress (Li et al., 2008), lower levels of glutamate (as measured by ^{1}H MRS) and glutamate metabolism/neurotransmission (as measured by ^{13}C MRS) in the PFC have been reported. Similarly, a review of 16 ^{1}H MRS studies (281 patients and 301 healthy subjects) reported significantly lower glutamate and GLX (the combined signal of glutamate plus glutamine) levels in individuals with MDD (Luykx et al., 2012), although a smaller number of studies reported higher levels of glutamate in MDD (Abdallah, Mason, et al., 2017; Hashimoto, Sawa, & Iyo, 2007; Sanacora et al., 2004) including our own examination (Abdallah, Mason, et al., 2017). Historically, clinical ^{13}C MRS studies of neuronal energetics and glutamate neurotransmission are far less common than ^{1}H MRS studies of glutamate and GLX levels. A 2014 study (Abdallah et al., 2014) using ^{13}C MRS did not detect differences in glutamate neurotransmission at the occipital-parietal lobe between control and MDD individuals. Given the recent rise in techniques and modeling methods (Harris, Saleh, & Edden, 2017; Rothman et al., 2019), and our ability to collect ^{13}C MRS data in the frontal cortical regions, it is likely that future neurotransmission studies will enhance our understanding of glutamate's role in the etiology and treatment of mood disorders. For example, Abdallah and colleagues (Abdallah et al., 2018) demonstrated an increase in prefrontal cortex glutamate–glutamine cycling following the ketamine (a noncompetitive NMDAR antagonist) infusion, as well as a significant correlation between the ratio of ^{13}C glutamate/glutamine enrichments (a putative measure of neurotransmission strength) and ketamine-induced dissociative symptoms in a small group of participants (14 healthy adults and 7 participants with MDD). These clinical findings are in keeping with earlier studies employing similar ^{13}C MRS techniques to demonstrate that increased prefrontal glutamate cycling in is associated with the rapid onset of ketamine's antidepressant-like effects in rats (Chowdhury et al., 2012, 2016).

Similar to the variable findings regarding levels of glutamate and related markers—that is, reports of lower (Hasler et al., 2007; Luykx et al., 2012) and higher (Abdallah, Mason, et al., 2017; Hashimoto et al., 2007; Sanacora et al., 2004) levels—the directionality of mGlu5 availability in depression varies across studies. The first in vivo report of mGlu5 availability in MDD (using ^{11}C]ABP688 PET, MDD: $n = 11$, healthy subjects: $n = 11$) reported 10–20% lower mGlu5 availability in several brain regions using DVR (Deschwanden et al., 2011). We have previously reported differences in mGlu5 availability of a similar magnitude in MDD using the same tracer, a similar cohort size (MDD: $n = 13$, healthy subjects: $n = 13$) and the outcome measure V_T (Esterlis et al., 2018). More recently, in a larger sample and using ^{18}F]FPEB, we observed no difference in prefrontal mGlu5, as measured by V_T (MDD: $n = 29$, healthy subjects: $n = 29$), but did observe significant negative correlations between mGlu5 availability and mood symptoms, such that lower V_T was associated with total Profile of Mood State (POMS) scores (Davis et al., 2019). A ^{11}C]ABP688 PET study found prefrontal mGlu5 availability, as measured by BP_{ND}, to be lower in drug-naïve young adults with MDD ($n = 16$) vs healthy youth ($n = 15$) (Kim et al., 2019). In the case of geriatric depression, however, examination of mGlu5 availability using V_T or DVR, again using [^{11}C]ABP688 PET and a larger sample (MDD: $n = 20$, healthy subjects: $n = 22$), did not provide support for lower mGlu5 associated (DeLorenzo et al., 2015). Given that the currently available data from studies in healthy participants does not suggest age-related changes in mGlu5 in vivo in humans (DuBois et al., 2016; Mecca et al., 2017), the variability in the mGlu5 findings in MDD may be a reflection of the heterogeneity of the disorder, variability in comorbidity with anxiety-related disorders, the confound of chronicity of the disorder with age, or a reflection of the smaller sample sizes in the above-mentioned studies. Furthermore, the possibility of an age-by-diagnosis interaction is a possibility that has not been systematically evaluated. However, there is evidence suggesting an influence of sex or gender on PET measures of mGlu5. While earlier studies did not observe differences between male and female participants (DuBois et al., 2016), a 2019 study by Smart and colleagues (Smart et al., 2019) of 25 male and 49 female healthy adults found [^{11}C]ABP688 BP_{ND} to be significantly higher in men as compared with women, with whole-brain BP_{ND} being 17% higher in men. Interestingly, this study found no relationship between mGlu5 availability and menstrual phase the women. Therefore, sex could be contributing to variability in mGlu5 findings in MDD, and future studies should consider the possibility of differences in mGlu5 binding contributing to sex differences in MDD specifically, and neurocognitive function and neuropsychiatric disorders more generally.

Larger studies are needed to compare neurotransmission changes between healthy and depressed groups, the relationship between induction of glutamatergic cycling and mGlu5 availability, and in relation to changes in depressive symptoms. However, the relationship between glutamate levels, glutamate signaling, and mGlu5 availability is complex, and unraveling this relationship is made more complicated by the method of assessing each. While PET measures neural and glial mGlu5 levels in the perisynaptic or extrasynaptic space (Lavialle et al., 2011; Lujan et al., 1996, 1997), ^{1}H-MRS assesses the total amount of glutamate within a "box" (i.e., a voxel) containing neuronal tissue, non-neuronal tissue, and extracellular fluid (Blüml, 2013; Mason, 2003). Using [^{18}F]FPEB PET, we did not detect significant between-group differences in individuals with MDD ($n = 30$) vs healthy subjects ($n = 35$) in mGlu5 availability using V_T or DVR

(Abdallah, Mason, et al., 2017). However, addition of MRS allowed us to observe critical results—there was a negative association between mGlu5 availability and tissue glutamate, providing the first potential evidence in vivo for the hypothesized excitotoxicity of receptors under conditions of elevated glutamate levels (Popoli, Yan, McEwen, & Sanacora, 2012). As an alternate explanation, higher glutamate levels that can lead to mGlu5 internalization would result in lower mGlu5 available for radiotracer binding, in line with our work showing downregulation in mGlu5 upon rapid increases in glutamate (see explanation below and Fig. 1) (DeLorenzo et al., 2015; Esterlis et al., 2018). Although our in vivo demonstrates mGlu5 downregulation throughout the brain (including neocortex,

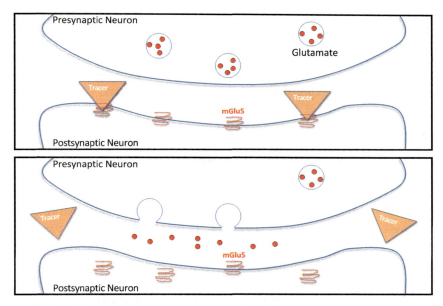

Fig. 1 mGlu5 availability is modulated with glutamate release in the synapse. Top: mGlu5 expressed at the cell surface with low levels of endogenous glutamate. The PET tracer is able to bind to the surface receptors. Bottom: Although the endogenous ligand, glutamate, and the PET tracer bind to different sites on mGlu5 (that is, the PET tracer binds to an allosteric site), when glutamate is released into the synapse, mGlu5 is downregulated/internalized, resulting in fewer receptors expressed on the cell surface. Because the PET tracer can only bind to receptors on the surface, tracer binding, and the PET signal, decreases. *Images adapted from Esterlis, I., Holmes, S. E., Sharma, P., Krystal, J. H., & DeLorenzo, C. (2018). Metabotropic glutamatergic receptor 5 and stress disorders: Knowledge gained from receptor imaging studies. Biological Psychiatry, 84(2), 95–105.*

striatum, hippocampus, and cerebellum) in response to a ketamine-induced glutamate challenge in both healthy (DeLorenzo, DellaGioia, et al., 2015) and depressed (Esterlis et al., 2018) human subjects, in vitro evaluations suggest regional differences in mGlu5 internalization related to endogenous differences in G protein-coupled receptor kinase 2 (GRK2) expression (Ribeiro et al., 2009).

While postmortem work is in line with the in vivo imaging studies suggesting lower or no differences in mGlu5 availability in MDD (Deschwanden et al., 2011; Fatemi, Folsom, Rooney, & Thuras, 2013; Matosin et al., 2014), preclinical studies repeatedly implicate reduced mGlu5 in the pathophysiology of depression (Palucha & Pilc, 2007). For example, mGlu5 knockout mice exhibit depressive-like behavior (Shin et al., 2015) and a number of studies (Kovacevic, Skelin, Minuzzi, Rosa-Neto, & Diksic, 2012; Wieronska et al., 2001), but not all (Mao & Wang, 2018), utilizing rat models of depression show lower mGlu5 protein and density in various brain regions. To our knowledge, no preclinical PET imaging studies have examined the relationships between mGlu5 availability and the expression of depressive-like behaviors.

3.2 Bipolar disorder

Glutamatergic dysfunction is also implicated in BD. In contrast to the majority of studies in MDD, higher Glx and glutamate levels are more commonly observed throughout the brain, particularly in frontal areas (dorsolateral prefrontal cortex and ACC) in BD (Cecil, DelBello, Morey, & Strakowski, 2002; Dager et al., 2004; Gigante et al., 2012; Sanacora et al., 2012; Taylor, 2014; Xu, Dydak, et al., 2013; Yüksel & Öngür, 2010). These constant findings of higher Glx and glutamate in BD may implicate and important difference in the pathophysiology of bipolar depression and MDD, or choice of regions examined. A particular challenge in evaluating studies and metanalyses, is BD groups are frequently inclusive of individuals across affective states that are not powered to model the effect of mood state on glutamate dynamics. Alternatively, studies may include just a single mood state (typically depressed), which would also miss dynamic glutamate fluctuations. Such fluctuation have been reported by Xu et al. (Xu, Dydak, et al., 2013; Xu, Zhu, et al., 2013), with higher glutamate levels in the left thalamus during depression and lower glutamate levels in the posterior cingulate cortex during mania. The chronicity/number of manic episodes has also been shown to be related to glutamate, with lower glutamate levels observed in the ACC of multi-episode relative to first episode BD subjects (Borgelt et al., 2019). However, these findings are difficult to interpret, given the above mentioned likely influence of mood state, as well as medication status (Scotti-Muzzi, Umla-Runge, & Soeiro-de-Souza, 2021).

Differences in affective state, medication status, and methodology further confound our understanding of the relationship between glutamate levels and mGlu5 in BD. Postmortem evaluation of mGlu5 demonstrated lower PFC (BA9) mRNA and protein expression as well as cerebellar mGlu5 protein expression in BD, as compared to MDD and controls (Fatemi et al., 2013). However, an autoradiography study did not observe significant differences in mGlu5 density in the ACC (Matosin et al., 2014). This discrepancy is likely influenced by the difficulty in accounting for mood state in postmortem work, as well as the difference in methodologies. Further, interestingly, a case report described symptoms of bipolar illness, including euphoria, impulsivity, psychomotor agitation and insomnia associated with deletion of the SHANK3 protein, a scaffold protein known to stabilize mGlu5 in the post-synaptic membrane of neurons (Fig. 2) (Guilmatre, Huguet, Delorme, & Bourgeron, 2014; Vucurovic et al., 2012). Further, in our recent PET study of mGlu5 availability (using $[^{18}F]$FPEB and V_T as the outcome measure) we found lower prefrontal mGlu5 availability in subjects with BD ($n = 17$ in a depressed state; $n = 10$ in a euthymic state) as compared with MDD ($n = 18$) and control groups ($n = 18$) (Holmes et al., 2023), adding further evidence of mGlu5 alterations in BD.

3.3 Treatment trials in mood disorders

Glutamatergic modulators with the greatest clinical efficacy in the treatment of MDD and BD to date are those targeting the ionotropic glutamate receptors (Henter, Park, & Zarate, 2021), specifically the noncompetitive NMDA receptor antagonist, ketamine [for recent metanalyses, see (Bahji, Zarate, & Vazquez, 2021; Kryst et al., 2020; McIntyre et al., 2020)], including in elderly patients (George et al., 2017). Ketamine leads to numerous neurochemical changes including a rapid surge in glutamate immediately following administration (Chowdhury et al., 2012, 2016; Moghaddam, Adams, Verma, & Daly, 1997). This glutamate surge increases AMPA receptor activation, which, coupled with ketamine's inhibition of extrasynaptic NMDA receptors, initiates and facilitates postsynaptic activation of

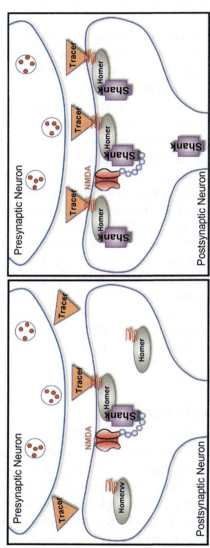

Fig. 2 mGlu5 radioligands bind to receptors only at the cell surface. Left: In a healthy individual, there is a balance of mGlu5 receptors expressed at the postsynaptic cell surface and internalized receptors. Right: In PTSD, we observed an increase in the expression of the scaffolding protein, Shank1, which helps anchor mGlu5 at the postsynaptic density. We hypothesize, despite no difference in mGlu5 protein expression between the PTSD and healthy brain, due to higher Shank1, there is a greater proportion of the total mGlu5 at the surface. Thus, quantification of mGlu5 via PET shows higher receptor density in PTSD. *Images adapted from Esterlis, I., DellaGioia, N., Pietrzak, R. H., Matuskey, D., Nabulsi, N., Abdallah, C. G., ... DeLorenzo, C. (2018). Ketamine-induced reduction in mGluR5 availability is associated with an antidepressant response: An [(11)C]ABP688 and PET imaging study in depression. Molecular Psychiatry, 23(4), 824–832. doi:10.1038/mp.2017.58.*

neuroplasticity-related signaling pathways, including those involving BDNF and mTOR (Abdallah et al., 2016; Lener et al., 2016; Li et al., 2010).

Modulation of mGlu5, via the glutamate surge, may work synergistically with other ketamine-induced changes to elicit the antidepressant response. We, therefore, established a PET paradigm to measure ketamine-induced mGlu5 changes (DeLorenzo, DellaGioia, et al., 2015). In this work, we found a downregulation of mGlu5 following ketamine [in both control subjects (DeLorenzo, DellaGioia, et al., 2015) and in subjects with MDD (Esterlis et al., 2018)], which we hypothesize to be related to the surge in glutamate (Fig. 1). This downregulation contributed to the antidepressant response (individuals with the greatest improvement in mood symptoms exhibited the greatest decrease in mGlu5 availability). Interestingly, reductions in receptor availability persisted 24 h after the ketamine infusion which may be among the persisting neuroadaptations that are associated with antidepressant effects beyond the glutamate surge (Duman et al., 2016). These sustained effects may rely on new protein synthesis and mTOR signaling pathways that lead to normalization of synaptic plasticity (Hou & Klann, 2004; Page et al., 2006) or pathways independent of mTOR (Yang et al., 2017). Although appearing counterintuitive given some postmortem and PET studies showing lower mGlu5 in individuals with MDD as compared with controls (Deschwanden et al., 2011; Fatemi & Folsom, 2014; Matosin et al., 2014), this substantial evidence from in vivo human studies suggests that mGlu5 availability does not differ as a function of MDD (Abdallah et al., 2017; DeLorenzo, Sovago, et al., 2015; Esterlis et al., 2022). Furthermore, the direction of ketamine-induced modulation needed for the antidepressant effect is consistent with preclinical studies in which mGlu5 antagonism and internalization that have also been associated with rapid, although not long lasting, antidepressant effects (Hughes et al., 2013; Li, Need, Baez, & Witkin, 2006; Palucha et al., 2005). It should also be noted that contrary to depression, studies related to anxiety show higher mGlu5 availability (preclinical and human), and mGlu5 agents have robust anxiolytic properties. Thus, given that our ketamine work in MDD shows most robust reductions with anxiety-related symptoms associated with mGlu5 downregulation, this work appears to be in line with previously published studies.

Preclinical evidence suggests that a metabolite of ketamine, (2S,6S;2R,6R)-hydroxynorketamine (HNK) may be necessary and sufficient for ketamine's antidepressant response (Zanos et al., 2016). The metabolite pathway is independent of NMDA receptors, acting through AMPA receptors to activate BDNF and mTOR pathways. Interestingly, even this independent pathway could affect mGlu5 downstream. Evidence from other studies suggests that AMPA receptor activation leads to changes in the mGlu5 signaling and facilitation of downstream effects (Kim et al., 2015). The eukaryotic elongation factor 2 (eEF2), part of the mGluR-eEF2-AMPAR pathway (Meyers et al., 2015), also effects plasticity independent of mTOR and has been implicated in the antidepressant actions of ketamine (Monteggia, Gideons, & Kavalali, 2013). Thus, it appears that mGlu5 alteration might be common to several mechanisms by which ketamine leads to rapid antidepressant effects.

Based on the above evidence, we propose that the downstream effects of ketamine may be enhanced by mGlu5 to produce the antidepressant response. Specifically, the activation of mGlu5 activates the inositol phosphate signaling cascade, raising levels of IP3. mGlu5 effects are linked to the endoplasmic reticulum via Homer and Shank (see Fig. 2), where IP3 causes the release of intracellular calcium stores, producing a variety of downstream effects (Sala, Roussignol, Meldolesi, & Fagni, 2005) including activation of mTOR (Li et al., 2010).

Metanalyses have also shown ketamine to be effective in bipolar depression, though as an augmentation therapy (Bahji et al., 2021; Lener, Kadriu, & Zarate, 2017). A randomized, placebo-controlled, double-blind, crossover, add-on study demonstrated that individuals with BD who were maintained on lithium or valproate and received an intravenous infusion of ketamine hydrochloride (0.5 mg/kg) showed an improvement in depressive symptoms as compared to those administered placebo (Zarate et al., 2012). That finding was recently replicated in a cohort of treatment–resistant BD patients (Lener et al., 2017).

Clinical trials of mGlu5-specific modulators in MDD have been undertaken, but thus far have failed to achieve the same efficacy as ketamine. A 6-week, phase II trial of AZD2066 (an mGlu5 antagonist) in 131 individuals with MDD reported that monotherapy (12–18 mg/day) with this drug was not superior to the SNRI duloxetine or placebo (AstraZeneca, 2012). Similarly, a phase IIb double blind randomized clinical trial of Basimglurant (an mGlu5 NAM with a modified release formulation) conducted in 333 adult patients with MDD in found at 0.5–1.5 mg in combination with ongoing antidepressant treatment showed no difference from placebo in the primary outcome measure of change in clinician-assessed depressive symptoms. However, the higher dose did induce greater improvements in patient-rated end points (Quiroz et al., 2016). The basimglurant trial highlights the importance of investigating mGlu5 modulators as a possible adjunct therapy and the consideration of dose, important factors in the future evaluation of these medications.

While mGlu5-specific compounds have failed to achieve the same anti-depressant efficacy as ketamine; this may be in part due to the synergistic effects of ketamine at targets other than the NMDAR and/or due to the level of downregulation of mGlu5 induced by ketamine treatment. For example, we estimated a $14 \pm 10\%$ decrease in surface expression of mGlu5 following ketamine in a depressed cohort (DeLorenzo, DellaGioia, et al., 2015). More recently, we additionally demonstrated a significant relationship between the decrease in depressive symptoms the reduction in mGlu5 availability immediately following ketamine infusion, leading us to hypothesize that ketamine-induced glutamate release moderates mGlu5 availability, which in turn appears to be related to antidepressant efficacy (Esterlis et al., 2018). Though multiple trials of mGlu5 specific drugs have been undertaken, none to date have evaluated the level of mGlu5 downregulation or receptor occupancy needed for clinical efficacy (which can both be done with PET). Thus, many important questions regarding dose–response, and related issues of optimal dosing rout and timing remain unclear. Studies on intermittent vs daily dosing are needed for glutamatergic agents, which will depend on route of delivery (Iadarola et al., 2015; Murrough, Abdallah, & Mathew, 2017).

4. Anxiety and trauma disorders

There is a paucity of evidence from preclinical and human studies implicating mGlu5 in anxiety, as well as obsessive-compulsive disorder (OCD) and post-traumatic stress disorder (PTSD), which were both formerly categorized as anxiety disorders before the introduction of the DSM-5 (American Psychiatric Association, 2000, 2013). In preclinical investigations, negative allosteric modulation of mGlu5 has shown promising anxiolytic effects. A range of behavioral paradigms (novelty suppressed feeding, elevated-plus maze, social interaction test, fear-potentiated startle, marble burying behavior, etc.) have been used to confirm the anxiolytic activity of mGlu5 NAMs in rodents, including MPEP (Ballard et al., 2005; Brodkin, Busse, Sukoff, & Varney, 2002; Iijima, Fukumoto, & Chaki, 2012; Mikulecká & Mares, 2009; Pérez de la Mora et al., 2006; Spooren & Gasparini, 2004; Varty et al., 2005), MTEP (Varty et al., 2005), fenobam (Porter et al., 2005), CTEP (Lindemann et al., 2011;

Peterlik et al., 2017), and others (Bates et al., 2014; Roppe et al., 2004). Together with the observation that the anxiolytic efficacy activity of MTEP was not diminished when co-administered with benzodiazepine antagonist, flumazenil, and a synergistic effect of non-effective MTEP doses was observed when combined with diazepam, suggests that mGlu5 antagonism may work independent of the GABA–benzodiazepine (BZD) receptor complex to elicit anxiolytic effects (Klodzinska et al., 2004). Furthermore, social isolation stress in mice is associated with increased mGlu5 expression and mGlu5-mediated neuronal excitability in the basolateral amygdala, and induces anxiety-like behaviors that can be rescued by MPEP administration (Lin et al., 2018).

Consistent with these preclinical findings, our clinical $[^{11}C]$ABP688 PET study (Esterlis et al., 2018) suggests that mGlu5 downregulation (decrease in V_T) is associated with reductions in anxiety symptoms among individuals with MDD. However, baseline anxiety symptoms in MDD (Deschwanden et al., 2011) were previously found to be negatively correlated with $[^{11}C]$ABP688 DVR. Furthermore, we also observed a negative correlation between trait anxiety and baseline mGlu5 availability, as measured by $[^{18}F]$FPEB V_T among individuals with MDD (but, interestingly, the negative correlation was not significant for the BD group) (Holmes et al., 2023). This supports the preclinical literature which suggests mGlu5 involvement in anxiety.

4.1 Posttraumatic stress disorder (PTSD)

There is overwhelming evidence that mGlu5 is a key modulator of synaptic plasticity within the corticolimbic circuitry that supports associative and adaptive learning (Fontanez-Nuin, Santini, Quirk, & Porter, 2011; Sethna & Wang, 2014, 2016; Simonyi, Schachtman, & Christoffersen, 2005; Xu, Zhu, Contractor, & Heinemann, 2009). More specifically, preclinical fear learning and fear extinction models of PTSD-relevant phenotypes implicate mGlu5 involvement (see Section 4.3, for an additional discussion of relevant preclinical findings). Furthermore, our recent work combining longitudinal in vivo $[^{18}F]$FPEB PET with a preclinical model of stress responsivity and fear learning indicates increased mGlu5 availability as a potential mediator of heightened behavioral responses and fear memory following to acute foot-shock stress in rats (Asch et al., 2022). However, it should also be noted that others have shown higher medial PFC and amygdala mGlu5 transcription levels (mRNA) post mortem in male rats exhibiting *resilience* to the predator scent stress model of PTSD as compared with susceptible or non-stressed control males (Schwendt et al., 2018; Shallcross, Wu, Wilkinson, Knackstedt, & Schwendt, 2021). Interestingly, in female rats, *susceptibility* to the predator scent stress model was associated with higher mGlu5 mRNA expression in the prelimbic PFC, basolateral amygdala, and nucleus reuniens (Blount, Dee, Wu, Schwendt, & Knackstedt, 2023). Collectively these findings highlight the complex role of mGlu5 receptor dynamics in fear learning and anxiety-like behavior, the regional and temporal specificity, and interactions with sex.

The increasing number of clinical neuroimaging studies have provided insight as to the role of glutamate, and mGlu5 specifically, in PTSD. However, studies utilizing MRS to measure glutamate, glutamine, or Glx levels among individuals with PTSD have yielded mixed results. Two studies report higher glutamate in the temporal cortex in subjects with PTSD relative to trauma exposed controls (Meyerhoff, Mon, Metzler, & Neylan, 2014; Pennington, Abé, Batki, & Johannes Meyerhoff, 2014). For the ACC, one study found lower Glx in the ACC of youth with PTSD relative to both participants in remission and trauma-exposed controls (Yang et al., 2015); but the majority of studies report no differences in ACC Glu or Glx levels (Meyerhoff et al., 2014; Michels et al., 2014; Pennington et al., 2014). In addition, a 2017 study by Harnett and colleagues found higher ACC Glx levels to be associated with greater acute and persistent posttraumatic diagnosis scale scores among trauma-exposed individuals ($n = 21$). Notably for this study, none of the trauma-exposed subjects met criteria for a clinical diagnosis, although there were no between-group differences in ACC metabolites relative to the no-trauma comparison group ($n = 18$) (Harnett et al., 2017). As mentioned above, it is only with recent technological advances that we have started to see clinical studies utilizing ^{13}C MRS to probe glutamate neurotransmission and neuroenergetics (Harris et al., 2017; Rothman et al., 2019). To data, the single reported clinical investigation of neuroenergetics in PTSD using ^{13}C-acetate MRS to estimate rates of neuronal oxidative energy production (V_{TCAn}) and glutamate cycling (V_{Cycle}), and their ratio (i.e., V_{TCAn}/V_{Cycle}, a putative biomarker related to glutamatergic synaptic strength) found a 28% reduction in the V_{TCAn}/V_{Cycle} ratio in the prefrontal cortex (Brodmann Area 10) of the subjects with PTSD as compared with healthy controls (Averill et al., 2022). Future multimodal neuroimaging studies will be needed to determine relationships between regional glutamate levels, glutamate metabolism, synaptic strength, mGlu5 availability, and PTSD symptoms and/or treatment response.

Regarding human subjects research using [^{18}F]FPEB PET, our group observed 20% higher mGlu5 availability (measured by V_T) in 16 individuals with PTSD compared to controls, and the extent of mGlu5 dysregulation was related to greater avoidance symptoms, suggesting that targeting mGlu5 in PTSD might be particularly beneficial to treating this symptom domain (Holmes et al., 2017). In a subsequent larger study, we again observed significantly higher mGlu5 availability, as measured by [^{18}F] FPEB V_T, in PTSD ($n=29$) relative to both HC ($n=29$) and MDD ($n=29$) individuals in frontolimbic brain regions (Davis et al., 2019). The greatest apparent upregulation was observed in the PTSD group with current suicide ideation (SI), as compared to PTSD with no SI, and MDD with/out SI. Moreover, interestingly, we observed opposite associations between mood symptoms and mGlu5 availability in MDD and PTSD groups. Whereas in the PTSD group, higher mGlu5 availability was associated with greater symptomatology, in the MDD group, lower mGlu5 availability was associated with such. Importantly, given the paucity of novel medication development specifically for PTSD, this work highlights the potential utility of mGlu5 as a biomarker and treatment target within for PTSD.

To help with the interpretation of in vivo findings, we conducted a postmortem examination in collaboration with the late Dr. Ron Duman: in samples of 19 individuals with PTSD and 19 matched controls, we found no differences in mGlu5 availability; however, we observed a 3.8-fold higher Shank1 and 3.5-fold lower FKBP5 (glucocorticoid regulating protein) in the PTSD group (Holmes et al., 2017). Shank1 acts as a scaffold, anchoring mGlu5 at the postsynaptic density (Tu et al., 1999), thus the observed PET finding of increased mGlu5 availability (i.e., cell surface expression) could in part reflect an upregulation of Shank1 (Fig. 2). Further, given the glucocorticoid system appears to interact with glutamatergic processes, including the observation of increased cortisol levels downregulating mGlu5 (Iyo et al., 2010), our postmortem observation could also be due to the chronic hypocortisol experience in PTSD. We propose that this change in mGlu5 localization might have an impact on psychological functioning.

4.2 Obsessive compulsive disorder (OCD)

Similarly, preclinical studies indicate promising anxiolytic effects of mGlu5 modulation. In rodent models of OCD, mGlu5 antagonists reduce stereotyped, repetitive behaviors that may be relevant to OCD (Bhattacharyya & Chakraborty, 2007; Mehta, Gandal, & Siegel, 2011; Pittenger, Krystal, & Coric, 2006; Spooren et al., 2000). Further, previous work indicates that when Sapap3 (also DLGAP3 or GKAP3), a postsynaptic scaffold protein that is involved in synaptic organization and neuronal cell signaling, is knocked-out in mice, they develop OCD-relevant phenotypes such as excessive self-grooming and anxiety-like behaviors, as well as defects in cortico-striatal synapses (Wan et al., 2014; Welch et al., 2007). Sapap3 KO mice also exhibit increased expression and activity of mGlu5, likely due to Sapap3 interaction with Shank and PSD-95. Interestingly, a recent study showed that chronic but not acute quinpirole administration, used to induce the quinpirole sensitization rat model for OCD, is associated with significant occupancy of mGlu5 (as measured by $[^{11}C]$ABP688) (Servaes, Glorie, Verhaeghe, Stroobants, & Staelens, 2017).

In the one $[^{11}C]$ABP688PET study investigating the mGlu5 in OCD in vivo in humans, no differences in mGlu5 availability (DVR) between 10 individuals with OCD and 10 matched controls were reported (Akkus et al., 2014). However, there was a positive correlation between mGlu5 availability and obsessive symptoms measured using Yale Brown Obsessive Compulsive Scale (Y-BOCS). These findings suggest that obsessions in particular may have an underlying glutamatergic pathology related to mGlu5, but further work is required.

4.3 Treatment trials in anxiety and trauma disorders

In a double-blind, placebo-controlled clinical trial, fenobam was comparable to anxiolytic diazepam with respect to its efficacy and onset of action, and was also shown to be useful in outpatients with severe anxiety symptoms (Pecknold, McClure, Appeltauer, Wrzesinski, & Allan, 1982). Although fenobam was considered promising agent in the treatment of anxiety with a good safety profile as early as in 1980, it was associated with troublesome side effects of a psychostimulant nature (Friedmann, Davis, Ciccone, & Rubin, 1980). Similarly, mGlu5 NAMs have not been successful thus far as clinical therapeutics, in part because of the induction of side effects, including psychotomimetic activity (Pecknold et al., 1982; Porter et al., 2005).

To our knowledge, there have been no treatment trials in humans using potential mGlu5 targets in the treatment of PTSD. However, mGlu5 activation has been shown to contribute to stress-induced fear conditioning (Tronson et al., 2010), mGlu5 antagonism using MPEP blocks the acquisition and expression of conditioned fear (Schulz et al., 2001), and ketamine (at 10 mg/kg but not 20 mg/kg) impaired fear memory reconsolidation (Duclot, Perez-Taboada, Wright, & Kabbaj, 2016), all suggesting that mGlu5 may play some role in trauma-related memories in PTSD. Although ketamine has many effects in addition to the recently implicated mGlu5 downregulation that likely contribute to its efficacy. Alternatively, MPEP has also been shown to prevent the extinction of contextual fear memory in mice (Sethna & Wang, 2016; Slattery, Neumann, Flor, & Zoicas, 2017) and rats (Fontanez-Nuin et al., 2011; Shallcross et al., 2021). Another study found that fear extinction was completely abolished in mGlu5 KO mice (Xu et al., 2009), suggesting that mGlu5 is required for adaptive learning in the context of fear extinction. Furthermore, administration of ADX47273 (a selective and potent mGlu5 PAM) enhanced extinction learning and improved reversal learning (Xu, Dydak, et al., 2013; Xu, Zhu, et al., 2013). Similarly, rodents identified as high novelty seekers [which is associated with higher mGlu5 V_T in humans (Leurquin-Sterk et al., 2016)] required fewer sessions to achieve fear extinction (Duclot et al., 2016) although equivocal evidence exists, as preclinical study related reduced mGlu5 in cortical neurons (due to ablation) with novelty seeking (Jew et al., 2013).

The above indicates differing roles for mGlu5 in fear-related processing relevant to PTSD. On the one hand, mGlu5 activity might underlie fear conditioning, and blocking mGlu5 disrupts the acquisition and expression of conditioned fear. However, blockade or knockout of mGlu5 also seems to interfere with fear extinction. Therefore, targeting the mGlu5 with an antagonist or NAM might reduce distress associated with PTSD, but might prolong the disorder through interference with fear extinction. Conversely, positive modulation of mGlu5 receptor signaling (agonist or PAM) might be useful in facilitating fear extinction learning, but might also provoke re-experiencing symptoms/reinstatement. A recent trial in mice partially validated this hypothesis by showing ketamine, administered prophylactically, reduced fear expression following fear conditioning; however, it was ineffective when administered following the conditioning or during extinction (McGowan et al., 2017). This may be similar to what is seen with benzodiazepines in treating PTSD, where these medications can have initial anxiolytic effects but are generally ineffective long-term (Guina, Rossetter, DeRHODES, Nahhas, & Welton, 2015). Further, this suggests that more study of mGlu5 is warranted, as the upregulation of surface mGlu5 we report in PTSD may exist prior to onset of the disorder, enhancing the potential for fear conditioning, as suggested by the rodent literature (consistent with an mGlu5 NAM or antagonist being effective in prevention of fear learning). Studies are under way in our group to examine some of these effects.

In the first reported randomized controlled trial for ketamine (Feder et al., 2014) (double-blind, crossover design in 41 individuals with chronic PTSD), a single intravenous infusion of ketamine (0.5 mg/kg) resulted in the significant reduction of PTSD symptoms, as measured 24-h after dosing, whereas the psychoactive control (midazolam 0.045 mg/kg) did not. However, the antidepressant effect did not persist to the 1-week post treatment timepoint. Thus, in subsequent study, the researchers tested the efficacy of 6 doses delivered over 2 weeks (Feder et al., 2021), finding that subjects receiving ketamine ($n=15$) greater symptom improvement relative to the midazolam group ($n=15$), with the median ketamine response lasting 27.5 days after the final dose. Although mGlu5 was likely downregulated upon ketamine administration, the cascade of other downstream effects associated with ketamine administration may have rescued the blockade of fear extinction. Further studies with mGlu5-specific medications are warranted.

In OCD, a small crossover double-blind study of a single ketamine dose was performed (Rodriguez et al., 2013). Significant decreases in OCD symptoms (in individuals meeting diagnostic criteria for OCD but not for comorbid MDD), were observed, with 50% of individuals showing treatment response 1-week post dose. This suggests that ketamine may have an anti-OCD effect regardless of potential improvements in depressive symptoms. However, a recent study of mavoglurant (an mGlu5 antagonist) as an add-on treatment to SSRIs in patients with primary OCD was terminated due to lack of efficacy compared to placebo (Rutrick et al., 2017). Further study is therefore warranted as potential benefits of mGlu5 modulation in preclinical models of OCD. For example, one study performed in Sapap3 KO model demonstrated that excessive mGlu5 signaling underlies OCD-like behaviors in these mice, and that a single systemic injection of the NAM MTEP rapidly reduced grooming activity in the Sapap3 KO mice but not wild-types (Ade et al., 2016), indicating that targeting mGlu5 can mitigate OCD-relevant behaviors in mice. Similarly, administration of the mGlu5 antagonist MPEP significantly reduced quinpirole-induced compulsive-like checking behavior in rats (Gök, Murat Demet, & Öztürk, 2017).

5. Substance use: Nicotine, cannabis, and alcohol

In this section on substance use disorder, we focus on the *legal* neuromodulatory substances nicotine, cannabis, and alcohol and the role of mGlu5. It is important to note that mGlu5 has also been implicated in the use and addiction to illicit drugs, specifically the psychostimulants cocaine and methamphetamine. However, rates of overall use and problematic use are substantially higher for alcohol, nicotine, and cannabis than they are for cocaine and methamphetamine (listed in order of highest to lowest rates of use in the United States among people age 12 and older in 2020) (Samhsa, 2021). Furthermore, mGlu5 action and the potential of mGlu5 modulation as a treatment within the context of cocaine (Hadizadeh et al., 2022; Niedzielska-Andres et al., 2021) and methamphetamine (Petzold, Szumlinski, & London, 2021) use have recently been reviewed.

5.1 Nicotine use disorder

Nicotine is among the most widely consumed psychoactive drugs in the world (Anderson, 2006; Cornelius, Wang, Jamal, Loretan, & Neff, 2020). Nicotine binds directly to nicotinic acetylcholine receptors (nAChRs), resulting in receptor activation and desensitization which both contribute to behavioral responses relevant to nicotine addiction and mood (Brody et al., 2006; Picciotto, Addy, Mineur, & Brunzell, 2008; Wooltorton, Pidoplichko, Broide, & Dani, 2003). Importantly, preclinical studies have demonstrated nicotine activation of nAChRs increases glutamatergic neurotransmission, particularly in the ventral tegmental area (VTA) (Mansvelder, Keath, & McGehee, 2002; Schilström et al., 2000) and nucleus accumbens (NAc) (Reid, Fox, Ho, & Berger, 2000; Toth, Vizi, & Lajtha, 1993), key regions in the mesocorticolimbic circuitry that has been extensively implicated in addiction. Furthermore, there is a growing body of evidence suggesting that alterations in glutamatergic neurotransmission, including mGlu5 signaling, play a significant role in pathophysiology of addiction (Akkus et al., 2020; Byrnes et al., 2009; Terbeck, Akkus, Chesterman, & Hasler, 2015), including nicotine specifically (Akkus et al., 2020, 2016; Paterson, Semenova, Gasparini, & Markou, 2003; Yararbas, Keser, Kanit, & Pogun, 2010).

There is abundant preclinical evidence that the reinforcing properties of nicotine are contingent on the activation of dopaminergic VTA projections to NAc (Balfour, 2015; Corrigall, Coen, & Adamson, 1994). However, nicotine-induced dopamine release can be completely blocked by pretreatment with an mGlu5 antagonist (MPEP, 5 mg/kg) (Tronci & Balfour, 2011). Additionally, MPEP treatment has been shown to reduce the rate of nicotine self-administration in rats (Tronci, Vronskaya, Montgomery, Mura, & Balfour, 2010). Similarly, MPEP administration attenuates expression of nicotine-induced conditioned place preference (CPP); interestingly, this study found that nicotine resulted in CPP in male, but not female rats (Yararbas et al., 2010). Overall, these data are consistent with the hypothesis that mGlu5 antagonism may attenuate reinforcing properties of nicotine via attenuation of mesocorticolimbic dopamine signaling (Chiamulera, Marzo, & Balfour, 2017).

A role for mGlu5 in nicotine use and addiction is further supported by human studies examining mGlu5 availability in nicotine users. Studies by Akkus and colleagues found 20% lower mGlu5 availability in smokers relative to non-smokers using $[^{11}C]ABP688$ PET, and DVR as the outcome measure (Akkus et al., 2013), with mGlu5 levels seemingly "recovering" following extended abstinence (>1.5 years) (Akkus et al., 2016). Recent work by our group further suggests differential relationships between mGlu5, executive function, and mood symptoms among nicotine users, such that psychiatric diagnosis had lower mGlu5 availability compared with controls, in line with the previous findings of Akkus and colleagues (Akkus et al., 2013, 2016), while nicotine users with PTSD had higher mGlu5 availability relative to non-users with PTSD (Baldassarri, et al., under revision). Furthermore, the relationships between mGlu5 and measures of nicotine cravings varied as a function of psychiatric diagnosis: nicotine users without other psychiatric disorders showed a negative correlation between mGlu5 and cravings (i.e., lower mGlu5 availability was associated with greater craving and relief from withdrawal symptoms), but the opposite correlation was observed among nicotine users carrying a PTSD diagnosis. In combination with preclinical findings, results from human imaging studies implicate mGlu5 in the pathology of nicotine addiction. Given the differential associations between mGlu5 and nicotine use as a function of comorbid psychiatric diagnosis, further study of mGlu5 as a potential treatment target in specific populations with nicotine addiction is warranted.

5.2 Cannabis use disorder (CUD)

CUD is broadly defined as the inability to stop consuming cannabis even when it is causing physical or psychological harm (American Psychiatric Association, 2013). With legalization of medical and recreational cannabis making this addictive drug more readily available, there is concern for increasing rates of cannabis use and CUD. Indeed, between 2002 and 2014 in the United States, past-month cannabis use among individuals 18–25 years increased from 17.3% to 19.6% (Azofeifa et al., 2016), and from 4.0% to 6.6% in those ≥26 years (Compton, Han, Jones, Blanco, & Hughes, 2016), with an estimated 10–30% of users becoming addicted or qualifying for CUD diagnosis (Hasin et al., 2015; Lopez-Quintero et al., 2011).

While the effects of chronic cannabis use and the transition from use to addiction and CUD have not been fully elucidated in humans, converging preclinical research indicates that acute cannabinoid exposure induces a dose-dependent increase in cortical and striatal extracellular glutamate (Galanopoulos, Polissidis, Papadopoulou-Daifoti, Nomikos, & Antoniou, 2011; Pistis et al., 2002), and chronic cannabis exposure can have a lasting impact on glutamatergic synapse development and homeostasis (Curran et al., 2016; Rubino et al., 2015). Furthermore, there is strong evidence implicating cannabinoid exposure alters glutamate homeostasis via mechanisms involving mGlu5 (Colizzi, McGuire, Pertwee, & Bhattacharyya, 2016; Cristino, Bisogno, & Di Marzo, 2020). Specifically, mGlu5 and the cannabinoid type 1 (CB1) receptor exhibit a close biochemical relationship: activation of postsynaptic mGlu5 enhances the synthesis of endogenous cannabinoids, namely 2-AG (Ohno-Shosaku, Shosaku, Tsubokawa, & Kano, 2002; Varma, Carlson, Ledent, & Alger, 2001). Activation of presynaptic CB1 receptors by (endogenous or exogenous) cannabinoids leads to decreased Ca2+ channel activity, thereby limiting glutamate release and downstream postsynaptic mGlu5 activation (Marsicano et al., 2003; Olmo, Ferreira-Vieira, & Ribeiro, 2016; Shen, Piser, Seybold, & Thayer, 1996; Wilson & Nicoll, 2001). Therefore, glutamate homeostasis is typically maintained via this presynaptic CB1/postsynaptic mGlu5 negative feed-back loop that is predicted to be disrupted by the consumption of exogenous cannabinoids (Colizzi et al., 2016).

The metabotropic glutamate receptor 5 as a biomarker

Studies of glutamatergic alterations in human populations have mixed results. Using ^{1}H-MRS, some studies have found lower glutamate in the ACC of adolescent cannabis users (Prescot, Locatelli, Renshaw, & Yurgelun-Todd, 2011), but not adults (Newman et al., 2019). However, a number of studies looking at other brain regions (hippocampus, parietal lobe, temporal lobe or frontal white matter) have found no differences in glutamate or glutamate metabolites between cannabis users and control subjects (Blest-Hopley et al., 2020; Muetzel et al., 2013; Silveri, Jensen, Rosso, Sneider, & Yurgelun-Todd, 2011; van de Giessen et al., 2017). Interestingly, a recent study by Watts and colleagues (Watts et al., 2020) did not detect metabolite difference between cannabis users and controls, but they did observe a positive correlation between past-year cannabis exposure (grams) and ACC glutamate and Glx in male, but not female, users. Thus, the effects of chronic cannabis use on glutamate and glutamate metabolism may be regionally specific and differ between males and females.

To date, no clinical PET studies have been published that examine mGlu5 availability within the context of cannabis use and CUD (although work in our lab is underway aimed at investigating this question). However, a recent investigation of mGlu5 availability (using $[^{11}C]$ABP688 BP_{ND}) in youth at risk for addiction, observed lower mGlu5 availability in the OFC, insula, and striatum specifically in the high-risk individuals that additionally reported high cannabis use (Cox et al., 2020). Significant work and longitudinal studies are needed to fully describe the role of mGlu5 in the transition from cannabis use to addiction, if and when receptor availability normalizes after a period of abstinence, and weather mGlu5 availability is related to relevant clinical measures.

5.3 Alcohol use disorder (AUD)

AUD is a major public health concern that affects over 10% of the US population, accruing great cost to individuals and society. Glutamate systems are altered by chronic alcohol use (Olive, Cleva, Kalivas, & Malcolm, 2012), including upregulation of NMDA receptors (Nelson, Ur, & Gruol, 2005) and excessive excitatory signaling and glutamate levels during withdrawal (Umhau et al., 2010). Evidence from preclinical studies consistently implicates a role for mGlu5 in behaviors relevant to AUD. Specifically, mGlu5 negative allosteric modulation, antagonism, or deletion have been shown to reduce alcohol self-administration (Besheer et al., 2010; Cowen, Djouma, & Lawrence, 2005; Gupta et al., 2008; Schroeder,

Overstreet, & Hodge, 2005) and diminish reinstatement (Gass et al., 2014; Sidhpura, Weiss, & Martin-Fardon, 2010; Sinclair, Cleva, Hood, Olive, & Gass, 2012). A 2019 study by de Laat and colleagues (de Laat et al., 2019) examining glutamate (using [^{1}H]MRS) and mGlu5 (using [^{18}F]FPEB) found that rats with higher NAc mGlu5 availability at baseline showed greater alcohol preference than those with lower baseline mGlu5; furthermore, exposure to alcohol reduced prefrontal glutamate, which then recovered to baseline levels after 8 days of alcohol withdrawal. Given these preclinical findings, mGlu5 negative allosteric modulation is a particularly attractive mechanism for AUD treatments that has motivated human laboratory studies evaluating safety and tolerability (Haass-Koffler et al., 2017) and alcohol effects on pharmacokinetics (Haass-Koffler et al., 2018) of negative allosteric modulators.

Evaluation of glutamate levels in clinical populations have yielded more variable results in terms of the directionality of findings. While most MRS studies in people with AUD report elevated extracellular glutamate levels (Hermann et al., 2012; Joffe, Winder, & Conn, 2021; Tsai et al., 1998; Umhau et al., 2010; Varodayan, Sidhu, Kreifeldt, Roberto, & Contet, 2018), a smaller number of studies have found lower glutamate signal in AUD (Mon, Durazzo, & Meyerhoff, 2012; Thoma et al., 2011). A recent meta-analysis of brain metabolite alterations related to alcohol use found no significant group differences for glutamate, glutamine, or Glx; the authors of this analysis cite demographic heterogeneity (particularly in the selection of participants in different phases of alcohol consumption, withdrawal, recovery, and relapse) and methodological discrepancies (magnet strength, pulse sequence, etc.) as likely contributors to the non-significant findings (Kirkland et al., 2022).

In clinical studies, multiple lines of evidence implicate associations between mGlu5 and alcohol use. First, genetic variations in mGlu5 are associated with AUD (Schumann et al., 2008). Second, studies using autoradiography in human *post-mortem* brain tissue have reported higher mGlu1/5 receptor binding in the brains of people with AUD as compared with controls (Kupila et al., 2013; Laukkanen, Kärkkäinen, Kautiainen, Tiihonen, & Storvik, 2019). Third, in vivo PET imaging studies have provided mixed results, not dissimilar from the variable [^{1}H]MRS findings in AUD. Two studies reported higher mGlu5 availability in cortical brain regions of people with AUD after 1–2 weeks abstinence compared to controls. The first used the radioligand [^{11}C]ABP688 in a completely nonsmoking population (Akkus et al., 2018), while the second larger study

The metabotropic glutamate receptor 5 as a biomarker   289

used [^{18}F]FPEB while carefully controlling for smoking status in participant recruitment and statistical analysis (Hillmer et al., 2021). Furthermore, in the latter study, there was no evidence for significant changes in mGlu5 availability over time (Hillmer et al., 2021). In contrast, a larger study using the radioligand [^{18}F]FPEB reported lower mGlu5 availability in limbic brain regions compared to controls (Leurquin-Sterk et al., 2018), which recovered after 6 months abstinence (Ceccarini et al., 2019). However, these later studied did not control for tobacco smoking, suggesting a potential interaction between tobacco smoking status and AUD status regarding mGlu5 availability during recovery, with further work being required to confirm this possibility. Overall, the human evidence for higher mGlu5 availability in people with AUD is consistent with extensive preclinical research, but careful attention to tobacco smoking status is needed for human studies.

5.4 Treatment trials in substance use disorders

For nicotine use and addiction, preclinical studies, such as those noted above, and as reviewed by Chiamulera et al. (Chiamulera et al., 2017), provide unequivocal evidence that mGlu5s play a pivotal role in the development and expression of nicotine dependence and in drug- and cue-induced relapse following extinction/abstinence. For clinical trials, Novartis's mGlu5 antagonist mavoglurant (AFQ056) has been tested in individuals with moderate to heavy degree of nicotine dependence, but with no expressed intention of quitting, and a primary outcome measure of self-report ratings of craving after 3 days of voluntary abstinence (Novartis, 2007). The study commenced in 2006 and completed recruitment, but no results have been disclosed (https://clinicaltrials.gov/ct2/show/study/NCT00414752, accessed July 2022). It is presumed the trial did not provide positive results. It has been postulated, given the observation that mGlu5 antagonism during early nicotine withdrawal may worsen reward processing impairment (Liechti & Markou, 2007), decreasing the negative affective state experienced during withdrawal is likely key, and could be a contributing factor as to why the mavoglurant trial was not successful (Barnes, Sheffler, Semenova, Cosford, & Bespalov, 2018).

While there are no glutamate pharmacotherapies currently approved for the treatment of AUD, a number of clinical treatment trials for AUD have looked at the efficacy of glutamate system modulators, including mGlu5 negative modulation. Ifenprodil is an NMDA receptor antagonist (although also acting as an inhibitor of G protein-coupled inwardly-rectifying potassium (GIRK) channels) approved for use in some countries as a neuroprotectant following brain infarction or hemorrhage (Morley et al., 2021; Sun et al., 2021). One randomized controlled trial in participants with alcohol dependence investigated the effects 12 weeks of ifenprodil treatment (60 mg daily; $n=25$) relative to placebo ($n=21$) and found alcohol use and rates of heavy drinking to be significantly lower in the ifenprodil group (Sugaya et al., 2018). Ketamine, also an NMDA receptor antagonist, has additionally produced promising results in the treatment of AUD. In combination with motivational enhancement therapy, a single ketamine infusion (0.71 mg/kg, $n=17$) was found to lower the likelihood of alcohol use over a 21-day period after infusion, as well as lower the likelihood of heavy alcohol use and longer time to relapse, relative to an active control (midazolam; 0.025 mg/kg, $n=23$) (Dakwar et al., 2020). This study's authors conclude that ketamine may be an effective pharmacotherapy that could work synergistically with psychotherapy to help initiate and sustain abstinence from alcohol. Interestingly, results of a study by Das et al. (Das et al., 2019) suggest that by blocking the consolidation and retrieval of maladaptive reward memories, a single ketamine infusion (350 ng/dL for 30 min, $n=11$) can produce both a rapid and long-lasting (up to 9 months) reduction in drinking. While these studies are an important first step and build on evidence implicating the glutamatergic system as a promising treatment target to promote recovery during alcohol abstinence, further investigation and larger studies incorporating imaging will be required to elucidate specific mechanisms of action for ifenprodil and ketamine within the context of AUD treatment, and the extent to which mGlu5 is involved.

Given the reduction in alcohol self-administration (Besheer et al., 2010; Cowen et al., 2005; Gupta et al., 2008; Schroeder et al., 2005) and diminished reinstatement (Gass et al., 2014; Sidhpura et al., 2010; Sinclair et al., 2012) observed in preclinical studies targeting mGlu5, the novel mGlu5 NAM, GET73 was developed for the treatment of AUD humans (Beggiato et al., 2018; Ferraro et al., 2013). Although GET73 has been well tolerated with minimal side effects in the early trials (Haass-Koffler et al., 2018, 2017), an inpatient, cross-over, randomized, double-blind, placebo-controlled study in non-treatment-seeking, alcohol-dependent individuals ($n=24$) found GET73 (300 mg, 3 times a day for 3 days) had no effect on alcohol cue-induced craving, or alcohol self-administration in the laboratory setting (Haass-Koffler et al., 2022). A longer dosing period in a treatment-seeking population, with additional clinically meaningful outcome measures, should be considered to assess the full efficacy of GET73, or other mGlu5 modulators, on alcohol consumption and/or craving in order to speak to any efficacy or make inferences about clinical relevance, beyond safety.

To date, mGlu5 modulators have not been tested for CUD.

While the evidence reviewed above focused on the application and efficacy of mGlu5 inhibition/negative modulation, there is preclinical body of work suggesting mGlu5 positive modulation as a viable treatment mechanism in substance use disorders (Johnson & Lovinger, 2020). For example, a recent report by Marszalek-Grabska and colleagues found that daily treatment with an mGlu5 PAM, VU-29 (30 mg/kg, i.p.), over a forced 10 days of abstinence significantly reduced previously acquired alcohol conditioned place preference (Marszalek-Grabska et al., 2020). Similarly, a single dose of the mGlu5 PAM CDPPB (30 mg/kg, s.c.) acutely attenuated drug-seeking in a context + cue relapse test after 45–80 days of abstinence (Gobin & Schwendt, 2020). Thus, the potential for mGlu5 PAM to attenuate maladaptive memories involved in relapse following withdrawal or enhance extinction learning during abstinence is an additional treatment mechanism that deserves further exploration.

6. Summary and commentary

The last two decades, or so, have seen an increased interest in the glutamatergic system, and mGlu5 in particular, as targets for treating psychiatric conditions, including mood disorders, anxiety and trauma-related disorders, and substance use disorders. For this reason, the current chapter focused on studies of mGlu5, using PET when available. However, previous reviews have explored the structure, pharmacology and treatment target implications of all three metabotropic glutamate receptor groups (Golubeva, Moloney, O'Connor, Dinan, & Cryan, 2016; Gregory, Dong, Meiler, & Conn, 2011; Niswender & Conn, 2010; Palucha & Pilc, 2007; Peterlik, Flor, & Uschold-Schmidt, 2016; Yin & Niswender, 2014).

Focusing on mGlu5, the largest exploration was performed in relation to depression, where rodent models of MDD (and some in vivo clinical MDD investigations) are associated with lower mGlu5 expression. Further, our work indicates changes in mGlu5 availability are related to the antidepressant response to ketamine in depression. We also demonstrated associations between mGlu5 availability and posttraumatic stress symptom severity (particularly avoidance symptoms and suicidal ideation).

However, it should be noted that mood and anxiety disorders are heterogeneous disorders and present in high comorbidity with other psychiatric disorders, including substance use disorders. Importantly, as reviewed in this chapter, drugs of abuse such as tobacco, cannabis, and alcohol may contribute to the mGlu5 dysregulation (Abdallah, Mason, et al., 2017; Akkus et al., 2013; Hulka et al., 2014; Martinez et al., 2014; Milella et al., 2014; Stoker, Olivier, & Markou, 2012). Although some of the earlier trials testing mGlu5 NAMs in substance use disorders have not produced significant results, to our knowledge modulation of mGlu5 availability in treatment trials, such as can be determined by PET, has not been tested and deserves consideration, as does potential for mGlu5 PAMs to attenuate maladaptive memories (such as reexperiencing symptoms in PTSD, or those involved in relapse following substance withdrawal) or enhance desired types of learning (fear extinction, for example). Furthermore, indirect mGlu5 modulation as part of the mechanism of treatments not specifically targeting mGlu5 (as seen with ketamine in depression, for example) within the context of substance use disorders is additionally worthy of future investigation.

Together, the studies and findings suggest that dysregulation of mGlu5 is evident in numerous psychiatric disorders, and normalization of glutamate neurotransmission via modulation of expression and/or signaling of mGlu5 may be a needed component in treating some psychiatric disorders or symptoms. Finally, PET remains an important tool to investigate mGlu5 in disease conditions and in response to therapy.

References

Abdallah, C. G., Jiang, L., De Feyter, H. M., Fasula, M., Krystal, J. H., Rothman, D. L., et al. (2014). Glutamate metabolism in major depressive disorder. *The American Journal of Psychiatry, 171*(12), 1320–1327. https://doi.org/10.1176/appi.ajp.2014.14010067.

Abdallah, C., Adams, T., Kelmendi, B., Esterlis, I., Sanacora, G., & Krystal, J. (2016). Ketamine's mechanism of action: A path to rapid-acting antidepressants. *Depression and Anxiety, 33*(8), 689–697.

Abdallah, C. G., Hannestad, J., Mason, G. F., Holmes, S. E., DellaGioia, N., Sanacora, G., et al. (2017). Metabotropic glutamate receptor 5 and glutamate involvement in major depressive disorder: A multimodal imaging study. *Biological Psychiatry: Cognitive Neuroscience and Neuroimaging, 2*(5), 449–456.

Abdallah, C., Mason, G., DellaGioia, N., Sanacora, G., Jiang, L., Matuskey, D., et al. (2017). mGluR5 and glutamate involvement in MDD: A multimodal imaging study. *Biological Psychiatry: Cognitive Neuroscience and Neuroimaging, 2*, 449–456.

Abdallah, C. G., De Feyter, H. M., Averill, L. A., Jiang, L., Averill, C. L., Chowdhury, G. M. I., et al. (2018). The effects of ketamine on prefrontal glutamate neurotransmission in healthy and depressed subjects. *Neuropsychopharmacology, 43*(10), 2154–2160. https://doi.org/10.1038/s41386-018-0136-3.

Ade, K. K., Wan, Y., Hamann, H. C., O'Hare, J. K., Guo, W., Quian, A., et al. (2016). Increased metabotropic glutamate receptor 5 signaling underlies obsessive-compulsive disorder-like behavioral and striatal circuit abnormalities in mice. *Biological Psychiatry, 80*(7), 522–533. https://doi.org/10.1016/j.biopsych.2016.04.023.

Akkus, F., Ametamey, S., Treyer, V., Burger, C., Johayem, A., Umbricht, D., et al. (2013). Marked global reduction in mGluR5 receptor binding in smokers and ex-smokers determined by [11C]ABP688 positron emission tomography. *Proceedings of the National Academy of Sciences of the United States of America, 110*(2), 737–742.

Akkus, F., Terbeck, S., Ametamey, S. M., Rufer, M., Treyer, V., Burger, C., et al. (2014). Metabotropic glutamate receptor 5 binding in patients with obsessive-compulsive disorder. *The International Journal of Neuropsychopharmacology, 17*(12), 1915–1922. https://doi.org/10.1017/s1461145714000716.

Akkus, F., Treyer, V., Johayem, A., Ametamey, S., Mancilla, B., Sovago, J., et al. (2016). Association of long-term nicotine abstinence with normal metabotropic glutamate receptor-5 binding. *Biological Psychiatry, 79,* 474–480.

Akkus, F., Mihov, Y., Treyer, V., Ametamey, S. M., Johayem, A., Senn, S., et al. (2018). Metabotropic glutamate receptor 5 binding in male patients with alcohol use disorder. *Translational Psychiatry, 8*(1), 17. https://doi.org/10.1038/s41398-017-0066-6 PMID - 29317611.

Akkus, F., Terbeck, S., Haggarty, C. J., Treyer, V., Dietrich, J. J., Hornschuh, S., et al. (2020). The role of the metabotropic glutamate receptor 5 in nicotine addiction. *CNS Spectrums, 1-6.* https://doi.org/10.1017/S1092852920001704.

American Psychiatric Association. (2000). *Diagnostic and statistical manual of mental disorders (DSM-IV-TR).* Washington, DC: American Psychiatric Publication.

American Psychiatric Association. (2013). *Diagnostic and statistical manual of mental disorders (DSM-5).* Washington, DC: American Psychiatric Publication.

Ametamey, S. M., Kessler, L. J., Honer, M., Wyss, M. T., Buck, A., Hintermann, S., et al. (2006). Radiosynthesis and preclinical evaluation of 11C-ABP688 as a probe for imaging the metabotropic glutamate receptor subtype 5. *Journal of Nuclear Medicine, 47*(4), 698–705. doi:47/4/698 [pii].

Anderson, P. (2006). Global use of alcohol, drugs and tobacco. *Drug and Alcohol Review, 25*(6), 489–502.

Asch, R. H., Pothula, S., Toyonaga, T., Fowles, K., Groman, S. M., Garcia-Milian, R., et al. (2022). Examining sex differences in responses to footshock stress and the role of the metabotropic glutamate receptor 5: An [(18)F]FPEB and positron emission tomography study in rats. *Neuropsychopharmacology.* https://doi.org/10.1038/s41386-022-01441-y.

AstraZeneca. (2012). *6-week Study Treatment to Evaluate the Safety and Effectiveness of AZD2066 in Patients With Major Depressive Disorder.* Retrieved from https://clinicaltrials.gov/ct2/show/study/NCT01145755.

Attucci, S., Carlà, V., Mannaioni, G., & Moroni, F. (2001). Activation of type 5 metabotropic glutamate receptors enhances NMDA responses in mice cortical wedges. *British Journal of Pharmacology, 132,* 799–806.

Averill, L. A., Jiang, L., Purohit, P., Coppoli, A., Averill, C. L., Roscoe, J., et al. (2022). Prefrontal glutamate neurotransmission in PTSD: A novel approach to estimate synaptic strength in vivo in humans. *Chronic Stress (Thousand Oaks), 6.* https://doi.org/10.1177/24705470221092734.

Awad, H., Hubert, G., Smith, Y., Levey, A., & Conn, P. (2000). Activation of metabotropic glutamate receptor 5 has direct excitatory effects and potentiates NMDA receptor currents in neurons of the subthalamic nucleus. *The Journal of Neuroscience, 20,* 7871–7879.

Azofeifa, A., Mattson, M. E., Schauer, G., McAfee, T., Grant, A., & Lyerla, R. (2016). National estimates of marijuana use and related indicators—National Survey on drug use and health, United States, 2002–2014. *Morbidity and Mortality Weekly Report. Surveillance Summaries, 65*(11), 1–25.

Bahji, A., Zarate, C. A., Jr., & Vazquez, G. H. (2021). Ketamine for bipolar depression: A systematic review. *International Journal of Neuropsychopharmacology, 24*(7), 535–541. https://doi.org/10.1093/ijnp/pyab023.

Balfour, D. J. (2015). The role of mesoaccumbens dopamine in nicotine dependence. *The Neuropharmacology of Nicotine Dependence*, 55–98.

Ballard, T. M., Woolley, M. L., Prinssen, E., Huwyler, J., Porter, R., & Spooren, W. (2005). The effect of the mGlu5 receptor antagonist MPEP in rodent tests of anxiety and cognition: A comparison. *Psychopharmacology, 179*(1), 218–229. https://doi.org/10.1007/s00213-005-2211-9.

Barnes, S. A., Sheffler, D. J., Semenova, S., Cosford, N. D., & Bespalov, A. (2018). Metabotropic glutamate receptor 5 as a target for the treatment of depression and smoking: Robust preclinical data but inconclusive clinical efficacy. *Biological Psychiatry, 83*(11), 955–962.

Bates, B. S., Rodriguez, A. L., Felts, A. S., Morrison, R. D., Venable, D. F., Blobaum, A. L., et al. (2014). Discovery of VU0431316: A negative allosteric modulator of mGlu5 with activity in a mouse model of anxiety. *Bioorganic & Medicinal Chemistry Letters, 24*(15), 3307–3314. https://doi.org/10.1016/j.bmcl.2014.06.003.

Beggiato, S., Borelli, A. C., Tomasini, M. C., Castelli, M. P., Pintori, N., Cacciaglia, R., et al. (2018). In vitro functional characterization of GET73 as possible negative allosteric modulator of metabotropic glutamate receptor 5. *Frontiers in Pharmacology, 9*, 327. https://doi.org/10.3389/fphar.2018.00327.

Besheer, J., Grondin, J. J., Cannady, R., Sharko, A. C., Faccidomo, S., & Hodge, C. W. (2010). Metabotropic glutamate receptor 5 activity in the nucleus accumbens is required for the maintenance of ethanol self-administration in a rat genetic model of high alcohol intake. *Biological Psychiatry, 67*(9), 812–822. https://doi.org/10.1016/j.biopsych.2009.09.016.

Bhattacharyya, S., & Chakraborty, K. (2007). Glutamatergic dysfunction—Newer targets for anti-obsessional drugs. *Recent Patents on CNS Drug Discovery, 2*(1), 47–55. https://doi.org/10.2174/157488907779561727.

Blest-Hopley, G., O'Neill, A., Wilson, R., Giampietro, V., Lythgoe, D., Egerton, A., et al. (2020). Adolescent-onset heavy cannabis use associated with significantly reduced glial but not neuronal markers and glutamate levels in the hippocampus. *Addiction Biology, 25*(6), e12827.

Blount, H. L., Dee, J., Wu, L., Schwendt, M., & Knackstedt, L. A. (2023). Stress resilience-associated behaviors following predator scent stress are accompanied by upregulated nucleus accumbens mGlu5 transcription in female Sprague Dawley rats. *Behavioural Brain Research, 436*, 114090. https://doi.org/10.1016/j.bbr.2022.114090.

Blüml, S. (2013). Magnetic resonance spectroscopy: Basics. In S. Blüml, & A. Panigrahy (Eds.), *MR spectroscopy of pediatric brain disorders* Springer Science+Business Media, LLC.

Borgelt, L., Strakowski, S. M., DelBello, M. P., Weber, W., Eliassen, J. C., Komoroski, R. A., et al. (2019). Neurophysiological effects of multiple mood episodes in bipolar disorder. *Bipolar Disorders, 21*(6), 503–513. https://doi.org/10.1111/bdi.12782.

Brodkin, J., Busse, C., Sukoff, S. J., & Varney, M. A. (2002). Anxiolytic-like activity of the mGluR5 antagonist MPEP: A comparison with diazepam and buspirone. *Pharmacology, Biochemistry, and Behavior, 73*(2), 359–366. https://doi.org/10.1016/s0091-3057(02)00828-6.

Brody, A. L., Mandelkern, M. A., London, E. D., Olmstead, R. E., Farahi, J., Scheibal, D., et al. (2006). Cigarette smoking saturates brain $\alpha 4\beta 2$ nicotinic acetylcholine receptors. *Archives of General Psychiatry, 63*(8), 907–914.

Byrnes, K. R., Stoica, B., Loane, D. J., Riccio, A., Davis, M. I., & Faden, A. I. (2009). Metabotropic glutamate receptor 5 activation inhibits microglial associated inflammation and neurotoxicity. *Glia, 57*(5), 550–560. https://doi.org/10.1002/glia.20783.

Ceccarini, J., Leurquin-Sterk, G., Crunelle, C. L., de Laat, B., Bormans, G., Peuskens, H., et al. (2019). Recovery of decreased metabotropic glutamate receptor 5 availability in abstinent alcohol-dependent patients. *Journal of Nuclear Medicine, 61*(2), 256–262. https://doi.org/10.2967/jnumed.119.228825. PMID - 31481578.

Cecil, K., DelBello, M., Morey, R., & Strakowski, S. (2002). Frontal lobe differences in bipolar disorder as determined by proton MR spectroscopy. *Bipolar Disorders, 4*, 357–365.

Chiamulera, C., Marzo, C. M., & Balfour, D. J. (2017). Metabotropic glutamate receptor 5 as a potential target for smoking cessation. *Psychopharmacology, 234*(9), 1357–1370.

Chowdhury, G., Behar, K., Cho, W., Thomas, M., Rothman, D., & Sanacora, G. (2012). ^{1}H-[^{13}C]-nuclear magnetic resonance spectroscopy measures of ketamine's effect on amino acid neurotransmitter metabolism. *Biological Psychiatry, 71*, 1022–1025.

Chowdhury, G., Zhang, J., Thomas, M., Banasr, M., Ma, X., Pittman, B., et al. (2016). Transiently increased glutamate cycling in rat PFC is associated with rapid onset of antidepressant-like effects. *Molecular Psychiatry, 22*(1), 120–126.

Colizzi, M., McGuire, P., Pertwee, R. G., & Bhattacharyya, S. (2016). Effect of cannabis on glutamate signalling in the brain: A systematic review of human and animal evidence. *Neuroscience and Biobehavioral Reviews, 64*, 359–381. https://doi.org/10.1016/j.neubiorev.2016.03.010.

Compton, W. M., Han, B., Jones, C. M., Blanco, C., & Hughes, A. (2016). Marijuana use and use disorders in adults in the USA, 2002–14: Analysis of annual cross-sectional surveys. *The Lancet Psychiatry, 3*(10), 954–964.

Conn, P. J., & Pin, J.-P. (1997). Pharmacology and functions of metabotropic glutamate receptors. *Annual Review of Pharmacology and Toxicology, 37*(1), 205–237.

Cornelius, M. E., Wang, T. W., Jamal, A., Loretan, C. G., & Neff, L. J. (2020). Tobacco product use among adults—United States, 2019. *Morbidity and Mortality Weekly Report, 69*(46), 1736.

Corrigall, W. A., Coen, K. M., & Adamson, K. L. (1994). Self-administered nicotine activates the mesolimbic dopamine system through the ventral tegmental area. *Brain Research, 653*(1–2), 278–284. https://doi.org/10.1016/0006-8993(94)90401-4.

Coutinho, V., & Knöpfel, T. (2002). Book review: Metabotropic glutamate receptors: Electrical and chemical signaling properties. *The Neuroscientist, 8*(6), 551–561.

Cowen, M. S., Djouma, E., & Lawrence, A. J. (2005). The metabotropic glutamate 5 receptor antagonist 3-[(2-methyl-1,3-thiazol-4-yl)ethynyl]-pyridine reduces ethanol self-administration in multiple strains of alcohol-preferring rats and regulates olfactory glutamatergic systems. *The Journal of Pharmacology and Experimental Therapeutics, 315*(2), 590–600. https://doi.org/10.1124/jpet.105.090449.

Cox, S. M. L., Tippler, M., Jaworska, N., Smart, K., Castellanos-Ryan, N., Durand, F., et al. (2020). mGlu5 receptor availability in youth at risk for addictions: Effects of vulnerability traits and cannabis use. *Neuropsychopharmacology, 45*(11), 1817–1825. https://doi.org/10.1038/s41386-020-0708-x.

Cristino, L., Bisogno, T., & Di Marzo, V. (2020). Cannabinoids and the expanded endocannabinoid system in neurological disorders. *Nature Reviews Neurology, 16*(1), 9–29. https://doi.org/10.1038/s41582-019-0284-z.

Curran, H. V., Freeman, T. P., Mokrysz, C., Lewis, D. A., Morgan, C. J., & Parsons, L. H. (2016). Keep off the grass? Cannabis, cognition and addiction. *Nature Reviews Neuroscience, 17*(5), 293–306.

Dager, S., Friedman, S., Parow, A., Demopulos, C., Stoll, A., Lyoo, I., et al. (2004). Brain metabolic alterations in medication-free patients with bipolar disorder. *Archives of General Psychiatry, 61*, 450–458.

Daggett, L. P., Sacaan, A. I., Akong, M., Rao, S. P., Hess, S. D., Liaw, C., et al. (1995). Molecular and functional characterization of recombinant human metabotropic glutamate receptor subtype 5. *Neuropharmacology, 34*(8), 871–886. Retrieved from http://www.ncbi.nlm.nih.gov/pubmed/8532169.

Dakwar, E., Levin, F., Hart, C. L., Basaraba, C., Choi, J., Pavlicova, M., et al. (2020). A single ketamine infusion combined with motivational enhancement therapy for alcohol use disorder: A randomized midazolam-controlled pilot trial. *The American Journal of Psychiatry, 177*(2), 125–133. https://doi.org/10.1176/appi.ajp.2019.19070684.

Das, R. K., Gale, G., Walsh, K., Hennessy, V. E., Iskandar, G., Mordecai, L. A., et al. (2019). Ketamine can reduce harmful drinking by pharmacologically rewriting drinking memories. *Nature Communications, 10*(1), 5187. https://doi.org/10.1038/s41467-019-13162-w.

Davis, M. T., Hillmer, A., Holmes, S. E., Pietrzak, R. H., DellaGioia, N., Nabulsi, N., et al. (2019). In vivo evidence for dysregulation of mGluR5 as a biomarker of suicidal ideation. *Proceedings of the National Academy of Sciences, 116*(23), 11490–11495.

de Laat, B., Weerasekera, A., Leurquin-Sterk, G., Gsell, W., Bormans, G., Himmelreich, U., et al. (2019). Effects of alcohol exposure on the glutamatergic system: A combined longitudinal 18F-FPEB and 1H-MRS study in rats. *Addiction Biology, 24*(4), 696–706.

DeLorenzo, C., Sovago, J., Gardus, J., Xu, J., Yang, J., Behrje, R., et al. (2015). Characterization of brain mGluR5 binding in a pilot study of late-life major depressive disorder using positron emission tomography and [^{11}C]ABP688. *Translational Psychiatry, 5*, e693.

DeLorenzo, C., DellaGioia, N., Bloch, M., Sanacora, G., Nabulsi, N., Abdallah, C., et al. (2015). In vivo ketamine-induced changes in [11C]ABP688 binding to metabotropic glutamate receptor subtype 5. *Biological Psychiatry, 77*, 266–275.

DeLorenzo, C., Gallezot, J., Gardus, J., Yang, J., Planeta, B., Nabulsi, N., et al. (2017). In vivo variation in same-day estimates of metabotropic glutamate receptor subtype 5 binding using. [11C]ABP688 and [18F]FPEB *JCBFM, 37*, 2716–2727.

Deschwanden, A., Karolewicz, B., Feyissa, A. M., Treyer, V., Ametamey, S. M., Johayem, A., et al. (2011). Reduced metabotropic glutamate receptor 5 density in major depression determined by [(11)C]ABP688 PET and postmortem study. *The American Journal of Psychiatry, 168*(7), 727–734. https://doi.org/10.1176/appi.ajp.2011.09111607.

DuBois, J., Rousset, O., Rowley, J., Porras-Betancourt, M., Reader, R., Labbe, A., et al. (2016). Characterization of age/sex and the regional distribution of mGluR5 availability in the healthy human brain measured by high-resolution [(11)C]ABP688 PET. *European Journal of Nuclear Medicine and Molecular Imaging, 43*, 152–162.

Duclot, F., Perez-Taboada, I., Wright, K. N., & Kabbaj, M. (2016). Prediction of individual differences in fear response by novelty seeking, and disruption of contextual fear memory reconsolidation by ketamine. *Neuropharmacology, 109*, 293–305. https://doi.org/10.1016/j.neuropharm.2016.06.022.

Duman, R., Aghajanian, G., Sanacora, G., & Krystal, J. (2016). Synaptic plasticity and depression: New insights from stress and rapid-acting antidepressants. *Nature Medicine, 22*, 238–249.

Esterlis, I., DellaGioia, N., Pietrzak, R. H., Matuskey, D., Nabulsi, N., Abdallah, C. G., et al. (2018). Ketamine-induced reduction in mGluR5 availability is associated with an antidepressant response: An [(11)C]ABP688 and PET imaging study in depression. *Molecular Psychiatry, 23*(4), 824–832. https://doi.org/10.1038/mp.2017.58.

Esterlis, I., DeBonee, S., Cool, R., Holmes, S., Baldassari, S. R., Maruff, P., et al. (2022). Differential role of mGluR5 in cognitive processes in posttraumatic stress disorder and major depression. *Chronic Stress, 6*.

Fatemi, S., & Folsom, T. (2014). Existence of monomer and dimer forms of mGluR5, under reducing conditions in studies of postmortem brain in various psychiatric disorders. *Schizophrenia Research, 158*, 270–271.

Fatemi, S., Folsom, T., Rooney, R., & Thuras, P. (2013). mRNA and protein expression for novel GABAA receptors θ and ρ2 are altered in schizophrenia and mood disorders; relevance to FMRP-mGluR5 signaling pathway. *Translational Psychiatry*, *3*, 1–13.

Feder, A., Parides, M., Murrough, J., Perez, A., Morgan, J., Saxena, S., et al. (2014). Efficacy of intravenous ketamine for treatment of chronic posttraumatic stress disorder: A randomized clinical trial. *JAMA Psychiatry*, *71*, 681–688.

Feder, A., Costi, S., Rutter, S. B., Collins, A. B., Govindarajulu, U., Jha, M. K., et al. (2021). A randomized controlled trial of repeated ketamine administration for chronic posttraumatic stress disorder. *The American Journal of Psychiatry*, *178*(2), 193–202. https://doi.org/10.1176/appi.ajp.2020.20050596.

Ferraro, L., Loche, A., Beggiato, S., Tomasini, M. C., Antonelli, T., Colombo, G., et al. (2013). The new compound GET73, N-[(4-trifluoromethyl)benzyl]4-methoxybutyramide, regulates hippocampal Aminoacidergic transmission possibly via an allosteric modulation of mGlu5 receptor. Behavioural evidence of its "anti-alcohol" and anxiolytic properties. *Current Medicinal Chemistry*, *20*(27), 3339–3357. https://doi.org/10.2174/09298673113209990167.

Fontanez-Nuin, D. E., Santini, E., Quirk, G. J., & Porter, J. T. (2011). Memory for fear extinction requires mGluR5-mediated activation of infralimbic neurons. *Cerebral Cortex*, *21*(3), 727–735. https://doi.org/10.1093/cercor/bhq147.

Friedmann, C., Davis, L., Ciccone, P., & Rubin, R. (1980). Phase II double blind controlled study of a new anxiolytic, fenobam (McN-3377) vs placebo. *Current Therapeutic Research*, *27*, 144–151.

Galanopoulos, A., Polissidis, A., Papadopoulou-Daifoti, Z., Nomikos, G. G., & Antoniou, K. (2011). Δ9-THC and WIN55,212-2 affect brain tissue levels of excitatory amino acids in a phenotype-, compound-, dose-, and region-specific manner. *Behavioural Brain Research*, *224*(1), 65–72. https://doi.org/10.1016/j.bbr.2011.05.018.

Gass, J. T., Trantham-Davidson, H., Kassab, A. S., Glen, W. B., Jr., Olive, M. F., & Chandler, L. J. (2014). Enhancement of extinction learning attenuates ethanol-seeking behavior and alters plasticity in the prefrontal cortex. *The Journal of Neuroscience*, *34*(22), 7562–7574. https://doi.org/10.1523/jneurosci.5616-12.2014.

George, D., Galvez, V., Martin, D., Kumar, D., Leyden, J., Hadzi-Pavlovic, D., et al. (2017). Pilot randomized controlled trial of titrated subcutaneous ketamine in older patients with treatment-resistant depression. *The American Journal of Geriatric Psychiatry*. https://doi.org/10.1016/j.jagp.2017.06.007.

Gerhard, D. M., Wohleb, E. S., & Duman, R. S. (2016). Emerging treatment mechanisms for depression: Focus on glutamate and synaptic plasticity. *Drug Discovery Today*, *21*(3), 454–464. https://doi.org/10.1016/j.drudis.2016.01.016.

Gigante, A., Bond, D., Lafer, B., Lam, R., Young, L., & Yatham, L. (2012). Brain glutamate levels measured by magnetic resonance spectroscopy in patients with bipolar disorder: A meta-analysis. *Bipolar Disorders*, *14*, 478–487.

Gobin, C., & Schwendt, M. (2020). The cognitive cost of reducing relapse to cocaine-seeking with mGlu5 allosteric modulators. *Psychopharmacology*, *237*(1), 115–125. https://doi.org/10.1007/s00213-019-05351-8.

Gök, Ş., Murat Demet, M., & Öztürk, Z. (2017). Glutamate mGlu5 receptor antagonist, MPEP, reduces the quinpirole-induced compulsive-like checking in rats. *Journal of Obsessive-Compulsive and Related Disorders*, *15*, 13–18. https://doi.org/10.1016/j.jocrd.2017.08.003.

Golubeva, A. V., Moloney, R. D., O'Connor, R. M., Dinan, T. G., & Cryan, J. F. (2016). Metabotropic glutamate receptors in central nervous system diseases. *Current Drug Targets*, *17*(5), 538–616. Retrieved from https://www.ncbi.nlm.nih.gov/pubmed/25777273.

Gregory, K. J., Dong, E. N., Meiler, J., & Conn, P. J. (2011). Allosteric modulation of metabotropic glutamate receptors: Structural insights and therapeutic potential. *Neuropharmacology, 60*(1), 66–81. https://doi.org/10.1016/j.neuropharm.2010.07.007.

Guilmatre, A., Huguet, G., Delorme, R., & Bourgeron, T. (2014). The emerging role of SHANK genes in neuropsychiatric disorders. *Developmental Neurobiology, 74*, 113–122.

Guina, J., Rossetter, S. R., DeRHODES, B. J., Nahhas, R. W., & Welton, R. S. (2015). Benzodiazepines for PTSD: A systematic review and meta-analysis. *Journal of Psychiatric Practice, 21*(4), 281–303.

Gupta, T., Syed, Y. M., Revis, A. A., Miller, S. A., Martinez, M., Cohn, K. A., et al. (2008). Acute effects of acamprosate and MPEP on ethanol drinking-in-the-dark in male C57BL/6J mice. *Alcoholism, Clinical and Experimental Research, 32*(11), 1992–1998. https://doi.org/10.1111/j.1530-0277.2008.00787.x.

Haass-Koffler, C. L., Goodyear, K., Long, V. M., Tran, H. H., Loche, A., Cacciaglia, R., et al. (2017). A phase I randomized clinical trial testing the safety, tolerability and preliminary pharmacokinetics of the mGluR5 negative allosteric modulator GET 73 following single and repeated doses in healthy volunteers. *European Journal of Pharmaceutical Sciences, 109*, 78–85.

Haass-Koffler, C. L., Goodyear, K., Loche, A., Long, V. M., Lobina, C., Tran, H. H., et al. (2018). Administration of the metabotropic glutamate receptor subtype 5 allosteric modulator GET 73 with alcohol: A translational study in rats and humans. *Journal of Psychopharmacology, 32*(2), 163–173. https://doi.org/10.1177/0269881117746904.

Haass-Koffler, C. L., Perciballi, R., Magill, M., Loche, A., Cacciaglia, R., Leggio, L., et al. (2022). An inpatient human laboratory study assessing the safety and tolerability, pharmacokinetics, and biobehavioral effect of GET 73 when co-administered with alcohol in individuals with alcohol use disorder. *Psychopharmacology, 239*(1), 35–46. https://doi.org/10.1007/s00213-021-06008-1.

Hadizadeh, H., Flores, J. M., Mayerson, T., Worhunsky, P. D., Potenza, M. N., & Angarita, G. A. (2022). Glutamatergic agents for the treatment of cocaine use disorder. *Current Behavioral Neuroscience Reports*, 1–12.

Harnett, N. G., Wood, K. H., Ference, E. W., Reid, M. A., Lahti, A. C., Knight, A. J., et al. (2017). Glutamate/glutamine concentrations in the dorsal anterior cingulate vary with post-traumatic stress disorder symptoms. *Journal of Psychiatric Research, 91*, 169–176. https://doi.org/10.1016/j.jpsychires.2017.04.010.

Harris, A. D., Saleh, M. G., & Edden, R. A. (2017). Edited 1 H magnetic resonance spectroscopy in vivo: Methods and metabolites. *Magnetic Resonance in Medicine, 77*(4), 1377–1389. https://doi.org/10.1002/mrm.26619.

Hashimoto, K., Sawa, A., & Iyo, M. (2007). Increased levels of glutamate in brains from patients with mood disorders. *Biological Psychiatry, 62*(11), 1310–1316. https://doi.org/10.1016/j.biopsych.2007.03.017.

Hasin, D. S., Saha, T. D., Kerridge, B. T., Goldstein, R. B., Chou, S. P., Zhang, H., et al. (2015). Prevalence of marijuana use disorders in the United States between 2001-2002 and 2012-2013. *JAMA Psychiatry, 72*(12), 1235–1242. https://doi.org/10.1001/jamapsychiatry.2015.1858.

Hasler, G., van der Veen, J. W., Tumonis, T., Meyers, N., Shen, J., & Drevets, W. C. (2007). Reduced prefrontal glutamate/glutamine and gamma-aminobutyric acid levels in major depression determined using proton magnetic resonance spectroscopy. *Archives of General Psychiatry, 64*(2), 193–200. https://doi.org/10.1001/archpsyc.64.2.193.

Henter, I. D., Park, L. T., & Zarate, C. A. (2021). Novel glutamatergic modulators for the treatment of mood disorders: Current status. *CNS Drugs, 35*(5), 527–543. https://doi.org/10.1007/s40263-021-00816-x.

Hermann, D., Weber-Fahr, W., Sartorius, A., Hoerst, M., Frischknecht, U., Tunc-Skarka, N., et al. (2012). Translational magnetic resonance spectroscopy reveals excessive central glutamate levels during alcohol withdrawal in humans and rats. *Biological Psychiatry, 71*(11), 1015–1021. https://doi.org/10.1016/j.biopsych.2011.07.034.

Hillmer, A. T., Angarita, G. A., Esterlis, I., Anderson, J. M., Nabulsi, N., Lim, K., et al. (2021). Longitudinal imaging of metabotropic glutamate 5 receptors during early and extended alcohol abstinence. *Neuropsychopharmacology, 46*(2), 380–385. https://doi.org/10.1038/s41386-020-00856-9.

Holmes, S., Girgenti, M., Davis, M., Pietrzak, R., DellaGioia, N., Nabulsi, N., et al. (2017). Altered metabotropic glutamate receptor 5 markers in PTSD: In-vivo and postmortem evidence. *PNAS, 114*, 8390–8839.

Holmes, S.E., R.H. Asch, M.T. Davis, N. DellaGioia, N. Pashankar, J-D. Gallezot, et al., (2023). Differences in quantification of the metabotropic glutamate receptor 5 (mGluR5) across bipolar disorder and major depressive disorder. Biological Psychiatry. In press.

Hou, L., & Klann, E. (2004). Activation of the phosphoinositide 3-kinase-Akt-mammalian target of rapamycin signaling pathway is required for metabotropic glutamate receptor-dependent long-term depression. *The Journal of Neuroscience, 24*, 6352–6361.

Hubert, G., Paquet, M., & Smith, Y. (2001). Differential subcellular localization of mGluR1a and mGluR5 in the rat and monkey substantia nigra. *The Journal of Neuroscience, 21*, 1838–1847.

Hughes, Z. A., Neal, S. J., Smith, D. L., Sukoff Rizzo, S. J., Pulicicchio, C. M., Lotarski, S., et al. (2013). Negative allosteric modulation of metabotropic glutamate receptor 5 results in broad spectrum activity relevant to treatment resistant depression. *Neuropharmacology, 66*, 202–214. https://doi.org/10.1016/j.neuropharm.2012.04.007.

Hulka, L., Treyer, V., Scheidegger, M., Preller, K., Vonmoos, M., Baumgartner, M., et al. (2014). Smoking but not cocaine use is associated with lower cerebral metabotropic glutamate receptor 5 density in humans. *Molecular Psychiatry, 19*, 625–632.

Iadarola, N. D., Niciu, M. J., Richards, E. M., Vande Voort, J. L., Ballard, E. D., Lundin, N. B., et al. (2015). Ketamine and other N-methyl-D-aspartate receptor antagonists in the treatment of depression: A perspective review. *Therapeutic Advances in Chronic Disease, 6*(3), 97–114. https://doi.org/10.1177/2040622315579059.

Iijima, M., Fukumoto, K., & Chaki, S. (2012). Acute and sustained effects of a metabotropic glutamate 5 receptor antagonist in the novelty-suppressed feeding test. *Behavioural Brain Research, 235*(2), 287–292. https://doi.org/10.1016/j.bbr.2012.08.016.

Iyo, A. H., Feyissa, A. M., Chandran, A., Austin, M. C., Regunathan, S., & Karolewicz, B. (2010). Chronic corticosterone administration down-regulates metabotropic glutamate receptor 5 protein expression in the rat hippocampus. *Neuroscience, 169*(4), 1567–1574.

Jew, C. P., Wu, C. S., Sun, H., Zhu, J., Huang, J. Y., Yu, D., et al. (2013). mGluR5 ablation in cortical glutamatergic neurons increases novelty-induced locomotion. *PLoS One, 8*(8), e70415. https://doi.org/10.1371/journal.pone.0070415.

Jin, D., Xue, B., Mao, L., & Wang, J. (2015). Metabotropic glutamate receptor 5 upregulates surface NMDA receptor expression in striatal neurons via CaMKII. *Brain Research, 1624*, 414–423.

Joffe, M. E., Winder, D. G., & Conn, P. J. (2021). Increased synaptic strength and mGlu2/3 receptor plasticity on mouse prefrontal cortex intratelencephalic pyramidal cells following intermittent access to ethanol. *Alcoholism: Clinical and Experimental Research, 45*(3), 518–529. https://doi.org/10.1111/acer.14546 PMID - 33434325.

Johnson, K. A., & Lovinger, D. M. (2020). Allosteric modulation of metabotropic glutamate receptors in alcohol use disorder: Insights from preclinical investigations. *Advances in Pharmacology, 88*, 193–232. https://doi.org/10.1016/bs.apha.2020.02.002.

Kim, H., Lee, K., Lee, D., Han, Y., Lee, S., Sohn, J., et al. (2015). Costimulation of AMPA and metabotropic glutamate receptors underlies phospholipase C activation by glutamate in hippocampus. *The Journal of Neuroscience, 35*, 6401–6412.

Kim, J.-H., Joo, Y.-H., Son, Y.-D., Kim, J.-H., Kim, Y.-K., Kim, H.-K., et al. (2019). In vivo metabotropic glutamate receptor 5 availability-associated functional connectivity alterations in drug-naïve young adults with major depression. *European Neuropsychopharmacology, 29*(2), 278–290.

Kirkland, A. E., Browning, B. D., Green, R., Leggio, L., Meyerhoff, D. J., & Squeglia, L. M. (2022). Brain metabolite alterations related to alcohol use: A meta-analysis of proton magnetic resonance spectroscopy studies. *Molecular Psychiatry*. https://doi.org/10.1038/s41380-022-01594-8.

Klodzinska, A., Tatarczyńska, E., Chojnacka-Wójcik, E., Nowak, G., Cosford, N. D., & Pilc, A. (2004). Anxiolytic-like effects of MTEP, a potent and selective mGlu5 receptor agonist does not involve GABAA signaling. *Neuropharmacology, 47*(3), 342–350.

Kovacevic, T., Skelin, I., Minuzzi, L., Rosa-Neto, P., & Diksic, M. (2012). Reduced metabotropic glutamate receptor 5 in the Flinders sensitive line of rats, an animal model of depression: An autoradiographic study. *Brain Research Bulletin, 87*(4–5), 406–412. https://doi.org/10.1016/j.brainresbull.2012.01.010.

Kryst, J., Kawalec, P., Mitoraj, A. M., Pilc, A., Lason, W., & Brzostek, T. (2020). Efficacy of single and repeated administration of ketamine in unipolar and bipolar depression: A meta-analysis of randomized clinical trials. *Pharmacological Reports, 72*(3), 543–562. https://doi.org/10.1007/s43440-020-00097-z.

Kupila, J., Karkkainen, O., Laukkanen, V., Tupala, E., Tiihonen, J., & Storvik, M. (2013). mGluR1/5 receptor densities in the brains of alcoholic subjects: A whole-hemisphere autoradiography study. *Psychiatry Research, 212*(3), 245–250. https://doi.org/10.1016/j.pscychresns.2012.04.003.

Laukkanen, V., Kärkkäinen, O., Kautiainen, H., Tiihonen, J., & Storvik, M. (2019). Increased [^{3}H]quisqualic acid binding density in the dorsal striatum and anterior insula of alcoholics: A post-mortem whole-hemisphere autoradiography study. *Psychiatry Research: Neuroimaging, 287*, 63–69. https://doi.org/10.1016/j.pscychresns.2019.04.002.

Lavialle, M., Aumann, G., Anlauf, E., Prols, F., Arpin, M., & Derouiche, A. (2011). Structural plasticity of perisynaptic astrocyte processes involves ezrin and metabotropic glutamate receptors. *Proceedings of the National Academy of Sciences of the United States of America, 108*(31), 12915–12919. https://doi.org/10.1073/pnas.1100957108.

Lener, M. S., Niciu, M. J., Ballard, E. D., Park, M., Park, L. T., Nugent, A. C., et al. (2016). Glutamate and gamma-aminobutyric acid Systems in the Pathophysiology of major depression and antidepressant response to ketamine. *Biological Psychiatry*. https://doi.org/10.1016/j.biopsych.2016.05.005.

Lener, M. S., Kadriu, B., & Zarate, C. A., Jr. (2017). Ketamine and beyond: Investigations into the potential of glutamatergic agents to treat depression. *Drugs, 77*(4), 381–401. https://doi.org/10.1007/s40265-017-0702-8.

Leurquin-Sterk, G., Van den Stock, J., Crunelle, C. L., de Laat, B., Weerasekera, A., Himmelreich, U., et al. (2016). Positive association between limbic metabotropic glutamate receptor 5 availability and novelty-seeking temperament in humans: An 18F-FPEB PET study. *Journal of Nuclear Medicine, 57*(11), 1746–1752. https://doi.org/10.2967/jnumed.116.176032.

Leurquin-Sterk, G., Ceccarini, J., Crunelle, C. L., de Laat, B., Verbeek, J., Deman, S., et al. (2018). Lower limbic metabotropic glutamate receptor 5 availability in alcohol dependence. *Journal of Nuclear Medicine, 59*(4), 682–690. https://doi.org/10.2967/jnumed.117.199422 PMID - 29348321.

Li, C.-X., Wang, Y., Gao, H., Pan, W.-J., Xiang, Y., Huang, M., et al. (2008). Cerebral metabolic changes in a depression-like rat model of chronic forced swimming studied by ex vivo high resolution 1H magnetic resonance spectroscopy. *Neurochemical Research, 33*(11), 2342–2349.

Li, N., Lee, B., Liu, R., Banasr, M., Dwyer, J., Iwata, M., et al. (2010). mTOR-dependent synapse formation underlies the rapid antidepressant effects of NMDA antagonists. *Science, 329*, 959–964.

Li, X., Need, A. B., Baez, M., & Witkin, J. M. (2006). Metabotropic glutamate 5 receptor antagonism is associated with antidepressant-like effects in mice. *The Journal of Pharmacology and Experimental Therapeutics, 319*(1), 254–259. https://doi.org/10.1124/jpet.106.103143.

Liechti, M. E., & Markou, A. (2007). Interactive effects of the mGlu5 receptor antagonist MPEP and the mGlu2/3 receptor antagonist LY341495 on nicotine self-administration and reward deficits associated with nicotine withdrawal in rats. *European Journal of Pharmacology, 554*(2–3), 164–174.

Lin, X., Donthamsetti, P., Skinberg, M., Slifstein, M., Abi-Dargham, A., & Javitch, J. (2015). FPEB and ABP688 cannot access internalized mGluR5 receptors. In *MGluR5 workshop* Columbia University.

Lin, S., Li, X., Chen, Y.-H., Gao, F., Chen, H., Hu, N.-Y., et al. (2018). Social isolation during adolescence induces anxiety behaviors and enhances firing activity in BLA pyramidal neurons via mGluR5 upregulation. *Molecular Neurobiology, 55*(6), 5310–5320. https://doi.org/10.1007/s12035-017-0766-1.

Lindemann, L., Jaeschke, G., Michalon, A., Vieira, E., Honer, M., Spooren, W., et al. (2011). CTEP: A novel, potent, long-acting, and orally bioavailable metabotropic glutamate receptor 5 inhibitor. *The Journal of Pharmacology and Experimental Therapeutics, 339*(2), 474–486. https://doi.org/10.1124/jpet.111.185660.

López-Bendito, G., Shigemoto, R., Fairén, A., & Luján, R. (2002). Differential distribution of group I metabotropic glutamate receptors during rat cortical development. *Cerebral Cortex, 12*, 625–638.

Lopez-Quintero, C., Pérez de los Cobos, J., Hasin, D. S., Okuda, M., Wang, S., Grant, B. F., et al. (2011). Probability and predictors of transition from first use to dependence on nicotine, alcohol, cannabis, and cocaine: Results of the National Epidemiologic Survey on alcohol and related conditions (NESARC). *Drug and Alcohol Dependence, 115*(1–2), 120–130. https://doi.org/10.1016/j.drugalcdep.2010.11.004.

Lujan, R., Nusser, Z., Roberts, J., Shigemoto, R., & Somogyi, P. (1996). Perisynaptic location of metabotropic glutamate receptors mGluR1 and mGluR5 on dendrites and dendritic spines in the rat hippocampus. *The European Journal of Neuroscience, 8*, 1488–1500.

Luján, R., Roberts, J., Shigemoto, R., Ohishi, H., & Somogyi, P. (1997). Differential plasma membrane distribution of metabotropic glutamate receptors mGluR1 alpha, mGluR2 and mGluR5, relative to neurotransmitter release sites. *Journal of Chemical Neuroanatomy, 13*(4), 219–241. Retrieved from https://www.ncbi.nlm.nih.gov/pubmed/9412905.

Luykx, J. J., Laban, K. G., van den Heuvel, M. P., Boks, M. P., Mandl, R. C., Kahn, R. S., et al. (2012). Region and state specific glutamate downregulation in major depressive disorder: A meta-analysis of (1)H-MRS findings. *Neuroscience and Biobehavioral Reviews, 36*(1), 198–205. https://doi.org/10.1016/j.neubiorev.2011.05.014.

Mansvelder, H. D., Keath, J. R., & McGehee, D. S. (2002). Synaptic mechanisms underlie nicotine-induced excitability of brain reward areas. *Neuron, 33*(6), 905–919.

Mao, L. M., & Wang, J. Q. (2018). Alterations in mGlu5 receptor expression and function in the striatum in a rat depression model. *Journal of Neurochemistry, 145*(4), 287–298. https://doi.org/10.1111/jnc.14307.

Marsicano, G., Goodenough, S., Monory, K., Hermann, H., Eder, M., Cannich, A., et al. (2003). CB1 cannabinoid receptors and on-demand defense against excitotoxicity. *Science, 302*(5642), 84–88.

Marszalek-Grabska, M., Gawel, K., Matosiuk, D., Gibula-Tarlowska, E., Listos, J., & Kotlinska, J. H. (2020). Effects of the positive allosteric modulator of metabotropic glutamate receptor 5, VU-29, on maintenance association between environmental cues and rewarding properties of ethanol in rats. *Biomolecules, 10*(5), 793. doi:ARTN 793 https://doi.org/10.3390/biom10050793.

Martinez, D., Slifstein, M., Nabulsi, N., Grassetti, A., Urban, N., Perez, A., et al. (2014). Imaging glutamate homeostasis in cocaine addiction with the metabotropic glutamate receptor 5 positron emission tomography radiotracer [(11)C]ABP688 and magnetic resonance spectroscopy. *Biological Psychiatry, 75*, 165–171.

Mason, G. (2003). Magnetic resonance spectroscopy for studies of neurotransmission in vivo. *Psychopharmacology Bulletin, 37*, 26–40.

Matosin, N., Fernandez-Enright, F., Frank, E., Deng, C., Wong, J., Huang, X. F., et al. (2014). Metabotropic glutamate receptor mGluR2/3 and mGluR5 binding in the anterior cingulate cortex in psychotic and nonpsychotic depression, bipolar disorder and schizophrenia: Implications for novel mGluR-based therapeutics. *Journal of Psychiatry & Neuroscience, 39*(6), 407–416. Retrieved from http://www.ncbi.nlm.nih.gov/pubmed/24949866.

McGowan, J. C., LaGamma, C. T., Lim, S. C., Tsitsiklis, M., Neria, Y., Brachman, R. A., et al. (2017). Prophylactic ketamine attenuates learned fear. *Neuropsychopharmacology.* https://doi.org/10.1038/npp.2017.19.

McIntyre, R. S., Carvalho, I. P., Lui, L. M., Majeed, A., Masand, P. S., Gill, H., et al. (2020). The effect of intravenous, intranasal, and oral ketamine in mood disorders: A meta-analysis. *Journal of Affective Disorders, 276*, 576–584.

Mecca, A., Rogers, K., Jacobs, Z., McDonald, J., Michalak, H., DellaGioia, N., et al. (2017). Investigating age related associations of metabotropic glutamate receptor 5 density using [18F]FPEB and PET. In *Paper presented at the AAGP, Dallas, Texas.*

Mehta, M. V., Gandal, M. J., & Siegel, S. J. (2011). mGluR5-antagonist mediated reversal of elevated stereotyped, repetitive behaviors in the VPA model of autism. *PLoS One, 6*(10), e26077. https://doi.org/10.1371/journal.pone.0026077.

Meyerhoff, D. J., Mon, A., Metzler, T., & Neylan, T. C. (2014). Cortical gamma-aminobutyric acid and glutamate in posttraumatic stress disorder and their relationships to self-reported sleep quality. *Sleep, 37*(5), 893–900. https://doi.org/10.5665/sleep.3654.

Meyers, J., Salling, M., Almli, L., Ratanatharathorn, A., Uddin, M., Galea, S., et al. (2015). Frequency of alcohol consumption in humans; the role of metabotropic glutamate receptors and downstream signaling pathways. *Translational Psychiatry, 5*, e586.

Michels, L., Schulte-Vels, T., Schick, M., O'Gorman, R. L., Zeffiro, T., Hasler, G., et al. (2014). Prefrontal GABA and glutathione imbalance in posttraumatic stress disorder: Preliminary findings. *Psychiatry Research: Neuroimaging, 224*(3), 288–295. https://doi.org/10.1016/j.pscychresns.2014.09.007.

Mikulecká, A., & Mares, P. (2009). Effects of mGluR5 and mGluR1 antagonists on anxiety-like behavior and learning in developing rats. *Behavioural Brain Research, 204*(1), 133–139. https://doi.org/10.1016/j.bbr.2009.05.032.

Milella, M., Marengo, L., Larcher, K., Fotros, A., Dagher, A., Rosa-Neto, P., et al. (2014). Limbic system mGluR5 availability in cocaine dependent subjects: A high-resolution PET [(11)C]ABP688 study. *NeuroImage, 98*, 195–202.

Mishra, P. K., Adusumilli, M., Deolal, P., Mason, G. F., Kumar, A., & Patel, A. B. (2020). Impaired neuronal and astroglial metabolic activity in chronic unpredictable mild stress model of depression: Reversal of behavioral and metabolic deficit with lanicemine. *Neurochemistry International, 137*, 104750.

Mishra, P. K., Kumar, A., Behar, K. L., & Patel, A. B. (2018). Subanesthetic ketamine reverses neuronal and astroglial metabolic activity deficits in a social defeat model of depression. *Journal of Neurochemistry, 146*(6), 722–734. https://doi.org/10.1111/jnc.14544.

Moghaddam, B., Adams, B., Verma, A., & Daly, D. (1997). Activation of glutamatergic neurotransmission by ketamine: A novel step in the pathway from NMDA receptor blockade to dopaminergic and cognitive disruptions associated with the prefrontal cortex. *The Journal of Neuroscience, 17*, 2921–2927.

Mon, A., Durazzo, T. C., & Meyerhoff, D. J. (2012). Glutamate, GABA, and other cortical metabolite concentrations during early abstinence from alcohol and their associations with neurocognitive changes. *Drug and Alcohol Dependence, 125*(1–2), 27–36.

Monteggia, L., Gideons, E., & Kavalali, E. (2013). The role of eukaryotic elongation factor 2 kinase in rapid antidepressant action of ketamine. *Biological Psychiatry, 73*, 1199–1203.

Morley, K. C., Perry, C. J., Watt, J., Hurzeler, T., Leggio, L., Lawrence, A. J., et al. (2021). New approved and emerging pharmacological approaches to alcohol use disorder: A review of clinical studies. *Expert Opinion on Pharmacotherapy, 22*(10), 1291–1303.

Muetzel, R. L., Marjańska, M., Collins, P. F., Becker, M. P., Valabrègue, R., Auerbach, E. J., et al. (2013). In vivo 1H magnetic resonance spectroscopy in young-adult daily marijuana users. *NeuroImage: Clinical, 2*, 581–589.

Murrough, J. W., Abdallah, C. G., & Mathew, S. J. (2017). Targeting glutamate signalling in depression: Progress and prospects. *Nature Reviews. Drug Discovery, 16*(7), 472–486. https://doi.org/10.1038/nrd.2017.16.

Nelson, T. E., Ur, C. L., & Gruol, D. L. (2005). Chronic intermittent ethanol exposure enhances NMDA-receptor-mediated synaptic responses and NMDA receptor expression in hippocampal CA1 region. *Brain Research, 1048*(1–2), 69–79. https://doi.org/10.1016/j.brainres.2005.04.041.

Newman, S. D., Cheng, H., Schnakenberg Martin, A., Dydak, U., Dharmadhikari, S., Hetrick, W., et al. (2019). An investigation of neurochemical changes in chronic cannabis users. *Frontiers in Human Neuroscience, 13*, 318.

Niedzielska-Andres, E., Pomierny-Chamioło, L., Andres, M., Walczak, M., Knackstedt, L. A., Filip, M., et al. (2021). Cocaine use disorder: A look at metabotropic glutamate receptors and glutamate transporters. *Pharmacology & Therapeutics, 221*, 107797.

Niswender, C. M., & Conn, P. J. (2010). Metabotropic glutamate receptors: Physiology, pharmacology, and disease. *Annual Review of Pharmacology and Toxicology, 50*, 295–322. https://doi.org/10.1146/annurev.pharmtox.011008.145533.

Novartis. (2007). *Effects of AFQ056 and nicotine in reducing cigarette smoking.* US National Library of Medicine.

O'Malley, K., Jong, Y., Gonchar, Y., Burkhalter, A., & Romano, C. (2003). Activation of metabotropic glutamate receptor mGlu5 on nuclear membranes mediates intranuclear Ca2+ changes in heterologous cell types and neurons. *The Journal of Biological Chemistry, 278*, 28210–28219.

Ohno-Shosaku, T., Shosaku, J., Tsubokawa, H., & Kano, M. (2002). Cooperative endocannabinoid production by neuronal depolarization and group I metabotropic glutamate receptor activation. *European Journal of Neuroscience, 15*(6), 953–961.

Olive, M. F., Cleva, R. M., Kalivas, P. W., & Malcolm, R. J. (2012). Glutamatergic medications for the treatment of drug and behavioral addictions. *Pharmacology Biochemistry and Behavior, 100*(4), 801–810. https://doi.org/10.1016/j.pbb.2011.04.015.

Olmo, I. G., Ferreira-Vieira, T. H., & Ribeiro, F. M. (2016). Dissecting the signaling pathways involved in the crosstalk between metabotropic glutamate 5 and cannabinoid type 1 receptors. *Molecular Pharmacology, 90*(5), 609. https://doi.org/10.1124/mol.116.104372.

Page, G., Khidir, F., Pain, S., Barrier, L., Fauconneau, B., Guillard, O., et al. (2006). Group I metabotropic glutamate receptors activate the p70S6 kinase via both mammalian target of rapamycin (mTOR) and extracellular signal-regulated kinase (ERK 1/2) signaling pathways in rat striatal and hippocampal synaptoneurosomes. *Neurochemistry International, 49*, 413–421.

Palucha, A., & Pilc, A. (2007). Metabotropic glutamate receptor ligands as possible anxiolytic and antidepressant drugs. *Pharmacology & Therapeutics, 115*(1), 116–147. https://doi.org/10.1016/j.pharmthera.2007.04.007.

Palucha, A., Branski, P., Szewczyk, B., Wieronska, J. M., Klak, K., & Pilc, A. (2005). Potential antidepressant-like effect of MTEP, a potent and highly selective mGluR5 antagonist. *Pharmacology, Biochemistry, and Behavior, 81*(4), 901–906. https://doi.org/10.1016/j.pbb.2005.06.015.

Patel, S., Hamill, T. G., Connolly, B., Jagoda, E., Li, W., & Gibson, R. E. (2007). Species differences in mGluR5 binding sites in mammalian central nervous system determined using in vitro binding with [18F]F-PEB. *Nuclear Medicine and Biology, 34*(8), 1009–1017. https://doi.org/10.1016/j.nucmedbio.2007.07.009.

Paterson, N. E., Semenova, S., Gasparini, F., & Markou, A. (2003). The mGluR5 antagonist MPEP decreased nicotine self-administration in rats and mice. *Psychopharmacology, 167*(3), 257–264.

Pecknold, J. C., McClure, D. J., Appeltauer, L., Wrzesinski, L., & Allan, T. (1982). Treatment of anxiety using fenobam (a nonbenzodiazepine) in a double-blind standard (diazepam) placebo-controlled study. *Journal of Clinical Psychopharmacology, 2*(2), 129–133.

Pennington, D. L., Abé, C., Batki, S. L., & Johannes Meyerhoff, D. (2014). A preliminary examination of cortical neurotransmitter levels associated with heavy drinking in post-traumatic stress disorder. *Psychiatry Research: Neuroimaging, 224*(3), 281–287. https://doi.org/10.1016/j.pscychresns.2014.09.004.

Pérez de la Mora, M., Lara-García, D., Jacobsen, K. X., Vázquez-García, M., Crespo-Ramírez, M., Flores-Gracia, C., et al. (2006). Anxiolytic-like effects of the selective metabotropic glutamate receptor 5 antagonist MPEP after its intra-amygdaloid microinjection in three different non-conditioned rat models of anxiety. *The European Journal of Neuroscience, 23*(10), 2749–2759. https://doi.org/10.1111/j.1460-9568.2006.04798.x.

Peterlik, D., Stangl, C., Bauer, A., Bludau, A., Keller, J., Grabski, D., et al. (2017). Blocking metabotropic glutamate receptor subtype 5 relieves maladaptive chronic stress consequences. *Brain, Behavior, and Immunity, 59*, 79–92. https://doi.org/10.1016/j.bbi.2016.08.007.

Peterlik, D., Flor, P. J., & Uschold-Schmidt, N. (2016). The emerging role of metabotropic glutamate receptors in the pathophysiology of chronic stress-related disorders. *Current Neuropharmacology, 14*(5), 514–539. Retrieved from https://www.ncbi.nlm.nih.gov/pubmed/27296643.

Petzold, J., Szumlinski, K. K., & London, E. D. (2021). Targeting mGlu5 for methamphetamine use disorder. *Pharmacology & Therapeutics, 224*, 107831.

Picciotto, M. R., Addy, N. A., Mineur, Y. S., & Brunzell, D. H. (2008). It is not "either/or": Activation and desensitization of nicotinic acetylcholine receptors both contribute to behaviors related to nicotine addiction and mood. *Progress in Neurobiology, 84*(4), 329–342. https://doi.org/10.1016/j.pneurobio.2007.12.005.

Piers, T. M., Kim, D. H., Kim, B. C., Regan, P., Whitcomb, D. J., & Cho, K. (2012). Translational concepts of mGluR5 in synaptic diseases of the brain. *Frontiers in Pharmacology, 3*, 199.

Pisani, A., Gubellini, P., Bonsi, P., Conquet, F., Picconi, B., Centonze, D., et al. (2001). Metabotropic glutamate receptor 5 mediates the potentiation of N-methyl-D-aspartate responses in medium spiny striatal neurons. *Neuroscience, 106*, 579–587.

The metabotropic glutamate receptor 5 as a biomarker 305

Pistis, M., Ferraro, L., Pira, L., Flore, G., Tanganelli, S., Gessa, G. L., et al. (2002). Delta(9)-tetrahydrocannabinol decreases extracellular GABA and increases extracellular glutamate and dopamine levels in the rat prefrontal cortex: An in vivo microdialysis study. *Brain Research, 948*(1–2), 155–158. https://doi.org/10.1016/s0006-8993(02)03055-x.

Pittenger, C., Krystal, J. H., & Coric, V. (2006). Glutamate-modulating drugs as novel pharmacotherapeutic agents in the treatment of obsessive-compulsive disorder. *NeuroRx: The Journal of the American Society for Experimental NeuroTherapeutics, 3*(1), 69–81. https://doi.org/10.1016/j.nurx.2005.12.006.

Popoli, M., Yan, Z., McEwen, B. S., & Sanacora, G. (2012). The stressed synapse: The impact of stress and glucocorticoids on glutamate transmission. *Nature Reviews. Neuroscience, 13*(1), 22–37. https://doi.org/10.1038/nrn3138.

Porter, R. H., Jaeschke, G., Spooren, W., Ballard, T. M., Buttelmann, B., Kolczewski, S., et al. (2005). Fenobam: A clinically validated nonbenzodiazepine anxiolytic is a potent, selective, and noncompetitive mGlu5 receptor antagonist with inverse agonist activity. *The Journal of Pharmacology and Experimental Therapeutics, 315*(2), 711–721. https://doi.org/10.1124/jpet.105.089839.

Prescot, A. P., Locatelli, A. E., Renshaw, P. F., & Yurgelun-Todd, D. A. (2011). Neurochemical alterations in adolescent chronic marijuana smokers: A proton MRS study. *NeuroImage, 57*(1), 69–75.

Purgert, C., Izumi, Y., Jong, Y., Kumar, V., Zorumski, C., & O'Malley, K. (2014). Intracellular mGluR5 can mediate synaptic plasticity in the hippocampus. *The Journal of Neuroscience, 34*, 4589–4598.

Quiroz, J. A., Tamburri, P., Deptula, D., Banken, L., Beyer, U., Rabbia, M., et al. (2016). Efficacy and safety of basimglurant as adjunctive therapy for major depression: A randomized clinical trial. *JAMA Psychiatry, 73*(7), 675–684. https://doi.org/10.1001/jamapsychiatry.2016.0838.

Reid, M. S., Fox, L., Ho, L. B., & Berger, S. P. (2000). Nicotine stimulation of extracellular glutamate levels in the nucleus accumbens: Neuropharmacological characterization. *Synapse, 35*(2), 129–136.

Ribeiro, F. M., Ferreira, L. T., Paquet, M., Cregan, T., Ding, Q., Gros, R., et al. (2009). Phosphorylation-independent regulation of metabotropic glutamate receptor 5 desensitization and internalization by G protein-coupled receptor kinase 2 in neurons. *The Journal of Biological Chemistry, 284*(35), 23444–23453. https://doi.org/10.1074/jbc.M109.000778.

Ritzén, A., Mathiesen, J. M., & Thomsen, C. (2005). Molecular pharmacology and therapeutic prospects of metabotropic glutamate receptor allosteric modulators. *Basic & Clinical Pharmacology & Toxicology, 97*(4), 202–213.

Rocher, J.-P., Bonnet, B., Bolea, C., Lutjens, R., Le Poul, E., Poli, S., et al. (2011). mGluR5 negative allosteric modulators overview: A medicinal chemistry approach towards a series of novel therapeutic agents. *Current Topics in Medicinal Chemistry, 11*(6), 680–695.

Rodriguez, C., Kegeles, L., Levinson, A., Feng, T., Marcus, S., Vermes, D., et al. (2013). Randomized controlled crossover trial of ketamine in obsessive-compulsive disorder: Proof-of-concept. *Neuropsychopharmacology, 38*, 2475–2483.

Roppe, J. R., Wang, B., Huang, D., Tehrani, L., Kamenecka, T., Schweiger, E. J., et al. (2004). 5-[(2-Methyl-1,3-thiazol-4-yl)ethynyl]-2,3′-bipyridine: A highly potent, orally active metabotropic glutamate subtype 5 (mGlu5) receptor antagonist with anxiolytic activity. *Bioorganic & Medicinal Chemistry Letters, 14*(15), 3993–3996. https://doi.org/10.1016/j.bmcl.2004.05.037.

Rothman, D. L., de Graaf, R. A., Hyder, F., Mason, G. F., Behar, K. L., & De Feyter, H. M. (2019). In vivo 13C and 1H-[13C] MRS studies of neuroenergetics and neurotransmitter cycling, applications to neurological and psychiatric disease and brain cancer. *NMR in Biomedicine, 32*(10), e4172.

Rubino, T., Prini, P., Piscitelli, F., Zamberletti, E., Trusel, M., Melis, M., et al. (2015). Adolescent exposure to THC in female rats disrupts developmental changes in the prefrontal cortex. *Neurobiology of Disease, 73*, 60–69. https://doi.org/10.1016/j.nbd.2014.09.015.

Rutrick, D., Stein, D. J., Subramanian, G., Smith, B., Fava, M., Hasler, G., et al. (2017). Mavoglurant augmentation in OCD patients resistant to selective serotonin reuptake inhibitors: A proof-of-concept, randomized, placebo-controlled, phase 2 study. *Advances in Therapy, 34*(2), 524–541. https://doi.org/10.1007/s12325-016-0468-5.

Sala, C., Roussignol, G., Meldolesi, J., & Fagni, L. (2005). Key role of the postsynaptic density scaffold proteins Shank and Homer in the functional architecture of Ca2+ homeostasis at dendritic spines in hippocampal neurons. *The Journal of Neuroscience, 25*, 4587–4592.

Samhsa. (2021). *Key substance use and mental health indicators in the United States: Results from the 2020 National Survey on Drug Use and Health*. Retrieved from https://www.samhsa.gov/.

Sanacora, G., Gueorguieva, R., Epperson, C., Wu, Y.-T., Appel, M., Rothman, D., et al. (2004). Subtype-specific alterations of GABA and glutamate in major depression. *Archives of General Psychiatry, 61*, 705–713.

Sanacora, G., Rothman, D. L., Mason, G., & Krystal, J. H. (2003). Clinical studies implementing glutamate neurotransmission in mood disorders. *Annals of the New York Academy of Sciences, 1003*(1), 292–308. https://doi.org/10.1196/annals.1300.018.

Sanacora, G., Treccani, G., & Popoli, M. (2012). Towards a glutamate hypothesis of depression: An emerging frontier of neuropsychopharmacology for mood disorders. *Neuropharmacology, 62*, 63–77.

Schilström, B., Fagerquist, M., Zhang, X., Hertel, P., Panagis, G., Nomikos, G., et al. (2000). Putative role of presynaptic $\alpha 7^*$ nicotinic receptors in nicotine stimulated increases of extracellular levels of glutamate and aspartate in the ventral tegmental area. *Synapse, 38*(4), 375–383.

Schroeder, J. P., Overstreet, D. H., & Hodge, C. W. (2005). The mGluR5 antagonist MPEP decreases operant ethanol self-administration during maintenance and after repeated alcohol deprivations in alcohol-preferring (P) rats. *Psychopharmacology, 179*(1), 262–270. https://doi.org/10.1007/s00213-005-2175-9. PMID - 15717208.

Schulz, B., Fendt, M., Gasparini, F., Lingenhöhl, K., Kuhn, R., & Koch, M. (2001). The metabotropic glutamate receptor antagonist 2-methyl-6-(phenylethynyl)-pyridine (MPEP) blocks fear conditioning in rats. *Neuropharmacology, 41*(1), 1–7.

Schumann, G., Johann, M., Frank, J., Preuss, U., Dahmen, N., Laucht, M., et al. (2008). Systematic analysis of glutamatergic neurotransmission genes in alcohol dependence and adolescent risky drinking behavior. *Archives of General Psychiatry, 65*(7), 826–838. https://doi.org/10.1001/archpsyc.65.7.826.

Schwendt, M., Shallcross, J., Hadad, N. A., Namba, M. D., Hiller, H., Wu, L., et al. (2018). A novel rat model of comorbid PTSD and addiction reveals intersections between stress susceptibility and enhanced cocaine seeking with a role for mGlu5 receptors. *Translational Psychiatry, 8*(1), 209. https://doi.org/10.1038/s41398-018-0265-9.

Scotti-Muzzi, E., Umla-Runge, K., & Soeiro-de-Souza, M. G. (2021). Anterior cingulate cortex neurometabolites in bipolar disorder are influenced by mood state and medication: A meta-analysis of (1)H-MRS studies. *European Neuropsychopharmacology, 47*, 62–73. https://doi.org/10.1016/j.euroneuro.2021.01.096.

Servaes, S., Glorie, D., Verhaeghe, J., Stroobants, S., & Staelens, S. (2017). Preclinical molecular imaging of glutamatergic and dopaminergic neuroreceptor kinetics in obsessive compulsive disorder. *Progress in Neuro-Psychopharmacology & Biological Psychiatry, 77*, 90–98. https://doi.org/10.1016/j.pnpbp.2017.02.027.

Sethna, F., & Wang, H. (2014). Pharmacological enhancement of mGluR5 facilitates contextual fear memory extinction. *Learning & Memory, 21*(12), 647–650.

Sethna, F., & Wang, H. (2016). Acute inhibition of mGluR5 disrupts behavioral flexibility. *Neurobiology of Learning and Memory, 130*, 1–6. https://doi.org/10.1016/j.nlm.2016.01.004.

Shallcross, J., Wu, L., Wilkinson, C. S., Knackstedt, L. A., & Schwendt, M. (2021). Increased mGlu5 mRNA expression in BLA glutamate neurons facilitates resilience to the long-term effects of a single predator scent stress exposure. *Brain Structure & Function, 226*(7), 2279–2293. https://doi.org/10.1007/s00429-021-02326-4.

Shen, M., Piser, T. M., Seybold, V. S., & Thayer, S. A. (1996). Cannabinoid receptor agonists inhibit glutamatergic synaptic transmission in rat hippocampal cultures. *Journal of Neuroscience, 16*(14), 4322–4334.

Shigemoto, R., Kinoshita, A., Wada, E., Nomura, S., Ohishi, H., Takada, M., et al. (1997). Differential presynaptic localization of metabotropic glutamate receptor subtypes in the rat hippocampus. *The Journal of Neuroscience, 17*, 7503–7522.

Shin, S., Kwon, O., Kang, J. I., Kwon, S., Oh, S., Choi, J., et al. (2015). mGluR5 in the nucleus accumbens is critical for promoting resilience to chronic stress. *Nature Neuroscience, 18*(7), 1017–1024. https://doi.org/10.1038/nn.4028.

Sidhpura, N., Weiss, F., & Martin-Fardon, R. (2010). Effects of the mGlu2/3 agonist LY379268 and the mGlu5 antagonist MTEP on ethanol seeking and reinforcement are differentially altered in rats with a history of ethanol dependence. *Biological Psychiatry, 67*(9), 804–811. https://doi.org/10.1016/j.biopsych.2010.01.005.

Silveri, M. M., Jensen, J. E., Rosso, I. M., Sneider, J. T., & Yurgelun-Todd, D. A. (2011). Preliminary evidence for white matter metabolite differences in marijuana-dependent young men using 2D J-resolved magnetic resonance spectroscopic imaging at 4 Tesla. *Psychiatry Research: Neuroimaging, 191*(3), 201–211.

Simonyi, A., Schachtman, T. R., & Christoffersen, G. R. (2005). The role of metabotropic glutamate receptor 5 in learning and memory processes. *Drug News & Perspectives, 18*(6), 353–361. https://doi.org/10.1358/dnp.2005.18.6.927927.

Sinclair, C. M., Cleva, R. M., Hood, L. E., Olive, M. F., & Gass, J. T. (2012). mGluR5 receptors in the basolateral amygdala and nucleus accumbens regulate cue-induced reinstatement of ethanol-seeking behavior. *Pharmacology Biochemistry and Behavior, 101*(3), 329–335.

Sistiaga, A., Herrero, I., Conquet, F., & Sánchez-Prieto, J. (1998). The metabotropic glutamate receptor 1 is not involved in the facilitation of glutamate release in cerebrocortical nerve terminals. *Neuropharmacology, 37*, 1485–1492.

Slattery, D. A., Neumann, I. D., Flor, P. J., & Zoicas, I. (2017). Pharmacological modulation of metabotropic glutamate receptor subtype 5 and 7 impairs extinction of social fear in a time-point-dependent manner. *Behavioural Brain Research, 328*, 57–61. https://doi.org/10.1016/j.bbr.2017.04.010.

Smart, K., Cox, S. M. L., Scala, S. G., Tippler, M., Jaworska, N., Boivin, M., et al. (2019). Sex differences in [(11)C]ABP688 binding: A positron emission tomography study of mGlu5 receptors. *European Journal of Nuclear Medicine and Molecular Imaging, 46*(5), 1179–1183. https://doi.org/10.1007/s00259-018-4252-4.

Spooren, W., & Gasparini, F. (2004). mGlu5 receptor antagonists: A novel class of anxiolytics? *Drug News & Perspectives, 17*(4), 251–257. https://doi.org/10.1358/dnp.2004.17.4.829052.

Spooren, W. P., Vassout, A., Neijt, H. C., Kuhn, R., Gasparini, F., Roux, S., et al. (2000). Anxiolytic-like effects of the prototypical metabotropic glutamate receptor 5 antagonist 2-methyl-6-(phenylethynyl)pyridine in rodents. *The Journal of Pharmacology and Experimental Therapeutics, 295*(3), 1267–1275.

Stoker, A., Olivier, B., & Markou, A. (2012). Involvement of metabotropic glutamate receptor 5 in brain reward deficits associated with cocaine and nicotine withdrawal and somatic signs of nicotine withdrawal. *Psychopharmacology, 221*, 317–327.

Sugaya, N., Ogai, Y., Aikawa, Y., Yumoto, Y., Takahama, M., Tanaka, M., et al. (2018). A randomized controlled study of the effect of ifenprodil on alcohol use in patients with alcohol dependence. *Neuropsychopharmacology Reports, 38*(1), 9–17. https://doi.org/10.1002/npr2.12001.

Sun, J. Y., Zhao, S. J., Wang, H. B., Hou, Y. J., Mi, Q. J., Yang, M. F., et al. (2021). Ifenprodil improves long-term neurologic deficits through antagonizing glutamate-induced excitotoxicity after experimental subarachnoid hemorrhage. *Translational Stroke Research, 12*(6), 1067–1080. https://doi.org/10.1007/s12975-021-00906-4.

Takumi, Y., Matsubara, A., Rinvik, E., & Ottersen, O. (1999). The arrangement of glutamate receptors in excitatory synapses. *Annals of the New York Academy of Sciences, 868*, 474–482.

Taylor, M. (2014). Could glutamate spectroscopy differentiate bipolar depression from unipolar? *Journal of Affective Disorders, 167*, 80–84.

Terbeck, S., Akkus, F., Chesterman, L. P., & Hasler, G. (2015). The role of metabotropic glutamate receptor 5 in the pathogenesis of mood disorders and addiction: Combining preclinical evidence with human Positron Emission Tomography (PET) studies. *Frontiers in Neuroscience, 9*, 86. https://doi.org/10.3389/fnins.2015.00086.

Thoma, R., Mullins, P., Ruhl, D., Monnig, M., Yeo, R. A., Caprihan, A., et al. (2011). Perturbation of the glutamate–glutamine system in alcohol dependence and remission. *Neuropsychopharmacology, 36*(7), 1359–1365.

Toth, E., Vizi, E., & Lajtha, A. (1993). Effect of nicotine on levels of extracellular amino acids in regions of the rat brain in vivo. *Neuropharmacology, 32*(8), 827–832.

Tronci, V., & Balfour, D. J. (2011). The effects of the mGluR5 receptor antagonist 6-methyl-2-(phenylethynyl)-pyridine (MPEP) on the stimulation of dopamine release evoked by nicotine in the rat brain. *Behavioural Brain Research, 219*(2), 354–357.

Tronci, V., Vronskaya, S., Montgomery, N., Mura, D., & Balfour, D. J. (2010). The effects of the mGluR5 receptor antagonist 6-methyl-2-(phenylethynyl)-pyridine (MPEP) on behavioural responses to nicotine. *Psychopharmacology, 211*(1), 33–42.

Tronson, N. C., Guzman, Y. F., Guedea, A. L., Huh, K. H., Gao, C., Schwarz, M. K., et al. (2010). Metabotropic glutamate receptor 5/Homer interactions underlie stress effects on fear. *Biological Psychiatry, 68*(11), 1007–1015. https://doi.org/10.1016/j.biopsych.2010.09.004.

Tsai, G. E., Ragan, P., Chang, R., Chen, S., Linnoila, V. M., & Coyle, J. T. (1998). Increased glutamatergic neurotransmission and oxidative stress after alcohol withdrawal. *The American Journal of Psychiatry, 155*(6), 726–732.

Tu, J. C., Xiao, B., Naisbitt, S., Yuan, J. P., Petralia, R. S., Brakeman, P., et al. (1999). Coupling of mGluR/Homer and PSD-95 complexes by the Shank family of postsynaptic density proteins. *Neuron, 23*(3), 583–592. https://doi.org/10.1016/S0896-6273(00)80810-7.

Umhau, J. C., Momenan, R., Schwandt, M. L., Singley, E., Lifshitz, M., Doty, L., et al. (2010). Effect of acamprosate on magnetic resonance spectroscopy measures of central glutamate in detoxified alcohol-dependent individuals: A randomized controlled experimental medicine study. *Archives of General Psychiatry, 67*(10), 1069–1077. https://doi.org/10.1001/archgenpsychiatry.2010.125.

van de Giessen, E., Weinstein, J. J., Cassidy, C. M., Haney, M., Dong, Z., Ghazzaoui, R., et al. (2017). Deficits in striatal dopamine release in cannabis dependence. *Molecular Psychiatry, 22*(1), 68–75.

Varma, N., Carlson, G. C., Ledent, C., & Alger, B. E. (2001). Metabotropic glutamate receptors drive the endocannabinoid system in hippocampus. *Journal of Neuroscience, 21*(24). RC188-RC188.

Varodayan, F. P., Sidhu, H., Kreifeldt, M., Roberto, M., & Contet, C. (2018). Morphological and functional evidence of increased excitatory signaling in the prelimbic cortex during ethanol withdrawal. *Neuropharmacology*, *133*, 470–480. https://doi.org/10.1016/j.neuropharm.2018.02.014.

Varty, G. B., Grilli, M., Forlani, A., Fredduzzi, S., Grzelak, M. E., Guthrie, D. H., et al. (2005). The antinociceptive and anxiolytic-like effects of the metabotropic glutamate receptor 5 (mGluR5) antagonists, MPEP and MTEP, and the mGluR1 antagonist, LY456236, in rodents: A comparison of efficacy and side-effect profiles. *Psychopharmacology*, *179*(1), 207–217. https://doi.org/10.1007/s00213-005-2143-4.

Veeraiah, P., Noronha, J. M., Maitra, S., Bagga, P., Khandelwal, N., Chakravarty, S., et al. (2014). Dysfunctional glutamatergic and γ-aminobutyric acidergic activities in prefrontal cortex of mice in social defeat model of depression. *Biological Psychiatry*, *76*(3), 231–238.

Vucurovic, K., Landais, E., Delahaigue, C., Eutrope, J., Schneider, A., Leroy, C., et al. (2012). Bipolar affective disorder and early dementia onset in a male patient with SHANK3 deletion. *European Journal of Medical Genetics*, *55*(11), 625–629. https://doi.org/10.1016/j.ejmg.2012.07.009.

Wan, Y., Ade, K. K., Caffall, Z., Ilcim Ozlu, M., Eroglu, C., Feng, G., et al. (2014). Circuit-selective striatal synaptic dysfunction in the Sapap3 knockout mouse model of obsessive-compulsive disorder. *Biological Psychiatry*, *75*(8), 623–630. https://doi.org/10.1016/j.biopsych.2013.01.008.

Watts, J. J., Garani, R., Da Silva, T., Lalang, N., Chavez, S., & Mizrahi, R. (2020). Evidence that Cannabis exposure, abuse, and dependence are related to glutamate metabolism and glial function in the anterior cingulate cortex: A (1)H-magnetic resonance spectroscopy study. *Frontiers in Psychiatry*, *11*, 764. https://doi.org/10.3389/fpsyt.2020.00764.

Welch, J. M., Lu, J., Rodriguiz, R. M., Trotta, N. C., Peca, J., Ding, J. D., et al. (2007). Cortico-striatal synaptic defects and OCD-like behaviours in Sapap3-mutant mice. *Nature*, *448*(7156), 894–900. https://doi.org/10.1038/nature06104.

Wieronska, J. M., Branski, P., Szewczyk, B., Palucha, A., Papp, M., Gruca, P., et al. (2001). Changes in the expression of metabotropic glutamate receptor 5 (mGluR5) in the rat hippocampus in an animal model of depression. *Polish Journal of Pharmacology*, *53*(6), 659–662. Retrieved from http://www.ncbi.nlm.nih.gov/pubmed/11985342.

Wilson, R. I., & Nicoll, R. A. (2001). Endogenous cannabinoids mediate retrograde signalling at hippocampal synapses. *Nature*, *410*(6828), 588–592.

Wooltorton, J. R., Pidoplichko, V. I., Broide, R. S., & Dani, J. A. (2003). Differential desensitization and distribution of nicotinic acetylcholine receptor subtypes in midbrain dopamine areas. *Journal of Neuroscience*, *23*(8), 3176–3185.

Xu, J., Zhu, Y., Kraniotis, S., He, Q., Marshall, J. J., Nomura, T., et al. (2013). Potentiating mGluR5 function with a positive allosteric modulator enhances adaptive learning. *Learning & Memory (Cold Spring Harbor, N.Y.)*, *20*(8), 438–445. https://doi.org/10.1101/lm.031666.113.

Xu, J., Dydak, U., Harezlak, J., Nixon, J., Dzemidzic, M., Gunn, A., et al. (2013). Neurochemical abnormalities in unmedicated bipolar depression and mania: A 2D 1H MRS investigation. *Psychiatry Research*, *213*, 235–241.

Xu, J., Zhu, Y., Contractor, A., & Heinemann, S. F. (2009). mGluR5 has a critical role in inhibitory learning. *The Journal of Neuroscience*, *29*(12), 3676–3684. https://doi.org/10.1523/jneurosci.5716-08.2009.

Yang, Z. Y., Quan, H., Peng, Z. L., Zhong, Y., Tan, Z. J., & Gong, Q. Y. (2015). Proton magnetic resonance spectroscopy revealed differences in the glutamate+ glutamine/

creatine ratio of the anterior cingulate cortex between healthy and pediatric post-traumatic stress disorder patients diagnosed after 2008 W enchuan earthquake. *Psychiatry and Clinical Neurosciences*, *69*(12), 782–790.

Yang, C., Ren, Q., Qu, Y., Zhang, J. C., Ma, M., Dong, C., et al. (2017). Mechanistic target of rapamycin-independent antidepressant effects of (R)-ketamine in a social defeat stress model. *Biological Psychiatry*. https://doi.org/10.1016/j.biopsych.2017.05.016.

Yararbas, G., Keser, A., Kanit, L., & Pogun, S. (2010). Nicotine-induced conditioned place preference in rats: Sex differences and the role of mGluR5 receptors. *Neuropharmacology*, *58*(2), 374–382.

Yin, S., & Niswender, C. M. (2014). Progress toward advanced understanding of metabotropic glutamate receptors: Structure, signaling and therapeutic indications. *Cellular Signalling*, *26*(10), 2284–2297. https://doi.org/10.1016/j.cellsig.2014.04.022.

Yüksel, C., & Öngür, D. (2010). Magnetic resonance spectroscopy studies of glutamate-related abnormalities in mood disorders. *Biological Psychiatry*, *68*, 785–794.

Zanos, P., Moaddel, R., Morris, P. J., Georgiou, P., Fischell, J., Elmer, G. I., et al. (2016). NMDAR inhibition-independent antidepressant actions of ketamine metabolites. *Nature*, *533*(7604), 481–486. https://doi.org/10.1038/nature17998.

Zarate, C. J., Brutsche, N., Ibrahim, L., Franco-Chaves, J., Diazgranados, N., Cravchik, A., et al. (2012). Replication of ketamine's antidepressant efficacy in bipolar depression: A randomized controlled add-on trial. *Biological Psychiatry*, *71*, 939–946.

CHAPTER SEVEN

Sex differences and hormonal regulation of metabotropic glutamate receptor synaptic plasticity

Carly B. Fabian[a,b,c], Marianne L. Seney[a,b,c], and Max E. Joffe[a,b,c,*]

[a]Center for Neuroscience, University of Pittsburgh, Pittsburgh, PA, United States
[b]Department of Psychiatry, University of Pittsburgh, Pittsburgh, PA, United States
[c]Translational Neuroscience Program, University of Pittsburgh, Pittsburgh, PA, United States
*Corresponding author: e-mail address: joffeme@upmc.edu

Contents

1. Introduction	312
2. Sex differences in mGlu receptor function	314
2.1 Regional distribution and receptor expression	314
2.2 Receptor trafficking	315
2.3 Signaling and synaptic plasticity	316
2.4 Sex hormone regulation	318
3. Insight gained from behavioral pharmacology, transgenic mice, and disease models	321
3.1 Cognition	322
3.2 Social behaviors	323
3.3 Alcohol use	324
3.4 Drug use	326
3.5 Anxiety-like and affective behavior	329
3.6 Neurodegeneration	333
4. Conclusions	334
Acknowledgments	336
Conflict of Interest	336
References	336

Abstract

Striking sex differences exist in presentation and incidence of several psychiatric disorders. For example, major depressive disorder is more prevalent in women than men, and women who develop alcohol use disorder progress through drinking milestones more rapidly than men. With regards to psychiatric treatment responses, women respond more favorably to selective serotonin reuptake inhibitors than men, whereas men have better outcomes when prescribed tricyclic antidepressants. Despite such

---

International Review of Neurobiology, Volume 168
ISSN 0074-7742
https://doi.org/10.1016/bs.irn.2022.10.002

Copyright © 2023 Elsevier Inc.
All rights reserved.

well-documented biases in incidence, presentation, and treatment response, sex as a biological variable has long been neglected in preclinical and clinical research. An emerging family of druggable targets for psychiatric diseases, metabotropic glutamate (mGlu) receptors are G-protein coupled receptors broadly distributed throughout the central nervous system. mGlu receptors confer diverse neuromodulatory actions of glutamate at the levels of synaptic plasticity, neuronal excitability, and gene transcription. In this chapter, we summarize the current preclinical and clinical evidence for sex differences in mGlu receptor function. We first highlight basal sex differences in mGlu receptor expression and function and proceed to describe how gonadal hormones, notably estradiol, regulate mGlu receptor signaling. We then describe sex-specific mechanisms by which mGlu receptors differentially modulate synaptic plasticity and behavior in basal states and models relevant for disease. Finally, we discuss human research findings and highlight areas in need of further research. Taken together, this review emphasizes how mGlu receptor function and expression can differ across sex. Gaining a more complete understanding of how sex differences in mGlu receptor function contribute to psychiatric diseases will be critical in the development of novel therapeutics that are effective in all individuals.

1. Introduction

Sex differences in the incidence, presentation, and treatment responses of several psychiatric and substance use disorders are well-known. Major depressive disorder (MDD), for instance, exhibits sex differences across several levels. Women are approximately twice as likely to have MDD than men, exhibit more severe symptoms, and often experience different symptomatology (Angst & Dobler-Mikola, 1984; Frank, Carpenter, & Kupfer, 1988; Kessler et al., 2005; Kornstein et al., 2000a; Najt, Fusar-Poli, & Brambilla, 2011; Silverstein, 1999; Young et al., 1990). Additionally, evidence suggests that women and men, on average, have superior treatment responses to different antidepressant medications. For instance, women have statistically superior responses to monoamine oxidase inhibitors and selective serotonin reuptake inhibitors than men; conversely, men respond more favorably to tricyclic antidepressants (Hamilton, Grant, & Jensvold, 1996; Kornstein et al., 2000b; Quitkin et al., 2002). In contrast, many reported sex differences in schizophrenia are weighted toward men. For example, men have an earlier age of onset of schizophrenia, with symptoms emerging in the early twenties in men, while symptom emergence is in the late twenties in women (Angermeyer, Goldstein, & Kuehn, 1989; Goldstein, Tsuang, & Faraone, 1989; Hafner et al., 1989; Hambrecht, Maurer, & Hafner, 1992; Sartorius et al., 1986). Men with schizophrenia also exhibit higher incidence of negative symptoms and poorer course compared to women with schizophrenia (Goldstein et al., 1998; Goldstein & Link, 1988; Huber et al., 1980; Salokangas, 1983). Similarly, sex-dependent vulnerabilities to specific disease symptoms have been observed in alcohol use disorders (AUD) and several concerning findings indicate that women are at greater risks for detrimental outcomes. While AUDs are more prevalent in men overall, women are more sensitive to peripheral diseases and cognitive disturbances that can stem from chronic alcohol consumption (Nixon, Tivis, & Parsons, 1995; Tuyns & Pequignot, 1984; Urbano-Marquez et al., 1995). Women who develop AUD also progress through disease milestones more rapidly than men (Diehl et al., 2007; Lewis, Hoffman, & Nixon, 2014; Randall et al., 1999), and women are disproportionately affected by consequences of acute intoxication (Gross & Billingham, 1998). Clearly, many psychiatric diseases display remarkable sex differences in presentation and etiology. With the goal of delivering optimal treatment outcomes to all, understanding the manifestation of these sex differences should be a top priority of biological psychiatry research.

Despite the well-documented sex differences in the epidemiology of psychiatric disorders, it is only relatively recently that researchers have begun to delve into potential biological underpinnings for these observed sex differences. A major driving force behind these recent studies is the National Institutes of Health mandate that sex as a biological variable (SABV) will be factored into research designs, analyses, and reporting in vertebrate animal and human studies. Prior to the mandate, many researchers argued that the circulating gonadal hormones associated with the estrous cycle would introduce too much variability into experimental measures; the result is that many previous basic and preclinical studies excluded female subjects and cells. Thus, much of what is known about biological mechanisms of psychiatric disorders is based on studies in males. The NIMH mandate to include SABV has resulted in an explosion of research documenting sex differences in cell signaling, synaptic plasticity, emotional behaviors, cognition, and social function. In this chapter, we describe and summarize the known sex differences in the functions of metabotropic glutamate (mGlu) receptors. Consistent with the broad and rich neuroendocrinology literature documenting acute effects of gonadal hormones on brain function, significant evidence suggests that sex hormones regulate a variety of mGlu receptor functions through proximal receptor-receptor interactions. However, much less is known about how other factors, including developmental hormone exposure and sex chromosome complement, might influence sex-specific behavioral and neurobiological outcomes. In addition to synthesizing key themes from the existing literature, we hope this chapter will be useful in identifying and highlighting the important gaps in our understanding of sex differences in mGlu receptor function.

2. Sex differences in mGlu receptor function

Metabotropic glutamate (mGlu) receptors are a class of type C G protein-coupled receptors widely distributed in the nervous system. Activation of mGlu receptors via extracellular binding of glutamate triggers intracellular signaling cascades that induce short- and long-term adaptations in synaptic physiology, neuron excitability, and gene expression (Niswender & Conn, 2010). The eight mGlu receptor subtypes are grouped into three families (Groups 1–3) based on sequence homology, cell signaling, and pharmacology (Conn & Pin, 1997).

2.1 Regional distribution and receptor expression

The Group I mGlu receptors, comprised of subtypes $mGlu_1$ and $mGlu_5$, are important regulators of excitatory glutamate transmission and synaptic plasticity. $mGlu_1$ and $mGlu_5$ receptors are primarily localized to postsynaptic compartments, where they signal through Gq proteins and associated downstream effectors (Hermans & Challiss, 2001; Niswender & Conn, 2010). Our current understanding of the regional distribution and localization of Group 1 receptors in the CNS is largely based on gene-cloning and immunoreactivity studies from preclinical models (Martin et al., 1992; Romano et al., 1995; Van den Pol, 1994; van den Pol, Romano, & Ghosh, 1995). $mGlu_1$ and $mGlu_5$ receptors have widespread, generally overlapping distributions throughout most of the brain, with a key exception being the cerebellum, where high levels of $mGlu_1$ (but not $mGlu_5$) are expressed in mature Purkinje cells (Baude et al., 1993; Casabona et al., 1997; Fazio et al., 2008). An important caveat to this literature is that most studies investigating the regional distribution of Group 1 receptors were restricted to male rodents or not statistically powered to test for interactions across sexes. As such, the potential for intrinsic sex differences in the regional distribution of $mGlu_{1/5}$ receptors remains largely unexplored.

Although major sex differences in the regional distribution of Group 1 receptors have not been reported (Wickens, Bangasser, & Briand, 2018), preclinical studies investigating the potential sex differences in levels of mGlu receptor protein expression have revealed conflicting results. In a

Western blot analysis of mGlu receptor subunit protein expression, adult female rats exhibited higher levels of $mGlu_5$ expression in the prefrontal cortex (PFC) and hippocampus than males, with no sex differences in $mGlu_1$ receptor protein expression in the PFC or hippocampus (Wang et al., 2015). These findings are in direct contrast to results from a recent study published in 2017, in which female rats had significantly lower levels of $mGlu_1$ receptor protein expression in the hippocampus than males (Ning et al., 2018). Both of these studies used Sprague-Dawley rats, suggesting that differences in animal strain do not account for the divergent findings. Interestingly, however, these studies employed antibodies raised against different portions of the intracellular C-terminal tail; therefore, it is possible that changes in the proportion of splice variants or post-translational modifications might account for the discrepancies in the literature. Additional studies examining cell type-specific expression of $mGlu_1$ and $mGlu_5$, as well as the relative expression of common splice variants and modifications, appear duly warranted.

There are limited published accounts describing sex differences in Group 2 and Group 3 mGlu receptor expression and function. As such, little is known about how these receptor subtypes endogenously differ in male and female brains. Higher levels of $mGlu_{2/3}$ receptor expression (assessed with a non-specific antibody) were found in the hippocampus of wild-type female rats in comparison to males, and no sex differences were detected in PFC or amygdala (Wang et al., 2015). In addition, a study in human temporal cortex found higher expression of $mGlu_3$ receptors in women relative to men (García-Bea et al., 2016). To our knowledge, sex differences in baseline Group 2 receptor expression have not been thoroughly explored in other brain regions. Similarly, to our knowledge, potential sex differences in regional or synaptic localization of Group 3 receptors have not been thoroughly investigated.

2.2 Receptor trafficking

Group 1 receptors are localized to postsynaptic densities and anchored to nearby receptors through interactions with a multitude of scaffolding proteins, most notably including Homer, Shank, and PSD95 (Tu et al., 1999). There are three genes that comprise the Homer family proteins (Homer1–3) (Brakeman et al., 1997; Xiao et al., 1998) each of which has multiple alternative short and long form splice variants involved in the trafficking and signaling of Group 1 receptors (Kammermeier et al., 2000).

Long form splice variants of Group 1 Homer proteins (Homer1b and Homer1c) differentially regulate mGlu$_{1/5}$ receptor trafficking on the post-synaptic membrane. In cultures from male rodents, Homer1b inhibits mGlu$_5$ receptor trafficking to the cell surface (Roche et al., 1999), whereas co-expression of Homer1c increases surface expression of mGlu$_1$ receptors (Ciruela et al., 2000; Xiao, Gustafsson, & Niu, 2006). To our knowledge, these findings have not been replicated in cultures from female animals. Nonetheless, there is little evidence to suggest basal sex differences in Homer protein regulation of Group 1 receptors, and neither Homer1 nor Homer2 knockout mice present with sex-dependent phenotypes in unconditioned behaviors (Szumlinski et al., 2004, 2005). Through their interactions with Homer proteins, the scaffolding proteins Shank and PSD-95 anchor Group 1 mGlu receptors within postsynaptic densities (Tu et al., 1999). Shank proteins are linked to NMDA receptors through PSD-95 associated complexes (Alagarsamy, Sorensen, & Conn, 2001; Xiao et al., 1998). In turn, Shank proteins enable mGlu$_{1/5}$ receptors to associate and couple with NMDA receptor complexes (Bertaso et al., 2010). Analysis of the baseline expression of PSD-95 in male and female rats showed no sex differences in protein levels in the hippocampus, amygdala, or PFC (Wang et al., 2015). Taken together, the existing cell signaling literature has not identified major sex differences in trafficking or scaffolding of mGlu receptors.

2.3 Signaling and synaptic plasticity

Throughout the forebrain, mGlu receptors are generally conceptualized as modulatory components that regulate fast neurotransmission through glutamate or GABA receptors. Canonical signaling downstream of Group 1 mGlu receptors proceeds through activation of Gq proteins, phospholipase C (PLC), Protein Kinase C (PKC), and the release of intracellular calcium (Conn & Pin, 1997; De Blasi et al., 2001; Valenti, Conn, & Marino, 2002). In addition to PKC activation, Group 1 receptors can activate a myriad of downstream effectors, including the mitogen-activated protein kinase/extracellular receptor kinase (MAPK/ERK) pathway (Mao & Wang, 2016). mGlu$_1$ and mGlu$_5$ receptors can also facilitate the production of endocannabinoids, such as 2-arachadonylglycerol and anandamide (Ohno–Shosaku et al., 2002; Varma et al., 2001), which can attenuate pre-synaptic release probability through CB1 receptors and also regulate post-synaptic plasticity through the transient receptor potential cation channel

TRPV1 (Chávez, Chiu, & Castillo, 2010; Grueter, Brasnjo, & Malenka, 2010). A recent study found that, in the rat PFC, 10 Hz electrical stimulation evoked $mGlu_5$ receptor-dependent LTD with distinct latent mechanisms across sexes (Bara et al., 2018). While LTD was mediated by $mGlu_5$ receptors and endocannabinoid signaling in both sexes, LTD was blocked by the CB1 antagonists in male rats but TRPV1 antagonists blocked LTD in female rats. These findings highlight the tremendous need to revisit basic mechanisms to probe for potential latent or overt differences in mGlu receptor synaptic plasticity.

In addition to their ability to regulate synaptic plasticity, in some nuclei and cell types, $mGlu_{1/5}$ receptor activation can elicit excitatory postsynaptic currents through more proximal actions on nonselective cation channels. For example, in several types of neurons within the cerebellum, $mGlu_1/5$ receptor activation can elicit slow excitatory postsynaptic currents by activating transient receptor potential TRPC3 channels (Hartmann, Henning, & Konnerth, 2011). Interestingly, there are sex differences in cerebellar nuclei firing rates, with cerebellar nuclei cells from female mice having larger $mGlu_{1/5}$ receptor currents than males, suggesting sex differences in $mGlu_{1/5}$ receptor function (Mercer et al., 2016). However, there were no differences in Group 1 receptor protein levels between male and female mice, but blockade of glutamate transport increased the magnitude of $mGlu_{1/5}$-dependent currents in male but not female mice. These data suggest that basal sex differences in cerebellar nuclei $mGlu_{1/5}$ currents may not stem from receptor availability but are instead related to the access of glutamate to Group 1 receptors.

Group 2 and 3 mGlu receptors are generally linked to Gi/o proteins, which inhibit adenylyl cyclase to decrease cAMP production. Group 2 and 3 receptors can also initiate a variety of downstream signaling pathways through the liberation of $G\beta\gamma$ subunits, notably including the inhibition of calcium channels and facilitation of potassium channels. Similar to Group 1 receptors, Group 2 and 3 mGlu receptors can also modulate kinase signaling, such as MAPK and phosphatidylinositol 3-kinase (PI3K) pathways. Despite similarities in sequence homology and pharmacology, $mGlu_2$ and $mGlu_3$ receptors regulate neurotransmission through distinct mechanisms of action. While both Group 2 receptor subtypes act as presynaptic autoreceptors to reduce glutamate release probability, $mGlu_3$ receptors are also heavily expressed at postsynaptic sites, where they can induce LTD through a postsynaptic signaling cascade involving $mGlu_5$ receptors, Homer proteins, PI3K pathway signaling, and the internalization of

AMPA receptors. Thus, any of the experience-dependent sex differences in $mGlu_5$ receptor and/or Homer signaling (described in later sections), might conceivably have ramifications for $mGlu_3$ receptor function as well.

2.4 Sex hormone regulation

Cellular pools of estrogen and other steroid hormones bind to both nuclear and membrane localized receptors, with varying affinity and response (Fuentes & Silveyra, 2019; Levin, 2009; Watson, Jeng, & Kochukov, 2008). In regions of the hippocampus, PFC, dorsal striatum, and nucleus accumbens, estrogen binds to membrane-localized receptors to activate second messenger systems and affect synaptic function (Milner et al., 2001; Razandi et al., 2004). Notably, several central effects of estradiol (E2) are linked to $mGlu_{1/5}$ receptors. E2 alters neuronal excitability and synaptic plasticity through stimulation of Group 1 receptor machinery in both a sex-dependent and independent manner (McEwen et al., 2001; Tonn Eisinger et al., 2018). Membrane-localized estrogen receptors, estrogen receptor α (ERα) and ERβ, are functionally coupled to $mGlu_{1/5}$ receptors expressed on the cell surface. Binding of E2 to these ER/mGlu complexes activates G proteins to trigger a glutamate-independent intracellular signaling cascade (Meitzen & Mermelstein, 2011). It is important to note that ER/mGlu complexes are present in both male and female brains, without major sex differences in overall expression. For example, immunocytochemistry studies targeting ERs in rat hippocampus (Weiland et al., 1997) and mouse amygdala (Bender et al., 2017) found no sex differences in ERα or ERβ expression levels. A more recent study separating mouse hippocampal ERs into nuclear and membrane populations detected greater extranuclear ERα in tissue from female mice, but no sex differences in ERβ expression (Mitterling et al., 2010). Moreover, ERα expression was highest in the CA1 region of mouse hippocampus during diestrus, while expression of ERβ in CA1 pyramidal cell layers was relatively high across the estrous cycle. We are not aware of any published findings describing Group 1 receptor expression throughout the estrous cycle.

Sex differences in ER/mGlu receptor localization and have been identified in the hippocampus. Co-immunoprecipitation studies indicate that ERα colocalizes with $mGlu_1$ and $mGlu_5$ receptors in both males and females (Boulware, Heisler, & Frick, 2013; Tabatadze et al., 2015). Interestingly, E2 treatment selectively increases the co-localization of ERα with $mGlu_1$ receptors in females, without affecting co-localization with $mGlu_1$ receptors in males or with $mGlu_5$ receptors in either sex. Other studies report sex differences in expression of proteins associated with the organization and signaling of ERα/mGlur1 complexes. Transcripts for caveolin 1 and the palmitoylacyltransferase DHHC-7, were expressed at lower levels in the hippocampus of adult female rats compared to males (Meitzen et al., 2019), suggesting that posttranslational modifications may also guide sex differences in ERα/mGlu receptor co-localization. Interestingly, no sex differences in gene expression were detected in hippocampal samples from neonates, suggesting a possible developmental sex-dependent regulation of ERα/mGlu receptor complexes. Taken together, ER signaling promotes the surface expression of $mGlu_{1/5}$ receptors to a greater extent in female individuals. Some evidence suggests the converse relationship holds as well (i.e., that mGlu receptors promote the expression of ERα). In ventromedial hypothalamus, genetic deletion of $mGlu_5$ receptors decreased ERα expression in female mice but increased ERα expression in males (Fagan et al., 2020). In this study, $mGlu_5$ receptors did not appear to regulate ERβ expression and $mGlu_1$ receptor function was not addressed. Taken together, the bidirectional relationship between ERα and $mGlu_{1/5}$ receptors raises the possibility that sex differences in downstream signaling may govern $mGlu_{1/5}$ receptor localization and function.

There are both sex and region differences in the pairing of ER/mGlu receptor subtype coupling and their downstream effects. E2 facilitates IP3 production in hippocampal slices from female rats to a much greater extent than in male rats (Tabatadze et al., 2015). E2-driven IP3 production was abolished by JNJ16259685, indicating an $mGlu_1$ receptor-dependent mechanism. Importantly, no difference in IP3 production was observed during pharmacological activation of $mGlu_1$ with the Group 1 agonist DHPG (Tabatadze et al., 2015), suggesting that the sex differences in signaling is manifested upstream of mGlu receptors. Other studies have yielded similar results and identified additional signaling nodes. In hippocampal cultures from female rats, E2 facilitates CREB phosphorylation through a pathway downstream of $mGlu_1$ receptors, PLC, PKC, IP3 receptors, and MAPK (Boulware et al., 2005). In these studies, an $mGlu_5$ negative allosteric modulator (NAM) had no effect, again indicating specific coupling between ERα and $mGlu_1$ receptors. Furthermore, ERα/$mGlu_1$ receptor-dependent CREB phosphorylation was specific to female hippocampal cultures, whereas similar E2 treatment in male cultures had no effect on CREB phosphorylation (Boulware et al., 2005). Together, these studies indicate that E2 signals through $mGlu_1$ receptor and canonical G_q signaling pathways in the female hippocampus. By contrast, E2 activates $mGlu_5$ receptor signaling in striatum. In female striatal cultures, the rapid actions of E2 lead to MAPK-dependent CREB phosphorylation through the activation of $ER\alpha/mGlu_5$ receptor signaling complexes (Grove-Strawser, Boulware, & Mermelstein, 2010). Similar to the studies in hippocampus cultures, male striatal cultures did not respond to E2 treatment. Thus, while ER-$mGlu_{1/5}$ receptor interactions are potentiated in the brains of female individuals, the mGlu receptor subtype involved varies based on region.

Functional studies also support the conclusion that there are sex differences in $ER\alpha/mGlu_{1/5}$ receptor regulation of neuronal physiology and synaptic plasticity. In ventromedial hypothalamus, genetic deletion of $mGlu_5$ receptors decreases current-evoked spike-firing in female but not male mice (Fagan et al., 2020). These actions were linked with E2 signaling, as the decreased excitability in $mGlu_5$ receptor knockouts was occluded by ovariectomy in control mice and restored by E2. Neurons within cerebellar nuclei also display larger $mGlu_{1/5}$ slow excitatory currents in female mice relative to males (Mercer et al., 2016). In addition to basic physiology and excitability, a growing literature now indicates that gonadal hormones regulate $mGlu_{1/5}$ receptor-dependent synaptic plasticity. While E2 is synthesized in the hippocampus of both males and females, the hormone exerts sex-specific effects on synaptic plasticity in CA1. In female hippocampal pyramidal cells, E2 suppresses inhibitory synapses through activation of post-synaptic $ER\alpha/mGlu$ receptor complexes, retrograde signaling through anandamide and CB1 receptors, and the attenuation of GABA release probability (Huang & Woolley, 2012). E2 also potentiates excitatory transmission in hippocampus through distinct mechanisms of action in females and males (Oberlander & Woolley, 2016), but it remains unclear whether mGlu receptor signaling is involved in these phenomena.

Naturally occurring fluctuations in apical dendritic spine density of CA1 hippocampal pyramidal cells are mediated, in part, by fluctuating E2 levels across the estrous cycle in female rats (Woolley et al., 1990; Woolley & McEwen, 1993). Studies in nucleus accumbens suggest that these endogenous cycles are regulated by $mGlu_1$ and $mGlu_5$. Consistent with the results from functional studies, E2 and other gonadal hormones alter dendritic spine plasticity in medium spiny neurons of the nucleus accumbens through the activation of Group 1 receptors (Gross et al., 2016, 2018). Though $mGlu_{1/5}$ receptors are structurally similar and tend to signal through conserved downstream pathways, evidence suggests that $mGlu_1$ and $mGlu_5$ receptors regulate spine density through distinct mechanisms. For example, systemic administration of selective mGlu$_1$ and mGlu$_5$ positive allosteric modulators increase and decrease spine density, respectively, in the nucleus accumbens of female rats (Gross et al., 2016). In addition, this structural plasticity is linked to gonadal hormones. For instance, E2 decreases spine density in nucleus accumbens core medium spiny neurons, but not dorsal striatum, and this effect was blocked by MPEP (Peterson, Mermelstein, & Meisel, 2015). Similar to E2, the male hormone dihydrotestosterone (DHT) can also influence dendritic plasticity through Group 1 activation. In castrated males, DHT administration decreased dendritic spine density on MSNs in the nucleus accumbens shell, but not core, of castrated male rats and this effect was blocked by pretreatment with MPEP (Gross et al., 2018). Thus, similar to their dissociable roles in mediating gonadal hormone effects on neuronal physiology, mGlu$_1$, and mGlu$_5$ receptors differentially regulate hormonal effects on structural plasticity in a region-specific manner.

Group 2 receptor signaling and female sex hormone signaling converge in the regulation of ion channels and activity-dependent gene transcription. Initial studies independently show that Group 2 receptor agonists (Chavis et al., 1994) and E2 each lead to the inhibition of L-type calcium channel currents through either ERα or ERβ (Chaban et al., 2003; Mermelstein, Becker, & Surmeier, 1996). LY341495 (at concentrations selective for mGlu$_{2/3}$ receptors) blocks the ability of E2 to attenuate calcium channel currents and CREB phosphorylation (Boulware et al., 2005). Though the mGlu$_{2/3}$ agonists suppressed CREB phosphorylation similarly in hippocampal cultures from male rats, E2's effects were specific to cultures from female rats. Additional studies have begun to tease apart the relative contribution of mGlu$_2$ vs mGlu$_3$ receptors. In striatal cultures from female rats, studies using subtype-specific interfering RNA revealed that knockdown of mGlu$_3$ receptors, but not mGlu$_2$ receptors, prevented E2 from attenuating CREB phosphorylation (Grove-Strawser et al., 2010). In addition, neither mGlu$_2$ nor mGlu$_3$ receptor expression appears to be altered by estrus cycle. Comparable receptor levels were detected in the nucleus accumbens when cocaine/vehicle self-administration rats were killed in the estrus vs. non-estrus phases (Logan et al., 2020). It is important to note, however, that these findings may or may not generalize to other brain areas.

3. Insight gained from behavioral pharmacology, transgenic mice, and disease models

Pharmacological and genetic manipulation of mGlu receptor subtypes in preclinical research have revealed further insights into their biological functions. Over the past two decades, numerous studies have used highly selective pharmacological ligands and transgenic animals to characterize the function of mGlu receptor functions in male preclinical models (each of which have been extensively detailed in (Niswender & Conn, 2010)). In stark comparison, only a handful of studies have addressed mGlu$_{1/5}$ receptor function in females, and even fewer studies have made direct comparisons to probe for potential sex differences. Nonetheless, the results of those few studies are consistent with significant sex differences in mGlu receptor functions in basal and altered states.

3.1 Cognition

mGlu$_1$ (Gil-Sanz et al., 2008) and mGlu$_5$ (Simonyi, Schachtman, & Christoffersen, 2005; Xu et al., 2009) receptors have been widely implicated in associative learning based on studies performed in male rodents (Luscher & Huber, 2010). However, it remains unclear whether similar contributions of mGlu$_1$ and mGlu$_5$ receptors guide cognition in female individuals. In a study using mice with selective deletion of mGlu$_5$ receptors from parvalbumin-expressing interneurons, there was a sex difference in perseverative behavior on the Barnes maze (Barnes et al., 2015). Male mice lacking mGlu$_5$ receptors in parvalbumin cells took longer to escape the Barnes maze and displayed more preservative errors than controls. By contrast, while female knockout mice also performed worse than female controls, changes in perseveration were not closely tied to the cognitive disruption. In addition, a genotype by sex interaction was observed with respect to auditory event related potentials, suggesting that mGlu$_5$ receptor signaling in cortical parvalbumin interneurons may differentially regulate rhythmic activity and behavioral flexibility across sexes. Considering that other studies observed no differences in PFC-dependent cognitive behaviors in models where mGlu$_5$ receptor function on pyramidal cells was disrupted (Bara et al., 2018), these findings reinforce that mGlu receptor function can regulate behavior in a cell type-specific manner.

Studies investigating sex hormone regulation of learning and memory also provide clear indications that mGlu$_{1/5}$ receptor signaling may guide latent sex differences in the molecular mechanisms of cognition. For example, E2 delivery to the dorsal hippocampus of female mice enhances novel object recognition (Fernandez et al., 2008; Zhao et al., 2012; Zhao, Fan, & Frick, 2010), an effect blocked by co-administration of LY367385 or the MEK inhibitor U0126 (Boulware et al., 2013). Considering that the female hippocampus displays enhanced ERK-dependent ER/mGlu$_1$ receptor signaling (see earlier section), a compelling hypothesis for future research is that mGlu$_1$ receptors differentially contribute to hippocampal-dependent cognition across sexes. Similarly, some evidence suggests sex-differences in the contribution upregulation of hippocampal mGlu$_5$ receptors to cognitive processing. In a model of phenylketonuria, male but not female mice, displayed increased mGlu$_5$ receptor protein levels in the hippocampus that coincided with a deficit in object location recognition (Nardecchia et al., 2018). The increased mGlu$_5$ receptor expression occurred at the protein, but not transcript, level and was also associated with increased expression of long Homer isoforms. Additionally, administration of MPEP to mutant male mice significantly improved behavioral abnormalities. Taken together, the preclinical literature suggests that potential mGlu$_1$ and mGlu$_5$ modulators may display sex differences in pro-cognitive efficacy.

Genetic deletion of mGlu$_8$ receptors alters hippocampal-dependent learning and memory (Duvoisin et al., 2010). Though mGlu$_8$ receptor deletion impaired novel object recognition and spatial memory retention in the Morris water maze in both males and females, the effects on memory retention were more pronounced in females. Later studies from the same group found impaired contextual fear learning in female knockout mice relative to wildtypes, whereas males were not affected (Torres et al., 2018). Together, this literature raises the possibility that hippocampal circuits in female individuals are particularly sensitive to manipulations in mGlu$_8$ receptor function. Transgenic mouse studies have also identified that mGlu$_7$ receptors are involved in working memory (Hölscher et al., 2004), fear and extinction learning (Goddyn et al., 2008), anxiety-like behaviors (Cryan et al., 2003), and social behaviors (Fisher et al., 2020, 2021). In each of these studies, female mice and male mice were included, similar effects were observed across sexes, and the datasets were collapsed for subsequent analysis and presentation. Thus, there is minimal evidence to suggest notable sex differences in mGlu$_7$ receptor function regulate cognition or other behaviors.

3.2 Social behaviors

As mentioned previously, Group 1 receptors drive endocannabinoid signaling to regulate synaptic plasticity and behavior (Varma et al., 2001). Previous studies involving prenatal cannabinoid exposure have linked adaptations to PFC mGlu$_{1/5}$ receptor function and endocannabinoids to social behavior in a sex-dependent manner. WIN-55,212 administration during gestation causes decreases in adult social exploration and play behaviors in male but not female rats (Bara et al., 2018). There were no differences in the elevated plus maze or in an order recognition task following gestational cannabinoid exposure, suggesting changes were specific to social behaviors in male individuals. In addition, prenatal cannabinoid exposure eliminated endocannabinoid LTD induced by 10-Hz stimulation in layer 5 pyramidal neurons of the PFC in male but not female rats. Previous studies have found this 10-Hz LTD is mediated by mGlu$_5$ receptors (Luscher & Huber, 2010), and the mGlu$_5$ PAM CDPPB restored LTD in male rats exposure to prenatal cannabinoids (Bara et al., 2018). Finally, the authors linked the sex-specificity of these phenomena to the latent mechanisms mediating PFC LTD, providing evidence that LTD in male rats is mediated by CB1 receptors while TRPV1 receptors mediate LTD in female rats (Bara et al., 2018). Systemic treatment with either the mGlu$_5$ PAM or a fatty acid amide hydrolase inhibitor URB597 rescued behavioral deficits in affected male rats, providing a sex-dependent link between mGlu$_5$ receptor plasticity and social interaction.

Evidence linking mGlu$_5$ receptor function with social behaviors has also emerged from transgenic animal studies investigating Shank3. Deficits in Shank3, a scaffolding protein that coordinates mGlu$_5$/Homer intracellular signaling, are associated with the pathogenesis of autism spectrum disorder (Uchino & Waga, 2013; Verpelli et al., 2011; Zoicas & Kornhuber, 2019). Mutant Shank3 mice exhibit significant impairments in social behaviors and hippocampal mGlu$_5$ receptor signaling, without obvious sex differences (Matas et al., 2021; Peixoto et al., 2016; Wang et al., 2011). While social behaviors related to Shank3 disruption appear comparable in both sexes, Shank3 mutant mice also exhibit changes in repetitive behaviors and motor control, with evidence of sex differences in effects. Male Shank3 mutants displayed greater impairments in several gait parameters than littermate female mutants (Wang et al., 2011), and this phenotype has been linked with lower expression of mGlu$_5$ receptors in cerebellum in male mice (Matas et al., 2021). Together, these studies suggest that Shank3 regulation of mGlu$_5$ receptor function exhibits sex differences in specific brain areas (i.e., cerebellum but not hippocampus) and indicate that not all insults to mGlu$_5$ receptor signaling affect social play in a sex-dependent manner.

3.3 Alcohol use

Genetic and pharmacological manipulations of mGlu$_{1/5}$ receptors have revealed a clear association with alcohol seeking behaviors (Joffe et al., 2018). The link between ethanol exposure and mGlu$_{1/5}$ receptor function is bidirectional. Acute and chronic applications of ethanol alter $mGlu_{1/5}$ receptor function in cultured oocytes and rat cerebellar slices (Belmeguenai et al., 2008; Carta, Mameli, & Valenzuela, 2006; Minami et al., 1998; Netzeband & Gruol, 1995). On the other hand, Group 1 antagonists, such as the prototypical $mGlu_5$ NAM MPEP, reduce ethanol-seeking and relapse-like behaviors in a broad variety of models and background (Backstrom et al., 2004; Besheer et al., 2008; Hodge et al., 2006; Lominac et al., 2006). While most of these studies were limited to male subjects, a few have found comparable effects in females. In a recent study directly assessing sex differences in binge drinking, MPEP displayed a similar dose-response curve in reducing binge alcohol consumption in male and female mice (Huang, Thompson, & Taylor, 2021). Similarly, MTEP administration decreased binge ethanol intake in male and female mice (Cozzoli et al., 2014). Interestingly, in that study, female but not male mice displayed increased ethanol consumption 24 h after MTEP exposure, suggesting that females may be uniquely sensitive to compensatory or rebound adaptations following $mGlu_5$ inhibition.

Alterations in $mGlu_{1/5}$ receptor signaling in several brain areas have been implicated in binge drinking models. Several studies have implicated Homer2 in the regulation of ethanol sensitivity and ethanol-induced synaptic plasticity (Szumlinski et al., 2005, 2008; Szumlinski, Ary, & Lominac, 2008), including initial studies using mixed-sex cohorts. Homer2 knockout mice of both sexes resist drinking high concentrations of ethanol and do not express ethanol locomotor sensitization or conditioned place preference. By contrast, binge-drinking upregulates $mGlu_5$ receptors, Homer2, and PI3K signaling in male mice (Cozzoli et al., 2009). Furthermore, both male and female transgenic mice that express a point mutation in $mGlu_5$ receptors rendering them insensitive to ERK phosphorylation exhibit decreased sensitivity to the aversive properties of high dose ethanol (Campbell et al., 2019). Taken together, this literature indicates that upregulation of Homer2/$mGlu_5$ receptor expression and activation of ERK-dependent phosphorylation function as adaptive compensatory mechanisms to limit alcohol consumption in both sexes.

Women are more likely than men to drink alcohol to regulate negative emotional state or stress (Peltier et al., 2019). In recent years, the sexually dimorphic bed nucleus of the stria terminalis (BNST) has been identified as a point of convergence for early stress- and alcohol-related disorders in modulating negative affect behaviors (Kasten et al., 2021; Lebow & Chen, 2016). This observation from AUD has been modeled by inducing dependence to ethanol during adolescence in rodents

(Crews et al., 2016). In adolescent female mice, acute withdrawal from chronic intermittent ethanol disrupted mGlu$_5$ receptor LTD in the BNST (Kasten et al., 2020). By contrast, male mice in acute withdrawal exhibited no differences in mGlu$_5$ receptor LTD, but instead displayed enhanced NMDA receptor functions (Carzoli et al., 2019). Interestingly, these sex-dependent changes in BNST plasticity mapped onto distinct behavioral phenotypes associated with withdrawal. Adolescent female mice showed enhanced anxiety/anhedonia-like behavior in a novelty-induced hypophagia task, whereas male mice in withdrawal displayed elevated freezing during fear conditioning paradigms (Kasten et al., 2020). Two pieces of convergent evidence link the adaptations to BNST mGlu$_5$ receptor plasticity with female-specific affective behaviors. First, systemic delivery of MTEP to female mice reversed withdrawal-induced increases in latency to each in the novelty-induced hypophagia assay (Kasten et al., 2020). Second, mGlu$_5$ receptor knockdown in the BNST reproduces anhedonia-like behavior in the novelty-induced hypophagia task in female but not male mice (Kasten et al., 2021). Consistent with the findings by Kasten et al. (2020) in mice, Chandler et al. also observed that adolescent chronic intermittent ethanol exposure enhanced contextual fear learning and spontaneous recovery of freezing behavior in male but not female rats (Chandler, Vaughan, & Gass, 2022). Interestingly, the mGlu$_5$ PAM CDPPB reversed the withdrawal-induced phenotypes in male rats, although it is not clear from these studies how female individuals would respond to enhanced mGlu$_5$ receptor function. Together, these findings indicate that mGlu$_5$ receptors influence behavioral phenotypes relevant to alcohol dependence in adolescence in both males and females, however there is a clear impact of sex on resulting behaviors (Table 1).

3.4 Drug use

Compared to males, female rodents acquire drug self-administration much more quickly, rapidly progress through milestones of drug use, and exhibit greater levels of drug reinstatement, mediated in part by the estrous cycle (Anker & Carroll, 2011; Carroll et al., 2002; Doncheck et al., 2018; Feltenstein & See, 2007). A significant literature, albeit generally restricted to males, indicates that Group 1 mGlu receptors contribute to drug-seeking and related behaviors. In general, mGlu$_1$ antagonists and NAMs block some cocaine-seeking behaviors (Xie et al., 2010; Yu et al., 2013), whereas mGlu$_1$ PAMs may prevent relapse-like behaviors (Loweth et al., 2014).

Table 1 Sex differences in mGlu receptor regulation of affective behaviors.

Domain	Family	Subtype	Species/Strain	Sex Diff	Key findings	First author (year)
Anxiety-like and affective behavior	Group I	mGlu$_{1/5}$	Rat, SD	Y	BLA infusion of the mGlu$_{1/5}$ agonist DHPG increased punished drinking in VCT in OVX ♀. BLA infusion of the DHPG decreased punished drinking in VCT in ♂ (De Jesús-Burgos, Torres-Llenza, & Pérez-Acevedo, 2012)	De Jesús-Burgos (2016)
		mGlu$_5$	Mouse, C57BL/6J, *Grm5$^{(AA/AA)}$*	N	Mice with mGlu$_5$ that cannot be phosphorylated by ERK displayed decreased float time in FST and increased open arm time in the EPM (Cozzoli et al., 2009)	Campbell (2019)
			Mouse, C57BL/6J, *Grm5^{loxp}*	Y	mGlu$_5$ KD in BNST prolonged latency to eat in NIH in ♀ but not ♂. mGlu$_5$ KD in BNST decreased center time in open field in ♂ but not ♀ (Lebow & Chen, 2016)	Kasten (2021)
			Mouse, C57BL/6J, *Grm5^{loxp}, SST-Cre*	N	SST-mGlu$_5$-KO mice resisted stress–induced alterations in PFC inhibitory transmission and spontaneous alternation on the Y maze (Di Menna et al., 2018)	Joffe (2022)
			Mouse, C57BL/6J, *Grm5^{loxp}, PV-Cre*	Y	♂ PV-mGlu$_5$-KO mice displayed more perseverative errors on the Barnes maze. ♀ KO mice did not display more perseverative errors than controls (Luscher & Huber, 2010)	Barnes (2015)
	Group 2	mGlu$_3$	Mouse, C57BL6/J x C57BL6/Ntac, *Grm3^{loxp}*	N	mGlu$_3$ KD in PFC increased open arm time in EZM and decreased immobility in TST/FST (Maksymetz & Joffe, 2021)	Joffe (2020)

Continued

Table 1 Sex differences in mGlu receptor regulation of affective behaviors.—cont'd

Domain	Family	Subtype	Species/Strain	Sex Diff	Key findings	First author (year)
	Group 3	mGlu$_4$	Mouse, C57BL6/J, $Grm4^{-/-}$	Y	♀ mGlu$_4$ KO mice displayed decreased open arm/center time in EZM and open field. ♂ mGlu$_4$ KO mice displayed increased open arm/center time in EZM and open field (Joffe et al., 2021)	Davis (2012)
		mGlu$_8$	Mouse, C57BL6/J, $Grm8^{-/-}$	Y	♀ mGlu$_8$ KO mice displayed increased open arm/center time in EZM and open field, and increased startle response. ♂ mGlu$_8$ KO mice displayed decreased open arm/center time in EZM and open field, and increased startle response. (Joffe et al., 2021)	Duvoisin (2010)
Alcohol Use	Group I	mGlu$_5$	Mouse, C57BL/6J	Y	Acute withdrawal from AIE disrupts mGlu$_5$ LTD in the BNST in ♀ but not ♂ (Crews et al., 2016)	Kasten (2020)
			Mouse, C57BL/6NCrl	N	mGlu$_5$ NAM MPEP (30 mg/kg) decreases binge drinking (Besheer et al., 2008)	Huang (2021)
			Mouse, C57BL/6J	Y	mGlu$_5$ NAM MTEP (20 mg/kg) decreases binge drinking. One month later, increased drinking was observed in ♀ but not ♂ (Huang et al., 2021)	Cozzoli (2014)
			Mouse, C57BL/6J, $Grm5^{(AA/AA)}$	N	Transgenic mice with mGlu$_5$ that cannot be phosphorylated by ERK display increased drinking (Cozzoli et al., 2009)	Campbell (2019)

All studies in this table employed mixed-sex cohorts

AIE, adolescent intermittent ethanol vapor; BLA, basolateral amygdala; BNST, bed nucleus of the stria terminalis; EPM, elevated plus-maze; ERK, extracellular signal-regulated kinase; EZM, elevated zero-maze; FST, forced swim test; KD, knock-down; KO, knock-out; LTD, long-term depression; NAM, negative allosteric modulator; OVX, ovariectomized; PFC, prefrontal cortex; PV, parvalbumin; SD, Sprague Dawley; SST, somatostatin; TST, tail suspension test; VCT, Vogel conflict test.

Metabotropic glutamate receptors in psychiatric and neurological disorders   329

While these studies allude to the involvement $mGlu_1$ receptors in the adaptations to chronic drug use, there is still limited information on how sex may influence $mGlu_1$ activity in SUD. In comparison to $mGlu_1$, we have a better understanding how sex differences in $mGlu_5$ receptor signaling guide behaviors related to drug use. Some of this evidence comes from studies examining sex hormone contributions to drug-seeking. In female individuals, the subjective effects of smoked cocaine fluctuate over the course of the menstrual cycle (Evans, Haney, & Foltin, 2002). Similar findings have been observed in several rodent models and laboratories: female rodents show increased cocaine-induced hyperlocomotion, enhanced cocaine cue and reward learning, increased motivation to self-administer, increased incubation of craving, and estrous cycle-dependent cocaine reinforcement (Calipari et al., 2017; Jackson, Robinson, & Becker, 2006; Johnson et al., 2019; Roberts, Bennett, & Vickers, 1989). In addition, in female rodent ovariectomy studies, E2 treatment facilitates cocaine self-administration (Hu & Becker, 2003; Lynch et al., 2001), related effects on hyperlocomotion and self-administration were blocked by MPEP administration (Gross et al., 2016; Martinez et al., 2014). Taken together, this literature suggests that changes in drug-seeking and cue attending during female hormonal cycles may be dependent on surges in E2 and potentiated signaling through downstream $mGlu_5$ receptors (Table 2).

3.5 Anxiety-like and affective behavior

Affective disorders, such as major depressive disorder and anxiety-based disorders, including post-traumatic stress and generalized anxiety disorders, are more prevalent in women than in men (Altemus, Sarvaiya, & Neill Epperson, 2014; Bangasser & Cuarenta, 2021; Kessler et al., 2012). Though monoaminergic systems have historically been the predominant focus for therapeutic targets, growing evidence suggests a functional role for the glutamate system in the treatment of anxiety and affective disorders (Joffe et al., 2019; Mathews, Henter, & Zarate, 2012; Palucha & Pilc, 2007). Indeed, several studies have reported the robust anxiolytic-like effects in preclinical models following pharmacological manipulations of Group 1 receptors (Ballard et al., 2005; Busse et al., 2004; Pietraszek et al., 2005), although this literature is heavily biased toward male subjects.

Stress, a major contributing factor to the development of anxiety and affective disorders, dynamically alters Group 1 receptor function. In the PFC of male rats, exposure to 2-days forced swim stress decreased

Table 2 Sex hormone regulation of mGlu receptor synaptic plasticity.

Family	Subtype	Region	Species/Strain	Sexes used	Sex diff	Key findings	First author (Year)
Group I	mGlu$_1$	HPC	Rat[a]	Both	Y	E2 facilitates CREB phosphorylation through ERα/mGlu$_1$, PLC, PKC, and IP3R in ♀ but not ♂ (Boulware et al., 2005)	Boulware (2005)
			Rat, SD	Both (OVX)	Y	E2 suppresses GABA release probability through ERα/mGlu$_1$, anandamide, and CB1R in ♀ but not ♂ (Huang & Woolley, 2012)	Huang (2012)
			Rat, SD	Both (OVX/CAS)	Y	E2 suppression of GABA release probability coincides with greater IP3 production in ♀ than ♂. ERα/mGlu$_1$ complexes in exist in ♂ but are only regulated by E2 in ♀ (Tabatadze et al., 2015)	Tabatadze (2015)
			Rat, SD	♀ (OVX)	n.a.	E2 increases dendritic spine density through activation of mGlu$_1$ (Peterson et al., 2015)	Peterson (2015)
	mGlu$_5$	Striatum	Rat[a]	♀	n.a.	E2 facilitates CREB phosphorylation through ERα/mGlu$_5$ and MAPK (Grove-Strawser et al., 2010)	Grove-Strawser (2010)
		NAC core	Rat, SD	♀ (OVX)	n.a.	E2 decreases dendritic spine density through activation of mGlu$_5$ (Peterson et al., 2015)	Peterson (2015)
		NAC shell	Rat, SD	♂ (CAS)	n.a.	DHT decreases dendritic spine density in the NAC shell, but not core, through mGlu$_5$ (Gross et al., 2018)	Gross (2018)
Group II	mGlu$_{2/3}$	HPC	Rat[a]	Both	Y	E2 decreases CREB phosphorylation through mGlu$_2$ and/or mGlu$_3$ inhibition of L-type calcium channels in ♀ but not ♂ (Boulware et al., 2005)	Boulware (2005)
	mGlu$_3$	Striatum	Rat[a]	♀	n.a.	E2 decreases CREB phosphorylation through mGlu$_3$ inhibition of L-type calcium channels (Grove-Strawser et al., 2010)	Grove-Strawser (2010)

[a]Pyramidal neurons cultured from P1-P2 rat pups, experiments performed at 9 d.i.v.; strain not specified

CAS, castrated; CB1R, cannabinoid receptor type 1; CREB, cAMP response element-binding protein; DHT, dihydrotestosterone; GABA, gamma-aminobutyric acid; E2, estradiol; ER, estrogen receptor; HPC, hippocampus; IP3R, inositol triphosphate receptor; MAPK, mitogen-activated protein kinase; n.a., not addressed; NAC, nucleus accumbens; OVX, ovariectomized; PLC, phospholipase C; PKC, protein kinase C; SD, Sprague Dawley mGlu$_1$ and increased mGlu$_5$ receptor protein expression (Wang et al., 2015). Consistent with a stress–induced dysregulation of PFC mGlu$_5$ receptor signaling, 20-min of restraint stress in male mice impairs the induction of mGlu$_3$ LTD that is dependent on mGlu$_5$ receptor signaling (Di Menna et al., 2018; Joffe et al., 2019). By contrast, female rats exposed to forced swim stress displayed intact mGlu$_1$ and decreased mGlu$_5$ receptor expression (Wang et al., 2015). Continued research should account for cell type-specific differences in mGlu$_5$ receptor signaling, as we recently found that restraint stress potentiates somatostatin interneuron activity through a mGlu$_5$ receptor-dependent mechanism in both male and female mice (Joffe et al., 2022). Together, these results suggest that stress–induced alterations in mGlu$_{1/5}$ receptor signaling may mediate some effects of stress and potentially increase vulnerability to affective behaviors in a sex-specific manner.

In female rodents, E2 can reduce anxiety-like behaviors (Hill, Karacabeyli, & Gorzalka, 2007; Mora, Dussaubat, & Díaz-Véliz, 1996); these effects are mediated by activation of Group 1 mGlu receptors (De Jesús-Burgos et al., 2012) and endocannabinoid mobilization (Hill et al., 2007). Studies from ovariectomized female rats suggest that the ability of mGlu$_{1/5}$ receptors to regulate anxiety-like behavior may vary across the estrous cycle. In E2-treated ovariectomized female rats, direct infusion of DHPG into the basolateral amygdala decreased anxiety-related behaviors in the elevated plus maze (De Jesús-Burgos et al., 2012) and Vogel conflict tests (De Jesús-Burgos et al., 2016). By contrast, this manipulation had no effect on non–E2 treated ovariectomized females in either assay or in males in the plus maze. Furthermore, stimulation of mGlu$_{1/5}$ receptors within the BLA *increased* anxiety-like behavior in male rats on the Vogel conflict test (De Jesús-Burgos et al., 2016). It is tempting to speculate that these differences could be related to differential contributions of CB1 receptor signaling to anxiety-like behavior (Bowers & Ressler, 2016), but more studies are needed to fully evaluate that hypothesis. In any case, the results from these studies are consistent with a sex-specific role of amygdalar Group 1 receptors in the attenuation of anxiety and depressive-type behaviors.

Antidepressant-like and anxiolytic-like phenotypes have been observed in mGlu$_2$ and mGlu$_3$ receptor knockout mice and following administration of compounds that inhibit mGlu$_2$ and/or mGlu$_3$ receptors (Highland et al., 2019; Joffe et al., 2020; Maksymetz & Joffe, 2021). Unfortunately, most of these studies were restricted to male mice. One exception is our recent study reporting a panel of emotional behavioral studies assessing frontal cortex deletion of mGlu$_3$ receptors (Joffe et al., 2021). We observed that frontal cortex knockdown of mGlu$_3$ receptors decreased immobility in the forced swim and tail suspension tests and increased open-arm time in the elevated-zero maze. We found no evidence for an interaction between mGlu$_3$ receptor knockout and sex; however, it is unclear to what extent these findings may generalize to additional behaviors, mGlu$_3$ receptor function in other brain areas, or acute pharmacological receptor inhibition. Additional studies examining potential sex-dependent behavioral effects of mGlu$_2$ and mGlu$_3$ modulators are certainly warranted.

Studies using selective pharmacology and knockout mice implicate Group 3 receptors in anxiety-like behaviors. Systemic treatment with the mGlu$_4$ PAM ADX88178 increased open-arm entries on the elevated plus maze in both male and female mice (Kalinichev et al., 2014), suggesting that mGlu$_4$ receptor signaling can reduce anxiety-like behavior in both sexes. Consistent with this hypothesis, male mGlu$_4$ receptor knockout mice display increased anxiety-like behavior in the open field and elevated plus maze that emerge at age 12 months (Davis et al., 2012). In striking contrast, however, female mGlu$_4$ receptor knockouts display reduced anxiety-like behaviors in both assays at 6 and 12 months. The authors observed similar sex- and age-dependent effects on sensorimotor function assessed on the accelerating rotarod. Male knockout mice at 12 months, but not 6 months, fell from the rotarod more quickly than wild-type controls, whereas female knockout mice at both ages displayed longer latencies to fall than female wildtypes. Interestingly, in the same studies, mGlu$_4$ knockout mice displayed altered cued fear learning across all ages and sexes. Thus, the sex- and age-dependencies of mGlu$_4$ receptor regulation of anxiety-like and sensorimotor behaviors likely occur through actions within one or more distinct neural circuits, rather than at the cellular level of receptor expression or function. Considering that mGlu$_4$ receptors are highly expressed in corticostriatal and thalamostriatal circuits—which are heavily implicated in anxiety and motor function but less so in fear learning—it is tempting to speculate that these receptor populations contribute to the observed sex- and age-dependent behavioral phenotypes in mGlu$_4$ knockout mice. Clearly, additional studies will be needed to address this speculative hypothesis.

Numerous studies suggest that mGlu$_4$ and mGlu$_8$ may have related and overlapping functions, particularly with respect to anxiety-like behaviors. Similar to the sex-dependent findings in mGlu4 receptor knockout mice, there is a sex difference in anxiety-like behavior in mGlu$_8$ receptor knockouts. In the elevated plus maze, open-arm time and entries were decreased in male knockouts but increased in female knockouts, respectively suggesting enhanced and reduced anxiety-like behaviors (Duvoisin et al., 2010). Consistent with this finding, the acoustic startle response was enhanced in male knockout mice but attenuated in female knockouts. An important caveat to these findings, however, is that mGlu$_8$ knockouts of both sexes exhibited less time in the center of an open field relative to matched controls. Therefore, the broad generalizability of the anxiety-like phenotype of female knockout mice should be interpreted with some caution.

3.6 Neurodegeneration

Alzheimer's disease, a type of dementia, is characterized as a progressive neurological disorder in which atrophy of the brain leads to cell death and subsequent loss of cognitive functions (Johnson et al., 2019). Though the incidence of Alzheimer's disease does not differ by sex, there are distinct sex-differences in the symptomatology, progression, and treatment of Alzheimer's disease (Ferretti et al., 2018). Aggregation of β-amyloid proteins in the formation of amyloid plaques is a pathological hallmark of Alzheimer's disease in both sexes (Murphy & LeVine 3rd, 2010). In fact, accumulation of oligomers induces abnormal clustering of mGlu$_5$ receptors at the cell membrane and, in turn, alters mGlu$_5$ glutamate transmission leading to synaptic degradation (Renner et al., 2010). In cortical tissue of wild-type male mice and human men, β-amyloid oligomers display a high-affinity (nanomolar range) interaction with mGlu$_5$ receptors, an effect not observed in female mice or in women (Abd-Elrahman et al., 2020; Abd-Elrahman & Ferguson, 2022). Furthermore, in primary neuronal cultures from male but not female mice, β-amyloid oligomers activate glycogen synthase kinase 3β signaling in an mGlu$_5$ receptor dependent manner. Consistent with these actions, systemic treatment with CTEP mGlu$_5$ inhibition reduced β-amyloid pathology and improved object recognition in male disease model mice but had no effect in females (Abd-Elrahman et al., 2020; Abd-Elrahman & Ferguson, 2022). An mGlu$_5$ silent allosteric modulator reversed deficits in hippocampal-dependent cognition in male Alzheimer's disease model mice, but cohorts of female mice were not included in that study (Haas et al., 2017). Taken together, these findings suggest a male-specific contribution of mGlu$_5$ receptor signaling in β-amyloid-driven pathology in Alzheimer's disease.

Evidence from Huntington's disease models similarly suggests that male individuals may be more sensitive to neurodegeneration related to mGlu$_5$ receptor signaling. MPEP administration in male mouse models of Huntington's disease inhibits disease progression (Schiefer et al., 2004) and mGlu$_5$ knockout improves motor behavior and decreases aggregate size of mutant huntingtin (Ribeiro et al., 2014). There is growing preclinical and clinical evidence to suggest that sex may influence the phenotype of Huntington's disease (Bode et al., 2008; Cao et al., 2019; Zielonka et al., 2013), and preclinical models are beginning to take this variation into account. A recent study examined CTEP administration in male and female Huntington's disease model mice; although both sexes displayed improvements in HD neuropathology following mGlu$_5$ blockage, female HD mice required longer CTEP treatment to show motor improvements (Li et al., 2022). Furthermore, mGlu$_5$ antagonism did not improve cognitive impairments or grip strength in female HD mice. These findings collectively suggest that males may be more sensitive to neurodegenerative processes related to aberrant mGlu$_5$ receptor signaling. One intriguing possible mechanism could be sex differences in coordination between mGlu$_3$ and mGlu$_5$ receptor signaling, as the same group observed that LY379268 exerts sex-dependent effects on Akt, GSK3β, and ERK1/2 phosphorylation in the striatum of Huntington's disease model mice (Li et al., 2022).

4. Conclusions

Despite preclinical evidence to suggest possible sex-dependent differences in mGlu receptor expression and function, few human studies have been adequately powered to make well-justified conclusions and it has been challenging to make clear comparisons across species. For example, positron emission tomography has enabled the quantification of available mGlu$_5$ receptors in the intact human brain, and the allosteric radiotracer [11C]-ABP688 has been used to assess sex differences in mGlu$_5$ receptor availability. While sex differences in mGlu$_5$ receptor availability were not detected in a small study (DuBois et al., 2016), a larger study found significantly higher rates of [11C]-ABP688 binding potential in men than women across several regions of the PFC, striatum, and hippocampus (Smart et al., 2019). The largest magnitude differences were observed in the orbitofrontal cortex and the dorsolateral PFC, in which binding potentials were 22% and 20% greater in men, respectively. Overall, a whole-brain comparison of mGlu$_5$ receptor availability found that [11C]-ABP688 binding potential was 17% greater in men. Consistent with these findings, an earlier human PET imaging study using [11C]-ABP688 reported 31% greater global mGlu$_5$ binding potential in non-smoking men when compared to non-smoking women (Akkus et al., 2013). Taken together, the larger availability of mGlu$_5$ receptors in women compared to men seems to conflict with the preclinical literature, as major sex differences in mGlu$_5$ receptor expression have not been consistently reported in rodents (Wickens et al., 2018) and available evidence points toward decreased cortical mGlu$_5$ receptor expression in male rats compared to females (Wang et al., 2015). Importantly, technical considerations impede direct comparisons between the clinical and preclinical literature. For one, studies using radioligands like [^{11}C]-ABP688 detect mGlu$_5$ receptors available on the cell surface, while Western blots with targeted antibodies can detect receptor populations in intracellular (Purgert et al., 2014) or nuclear (Jong et al., 2005) membranes. In addition, mGlu receptor antibodies were raised against sections of the intracellular C-terminal tail or the N-terminal extracellular domain, whereas allosteric ligands like [^{11}C]-ABP688 interact with the allosteric binding pocket within the transmembrane domain (Gregory & Conn, 2015). Based on this, sex-dependent changes in splicing, post-translational modifications, or interactions with other proteins (e.g., ER) could disproportionately affect one method of assessing mGlu$_5$ receptor expression. Moving forward, parallel cross-species studies should be designed to use identical techniques to assess potential sex differences in mGlu receptor expression. Ideal studies would examine expression of transcripts, protein, and available receptors for all mGlu receptor subtypes.

Here, we have summarized key pieces of evidence suggesting sex differences in mGlu receptor functions, highlighting literature showing that the gonadal hormone E2 has acute effects on these mGlu receptor functions. However, much less is known about how other factors, including acute effects of other gonadal hormones (e.g., testosterone, progesterone), developmental gonadal hormone effects (i.e., organizational effects of hormones), and sex chromosome complement might influence sex-specific behavioral and neurobiological outcomes. Indeed, there is a rich literature in neuroendocrinology and the sex differences field showing that these other factors influence behavior and neurobiology. For instance, rodent studies have revealed links between circulating testosterone and anxiety-like behaviors. Mice and rats with the testicular feminization mutation, which renders the androgen receptor insensitive to testosterone, exhibit higher anxiety-like behaviors compared to wildtypes (Chen et al., 2014; Zuloaga et al., 2008, 2011). We have also shown that testosterone exposure in adult mice reduces measures of anxiety−/depressive-like behavior after chronic stress exposure in both males and females (Seney et al., 2013a). Whether mGlu receptor signaling mediates the behavioral effects of androgens remains, by and large, an open question.

Additional sources of sex differences may originate in early developmental events and from sex chromosome genes. Exposure to gonadal hormones during sensitive periods of development ("organizational effects") is known to cause permanent sex differences in behavior and brain structure (Cooke et al., 1998; Handa et al., 1985; Phoenix et al., 1959). Moreover, sex chromosomes differ between males and females in the presence or absence of a Y and in the dosage of X chromosomes. The role of genetic sex has been difficult to investigate due to mosaicism in females due to X-linked inactivation and the relationship between genetic sex and gonadal sex. We and others have begun to parse genetic sex from gonadal sex by using the Four Core Genotypes (FCG) mice. In the FCG strain, the testes determining gene, *Sry*, was placed on an autosome after spontaneous deletion from the Y chromosome; thus, both XX mice and XY mice can have either ovaries or testes, dissociating genetic from gonadal sex (Arnold & Chen, 2009; McCarthy & Arnold, 2011). Using the FCG mice, we showed that genetic sex influences anxiety-like behavior (Seney et al., 2013a) and gene expression in mood-relevant brain regions (Barko et al., 2019; Paden et al., 2020; Puralewski, Vasilakis, & Seney, 2016; Seney et al., 2013b). Other studies in FCG mice have revealed sex chromosome influences on sex differences in aggressive, parental, social, cognitive, circadian, and substance use–associated behaviors (Aarde et al., 2021; Barker et al., 2010; Cox & Rissman, 2011; Gatewood et al., 2006; Kuljis et al., 2013; Martini et al., 2020; McPhie-Lalmansingh et al., 2008). Together, it is clear that factors other than circulating E2 influence sex-related outcomes, and that there is much work to be done related to how these factors may impact mGlu function.

Acknowledgments

This work was supported in part by the National Institute of Health AA027806 (MEJ) and MH120066 (MLS). CBF was supported by the Center for Neuroscience University of Pittsburgh.

Conflict of Interest

The authors declare no potential conflicts of interest.

References

Aarde, S. M., et al. (2021). Sex chromosome complement affects multiple aspects of reversal-learning task performance in mice. *Genes, Brain, and Behavior, 20*(1), e12685.

Abd-Elrahman, K. S., & Ferguson, S. S. G. (2022). Noncanonical metabotropic glutamate receptor 5 signaling in Alzheimer's disease. *Annual Review of Pharmacology and Toxicology, 62*, 235–254.

Abd-Elrahman, K. S., et al. (2020). Aβ oligomers induce pathophysiological mGluR5 signaling in Alzheimer's disease model mice in a sex-selective manner. *Science Signaling*, *13*(662), eabd2494.

Akkus, F., et al. (2013). Marked global reduction in mGluR5 receptor binding in smokers and ex-smokers determined by [11C]ABP688 positron emission tomography. *Proceedings of the National Academy of Sciences of the United States of America*, *110*(2), 737–742.

Alagarsamy, S., Sorensen, S. D., & Conn, P. J. (2001). Coordinate regulation of metabotropic glutamate receptors. *Current Opinion in Neurobiology*, *11*(3), 357–362.

Altemus, M., Sarvaiya, N., & Neill Epperson, C. (2014). Sex differences in anxiety and depression clinical perspectives. *Frontiers in Neuroendocrinology*, *35*(3), 320–330.

Angermeyer, M. C., Goldstein, J. M., & Kuehn, L. (1989). Gender differences in schizophrenia: Rehospitalization and community survival. *Psychological Medicine*, *19*(2), 365–382.

Angst, J., & Dobler-Mikola, A. (1984). Do the diagnostic criteria determine the sex ratio in depression? *Journal of Affective Disorders*, *7*(3–4), 189–198.

Anker, J. J., & Carroll, M. E. (2011). Females are more vulnerable to drug abuse than males: Evidence from preclinical studies and the role of ovarian hormones. *Current Topics in Behavioral Neurosciences*, *8*, 73–96.

Arnold, A. P., & Chen, X. (2009). What does the "four core genotypes" mouse model tell us about sex differences in the brain and other tissues? *Frontiers in Neuroendocrinology*, *30*(1), 1–9.

Backstrom, P., et al. (2004). mGluR5 antagonist MPEP reduces ethanol-seeking and relapse behavior. *Neuropsychopharmacology*, *29*(5), 921–928.

Ballard, T. M., et al. (2005). The effect of the mGlu5 receptor antagonist MPEP in rodent tests of anxiety and cognition: A comparison. *Psychopharmacology*, *179*(1), 218–229.

Bangasser, D. A., & Cuarenta, A. (2021). Sex differences in anxiety and depression: Circuits and mechanisms. *Nature Reviews. Neuroscience*, *22*(11), 674–684.

Bara, A., et al. (2018). Sex-dependent effects of in utero cannabinoid exposure on cortical function. *eLife*, *7*, e36234.

Barker, J. M., et al. (2010). Dissociation of genetic and hormonal influences on sex differences in alcoholism-related behaviors. *The Journal of Neuroscience*, *30*(27), 9140–9144.

Barko, K., et al. (2019). Sex-specific effects of stress on mood-related gene expression. *Molecular Neuropsychiatry*, *5*(3), 162–175.

Barnes, S. A., et al. (2015). Disruption of mGluR5 in parvalbumin-positive interneurons induces core features of neurodevelopmental disorders. *Molecular Psychiatry*, *20*(10), 1161–1172.

Baude, A., et al. (1993). The metabotropic glutamate receptor (mGluR1 alpha) is concentrated at perisynaptic membrane of neuronal subpopulations as detected by immunogold reaction. *Neuron*, *11*(4), 771–787.

Belmeguenai, A., et al. (2008). Alcohol impairs long-term depression at the cerebellar parallel fiber-Purkinje cell synapse. *Journal of Neurophysiology*, *100*(6), 3167–3174.

Bender, R. A., et al. (2017). Sex-dependent regulation of aromatase-mediated synaptic plasticity in the basolateral amygdala. *The Journal of Neuroscience*, *37*(6), 1532–1545.

Bertaso, F., et al. (2010). Homer1a-dependent crosstalk between NMDA and metabotropic glutamate receptors in mouse neurons. *PLoS One*, *5*(3), e9755.

Besheer, J., et al. (2008). Regulation of motivation to self-administer ethanol by mGluR5 in alcohol-preferring (P) rats. *Alcoholism, Clinical and Experimental Research*, *32*(2), 209–221.

Bode, F. J., et al. (2008). Sex differences in a transgenic rat model of Huntington's disease: Decreased 17beta-estradiol levels correlate with reduced numbers of DARPP32+ neurons in males. *Human Molecular Genetics*, *17*(17), 2595–2609.

Boulware, M. I., Heisler, J. D., & Frick, K. M. (2013). The memory-enhancing effects of hippocampal estrogen receptor activation involve metabotropic glutamate receptor signaling. *The Journal of Neuroscience*, *33*(38), 15184–15194.

Boulware, M. I., et al. (2005). Estradiol activates group I and II metabotropic glutamate receptor signaling, leading to opposing influences on cAMP response element-binding protein. *The Journal of Neuroscience, 25*(20), 5066–5078.

Bowers, M. E., & Ressler, K. J. (2016). Sex-dependence of anxiety-like behavior in cannabinoid receptor 1 (Cnr1) knockout mice. *Behavioural Brain Research, 300*, 65–69.

Brakeman, P. R., et al. (1997). Homer: A protein that selectively binds metabotropic glutamate receptors. *Nature, 386*(6622), 284–288.

Busse, C. S., et al. (2004). The behavioral profile of the potent and selective mGlu5 receptor antagonist 3-[(2-methyl-1,3-thiazol-4-yl)ethynyl]pyridine (MTEP) in rodent models of anxiety. *Neuropsychopharmacology, 29*(11), 1971–1979.

Calipari, E. S., et al. (2017). Dopaminergic dynamics underlying sex-specific cocaine reward. *Nature Communications, 8*, 13877.

Campbell, R. R., et al. (2019). Increased alcohol-drinking induced by manipulations of mGlu5 phosphorylation within the bed nucleus of the Stria terminalis. *The Journal of Neuroscience, 39*(14), 2745–2761.

Cao, J. K., et al. (2019). Sex-dependent impaired locomotion and motor coordination in the HdhQ200/200 mouse model of Huntington's disease. *Neurobiology of Disease, 132*, 104607.

Carroll, M. E., et al. (2002). Intravenous cocaine and heroin self-administration in rats selectively bred for differential saccharin intake: Phenotype and sex differences. *Psychopharmacology, 161*(3), 304–313.

Carta, M., Mameli, M., & Valenzuela, C. F. (2006). Alcohol potently modulates climbing fiber—>Purkinje neuron synapses: Role of metabotropic glutamate receptors. *The Journal of Neuroscience, 26*(7), 1906–1912.

Carzoli, K. L., et al. (2019). Regulation of NMDA receptor plasticity in the BNST following adolescent alcohol exposure. *Frontiers in Cellular Neuroscience, 13*, 440.

Casabona, G., et al. (1997). Expression and coupling to polyphosphoinositide hydrolysis of group I metabotropic glutamate receptors in early postnatal and adult rat brain. *The European Journal of Neuroscience, 9*(1), 12–17.

Chaban, V. V., et al. (2003). Estradiol inhibits atp-induced intracellular calcium concentration increase in dorsal root ganglia neurons. *Neuroscience, 118*(4), 941–948.

Chandler, L. J., Vaughan, D. T., & Gass, J. T. (2022). Adolescent alcohol exposure results in sex-specific alterations in conditioned fear learning and memory in adulthood. *Frontiers in Pharmacology, 13*, 837657.

Chávez, A. E., Chiu, C. Q., & Castillo, P. E. (2010). TRPV1 activation by endogenous anandamide triggers postsynaptic long-term depression in dentate gyrus. *Nature Neuroscience, 13*(12), 1511–1518.

Chavis, P., et al. (1994). The metabotropic glutamate receptor types 2/3 inhibit L-type calcium channels via a pertussis toxin-sensitive G-protein in cultured cerebellar granule cells. *The Journal of Neuroscience, 14*(11 Pt 2), 7067–7076.

Chen, C. V., et al. (2014). New knockout model confirms a role for androgen receptors in regulating anxiety-like behaviors and HPA response in mice. *Hormones and Behavior, 65*(3), 211–218.

Ciruela, F., et al. (2000). Homer-1c/Vesl-1L modulates the cell surface targeting of metabotropic glutamate receptor type 1alpha: Evidence for an anchoring function. *Molecular and Cellular Neurosciences, 15*(1), 36–50.

Conn, P. J., & Pin, J. P. (1997). Pharmacology and functions of metabotropic glutamate receptors. *Annual Review of Pharmacology and Toxicology, 37*, 205–237.

Cooke, B., et al. (1998). Sexual differentiation of the vertebrate brain: Principles and mechanisms. *Frontiers in Neuroendocrinology, 19*(4), 323–362.

Cox, K. H., & Rissman, E. F. (2011). Sex differences in juvenile mouse social behavior are influenced by sex chromosomes and social context. *Genes, Brain, and Behavior, 10*(4), 465–472.

Cozzoli, D. K., et al. (2009). Binge drinking upregulates accumbens mGluR5-Homer2-PI3K signaling: Functional implications for alcoholism. *The Journal of Neuroscience, 29*(27), 8655–8668.

Cozzoli, D. K., et al. (2014). The effect of mGluR5 antagonism during binge Drinkingon subsequent ethanol intake in C57BL/6J mice: Sex- and age-induced differences. *Alcoholism, Clinical and Experimental Research, 38*(3), 730–738.

Crews, F. T., et al. (2016). Adolescent alcohol exposure persistently impacts adult neurobiology and behavior. *Pharmacological Reviews, 68*(4), 1074–1109.

Cryan, J. F., et al. (2003). Antidepressant and anxiolytic-like effects in mice lacking the group III metabotropic glutamate receptor mGluR7. *The European Journal of Neuroscience, 17*(11), 2409–2417.

Davis, M. J., et al. (2012). Measures of anxiety, sensorimotor function, and memory in male and female mGluR4^{-}/$^{-}$ mice. *Behavioural Brain Research, 229*(1), 21–28.

De Blasi, A., et al. (2001). Molecular determinants of metabotropic glutamate receptor signaling. *Trends in Pharmacological Sciences, 22*(3), 114–120.

De Jesús-Burgos, M., Torres-Llenza, V., & Pérez-Acevedo, N. L. (2012). Activation of amygdalar metabotropic glutamate receptors modulates anxiety, and risk assessment behaviors in ovariectomized estradiol-treated female rats. *Pharmacology, Biochemistry, and Behavior, 101*(3), 369–378.

De Jesús-Burgos, M. I., et al. (2016). Amygdalar activation of group I metabotropic glutamate receptors produces anti- and pro-conflict effects depending upon animal sex in a sexually dimorphic conditioned conflict-based anxiety model. *Behavioural Brain Research, 302*, 200–212.

Di Menna, L., et al. (2018). Functional partnership between mGlu3 and mGlu5 metabotropic glutamate receptors in the central nervous system. *Neuropharmacology, 128*, 301–313.

Diehl, A., et al. (2007). Alcoholism in women: Is it different in onset and outcome compared to men? *European Archives of Psychiatry and Clinical Neuroscience, 257*(6), 344–351.

Doncheck, E. M., et al. (2018). 17β-estradiol potentiates the reinstatement of cocaine seeking in female rats: Role of the prelimbic prefrontal cortex and cannabinoid Type-1 receptors. *Neuropsychopharmacology, 43*(4), 781–790.

DuBois, J. M., et al. (2016). Characterization of age/sex and the regional distribution of mGluR5 availability in the healthy human brain measured by high-resolution [(11)C] ABP688 PET. *European Journal of Nuclear Medicine and Molecular Imaging, 43*(1), 152–162.

Duvoisin, R. M., et al. (2010). Sex-dependent cognitive phenotype of mice lacking mGluR8. *Behavioural Brain Research, 209*(1), 21–26.

Evans, S. M., Haney, M., & Foltin, R. W. (2002). The effects of smoked cocaine during the follicular and luteal phases of the menstrual cycle in women. *Psychopharmacology, 159*(4), 397–406.

Fagan, M. P., et al. (2020). Essential and sex-specific effects of mGluR5 in ventromedial hypothalamus regulating estrogen signaling and glucose balance. *Proceedings of the National Academy of Sciences of the United States of America, 117*(32), 19566–19577.

Fazio, F., et al. (2008). Switch in the expression of mGlu1 and mGlu5 metabotropic glutamate receptors in the cerebellum of mice developing experimental autoimmune encephalomyelitis and in autoptic cerebellar samples from patients with multiple sclerosis. *Neuropharmacology, 55*(4), 491–499.

Feltenstein, M. W., & See, R. E. (2007). Plasma progesterone levels and cocaine-seeking in freely cycling female rats across the estrous cycle. *Drug and Alcohol Dependence, 89*(2–3), 183–189.

Fernandez, S. M., et al. (2008). Estradiol-induced enhancement of object memory consolidation involves hippocampal extracellular signal-regulated kinase activation and membrane-bound estrogen receptors. *The Journal of Neuroscience, 28*(35), 8660–8667.

Ferretti, M. T., et al. (2018). Sex differences in Alzheimer disease—The gateway to precision medicine. *Nature Reviews. Neurology, 14*(8), 457–469.

Fisher, N. M., et al. (2020). Phenotypic profiling of mGlu(7) knockout mice reveals new implications for neurodevelopmental disorders. *Genes, Brain, and Behavior, 19*(7), e12654.

Fisher, N. M., et al. (2021). A GRM7 mutation associated with developmental delay reduces mGlu7 expression and produces neurological phenotypes. *JCI Insight, 6*(4), e143324.

Frank, E., Carpenter, L. L., & Kupfer, D. J. (1988). Sex differences in recurrent depression: Are there any that are significant? *The American Journal of Psychiatry, 145*(1), 41–45.

Fuentes, N., & Silveyra, P. (2019). Estrogen receptor signaling mechanisms. *Advances in Protein Chemistry and Structural Biology, 116*, 135–170.

García-Bea, A., et al. (2016). Metabotropic glutamate receptor 3 (mGlu3; mGluR3; GRM3) in schizophrenia: Antibody characterisation and a semi-quantitative western blot study. *Schizophrenia Research, 177*(1–3), 18–27.

Gatewood, J. D., et al. (2006). Sex chromosome complement and gonadal sex influence aggressive and parental behaviors in mice. *The Journal of Neuroscience, 26*(8), 2335–2342.

Gil-Sanz, C., et al. (2008). Involvement of the mGluR1 receptor in hippocampal synaptic plasticity and associative learning in behaving mice. *Cerebral Cortex, 18*(7), 1653–1663.

Goddyn, H., et al. (2008). Deficits in acquisition and extinction of conditioned responses in mGluR7 knockout mice. *Neurobiology of Learning and Memory, 90*(1), 103–111.

Goldstein, J. M., & Link, B. G. (1988). Gender and the expression of schizophrenia. *Journal of Psychiatric Research, 22*(2), 141–155.

Goldstein, J. M., Tsuang, M. T., & Faraone, S. V. (1989). Gender and schizophrenia: Implications for understanding the heterogeneity of the illness. *Psychiatry Research, 28*(3), 243–253.

Goldstein, J. M., et al. (1998). Are there sex differences in neuropsychological functions among patients with schizophrenia? *The American Journal of Psychiatry, 155*(10), 1358–1364.

Gregory, K. J., & Conn, P. J. (2015). Molecular insights into metabotropic glutamate receptor allosteric modulation. *Molecular Pharmacology, 88*(1), 188–202.

Gross, W. C., & Billingham, R. E. (1998). Alcohol consumption and sexual victimization among college women. *Psychological Reports, 82*(1), 80–82.

Gross, K. S., et al. (2016). Opposite effects of mGluR1a and mGluR5 activation on nucleus Accumbens medium spiny neuron dendritic spine density. *PLoS One, 11*(9), e0162755.

Gross, K. S., et al. (2018). mGluR5 mediates dihydrotestosterone-induced nucleus Accumbens structural plasticity, but not conditioned reward. *Frontiers in Neuroscience, 12*, 855.

Grove-Strawser, D., Boulware, M. I., & Mermelstein, P. G. (2010). Membrane estrogen receptors activate the metabotropic glutamate receptors mGluR5 and mGluR3 to bidirectionally regulate CREB phosphorylation in female rat striatal neurons. *Neuroscience, 170*(4), 1045–1055.

Grueter, B. A., Brasnjo, G., & Malenka, R. C. (2010). Postsynaptic TRPV1 triggers cell type-specific long-term depression in the nucleus accumbens. *Nature Neuroscience, 13*(12), 1519–1525.

Haas, L. T., et al. (2017). Silent allosteric modulation of mGluR5 maintains glutamate signaling while rescuing Alzheimer's mouse phenotypes. *Cell Reports, 20*(1), 76–88.

Hafner, H., et al. (1989). How does gender influence age at first hospitalization for schizophrenia? A transnational case register study. *Psychological Medicine, 19*(4), 903–918.

Hambrecht, M., Maurer, K., & Hafner, H. (1992). Gender differences in schizophrenia in three cultures. Results of the WHO collaborative study on psychiatric disability. *Social Psychiatry and Psychiatric Epidemiology, 27*(3), 117–121.

Hamilton, J. A., Grant, M., & Jensvold, M. F. (1996). Sex and treatment of depression: When does it matter? In M. F. Jensvold, U. Halbreich, & J. A. Hamilton (Eds.), *Psychopharmacology and women: Sex, gender, and hormones* (pp. 241–257). Washington, DC: American Psychiatric Press.

Metabotropic glutamate receptors in psychiatric and neurological disorders 341

Handa, R. J., et al. (1985). Differential effects of the perinatal steroid environment on three sexually dimorphic parameters of the rat brain. *Biology of Reproduction, 32*(4), 855–864.

Hartmann, J., Henning, H. A., & Konnerth, A. (2011). mGluR1/TRPC3-mediated synaptic transmission and calcium signaling in mammalian central neurons. *Cold Spring Harbor Perspectives in Biology, 3*(4), a006726.

Hermans, E., & Challiss, R. A. (2001). Structural, signalling and regulatory properties of the group I metabotropic glutamate receptors: Prototypic family C G-protein-coupled receptors. *The Biochemical Journal, 359*(Pt 3), 465–484.

Highland, J. N., et al. (2019). Group II metabotropic glutamate receptor blockade promotes stress resilience in mice. *Neuropsychopharmacology, 44*(10), 1788–1796.

Hill, M. N., Karacabeyli, E. S., & Gorzalka, B. B. (2007). Estrogen recruits the endocannabinoid system to modulate emotionality. *Psychoneuroendocrinology, 32*(4), 350–357.

Hodge, C. W., et al. (2006). The mGluR5 antagonist MPEP selectively inhibits the onset and maintenance of ethanol self-administration in C57BL/6J mice. *Psychopharmacology, 183*(4), 429–438.

Hölscher, C., et al. (2004). Lack of the metabotropic glutamate receptor subtype 7 selectively impairs short-term working memory but not long-term memory. *Behavioural Brain Research, 154*(2), 473–481.

Hu, M., & Becker, J. B. (2003). Effects of sex and estrogen on behavioral sensitization to cocaine in rats. *The Journal of Neuroscience, 23*(2), 693–699.

Huang, G., Thompson, S. L., & Taylor, J. R. (2021). MPEP lowers binge drinking in male and female C57BL/6 mice: Relationship with mGlu5/Homer2/Erk2 signaling. *Alcoholism, Clinical and Experimental Research, 45*(4), 732–742.

Huang, G. Z., & Woolley, C. S. (2012). Estradiol acutely suppresses inhibition in the hippocampus through a sex-specific endocannabinoid and mGluR-dependent mechanism. *Neuron, 74*(5), 801–808.

Huber, G., et al. (1980). Longitudinal studies of schizophrenic patients. *Schizophrenia Bulletin, 6*(4), 592–605.

Jackson, L. R., Robinson, T. E., & Becker, J. B. (2006). Sex differences and hormonal influences on acquisition of cocaine self-administration in rats. *Neuropsychopharmacology, 31*(1), 129–138.

Joffe, M. E., et al. (2018). Metabotropic glutamate receptors in alcohol use disorder: Physiology, plasticity, and promising pharmacotherapies. *ACS Chemical Neuroscience, 9*(9), 2188–2204.

Joffe, M. E., et al. (2019). Metabotropic glutamate receptor subtype 3 gates acute stress-induced dysregulation of amygdalo-cortical function. *Molecular Psychiatry, 24*(6), 916–927.

Joffe, M. E., et al. (2020). mGlu2 and mGlu3 negative allosteric modulators divergently enhance Thalamocortical transmission and exert rapid antidepressant-like effects. *Neuron, 105*(1), 46–59.e3.

Joffe, M. E., et al. (2021). Frontal cortex genetic ablation of metabotropic glutamate receptor subtype 3 (mGlu(3)) impairs postsynaptic plasticity and modulates affective behaviors. *Neuropsychopharmacology, 46*(12), 2148–2157.

Joffe, M., et al. (2022). Acute restraint stress redirects prefrontal cortex circuit function through mGlu5 receptor plasticity on somatostatin-expressing interneurons. *Neuron, 110*(6), 1068–1083.e5.

Johnson, A. R., et al. (2019). Cues play a critical role in estrous cycle-dependent enhancement of cocaine reinforcement. *Neuropsychopharmacology, 44*(7), 1189–1197.

Jong, Y. J., et al. (2005). Functional metabotropic glutamate receptors on nuclei from brain and primary cultured striatal neurons. Role of transporters in delivering ligand. *The Journal of Biological Chemistry, 280*(34), 30469–30480.

Kalinichev, M., et al. (2014). Characterization of the novel positive allosteric modulator of the metabotropic glutamate receptor 4 ADX88178 in rodent models of neuropsychiatric disorders. *The Journal of Pharmacology and Experimental Therapeutics, 350*(3), 495–505.

Kammermeier, P. J., et al. (2000). Homer proteins regulate coupling of group I metabotropic glutamate receptors to N-type calcium and M-type potassium channels. *The Journal of Neuroscience, 20*(19), 7238–7245.

Kasten, C. R., et al. (2020). Adolescent alcohol exposure produces sex differences in negative affect-like behavior and group I mGluR BNST plasticity. *Neuropsychopharmacology, 45*(8), 1306–1315.

Kasten, C. R., et al. (2021). BNST specific mGlu5 receptor knockdown regulates sex-dependent expression of negative affect produced by adolescent ethanol exposure and adult stress. *Translational Psychiatry, 11*(1), 178.

Kessler, R. C., et al. (2005). Lifetime prevalence and age-of-onset distributions of DSM-IV disorders in the National Comorbidity Survey Replication. *Archives of General Psychiatry, 62*(6), 593–602.

Kessler, R. C., et al. (2012). Twelve-month and lifetime prevalence and lifetime morbid risk of anxiety and mood disorders in the United States. *International Journal of Methods in Psychiatric Research, 21*(3), 169–184.

Kornstein, S. G., et al. (2000a). Gender differences in chronic major and double depression. *Journal of Affective Disorders, 60*(1), 1–11.

Kornstein, S. G., et al. (2000b). Gender differences in treatment response to sertraline versus imipramine in chronic depression. *The American Journal of Psychiatry, 157*(9), 1445–1452.

Kuljis, D. A., et al. (2013). Gonadal- and sex-chromosome-dependent sex differences in the circadian system. *Endocrinology, 154*(4), 1501–1512.

Lebow, M. A., & Chen, A. (2016). Overshadowed by the amygdala: The bed nucleus of the stria terminalis emerges as key to psychiatric disorders. *Molecular Psychiatry, 21*(4), 450–463.

Levin, E. R. (2009). Plasma membrane estrogen receptors. *Trends in Endocrinology and Metabolism, 20*(10), 477–482.

Lewis, B., Hoffman, L. A., & Nixon, S. J. (2014). Sex differences in drug use among polysubstance users. *Drug and Alcohol Dependence, 145*, 127–133.

Li, S. H., et al. (2022). Metabotropic glutamate receptor 5 antagonism reduces pathology and differentially improves symptoms in male and female heterozygous zQ175 Huntington's mice. *Frontiers in Molecular Neuroscience, 15*, 801757.

Logan, C. N., et al. (2020). Ceftriaxone and mGlu2/3 interactions in the nucleus accumbens core affect the reinstatement of cocaine-seeking in male and female rats. *Psychopharmacology, 237*(7), 2007–2018.

Lominac, K. D., et al. (2006). Behavioral and neurochemical interactions between group 1 mGluR antagonists and ethanol: Potential insight into their anti-addictive properties. *Drug and Alcohol Dependence, 85*(2), 142–156.

Loweth, J. A., et al. (2014). Synaptic depression via mGluR1 positive allosteric modulation suppresses cue-induced cocaine craving. *Nature Neuroscience, 17*(1), 73–80.

Luscher, C., & Huber, K. M. (2010). Group 1 mGluR-dependent synaptic long-term depression: Mechanisms and implications for circuitry and disease. *Neuron, 65*(4), 445–459.

Lynch, W. J., et al. (2001). Role of estrogen in the acquisition of intravenously self-administered cocaine in female rats. *Pharmacology, Biochemistry, and Behavior, 68*(4), 641–646.

Maksymetz, J., & Joffe, M. E. (2021). *mGlu* Receptor modulation in murine models of stress and affective disorders. In M. F. Olive, B. T. Burrows, & J. M. Leyrer-Jackson (Eds.), *Metabotropic glutamate receptor technologies* (pp. 259–296). New York, NY: Humana.

Mao, L. M., & Wang, J. Q. (2016). Regulation of group I metabotropic glutamate receptors by MAPK/ERK in neurons. *Journal of Nature and Science, 2*(12), e268.

Martin, L. J., et al. (1992). Cellular localization of a metabotropic glutamate receptor in rat brain. *Neuron, 9*(2), 259–270.

Martinez, L. A., et al. (2014). Estradiol facilitation of cocaine-induced locomotor sensitization in female rats requires activation of mGluR5. *Behavioural Brain Research, 271*, 39–42.

Martini, M., et al. (2020). Sex chromosome complement influences vulnerability to cocaine in mice. *Hormones and Behavior, 125*, 104821.

Matas, E., et al. (2021). Major motor and gait deficits with sexual dimorphism in a Shank3 mutant mouse model. *Molecular Autism, 12*(1), 2.

Mathews, D. C., Henter, I. D., & Zarate, C. A. (2012). Targeting the glutamatergic system to treat major depressive disorder: Rationale and progress to date. *Drugs, 72*(10), 1313–1333.

McCarthy, M. M., & Arnold, A. P. (2011). Reframing sexual differentiation of the brain. *Nature Neuroscience, 14*(6), 677–683.

McEwen, B., et al. (2001). Tracking the estrogen receptor in neurons: Implications for estrogen-induced synapse formation. *Proceedings of the National Academy of Sciences of the United States of America, 98*(13), 7093–7100.

McPhie-Lalmansingh, A. A., et al. (2008). Sex chromosome complement affects social interactions in mice. *Hormones and Behavior, 54*(4), 565–570.

Meitzen, J., & Mermelstein, P. G. (2011). Estrogen receptors stimulate brain region specific metabotropic glutamate receptors to rapidly initiate signal transduction pathways. *Journal of Chemical Neuroanatomy, 42*(4), 236–241.

Meitzen, J., et al. (2019). The expression of select genes necessary for membrane-associated estrogen receptor signaling differ by sex in adult rat hippocampus. *Steroids, 142*, 21–27.

Mercer, A. A., et al. (2016). Sex differences in cerebellar synaptic transmission and sex-specific responses to autism-linked Gabrb3 mutations in mice. *eLife, 5*, e07596.

Mermelstein, P. G., Becker, J. B., & Surmeier, D. J. (1996). Estradiol reduces calcium currents in rat neostriatal neurons via a membrane receptor. *The Journal of Neuroscience, 16*(2), 595–604.

Milner, T. A., et al. (2001). Ultrastructural evidence that hippocampal alpha estrogen receptors are located at extranuclear sites. *The Journal of Comparative Neurology, 429*(3), 355–371.

Minami, K., et al. (1998). Effects of ethanol and anesthetics on type 1 and 5 metabotropic glutamate receptors expressed in Xenopus laevis oocytes. *Molecular Pharmacology, 53*(1), 148–156.

Mitterling, K. L., et al. (2010). Cellular and subcellular localization of estrogen and progestin receptor immunoreactivities in the mouse hippocampus. *The Journal of Comparative Neurology, 518*(14), 2729–2743.

Mora, S., Dussaubat, N., & Díaz-Véliz, G. (1996). Effects of the estrous cycle and ovarian hormones on behavioral indices of anxiety in female rats. *Psychoneuroendocrinology, 21*(7), 609–620.

Murphy, M. P., & LeVine, H., 3rd. (2010). Alzheimer's disease and the amyloid-beta peptide. *Journal of Alzheimer's Disease, 19*(1), 311–323.

Najt, P., Fusar-Poli, P., & Brambilla, P. (2011). Co-occurring mental and substance abuse disorders: A review on the potential predictors and clinical outcomes. *Psychiatry Research, 186*(2–3), 159–164.

Nardecchia, F., et al. (2018). Targeting mGlu5 metabotropic glutamate receptors in the treatment of cognitive dysfunction in a mouse model of phenylketonuria. *Frontiers in Neuroscience, 12*, 154.

Netzeband, J. G., & Gruol, D. L. (1995). Modulatory effects of acute ethanol on metabotropic glutamate responses in cultured Purkinje neurons. *Brain Research, 688*(1–2), 105–113.

Ning, L. N., et al. (2018). Gender-related hippocampal proteomics study from Young rats after chronic unpredicted mild stress exposure. *Molecular Neurobiology, 55*(1), 835–850.

Niswender, C. M., & Conn, P. J. (2010). Metabotropic glutamate receptors: Physiology, pharmacology, and disease. *Annual Review of Pharmacology and Toxicology, 50*, 295–322.

Nixon, S. J., Tivis, R., & Parsons, O. A. (1995). Behavioral dysfunction and cognitive efficiency in male and female alcoholics. *Alcoholism, Clinical and Experimental Research, 19*(3), 577–581.

Oberlander, J. G., & Woolley, C. S. (2016). 17β-estradiol acutely potentiates glutamatergic synaptic transmission in the Hippocampus through distinct mechanisms in males and females. *The Journal of Neuroscience, 36*(9), 2677–2690.

Ohno-Shosaku, T., et al. (2002). Cooperative endocannabinoid production by neuronal depolarization and group I metabotropic glutamate receptor activation. *The European Journal of Neuroscience, 15*(6), 953–961.

Paden, W., et al. (2020). Sex differences in adult mood and in stress-induced transcriptional coherence across mesocorticolimbic circuitry. *Translational Psychiatry, 10*(1), 59.

Palucha, A., & Pilc, A. (2007). Metabotropic glutamate receptor ligands as possible anxiolytic and antidepressant drugs. *Pharmacology & Therapeutics, 115*(1), 116–147.

Peixoto, R. T., et al. (2016). Early hyperactivity and precocious maturation of corticostriatal circuits in Shank3B(−/−) mice. *Nature Neuroscience, 19*(5), 716–724.

Peltier, M. R., et al. (2019). Sex differences in stress-related alcohol use. *Neurobiology of Stress, 10*, 100149.

Peterson, B. M., Mermelstein, P. G., & Meisel, R. L. (2015). Estradiol mediates dendritic spine plasticity in the nucleus accumbens core through activation of mGluR5. *Brain Structure & Function, 220*(4), 2415–2422.

Phoenix, C. H., et al. (1959). Organizing action of prenatally administered testosterone propionate on the tissues mediating mating behavior in the female guinea pig. *Endocrinology, 65*, 369–382.

Pietraszek, M., et al. (2005). Anxiolytic-like effects of mGlu1 and mGlu5 receptor antagonists in rats. *European Journal of Pharmacology, 514*(1), 25–34.

Puralewski, R., Vasilakis, G., & Seney, M. L. (2016). Sex-related factors influence expression of mood-related genes in the basolateral amygdala differentially depending on age and stress exposure. *Biology of Sex Differences, 7*, 50.

Purgert, C. A., et al. (2014). Intracellular mGluR5 can mediate synaptic plasticity in the hippocampus. *The Journal of Neuroscience, 34*(13), 4589–4598.

Quitkin, F. M., et al. (2002). Are there differences between women's and men's antidepressant responses? *The American Journal of Psychiatry, 159*(11), 1848–1854.

Randall, C. L., et al. (1999). Telescoping of landmark events associated with drinking: A gender comparison. *Journal of Studies on Alcohol, 60*(2), 252–260.

Razandi, M., et al. (2004). Plasma membrane estrogen receptors exist and functions as dimers. *Molecular Endocrinology, 18*(12), 2854–2865.

Renner, M., et al. (2010). Deleterious effects of amyloid beta oligomers acting as an extracellular scaffold for mGluR5. *Neuron, 66*(5), 739–754.

Ribeiro, F. M., et al. (2014). Metabotropic glutamate receptor 5 knockout promotes motor and biochemical alterations in a mouse model of Huntington's disease. *Human Molecular Genetics, 23*(8), 2030–2042.

Roberts, D. C., Bennett, S. A., & Vickers, G. J. (1989). The estrous cycle affects cocaine self-administration on a progressive ratio schedule in rats. *Psychopharmacology, 98*(3), 408–411.

Roche, K. W., et al. (1999). Homer 1b regulates the trafficking of group I metabotropic glutamate receptors. *The Journal of Biological Chemistry, 274*(36), 25953–25957.

Romano, C., et al. (1995). Distribution of metabotropic glutamate receptor mGluR5 immunoreactivity in rat brain. *The Journal of Comparative Neurology, 355*(3), 455–469.

Salokangas, R. K. (1983). Prognostic implications of the sex of schizophrenic patients. *The British Journal of Psychiatry, 142*, 145–151.

Sartorius, N., et al. (1986). Early manifestations and first-contact incidence of schizophrenia in different cultures. A preliminary report on the initial evaluation phase of the WHO collaborative study on determinants of outcome of severe mental disorders. *Psychological Medicine, 16*(4), 909–928.

Schiefer, J., et al. (2004). The metabotropic glutamate receptor 5 antagonist MPEP and the mGluR2 agonist LY379268 modify disease progression in a transgenic mouse model of Huntington's disease. *Brain Research, 1019*(1–2), 246–254.

Seney, M. L., et al. (2013a). The role of genetic sex in affect regulation and expression of GABA-related genes across species. *Frontiers in Psychiatry, 4*, 104.

Seney, M. L., et al. (2013b). Sex chromosome complement regulates expression of mood-related genes. *Biology of Sex Differences, 4*(1), 20.

Silverstein, B. (1999). Gender difference in the prevalence of clinical depression: The role played by depression associated with somatic symptoms. *The American Journal of Psychiatry, 156*(3), 480–482.

Simonyi, A., Schachtman, T. R., & Christoffersen, G. R. (2005). The role of metabotropic glutamate receptor 5 in learning and memory processes. *Drug News & Perspectives, 18*(6), 353–361.

Smart, K., et al. (2019). Sex differences in [(11)C]ABP688 binding: A positron emission tomography study of mGlu5 receptors. *European Journal of Nuclear Medicine and Molecular Imaging, 46*(5), 1179–1183.

Szumlinski, K. K., Ary, A. W., & Lominac, K. D. (2008). Homers regulate drug-induced neuroplasticity: Implications for addiction. *Biochemical Pharmacology, 75*(1), 112–133.

Szumlinski, K. K., et al. (2004). Homer proteins regulate sensitivity to cocaine. *Neuron, 43*(3), 401–413.

Szumlinski, K. K., et al. (2005). Homer2 is necessary for EtOH-induced neuroplasticity. *The Journal of Neuroscience, 25*(30), 7054–7061.

Szumlinski, K. K., et al. (2008). Accumbens Homer2 overexpression facilitates alcohol-induced neuroplasticity in C57BL/6J mice. *Neuropsychopharmacology, 33*(6), 1365–1378.

Tabatadze, N., et al. (2015). Sex differences in molecular signaling at inhibitory synapses in the Hippocampus. *The Journal of Neuroscience, 35*(32), 11252–11265.

Tonn Eisinger, K. R., et al. (2018). Interactions between estrogen receptors and metabotropic glutamate receptors and their impact on drug addiction in females. *Hormones and Behavior, 104*, 130–137.

Torres, E. R. S., et al. (2018). Effects of sub-chronic MPTP exposure on behavioral and cognitive performance and the microbiome of wild-type and mGlu8 knockout female and male mice. *Frontiers in Behavioral Neuroscience, 12*, 140.

Tu, J. C., et al. (1999). Coupling of mGluR/Homer and PSD-95 complexes by the shank family of postsynaptic density proteins. *Neuron, 23*(3), 583–592.

Tuyns, A. J., & Pequignot, G. (1984). Greater risk of ascitic cirrhosis in females in relation to alcohol consumption. *International Journal of Epidemiology, 13*(1), 53–57.

Uchino, S., & Waga, C. (2013). SHANK3 as an autism spectrum disorder-associated gene. *Brain & Development, 35*(2), 106–110.

Urbano-Marquez, A., et al. (1995). The greater risk of alcoholic cardiomyopathy and myopathy in women compared with men. *JAMA, 274*(2), 149–154.

Valenti, O., Conn, P. J., & Marino, M. J. (2002). Distinct physiological roles of the Gq-coupled metabotropic glutamate receptors co-expressed in the same neuronal populations. *Journal of Cellular Physiology, 191*(2), 125–137.

Van den Pol, A. N. (1994). Metabotropic glutamate receptor mGluR1 distribution and ultrastructural localization in hypothalamus. *The Journal of Comparative Neurology, 349*(4), 615–632.

van den Pol, A. N., Romano, C., & Ghosh, P. (1995). Metabotropic glutamate receptor mGluR5 subcellular distribution and developmental expression in hypothalamus. *The Journal of Comparative Neurology, 362*(1), 134–150.

Varma, N., et al. (2001). Metabotropic glutamate receptors drive the endocannabinoid system in hippocampus. *The Journal of Neuroscience, 21*(24), Rc188.

Verpelli, C., et al. (2011). Importance of Shank3 protein in regulating metabotropic glutamate receptor 5 (mGluR5) expression and signaling at synapses. *The Journal of Biological Chemistry, 286*(40), 34839–34850.

Wang, X., et al. (2011). Synaptic dysfunction and abnormal behaviors in mice lacking major isoforms of Shank3. *Human Molecular Genetics, 20*(15), 3093–3108.

Wang, Y., et al. (2015). Prenatal chronic mild stress induces depression-like behavior and sex-specific changes in regional glutamate receptor expression patterns in adult rats. *Neuroscience, 301*, 363–374.

Watson, C. S., Jeng, Y. J., & Kochukov, M. Y. (2008). Nongenomic actions of estradiol compared with estrone and estriol in pituitary tumor cell signaling and proliferation. *The FASEB Journal, 22*(9), 3328–3336.

Weiland, N. G., et al. (1997). Distribution and hormone regulation of estrogen receptor immunoreactive cells in the hippocampus of male and female rats. *The Journal of Comparative Neurology, 388*(4), 603–612.

Wickens, M. M., Bangasser, D. A., & Briand, L. A. (2018). Sex differences in psychiatric disease: A focus on the glutamate system. *Frontiers in Molecular Neuroscience, 11*, 197.

Woolley, C. S., & McEwen, B. S. (1993). Roles of estradiol and progesterone in regulation of hippocampal dendritic spine density during the estrous cycle in the rat. *The Journal of Comparative Neurology, 336*(2), 293–306.

Woolley, C. S., et al. (1990). Naturally occurring fluctuation in dendritic spine density on adult hippocampal pyramidal neurons. *The Journal of Neuroscience, 10*(12), 4035–4039.

Xiao, M. Y., Gustafsson, B., & Niu, Y. P. (2006). Metabotropic glutamate receptors in the trafficking of ionotropic glutamate and GABA(A) receptors at central synapses. *Current Neuropharmacology, 4*(1), 77–86.

Xiao, B., et al. (1998). Homer regulates the association of group 1 metabotropic glutamate receptors with multivalent complexes of homer-related, synaptic proteins. *Neuron, 21*(4), 707–716.

Xie, X., et al. (2010). Effects of mGluR1 antagonism in the dorsal hippocampus on drug context-induced reinstatement of cocaine-seeking behavior in rats. *Psychopharmacology, 208*(1), 1–11.

Xu, J., et al. (2009). mGluR5 has a critical role in inhibitory learning. *The Journal of Neuroscience, 29*(12), 3676–3684.

Young, M. A., et al. (1990). Sex differences in the lifetime prevalence of depression: Does varying the diagnostic criteria reduce the female/male ratio? *Journal of Affective Disorders, 18*(3), 187–192.

Yu, F., et al. (2013). Metabotropic glutamate receptor I (mGluR1) antagonism impairs cocaine-induced conditioned place preference via inhibition of protein synthesis. *Neuropsychopharmacology, 38*(7), 1308–1321.

Zhao, Z., Fan, L., & Frick, K. M. (2010). Epigenetic alterations regulate estradiol-induced enhancement of memory consolidation. *Proceedings of the National Academy of Sciences of the United States of America, 107*(12), 5605–5610.

Zhao, Z., et al. (2012). Hippocampal histone acetylation regulates object recognition and the estradiol-induced enhancement of object recognition. *The Journal of Neuroscience, 32*(7), 2344–2351.

Zielonka, D., et al. (2013). The influence of gender on phenotype and disease progression in patients with Huntington's disease. *Parkinsonism & Related Disorders, 19*(2), 192–197.

Zoicas, I., & Kornhuber, J. (2019). The role of metabotropic glutamate receptors in social behavior in rodents. *International Journal of Molecular Sciences, 20*(6), 1412.

Zuloaga, D. G., et al. (2008). Mice with the testicular feminization mutation demonstrate a role for androgen receptors in the regulation of anxiety-related behaviors and the hypothalamic-pituitary-adrenal axis. *Hormones and Behavior, 54*(5), 758–766.

Zuloaga, D. G., et al. (2011). Male rats with the testicular feminization mutation of the androgen receptor display elevated anxiety-related behavior and corticosterone response to mild stress. *Hormones and Behavior, 60*(4), 380–388.

CHAPTER EIGHT

Roles of metabotropic glutamate receptor 8 in neuropsychiatric and neurological disorders

Li-Min Mao[a], Nirav Mathur[b], Karina Shah[a], and John Q. Wang[a,b,*]

[a]Department of Biomedical Sciences, University of Missouri-Kansas City, School of Medicine, Kansas City, MO, United States
[b]Department of Anesthesiology, University of Missouri-Kansas City, School of Medicine, Kansas City, MO, United States
*Corresponding author: e-mail address: wangjq@umkc.edu

Contents

1. Introduction	350
2. mGlu8 receptors	351
3. Anxiety	352
4. Epilepsy	354
5. Parkinson's disease	355
6. Drug addiction	356
7. Chronic pain	357
8. Alzheimer's disease	358
9. Conclusions	359
Author contributions	359
Funding	359
Conflict of interest	359
References	359

Abstract

Metabotropic glutamate (mGlu) receptors are G protein-coupled receptors. Among eight mGlu subtypes (mGlu1–8), mGlu8 has drawn increasing attention. This subtype is localized to the presynaptic active zone of neurotransmitter release and is among the mGlu subtypes with high affinity for glutamate. As a $G_{i/o}$-coupled autoreceptor, mGlu8 inhibits glutamate release to maintain homeostasis of glutamatergic transmission. mGlu8 receptors are expressed in limbic brain regions and play a pivotal role in modulating motivation, emotion, cognition, and motor functions. Emerging evidence emphasizes the increasing clinical relevance of abnormal mGlu8 activity. Studies using mGlu8 selective agents and knockout mice have revealed the linkage of mGlu8 receptors to multiple neuropsychiatric and neurological disorders, including anxiety, epilepsy, Parkinson's disease, drug addiction, and chronic pain. Expression and function of mGlu8 receptors in some limbic structures undergo long-lasting adaptive changes in animal models of these disorders, which may contribute to the remodeling of glutamatergic transmission critical for the pathogenesis and symptomatology of brain illnesses. This review summarizes the current understanding of mGlu8 biology and the possible involvement of the receptor in several common psychiatric and neurological disorders.

1. Introduction

Metabotropic glutamate (mGlu) receptors are a family of eight subtypes (mGlu1–8) that are subdivided into three functional groups (I–III) based on the receptor structure, signaling pathways, and physiological activity. All mGlu receptors are membrane-bound G protein–coupled receptors (GPCR). Group I receptors (mGlu1/5) are coupled to $G_{q/11}$ proteins, whereas group II (mGlu2/3) and III (mGlu4/6/7/8) receptors are coupled to $G_{i/o}$ proteins (Nicoletti et al., 2011; Niswender & Conn, 2010). Thus, activation of group I receptors causes phospholipase Cβ to hydrolyze phosphoinositide to form inositol-1,4,5-trisphosphate and diacylglycerol which triggers intracellular Ca^{2+} release and activates protein kinase C, respectively. On the other hand, activation of group II/III receptors inhibits adenylyl cyclase and subsequently reduces cytosolic cAMP formation. One group III subtype, mGlu8, has drawn increasing attention due to its emerging roles in various neuropsychiatric and neurological disorders. In addition to its canonical cAMP-dependent signaling pathway, mGlu8 receptors modulate other effectors, including specific ion channels (Nicoletti et al., 2011; Niswender & Conn, 2010).

mGlu8 receptors are distributed in broad brain regions, including the limbic structures such as the hippocampus, striatum, and amygdala (Ferraguti & Shigemoto, 2006; Robbins et al., 2007). Like other group II/III subtypes, mGlu8 receptors are predominantly presynaptic as opposed to group I receptors which are localized to postsynaptic sites (Nicoletti et al., 2011; Niswender & Conn, 2010). Due to their confined location in the active zone close to the glutamate release site, mGlu8 receptors respond to synaptic glutamate and act as autoreceptors to inhibit glutamate release. Besides, mGlu8 serves as heteroreceptors to inhibit nonglutamatergic transmitter release (Kogo et al., 2004; Marabese et al., 2005). The mGlu8 subtype is notably among the presynaptic mGlu receptors that bind glutamate with high affinity, as opposed to the mGlu7 subtype that has very low affinity (Nicoletti et al., 2011; Niswender & Conn, 2010). Thus, mGlu8 receptors actively modulate glutamatergic transmission and synaptic plasticity such as long-term depression (LTD).

As a key element at excitatory synapses, the mGlu8 receptor is linked to the pathogenesis and symptomatology of various neuropsychiatric and neurological disorders (Crupi, Impellizzeri, & Cuzzocrea, 2019). Accumulating pharmacological evidence shows that activation of mGlu8 receptors with an mGlu8 selective agonist negatively regulates the addictive property of drugs of abuse (Mao, Guo, Jin, Xue, & Wang, 2013) and alleviates motor symptoms of Parkinson's disease (PD). The mGlu8 agonist and/or positive allosteric modulators (PAM) also exhibit anxiolytic and anticonvulsant properties in animal models of anxiety and epilepsy, respectively (Raber & Duvoisin, 2015). In mGlu8 knockout mice, mGlu8 receptors show an involvement in determining the vulnerability to anxiety. In addition, long-lasting plastic changes in expression and function of mGlu8 receptors in the hippocampus occur in chronic animal models of epilepsy. These mGlu8-sensitive neuroadaptive events play either a promoting or a compensatory role in the remodeling of glutamatergic transmission critical for shaping synaptic plasticity and phenotypic behavioral outcomes.

2. mGlu8 receptors

Grm8 gene cloned from rats (Saugstad, Kinzie, Shinohara, Segerson, & Westbrook, 1997) and mice (Duvoisin, Zhang, & Ramonell, 1995) and *GRM8* gene, a human homolog (Scherer et al., 1996; Scherer, Soder, Duvoisin, Huizenga, & Tsui, 1997; Wu et al., 1998), encode the mGlu8 receptor (accession numbers: rat, NP_071538; mouse, NP_032200; human, NP_000836). Two alternative splice variants (mGlu8a and mGlu8b) have been reported in rats and humans, which differ in the last 16 amino acids of the C-terminal domain (Corti et al., 1998; Malherbe et al., 1999). An additional variant in humans, mGlu8c, is predicted to encode only the N-terminal region of the receptor (Malherbe et al., 1999). As a membrane-bound GPCR, mGlu8 possesses typical topology, i.e., seven transmembrane domains which give rise to a large extracellular N-terminus, three intracellular loops, and an intracellular C-terminus. The C-terminal tail is the key domain where protein–protein interactions and multiple types of posttranslational modifications often occur (Mao et al., 2011).

The canonical postreceptor signaling pathway of $G_{i/o}$-coupled mGlu8 receptors is inhibition of adenylyl cyclase, which in turn decreases cAMP production and protein kinase A activity (Niswender & Conn, 2010). In addition, mGlu8 receptors are coupled to other downstream effectors. For example, mGlu8 receptors activate G protein-coupled inwardly rectifying K^+ channels (Corti et al., 1998; Saugstad et al., 1997) and inhibit L- and N-type voltage-gated Ca^{2+} channels (Guo & Ikeda, 2005; Koulen, Kuhn, Wassle, & Brandstatter, 1999). Activation of mGlu8 also induces chemical LTD, a common form of synaptic plasticity, at lateral perforant path-dentate gyrus synapses in the hippocampus (Lodge et al., 2013).

mGlu8 receptors are primarily expressed in the CNS, while they are also detected in peripheral tissues such as the pancreas, testes, and osteoblasts (Hinoi, Fujimori, Nakamura, & Yoneda, 2001; Julio-Pieper, Flor, Dinan, & Cryan, 2011; Tong & Kirchgessner, 2003; Tong, Ouedraogo, & Kirchgessner, 2002). In addition to neurons in which mGlu8 receptors are primarily expressed, cultured rat microglial cells express functional mGlu8 (Taylor, Diemel, & Pocock, 2003). mGlu8 receptors are widely distributed in brain regions, including the olfactory bulb, hippocampus, striatum, amygdala, cerebellum, and cortical areas (Ferraguti & Shigemoto, 2006; Robbins et al., 2007). At the sub-synaptic level, mGlu8 receptors are predominantly presynaptic but have been identified in some postsynaptic locations (Lavreysen & Dautzenberg, 2008). As with other group III subtypes (mGlu4/7), mGlu8 receptors are localized to the active zone close to the glutamate release site. They can therefore be activated by synaptic glutamate to act as autoreceptors to inhibit glutamate release, noting that glutamate binds to mGlu4/8 with relatively high affinity and mGlu7 with very low affinity (Nicoletti et al., 2011; Niswender & Conn, 2010). Meanwhile, mGlu8 could play a heteroreceptor's role to inhibit nonglutamatergic transmitter release such as gamma-aminobutyric acid (GABA) (Kogo et al., 2004; Marabese et al., 2005). Of note, while mGlu8 like other mGlu subtypes functions primarily as mGlu8/8 homodimers, mGlu8 also likely heterodimerizes with one of group II (mGlu2/3)/III (mGlu4/7) but not group I subtypes to form functional heterodimers (Doumazane et al., 2011).

3. Anxiety

Genetic studies with the loss-of-function approach investigated a role of mGlu8 receptors in anxiety (Duvoisin, Villasana, Pfankuch, & Raber, 2010). In mice lacking the mGlu8 receptor, anxiety-like behavior was observed in an unconditioned behavioral model of anxiety, i.e., elevated plus maze (Duvoisin et al., 2005; Linden et al., 2002; Robbins et al., 2007). The enhanced anxiety behavior in mGlu8$^{(-/-)}$ mice was associated with an increase in c-Fos expression (a marker of neuronal activity) in the centromedial nucleus of the thalamus, indicating that loss of mGlu8 receptors reduces the threshold of neuronal activation in the stress-related brain region (Linden, Baez, Bergeron, & Schoepp, 2003). However, the anxiogenic-like phenotype of mGlu8-null mice was not consistently seen in the elevated plus maze (Fendt et al., 2010; Goddyn, Callaerts-Vegh, & D'Hooge, 2015) or elevated zero maze (Davis, Duvoisin, & Raber, 2013). Moreover, mGlu8 knockout mice displayed an anxiolytic-like phenotype in a contextual conditioned fear model (Fendt et al., 2010, 2013; Gerlai, Adams, Fitch, Chaney, & Baez, 2002). Future studies are needed to elucidate the exact role of mGlu8 receptors under the same experimental conditions and genetic background.

In pharmacological studies, an mGlu8 selective orthosteric agonist, (S)-3,4-dicarboxyphenylglycine [(S)-3,4-DCPG] (Thomas et al., 2001), with the subtype selectivity demonstrated in mGlu8 knockout mice in vivo (Linden, Bergeron, Baez, & Schoepp, 2003), was used to study the role of mGlu8 receptors. Acute intraperitoneal (i.p.) administration of (S)-3,4-DCPG (3 mg/kg) reduced the heightened level of anxiety-like behavior in the elevated zero maze test in male C57BL/6J mice, while (S)-3,4-DCPG had no much effect on baseline anxiety responses (Duvoisin, Pfankuch, et al., 2010). AZ12216052, an mGlu8 PAM, showed similar anxiolytic properties (Duvoisin, Pfankuch, et al., 2010). In the elevated plus maze test, injection of (S)-3,4-DCPG (10 μM) into the central nucleus of amygdala was anxiolytic in male arthritic Sprague–Dawley rats, although (S)-3,4-DCPG had no effect in normal rats (Palazzo, Fu, Ji, Maione, & Neugebauer, 2008). Systemic (S)-3,4-DCPG (30 mg/kg, i.p.) also exerted an anxiolytic effect in other paradigms of unconditioned conflict anxiety such as the open field and light-dark box tests in male C57BL/6J mice (Bahi, 2017) and altered neuronal excitability in mouse stress-related brain regions (Linden, Bergeron, et al., 2003). Bath application of (S)-3,4-DCPG depressed excitatory synaptic transmission in the bed nucleus of the stria terminalis, an integral component of the anxiety circuitry (Gosnell et al., 2011; Grueter & Winder, 2005). When injected into the lateral amygdala, (S)-3,4-DCPG (0.03–3 μM) inhibited acquisition and expression of conditioning fear in a fear-potentiated startle paradigm in male Sprague–Dawley rats (Schmid & Fendt, 2006). These results support a role of mGlu8 in the negative regulation of heightened or baseline anxiety-like behavior. However, no anxiolytic-like effects were observed in the conflict drinking Vogel test after local injection of (S)-3,4-DCPG (10, 50, and 100 nmol) into the basolateral amygdala or hippocampal CA1 region of male Wistar rats (Stachowicz, Klak, Pilc, & Chojnacka-Wojcik, 2005). An anxiogenic-like activity after intra-amygdala injection of (*S*)-3,4-DCPG (10 nmol) was observed during extinction of contextual fear in male C57BL/6J mice (Dobi et al., 2013). The different results obtained with (*S*)-3,4-DCPG may be due to differences in measures of anxiety, species used, injections (routes and sites), and doses of (*S*)-3,4-DCPG (Mercier et al., 2013). Future studies are warranted to define the exact role of mGlu8 in distinct anxiety-related brain sites at different stages and models of anxiety.

4. Epilepsy

Epilepsy is a progressive neurological disorder characterized by recurrent epileptic seizures. Ample evidence supports the role of glutamatergic transmission in this disorder (Barker-Haliski & White, 2015). In studies that aimed to define the role of mGlu8, (*R,S*)-4-phosphonophenylglycine (PPG), a group III mGlu agonist preferential for mGlu8 (Gasparini et al., 1999), was found to be anticonvulsive and neuroprotective after an intra-cerebroventricular (icv) injection in the maximal electroshock seizure model in mice (Barton, Peters, & Shannon, 2003; Gasparini et al., 1999). PPG administered icv or into the inferior colliculus also blocked sound-induced seizures in both mice and genetically epilepsy-prone rats (Chapman, Nanan, Yip, & Meldrum, 1999). These pharmacological results suggest a role of group III (likely mGlu8) receptors in the negative regulation of epilepsy. In support of an mGlu8 subtype-specific role, the mGlu8 agonist (*S*)-3,4-DCPG (icv) inhibited sound-induced seizures (Moldrich, Beart, Jane, Chapman, & Meldrum, 2001) and pilocarpine-induced status epilepticus (Jiang, Tang, Chia, Jay, & Tang, 2007) in mice. Of note, concurrent activation of mGlu8 and antagonism of AMPA receptors with the (*RS*)-3,4-DCPG racemate, an agent that activates mGlu8 receptors via the *S*-isomer [(*S*)-3,4-DCPG] and antagonizes AMPA receptors via the *R*-isomer [(*R*)-3,4-DCPG], produced synergistic inhibition of sound-induced seizures in DBA/2 mice (Moldrich et al., 2001). Inhibition of cortical excitatory amino acid release may underlie the antiepileptic effect of (*RS*)-3,4-DCPG, although this may not be the sole mechanism (Lee, Jane, & Croucher, 2003).

mGlu8 receptors serve as autoreceptors on lateral perforant path afferents to the dentate gyrus in the hippocampus (Shigemoto et al., 1997; Zhai et al., 2002). (*S*)-3,4-DCPG inhibited excitatory transmission in this pathway (Mercier et al., 2013). A reduced efficacy of PPG in presynaptically inhibiting field excitatory postsynaptic potentials at the lateral perforant path–dentate gyrus synapses was seen in rats that developed spontaneous recurrent seizures after pilocarpine-induced status epilepticus (Kral et al., 2003). A similar reduction in the efficacy of (S)-3,4-DCPG was seen in epileptic mice (Jiang et al., 2007). Thus, chronic limbic epilepsy downregulates the function of presynaptic mGlu8 autoreceptors on the lateral perforant path terminals, which is likely linked to the subsequent occurrence of spontaneous recurrent seizures. In agreement with this notion, mGlu8 proteins in the outer molecular layer of the dentate gyrus, where the lateral perforant path afferents terminate, were drastically reduced in rats following pilocarpine-induced epileptic seizures (Tang, Lee, Yang, Sim, & Ling, 2001). Unlike perforant path-dentate gyrus synapses, (S)-3,4-DCPG had no effect at Schaffer collateral-CA1 (SC-CA1) synapses in the normal adult rat hippocampus, but depressed SC-CA1 synaptic transmission in epileptic tissue via a presynaptic mechanism (Dammann et al., 2018). Thus, at SC-CA1 synapses, the presynaptic mGlu8 receptor on SC terminals seems to exert its inhibition of synaptic transmission primarily in the epileptic hippocampus. This is viewed as a compensatory response counteracting a hyperexcitable epileptic state and provides evidence supporting the potential of mGlu8 agonists as useful anticonvulsants (Dammann et al., 2018).

5. Parkinson's disease

PD is a chronic neurodegenerative disorder derived from the progressive loss of dopaminergic neurons in the substantia nigra pars compacta. Excessive glutamate activity at multiple levels of the basal ganglia is linked to PD pathology (Ribeiro, Vieira, Pires, Olmo, & Ferguson, 2017; Wang, Wang, Mao, & Qu, 2020; Zhang et al., 2019). As such, reducing the overactive glutamatergic transmission is thought to have a therapeutic value. Early attempts to pharmacologically oppose glutamate hyperactivity focused on ionotropic glutamate (iGlu) receptors (Duty, 2012; Johnson, Conn, & Niswender, 2009; Ossowska et al., 2007). However, iGlu antagonists had a limited success due to their non-motor side effects (Amalric et al., 2013; Starr, Starr, & Kaur, 1997). Recent attention has been drawn to mGlu receptors (Masilamoni & Smith, 2018; Sebastianutto & Cenci, 2018). As with group I mGlu receptor antagonism and groups II mGlu receptor activation, group III receptor activation is thought to improve motor symptoms of PD (Amalric, 2015; Amalric et al., 2013; Duty, 2010). Indeed, nonselective group III agonists show antiparkinsonian effects in animal models of PD (reviewed in Amalric et al., 2013).

Regarding mGlu8, moderate to high levels of mGlu8 immunoreactivity are present in the striatum and substantia nigra pars reticulata (SNr) (Duty, 2010). No changes in mGlu8 mRNA expression were observed in the rat basal ganglia following a 6-hydroxydopamine (6-OHDA) lesion (Messenger, Dawson, & Duty, 2002). Whether mGlu8 receptors undergo adaptive changes in protein expression, posttranslational modification, trafficking, and functions in response to dopamine depletion remains elusive. In pharmacological studies, the mGlu8 agonist (S)-3,4-DCPG was injected into different basal ganglia nuclei to evaluate the role of mGlu8 in acute vs prolonged models of PD. (S)-3,4-DCPG infused into the rat globus pallidus or SNr had no effect on haloperidol-induced catalepsy in an acute model of PD (Lopez et al., 2007). Intra-nigral injection of (S)-3,4-DCPG at high doses (30–300 nmol) reduced akinesia induced by prolonged reserpine treatment in rats, although the effect of the agonist was not readily blocked by a group III antagonist (Broadstock, Austin, Betts, & Duty, 2012). In another study, icv administration of (S)-3,4-DCPG (2.5 or 10 nmol) did not alleviate rat motor deficits induced by acute haloperidol or reserpine treatment, but it robustly reversed catalepsy and akinesia induced by prolonged haloperidol or reserpine treatment, respectively (Johnson et al., 2013). (S)-3,4-DCPG also ameliorated forelimb use asymmetry in 6-OHDA-lesioned rats. Thus, mGlu8 activation seems to exert a antiparkinsonian effect in prolonged rather than acute models of PD. The basal ganglia site(s) underlying the antiparkinsonian effect of icv (S)-3,4-DCPG are still unclear. Activation of mGlu8 in the globus pallidus is less likely to mediate the (S)-3,4-DCPG effect given that direct intrapallidal infusion of the agonist did not reverse 6-OHDA-induced motor deficits (Beurrier et al., 2009). Additionally, (S)-3,4-DCPG and the mGlu8 PAM (AZ12216052) were neuroprotective in an in vitro model of PD with the mitochondrial neurotoxin MPP(+) (Jantas et al., 2014). However, in the 1-methyl-4-phenyl-1,2,3,6-tetrahydropyridine (MPTP) model of PD, MPTP increased the time spent exploring objects in female and male wild-type but not mGlu8$^{(-/-)}$ mice and impaired cued fear memory in male wild-type but not mGlu8$^{(-/-)}$ mice (Torres et al., 2018), suggesting a potential protective role of mGlu8 deficiency in these symptoms of PD.

6. Drug addiction

In addition to dopamine, glutamatergic transmission is involved in the addictive properties of drugs of abuse (Chiamulera, Piva, & Abraham, 2020; Kalivas, 2009). In the striatum, a key structure in the reward circuitry, mGlu8 mRNAs are expressed (Messenger et al., 2002; Robbins et al., 2007).

In the dorsal striatum, presynaptic mGlu8 receptors are distributed on vesicular GABA transporter-expressing GABAergic terminals but not on vesicular glutamate transporter-1-expressing glutamatergic terminals and parvalbumin-expressing GABAergic interneurons (Rossi et al., 2014). Several pharmacological studies were conducted to define the role of mGlu8. While systemic (S)-3,4-DCPG did not alter motor responses to acute amphetamine in rats, icv (S)-3,4-DCPG reversed amphetamine-induced hyperactivity in mice (Robbins et al., 2007). Systemic (S)-3,4-DCPG also attenuated the reinforcing effect of ethanol in an operant self-administration paradigm and cue-induced reinstatement of ethanol seeking in rats (Bäckström & Hyytiä, 2005), indicating that mGlu8 receptors may exert the inhibitory modulation of addictive responses to ethanol. However, a decrease in spontaneous locomotor activity was observed following (S)-3,4-DCPG treatment (Bäckström & Hyytiä, 2005. Robbins et al., 2007), confounding the interpretation of the agonist's influence over behavioral responses to amphetamine and ethanol. Selective mGlu8 agents with less motor-suppressant side effects are thus preferred to define exact roles of mGlu8 in drug seeking behavior. Recently, systemic (S)-3,4-DCPG at a low dose that did not alter overall motor activity, total liquid consumption, and taste sensitivity reduced voluntary ethanol intake, ethanol preference, and the acquisition although not expression of ethanol-induced conditioned place preference in mice (Bahi, 2017). This supports a notion that activation of mGlu8 receptors attenuates the vulnerability to alcoholism.

Chronic amphetamine administration produced a long-lasting increase in striatal mGlu8 mRNA levels (Parelkar & Wang, 2008). Chronic ethanol use reduced and had no effect on mGlu8 mRNA expression in the rat prefrontal cortex and hippocampus, respectively (Pickering et al., 2015; Simonyi, Christian, Sun, & Sun, 2004). Thus, striatal and cortical mGlu8 mRNA expression is sensitive to chronic drug administration. It is currently unclear whether these changes at the mRNA level can translate to responses at protein and functional levels. It is possible that limbic mGlu8 receptors may undergo adaptive changes during chronic drug exposure, which may either contribute to or compensate for pathological changes in glutamatergic transmission related to drug seeking behavior.

7. Chronic pain

Supraspinal mGlu8 receptors modulate pain transmission and are potential targets for analgesics (reviewed in Chiechio, 2016; Palazzo et al., 2017; Boccella et al., 2020). Briefly, the role of mGlu8 receptors has been examined with (S)-3,4-DCPG. Systemic (S)-3,4-DCPG showed antinociceptive properties in both inflammatory and neuropathic pain models. Local infusion of the agonist into supraspinal centers of the pain neuraxis such as the amygdala, periaqueductal gray (Hosseini et al., 2020), or dorsal striatum produced a similar analgesic effect in various pain models. However, (S)-3,4-DCPG when infused into the nucleus tractus solitarius facilitated cardiac nociception. Thus, mGlu8 receptors appear to exert a region-specific modulation of pain responses.

8. Alzheimer's disease

Alzheimer's Disease (AD) is the most common cause of dementia in the elderly, yet the exact pathogenesis of the disease remains unclear. However, all forms of AD have demonstrated an overproduction or decreased clearance of β-amyloid (Aβ) protein peptides that form diffuse neuritic plaques. Another hallmark neuropathologic change associated with AD is intracellular hyperphosphorylated tau protein aggregates that form neurofibrillary tangles. Sabelhaus, Schröder, Breder, Henrich-Noack, and Reymann (2000) have sought to identify roles of group III mGlu receptors in neurodegeneration after a hypoxic insult to the rat hippocampus. Using an mGlu8 preferential agonist PPG, they found a significant improvement in recovery of hippocampal synaptic transmission, suggesting that mGlu8 preferential agonists might be beneficial for treating neurodegeneration involving excitotoxicity such as AD.

Furthermore, glutamate uptake and recycling are crucial for proper signaling processes, which might be impaired in AD (Wang & Reddy, 2017). The synaptic loss and neurodegeneration found in AD are a result of glutamatergic excitotoxicity (Lewerenz & Maher, 2015; Srivastava, Das, Yao, & Yan, 2020). When the glutamate receptors are activated beyond their threshold, dendrites, cell bodies, and other postsynaptic structures are lost. Likewise, this neuronal death leads to further glutamate release (Sabelhaus et al., 2000). mGlu8 agonists may provide therapeutic advantages in modulating this cycle of degenerative changes. Gerlai et al. (2002) found that mGlu8 knockout mice showed behavioral changes, including weakened acclimatization to novel environments, interrupted context-based freezing response to fear, and hyperactivity and anxiety. These changes coincide with the dysfunctional hippocampus, a commonly observed pathologic feature of AD. Thus, mGlu8 agonists could have the therapeutic potential for AD treatment (Srivastava et al., 2020). Future studies are warranted to explore the direct linkage of the individual mGlu8 subtype to AD.

9. Conclusions

As a $G_{i/o}$-coupled presynaptic receptor, mGlu8 inhibits glutamate and other transmitter release. Expression of mGlu8 in key limbic regions implicates the receptor in the regulation of motivation, mood, motor, and cognitive functions. Dysfunctional mGlu8 receptors are linked to the pathogenesis of various neuropsychiatric and neurological disorders. Activation of mGlu8 receptors modulates the addictive properties of drugs of abuse. The mGlu8 agonist or PAM when administered systemically or locally produces anxiolytic, antiepileptic, and antiparkinsonian effects, although the results sometimes depend on brain injection sites, animal models used, symptoms and stages of disease, and species. Expression and activity of mGlu8 receptors in the limbic circuitry are vulnerable to brain disorders and may undergo adaptive changes to either contribute to or compensate for the development of these diseases. Thus, the mGlu8 receptor represents an attractive therapeutic target for treating multiple neuropsychiatric and neurological disorders (Duty, 2010; Nicoletti, Bruno, Ngomba, Gradini, & Battaglia, 2015; Niswender & Conn, 2010; Raber & Duvoisin, 2015).

Author contributions

LMM and JQW planned and wrote the manuscript. NM and KS critically read, edited, and/or participated in the writing of the manuscript. All authors contributed to the article and approved the submitted version.

Funding

This study was supported by a grant from the NIH (R01-MH061469 to JQW).

Conflict of interest

The authors declare that the research was conducted in the absence of any commercial or financial relationships that could be construed as a potential conflict of interest.

References

Amalric, M. (2015). Targeting metabotropic glutamate receptors (mGluRs) in Parkinson's disease. *Current Opinion in Pharmacology, 20*, 29–34. https://doi.org/10.1016/j.coph.2014.11.001.

Amalric, M., Lopez, S., Goudet, C., Fisone, G., Battaglia, G., Nicocletti, F., et al. (2013). Group III and subtype 4 metabotropic glutamate receptor agonists: Discovery and pathophysiological applications in Parkinson's disease. *Neuropharmacology*, *66*, 53–64. https://doi.org/10.1016/j.neuropharm.2012.05.026.

Bäckström, P., & Hyytiä, P. (2005). Suppression of alcohol self-administration and cue-induced reinstatement of alcohol seeking by the mGlu2/3 receptor agonist LY379268 and the mGlu8 receptor agonist (S)-3,4-DCPG. *European Journal of Pharmacology*, *528*, 110–118. https://doi.org/10.1016/j.ejphar.2005.10.051.

Bahi, A. (2017). Decreased anxiety, voluntary ethanol intake and ethanol-induced CPP acquisition following activation of the metabotropic glutamate receptor 8 "mGluR8". *Pharmacology Biochemistry and Behavior*, *155*, 32–42. https://doi.org/10.1016/j.pbb.2017.03.004.

Barker-Haliski, M., & White, H. S. (2015). Glutamatergic mechanisms associated with seizures and epilepsy. *Cold Spring Harbor Perspectives in Medicine*, *5*, a022863. https://doi.org/10.1101/cshperspect.a022863.

Barton, M. E., Peters, S. C., & Shannon, H. E. (2003). Comparison of the effect of glutamate receptor modulators in the 6 Hz and maximal electroshock seizure models. *Epilepsy Research*, *56*, 17–26. https://doi.org/10.1016/j.eplepsyres.2003.08.001.

Beurrier, C., Lopez, S., Revy, D., Selvam, C., Goudet, C., Lherondel, M., et al. (2009). Electrophysiological and behavioral evidence that modulation of metabotropic glutamate receptor 4 with a new agonist reverses experimental parkinsonism. *FASEB Journal*, *23*, 3619–3628. https://doi.org/10.1096/fj.09-131789.

Boccella, S., Marabese, I., Guida, F., Luongo, L., Maione, S., & Palazzo, E. (2020). The modulation of pain by metabotropic glutamate receptors 7 and 8 in the dorsal striatum. *Current Neuropharmacology*, *18*, 34–50. https://doi.org/10.2174/1570159X17666190618121859.

Broadstock, M., Austin, P. J., Betts, M. J., & Duty, S. (2012). Antiparkinsonian potential of targeting group III metabotropic glutamate receptor subtypes in the rodent substantia nigra pars reticulate. *British Journal of Pharmacology*, *165*, 103401045. https://doi.org/10.1111/j.1476-5381.2011.01515.x.

Chapman, A. G., Nanan, K., Yip, P., & Meldrum, B. S. (1999). Anticonvulsant activity of a metabotropic glutamate receptor 8 preferential agonist, (R,S)-4-phosphonophenylglycine. *European Journal of Pharmacology*, *383*, 23–27. https://doi.org/10.1016/s0014-2999(99)00615-9.

Chiamulera, C., Piva, A., & Abraham, W. C. (2020). Glutamate receptors and metaplasticity in addiction. *Current Opinion in Pharmacology*, *56*, 39–45. https://doi.org/10.1016/j.coph.2020.09.005.

Chiechio, S. (2016). Modulation of chronic pain by metabotropic glutamate receptors. *Advances in Pharmacology*, *75*, 63–89. https://doi.org/10.1016/bs.apha.2015.11.001.

Corti, C., Restituito, S., Rimland, J. M., Brabet, I., Corsi, M., Pin, J. P., et al. (1998). Cloning and characterization of alternative mRNA forms for the rat metabotropic glutamate receptors mGluR7 and mGluR8. *European Journal of Neuroscience*, *10*, 3629–3641. https://doi.org/10.1046/j.1460-9568.1998.00371.x.

Crupi, R., Impellizzeri, D., & Cuzzocrea, S. (2019). Role of metabotropic glutamate receptors in neurological disorders. *Frontiers in Molecular Neuroscience*, *12*, 20. https://doi.org/10.3389/fnmol.2019.00020.

Dammann, F., Kirschstein, T., Guli, X., Muller, S., Porath, K., Rohde, M., et al. (2018). Bidirectional shift of group III metabotropic glutamate receptor-mediated synaptic depression in the epileptic hippocampus. *Epilepsy Research*, *139*, 157–163. https://doi.org/10.1016/j.eplepsyres.2017.12.002.

Davis, M. J., Duvoisin, R. M., & Raber, J. (2013). Related functions of mGlu4 and mGlu8. *Pharmacology Biochemistry and Behavior*, *111*, 11–16. https://doi.org/10.1016/j.pbb.2013.07.022.

Dobi, A., Sartori, S. B., Busti, D., van der Putten, H., Singewald, N., Shigemoto, R., et al. (2013). Neural substrates for the distinct effects of presynaptic group III metabotropic glutamate receptors on extinction of contextual fear conditioning in mice. *Neuropharmacology, 66*, 274–289. https://doi.org/10.1016/j.neuropharm.2012.05.025.

Doumazane, E., Scholler, P., Zwier, J. M., Trinquet, E., Rondard, P., & Pin, J. P. (2011). A new approach to analyze cell surface protein complexes reveals specific heterodimeric metabotropic glutamate receptors. *FASEB Journal, 25*, 66–77. https://doi.org/10.1096/fj.10-163147.

Duty, S. (2010). Therapeutic potential of targeting group III metabotropic glutamate receptors in the treatment of Parkinson's disease. *British Journal of Pharmacology, 161*, 271–287. https://doi.org/10.1111/j.1476-5381.2010.00882.x.

Duty, S. (2012). Targeting glutamate receptors to tackle the pathogenesis, clinical symptoms and levodopa-induced dyskinesia associated with Parkinson's disease. *CNS Drugs, 26*, 1017–1032. https://doi.org/10.1007/s40263-012-0016-z.

Duvoisin, R. M., Pfankuch, T., Wilson, J. M., Grabell, J., Chhajlani, V., Brown, D. G., et al. (2010). Acute pharmacological modulation of mGluR8 reduces measures of anxiety. *Behavioral Brain Research, 212*, 168–173. https://doi.org/10.1016/j.bbr.2010.04.006.

Duvoisin, R. M., Villasana, I., Pfankuch, T., & Raber, J. (2010). Sex-dependent cognitive phenotype of mice lacking mGluR8. *Behavioral Brain Research, 209*, 21–26. https://doi.org/10.1016/j.bbr.2010.01.006.

Duvoisin, R. M., Zhang, C., Pfankuch, T. F., O'Connor, H., Gayet-Primo, J., Quraishi, S., et al. (2005). Increased measures of anxiety and weight gain in mice lacking the group III metabotropic glutamate receptor mGluR8. *European Journal of Neuroscience, 22*, 425–436. https://doi.org/10.1111/j.1460-9568.2005.04210.x.

Duvoisin, R. M., Zhang, C., & Ramonell, K. (1995). A novel metabotropic glutamate receptor expressed in the retina and olfactory bulb. *Journal of Neuroscience, 15*, 3075–3083. https://doi.org/10.1523/JNEUROSCI.15-04-03075.1995.

Fendt, M., Burki, H., Imobersteg, S., van der Putten, H., McAllister, K., Leslie, J. C., et al. (2010). The effect of mGlu8 deficiency in animal models of psychiatric diseases. *Genes, Brain and Behavior, 9*, 33–44. https://doi.org/10.1111/j.1601-183X.2009.00532.x.

Fendt, M., Imobersteg, S., Peterlik, D., Chaperon, F., Mattes, C., Wittmann, C., et al. (2013). Differential roles of mGlu(7) and mGlu(8) in amygdala-dependent behavior and physiology. *Neuropharmacology, 72*, 215–223. https://doi.org/10.1016/j.neuropharm.2013.04.052.

Ferraguti, F., & Shigemoto, R. (2006). Metabotropic glutamate receptors. *Cell and Tissue Research, 326*, 483–504. https://doi.org/10.1007/s00441-006-0266-5.

Gasparini, F., Bruno, V., Battaglia, G., Lukic, S., Leonhardt, T., Inderbitzin, W., et al. (1999). (R,S)-4-phosphonophenylglycine, a potent and selective group III metabotropic glutamate receptor agonist, is anticonvulsisve and neuroprotective *in vivo*. *Journal of Pharmacology and Experimental Therapeutics, 289*, 1678–1687. 10336568.

Gerlai, R., Adams, B., Fitch, T., Chaney, S., & Baez, M. (2002). Performance deficits of mGluR8 knockout mice in learning tasks: The effects of null mutation and the background genotype. *Neuropharmacology, 43*, 235–249. https://doi.org/10.1016/s0028-3908(02)00078-3.

Goddyn, H., Callaerts-Vegh, Z., & D'Hooge, R. (2015). Functional dissociation of group III metabotropic glutamate receptors revealed by direct comparison between the behavioral profiles of knockout mouse lines. *International Journal of Neuropsychopharmacology, 18*, pyv053. https://doi.org/10.1093/ijnp/pyv053.

Gosnell, H. B., Silberman, Y., Grueter, B. A., Duvoisin, R. M., Raber, J., & Winder, D. G. (2011). mGluR8 modulates excitatory transmission in the bed nucleus of the stria terminalis in a stress-dependent manner. *Neuropsychopharmacoclogy, 36*, 1599–1607. https://doi.org/10.1038/npp.2011.40.

Grueter, B. A., & Winder, D. G. (2005). Group II and III metabotropic glutamate receptors suppress excitatory synaptic transmission in the dorsolateral bed nucleus of the stria terminalis. *Neuropsychopharmacology, 30*, 1302–1311. https://doi.org/10.1038/sj.npp.1300672.

Guo, J., & Ikeda, S. R. (2005). Coupling of metabotropic glutamate receptor 8 to N-type Ca^{2+} channels in rat sympathetic neurons. *Molecular Pharmacology, 67*, 1840–1851. https://doi.org/10.1124/mol.105.010975.

Hinoi, E., Fujimori, S., Nakamura, Y., & Yoneda, Y. (2001). Group III metabotropic glutamate receptors in rat cultured calvarial osteoblasts. *Biochemical and Biophysical Research Communications, 281*, 341–346. https://doi.org/10.1006/bbrc.2001.4355.

Hosseini, M., Parviz, M., Shabanzadeh, A. P., Zamani, E., Mohseni-Moghaddam, P., Gholami, L., et al. (2020). The inhibiting role of periaqueductal gray metabotropic glutamate receptor subtype 8 in a rat model of central neuropathic pain. *Neurological Research, 42*, 515–521. https://doi.org/10.1080/01616412.2020.1747730.

Jantas, D., Greda, A., Golda, S., Korostynski, M., Grygier, B., Roman, A., et al. (2014). Neuroprotective effects of metabotropic glutamate receptor group II and III activators against MPP(+)-induced cell death in human neuroblastoma SH-SY5Y cells: The impact of cell differentiation state. *Neuropharmacology, 83*, 36–53. https://doi.org/10.1016/j.neuropharm.2014.03.019.

Jiang, F. L., Tang, Y. C., Chia, S. C., Jay, T. M., & Tang, F. R. (2007). Anticonvulsive effect of a selective mGluR8 agonist (*S*)-3,4-dicarboxyphenylglycine (*S*-3,4-DCPG) in the mouse pilocarpine model of status epilepticus. *Epilepsia, 48*, 783–792. https://doi.org/10.1111/j.1528-1167.2007.01000.x.

Johnson, K. A., Conn, P. J., & Niswender, C. M. (2009). Glutamate receptors as therapeutic targets for Parkinson's disease. *CNS & Neurological Disorders—Drug Targets, 8*, 475–491. https://doi.org/10.2174/187152709789824606.

Johnson, K. A., Jones, C. K., Tantawy, M. N., Bubser, M., Marvanova, M., Ansari, M. S., et al. (2013). The metabotropic glutamate receptor 8 agonist (*S*)-3,4-DCPG reverses motor deficits in prolonged but not acute models of Parkinson's disease. *Neuropharmacology, 66*, 187–195. https://doi.org/10.1016/j.neuropharm.2012.03.029.

Julio-Pieper, M., Flor, P. J., Dinan, T. G., & Cryan, J. F. (2011). Exciting times beyond the brain: Metabotropic glutamate receptors in peripheral and non-neural tissues. *Pharmacological Reviews, 63*, 35–58. https://doi.org/10.1124/pr.110.004036.

Kalivas, P. W. (2009). The glutamate homeostasis hypothesis of addiction. *Nature Reviews Neuroscience, 10*(8), 561–572. https://doi.org/10.1038/nrn2515.

Kogo, N., Dalezios, Y., Capogna, M., Ferraguti, F., Shigemoto, R., & Somogyi, P. (2004). Depression of GABAergic input to identified hippocampal neurons by group III metabotropic glutamate receptors in the rat. *European Journal of Neuroscience, 19*, 2727–2740. https://doi.org/10.1111/j.0953-816X.2004.03394.x.

Koulen, P., Kuhn, R., Wassle, H., & Brandstatter, J. H. (1999). Modulation of the intracellular calcium concentration in photoreceptor terminals by a presynaptic metabotropic glutamate receptor. *Proceedings of the National Academy of Sciences of the United States of America, 96*, 9909–9914. https://doi.org/10.1073/pnas.96.17.9909.

Kral, T., Erdmann, E., Sochivko, D., Clusmann, H., Schramm, J., & Dietrich, D. (2003). Down-regulation of mGluR8 in pilocarpine epileptic rats. *Synapse, 47*, 278–284. https://doi.org/10.1002/syn.10178.

Lavreysen, H., & Dautzenberg, F. M. (2008). Therapeutic potential of group III metabotropic glutamate receptors. *Current Medicinal Chemistry, 15*, 671–684. https://doi.org/10.2174/092986708783885246.

Lee, J. J., Jane, D. E., & Croucher, M. J. (2003). Anticonvulsant dicarboxyphenylglycines differentially modulate excitatory amino acid release in the rat cerebral cortex. *Brain Research, 977*, 119–123. https://doi.org/10.1016/s0006-8993(03)02657-x.

Lewerenz, J., & Maher, P. (2015). Chronic glutamate toxicity in neurodegenerative diseases—what is the evidence? *Frontiers in Neuroscience, 9*, 469. https://doi.org/10.3389/fnins.2015.00469.

Linden, A. M., Baez, M., Bergeron, M., & Schoepp, D. D. (2003). Increased c-Fos expression in the centromedial nucleus of the thalamus in metabotropic glutamate 8 receptor knockout mice following the elevated plus maze test. *Neuroscience, 121*, 167–178. https://doi.org/10.1016/s0306-4522(03)00393-2.

Linden, A. M., Bergeron, M., Baez, M., & Schoepp, D. D. (2003). Systemic administration of the potent mGlu8 receptor agonist (S)-3,4-DCPG induces c-Fos in stress-related brain regions in wild-type, but not mGlu8 receptor knockout mice. *Neuropharmacology, 45*, 473–483. https://doi.org/10.1016/s0028-3908(03)00200-4.

Linden, A. M., Johnson, B. G., Peters, S. C., Shannon, H. E., Tian, M., Wang, Y., et al. (2002). Increased anxiety-related behavior in mice deficient for metabotropic glutamate 8 (mGlu8) receptor. *Neuropharmacology, 43*, 251–259. https://doi.org/10.1016/s0028-3908(02)00079-5.

Lodge, D., Tidball, P., Mercier, M. S., Lucas, S. J., Hanna, L., Ceolin, L., et al. (2013). Antagonists reversibly reverse chemical LTD induced by group I, group II and group III metabotropic glutamate receptors. *Neuropharmacology, 74*, 135–146. https://doi.org/10.1016/j.neuropharm.2013.03.011.

Lopez, S., Turle-Lorenzo, N., Acher, F., De Leonibus, E., Mele, A., & Amalric, M. (2007). Targeting group III metabotropic glutamate receptors produces complex behavioral effects in rodent models of Parkinson's disease. *Journal of Neuroscience, 27*, 6701–6711. https://doi.org/10.1523/JNEUROSCI.0299-07.2007.

Malherbe, P., Kratzersen, C., Lundstrom, K., Richards, J. G., Faull, R. L., & Mutel, V. (1999). Cloning and functional expression of alternative spliced variants of the human metabotropic glutamate receptor 8. *Molecular Brain Research, 67*, 201–210. https://doi.org/10.1016/s0169-328x(99)00050-9.

Mao, L. M., Guo, M. L., Jin, D. Z., Fibuch, E. E., Choe, E. S., & Wang, J. Q. (2011). Post-translational modification biology of glutamate receptors and drug addiction. *Frontiers in Neuroanatomy, 5*, 19. https://doi.org/10.3389/fnana.2011.00019.

Mao, L., Guo, M., Jin, D., Xue, B., & Wang, J. Q. (2013). Group III metabotropic glutamate receptors and drug addiction. *Frontiers in Medicine, 7*, 445–451. https://doi.org/10.1007/s11684-013-0291-1.

Marabese, I., de Novellis, V., Palazzo, E., Mariani, L., Siniscalco, D., Rodella, L., et al. (2005). Differential roles of mGlu8 receptors in the regulation of glutamate and gamma-aminobutyric acid release at periaqueductal grey level. *Neuropharmacology, 49*(Suppl. 1), 157–166. https://doi.org/10.1016/j.neuropharm.2005.02.006.

Masilamoni, G. J., & Smith, Y. (2018). Metabotropic glutamate receptors: Targets for neuroprotective therapies in Parkinson's disease. *Current Opinion in Pharmacology, 38*, 72–80. https://doi.org/10.1016/j.coph.2018.03.004.

Mercier, M. S., Lodge, D., Fang, G., Nicolas, C. S., Collett, V. J., Jane, D. E., et al. (2013). Characterization of an mGlu8 receptor-selective agonist and antagonist in the lateral and medial perforant path inputs to the dentate gyrus. *Neuropharmacology, 67*, 294–303. https://doi.org/10.1016/j.neuropharm.2012.11.020.

Messenger, M. J., Dawson, L. G., & Duty, S. (2002). Changes in metabotropic glutamate receptor 1-8 gene expression in the rodent basal ganglia motor loop following lesion of the nigrostriatal tract. *Neuropharmacology, 43*, 261–271. https://doi.org/10.1016/s0028-3908(02)00090-4.

Moldrich, R. X., Beart, P. M., Jane, D. E., Chapman, A. G., & Meldrum, B. S. (2001). Anticonvulsant activity of 3,4-dicarboxyphenylglycines in DBA/2 mice. *Neuropharmacology, 40*, 732–735. https://doi.org/10.1016/s0028-3908(01)00002-8.

Nicoletti, F., Bockaert, J., Collingridge, G. L., Conn, P. J., Ferraguti, F., Schoepp, D. D., et al. (2011). Metabotropic glutamate receptors: From the workbench to the bedside. *Neuropharmacology, 60*, 1017–1041. https://doi.org/10.1016/j.neuropharm.2010.10.022.

Nicoletti, F., Bruno, V., Ngomba, R. T., Gradini, R., & Battaglia, G. (2015). Metabotropic glutamate receptors as drug targets: What's new? *Current Opinion in Pharmacology, 20*, 89–94. https://doi.org/10.1016/j.coph.2014.12.002.

Niswender, C. M., & Conn, P. J. (2010). Metabotropic glutamate receptors: Physiology, pharmacology, and disease. *Annual Review of Pharmacology and Toxicology, 50*, 295–322. https://doi.org/10.1146/annurev.pharmtox.011008.145533.

Ossowska, K., Konieczny, J., Wardas, J., Pietraszek, M., Kuter, K., Wolfarth, S., et al. (2007). An influence of ligands of metabotropic glutamate receptor subtypes on parkinsonian-like symptoms and the striatopallidal pathway in rats. *Amino Acids, 32*, 179–188. https://doi.org/10.1007/s00726-006-0317-y.

Palazzo, E., Fu, Y., Ji, G., Maione, S., & Neugebauer, V. (2008). Group III mGluR7 and mGluR8 in the amygdala differentially modulate nocifensive and affective pain behaviors. *Neuropharmacology, 55*, 537–545. https://doi.org/10.1016/j.neuropharm.2008.05.007.

Palazzo, E., Marabese, I., Luongo, L., Guida, F., Novellis, V., & Maione, S. (2017). Nociception modulation by supraspinal group III metabotropic glutamate receptors. *Journal of Neurochemistry, 141*, 507–519. https://doi.org/10.1111/jnc.13725.

Parelkar, N. K., & Wang, J. Q. (2008). Upregulation of metabotropic glutamate receptor 8 mRNA expression in the rat forebrain after repeated amphetamine administration. *Neuroscience Letters, 433*(3), 250–254. https://doi.org/10.1016/j.neulet.2008.01.015.

Pickering, C., Alsio, J., Morud, J., Ericson, M., Robbins, T. W., & Soderpalm, B. (2015). Ethanol impairment of spontaneous alternation behaviour and associated changes in the middle prefrontal glutamatergic gene expression precede putative markers of dependence. *Pharmacology Biochemistry and Behavior, 132*, 63–70. https://doi.org/10.1016/j.pbb.2015.02.021.

Raber, J., & Duvoisin, R. M. (2015). Novel metabotropic glutamate receptor 4 and glutamate receptor 8 therapeutics for the treatment of anxiety. *Expert Opinion on Investigational Drugs, 24*, 519–528. https://doi.org/10.1517/13543784.2014.986264.

Ribeiro, F. M., Vieira, L. B., Pires, R. G. W., Olmo, R. P., & Ferguson, S. S. G. (2017). Metabotropic glutamate receptors and neurodegenerative diseases. *Pharmacological Research, 115*, 179–191. https://doi.org/10.1016/j.phrs.2016.11.013.

Robbins, M. J., Starr, K. R., Honey, A., Soffin, E. M., Rourke, C., Jones, G. A., et al. (2007). Evaluation of the mGlu8 receptor as a putative therapeutic target in schizophrenia. *Brain Research, 1152*, 215–227. https://doi.org/10.1016/j.brainres.2007.03.028.

Rossi, F., Marabese, I., De Chiaro, M., Boccella, S., Luongo, L., Guida, F., et al. (2014). Dorsal striatum metabotropic glutamate receptor 8 affects nocifensive responses and rostral ventromedial medulla cell activity in neuropathic pain conditions. *Journal of Neurophysiology, 111*, 2196–2209. https://doi.org/10.1152/jn.00212.2013.

Sabelhaus, C. F., Schröder, U. H., Breder, J., Henrich-Noack, P., & Reymann, K. G. (2000). Neuroprotection against hypoxic/hypoglycaemic injury after the insult by the group III metabotropic glutamate receptor agonist (R, S)-4-phosphonophenylglycine. *British Journal of Pharmacology, 131*, 655–658. https://doi.org/10.1038/sj.bjp.0703646.

Saugstad, J. A., Kinzie, J. M., Shinohara, M. M., Segerson, T. P., & Westbrook, G. L. (1997). Cloning and expression of rat metabotropic glutamate receptor 8 reveals a distinct pharmacological profile. *Molecular Pharmacology, 51*, 119–125. https://doi.org/10.1124/mol.51.1.119.

Scherer, S. W., Duvoisin, R. M., Kuhn, R., Heng, H. H., Belloni, E., & Tsui, L. C. (1996). Localization of two metabotropic glutamate receptor genes, GRM3 and GRM8, to human chromosome 7q. *Genomics, 31*, 230–233. https://doi.org/10.1006/geno.1996.0036.

Scherer, S. W., Soder, S., Duvoisin, R. M., Huizenga, J. J., & Tsui, L. C. (1997). The human metabotropic glutamate receptor 8 (GRM8) gene: A disproportionately large gene located at 7q31.3-q32.1. *Genomics, 44*, 232–236. https://doi.org/10.1006/geno.1997.4842.

Schmid, S., & Fendt, M. (2006). Effects of the mGluR8 agonist (S)-3,4-DCPG in the lateral amygdala on acquisition/expression of fear-potentiated startle, synaptic transmission, and plasticity. *Neuropharmacology*, *50*, 154–164. https://doi.org/10.1016/j.neuropharm. 2005.08.002.

Sebastianutto, I., & Cenci, M. A. (2018). mGlu receptors in the treatment of Parkinson's disease and L-DOPA-induced dyskinesia. *Current Opinion in Pharmacology*, *38*, 81–89. https://doi.org/10.1016/j.coph.2018.03.003.

Shigemoto, R., Kinoshita, A., Wada, E., Nomura, S., Ohishi, H., Takada, M., et al. (1997). Differential presynaptic localization of metabotropic glutamate receptor subtypes in the rat hippocampus. *Journal of Neuroscience*, *17*, 7503–7522. https://doi.org/10.1523/JNEUROSCI.17-19-07503.1997.

Simonyi, A., Christian, M. R., Sun, A. Y., & Sun, G. Y. (2004). Chronic ethanol-induced subtype- and subregion-specific decrease in the mRNA expression of metabotropic glutamate receptors in rat hippocampus. *Alcoholism: Clinical and Experimental Research*, *28*, 1419–1423. https://doi.org/10.1097/01.alc.0000139825.35438.a4.

Srivastava, A., Das, B., Yao, A. Y., & Yan, R. (2020). Metabotropic glutamate receptors in Alzheimer's disease synaptic dysfunction: Therapeutic opportunities and hope for the future. *Journal of Alzheimer's Disease*, *78*, 1345–1361. https://doi.org/10.3233/JAD-201146.

Stachowicz, K., Klak, K., Pilc, A., & Chojnacka-Wojcik, E. C. (2005). Lack of the antianxiety-like effect of (S)-3,4-DCPG, an mGlu8 receptor agonist, after central administration in rats. *Pharmacological Reports*, *57*, 856–860. 16382208.

Starr, M. S., Starr, B. S., & Kaur, S. (1997). Stimulation of basal and L-DOPA-induced motor activity by glutamate antagonists in animal models of Parkinson's disease. *Neuroscience and Biobehavioral Reviews*, *21*(4), 437–446. https://doi.org/10.1016/s0149-7634(96)00039-5.

Tang, F. R., Lee, W. L., Yang, J., Sim, M. K., & Ling, E. A. (2001). Metabotropic glutamate receptor 8 in the rat hippocampus after pilocarpine induced status epilepticus. *Neuroscience Letters*, *300*, 137–140. https://doi.org/10.1016/s0304-3940(01)01579-8.

Taylor, D. L., Diemel, L. T., & Pocock, J. M. (2003). Activation of microglial group III metabotropic glutamate receptors protects neurons against microglial neurotoxicity. *Journal of Neuroscience*, *23*, 2150–2160. https://doi.org/10.1523/JNEUROSCI.23-06-02150.2003.

Thomas, N. K., Wright, R. A., Howson, P. A., Kingston, A. E., Schoepp, D. D., & Jane, D. E. (2001). (S)-3,4-DCPG, a potent and selective mGlu8a receptor agonist, activates metabotropic glutamate receptors on primary afferent terminals in the neonatal rat spinal cord. *Neuropharmacology*, *40*, 311–318. https://doi.org/10.1016/s0028-3908(00)00169-6.

Tong, Q., & Kirchgessner, A. L. (2003). Localization and function of metabotropic glultamate receptor 8 in the enteric nervous system. *American Journal of Physiology. Gastrointestinal and Liver Physiology*, *285*, G992–1003. https://doi.org/10.1152/ajpgi.00118.2003.

Tong, Q., Ouedraogo, R., & Kirchgessner, A. L. (2002). Localization and function of group III metabotropic glutamate receptors in rat pancreatic islets. *American Journal of Physiology. Endocrinology and Metabolism*, *282*, E1324–E1333. https://doi.org/10.1152/ajpendo.00460.2001.

Torres, E. R. S., Akinyeke, T., Stagaman, K., Duvoisin, R. M., Meshul, C. K., Sharpton, T. J., et al. (2018). Effects of sub-chronic MPTP exposure on behavioral and cognitive performance and the microbiome of wild-type and mGlu8 knockout female and male mice. *Frontiers in Behavioral Neuroscience*, *12*, 140. https://doi.org/10.3389/fnbeh.2018.00140.

Wang, R., & Reddy, P. H. (2017). Role of glutamate and NMDA receptors in Alzheimer's disease. *Journal of Alzheimer's Disease*, *57*, 1041–1048. https://doi.org/10.3233/JAD-160763.

Wang, J., Wang, F., Mao, D., & Qu, S. (2020). Molecular mechanisms of glutamate toxicity in Parkinson's disease. *Frontiers in Neuroscience*, *14*, 585584. https://doi.org/10.3389/fnins.2020.585584.

Wu, S., Wright, R. A., Rockey, P. K., Burgett, S. G., Arnold, J. S., Jr, R., et al. (1998). Group III human metabotropic glutamate receptors 4, 7 and 8: Molecular cloning, functional expression, and comparison of pharmacological properties in RGT cells. *Molecular Brain Research*, *53*, 88–97. https://doi.org/10.1016/s0169-328x(97)00277-5.

Zhai, J., Tian, M. T., Wang, Y., Yu, J. L., Koster, A., Baez, M., et al. (2002). Modulation of lateral perforant path excitatory responses by metabotropic glutamate receptor 8 (mGlu8) receptors. *Neuropharmacology*, *43*, 223–230. https://doi.org/10.1016/s0028-3908(02)00087-4.

Zhang, Z., Zhang, S., Fu, P., Zhang, Z., Lin, K., Ko, J. K. S., et al. (2019). Roles of glutamate receptors in Parkinson's disease. *International Journal of Molecular Sciences*, *20*, 4391. https://doi.org/10.3390/ijms20184391.

> **CHAPTER NINE**

Metabotropic glutamate receptors and cognition: From underlying plasticity and neuroprotection to cognitive disorders and therapeutic targets

Brandon K. Hoglund[a], Vincent Carfagno[b], M. Foster Olive[c], and Jonna M. Leyrer-Jackson[a,*]

[a]Department of Medical Education, School of Medicine, Creighton University, Phoenix, AZ, United States
[b]School of Medicine, Midwestern University, Glendale, AZ, United States
[c]Department of Psychology, Arizona State University, Tempe, AZ, United States
*Corresponding author: e-mail address: jonnajackson1@creighton.edu

Contents

1.	Introduction	368
2.	Metabotropic glutamate receptor physiology	368
	2.1 Glutamatergic neurotransmission	368
	2.2 mGlu receptor families and signaling mechanisms	369
	2.3 Cellular and neuroanatomical distribution of mGlu receptors	370
3.	Cognition: An executive function	371
4.	Mediating cognition: The prefrontal cortex	372
5.	Synaptic plasticity and mGlu receptors	373
6.	mGlu receptors and cognition: Diseases and associations	375
	6.1 Fragile X syndrome	378
	6.2 Schizophrenia	380
	6.3 Parkinson's disease	384
	6.4 Alzheimer's disease	388
	6.5 Post-traumatic stress disorder (PTSD)	391
7.	Targeting mGlu receptors as a therapeutic for cognitive therapy	393
8.	Conclusion	398
Funding		398
References		398

Abstract

Metabotropic glutamate (mGlu) receptors are G protein-coupled receptors that play pivotal roles in mediating the activity of neurons and other cell types within the brain, communication between cell types, synaptic plasticity, and gene expression. As such, these receptors play an important role in a number of cognitive processes. In this chapter, we discuss the role of mGlu receptors in various forms of cognition and their underlying physiology, with an emphasis on cognitive dysfunction. Specifically, we highlight evidence that links mGlu physiology to cognitive dysfunction across brain disorders including Parkinson's disease, Alzheimer's disease, Fragile X syndrome, post-traumatic stress disorder, and schizophrenia. We also provide recent evidence demonstrating that mGlu receptors may elicit neuroprotective effects in particular disease states. Lastly, we discuss how mGlu receptors can be targeted utilizing positive and negative allosteric modulators as well as subtype specific agonists and antagonist to restore cognitive function across these disorders.

1. Introduction

Metabotropic glutamate receptors (mGlu) are G protein-coupled receptors activated by glutamate, the primary excitatory neurotransmitter within the central nervous system. These mGlu receptors play highly important roles in mediating excitatory neurotransmission, intercellular communication, synaptic plasticity, metaplasticity, fine-tuning of neuronal activity as well as gene expression. It is therefore unsurprising that mGlu receptors maintain cellular activity underlying a number of cognitive processes. In this chapter, we discuss the role of mGlu receptors in mediating the neural basis of cognition with an emphasis on mGlu-linked cognitive dysfunction. Given the unique phenotypes of cognitive dysfunction across various brain disorders, we have highlighted important evidence linking mGlu physiology to cognitive dysfunction in central nervous system pathologies including Parkinson's disease (PD), Alzheimer's Disease (AD), Fragile X syndrome (FXS), post-traumatic stress disorder (PTSD), and schizophrenia. Lastly, we review how mGlu receptors can be targeted pharmacologically utilizing positive and negative allosteric modulators as well as subtype specific agonists and antagonists to facilitate the restoration of cognitive function across these disorders. The information provided here lend insight into the important role mGlu receptors play in cognition and allows readers to understand and advance the development of mGlu targeted therapies for cognitive dysfunction.

2. Metabotropic glutamate receptor physiology
2.1 Glutamatergic neurotransmission

Chemical transmission in the brain is largely mediated by the excitatory amino acid glutamate, which accounts for as much as 70% of synaptic transmission within the central nervous system. Prior to release from presynaptic terminals, glutamate is packaged into synaptic vesicles via vesicular glutamate transporters (vGluTs) (Shigeri, Seal, & Shimamoto, 2004). Located on both presynaptic and postsynaptic membranes are a variety of types of receptors for glutamate. These receptors are primarily divided into ionotropic glutamate receptors (iGlu) which include N-methyl-D-aspartate (NMDA), α-amino-3-hydroxy-5-methylisoxazole-4-propionic acid (AMPA), and kainate receptors, which are ligand–gated ion channels that mediate fast excitatory neurotransmission. The other primary type of glutamate receptors are metabotropic glutamate (mGlu) receptors, which are seven transmembrane domain G protein–coupled receptors (GPCRs) that participate in slower, modulatory glutamatergic transmission. Following release into the extracellular fluid, glutamate is rapidly cleared by one or more excitatory amino acid transporters (EAATs), or other mechanisms such as the cystine–glutamate exchanger (x_c) (Vandenberg & Ryan, 2013).

2.2 mGlu receptor families and signaling mechanisms

Based on their pharmacological activation and signal transduction properties, mGlu receptors can be subdivided into three distinct groups— Group I which includes the mGlu1 and mGlu5 subtypes, Group II which is comprised of mGlu2 and mGlu3 subtypes, and Group III which includes mGlu4, mGlu6, mGlu7, and mGlu8. Group I mGlu receptors activate the $G\alpha_{q/11}$ class of G-proteins, which can stimulate one of several phospholipases (including phospholipase C) to induce phosphoinositol hydrolysis and formation of signaling intermediates such as inositol triphosphate (IP_3) and diacylglycerol (DAG) (Baude et al., 1993; Homayoun & Moghaddam, 2010; Lujan, Nusser, Roberts, Shigemoto, & Somogyi, 1996; Tanaka et al., 2000). These signaling intermediates in turn can activate other intracellular messengers including protein kinase C (PKC), and also liberate intracellular calcium from IP_3 receptor-mediated intracellular stores, which in turn activate other intracellular messengers such as calcium/calmodulin–dependent kinase II (CaMKII). Additional effector systems that may be activated by Group I mGlu receptors include mammalian target of rapamycin (mTOR) and mitogen-activated protein kinase/extracellular kinase (MAPK/ERK) (Hermans & Challiss, 2001; Nicoletti et al., 2011; Niswender & Conn, 2010).

As opposed to receptors in the Group I family, those in Groups II and III are predominantly inhibitory in nature, as they are coupled to the $G\alpha_i$ class of G-proteins which inhibit adenylyl cyclase activity to yield reductions in intracellular levels of cyclic adenosine monophosphate (cAMP). Group II mGlu receptor activation also promotes closure of voltage-gated calcium channels as well as the opening of voltage-activated potassium channels. Receptors in the Group III family are also coupled to inhibitory $G\alpha_i$ class of G-proteins. A signaling target of the mGlu6 subtype, which is expressed largely in the retina, is cyclic guanosine monophosphate (cGMP) phosphodiesterase (Hermans & Challiss, 2001; Nicoletti et al., 2011; Niswender & Conn, 2010).

2.3 Cellular and neuroanatomical distribution of mGlu receptors

Subcellular localization approaches have revealed that Group I mGlu receptors are primarily localized to postsynaptic elements such as the perisynaptic annulus (Lujan et al., 1996), though these receptors can be found on presynaptic terminals, albeit rarely. On the postsynaptic side, Group I mGlu receptors can be structurally linked to NMDA receptors via a variety of scaffolding proteins such as Homer, Shank and others. Group II and III mGlu receptors are predominantly found on presynaptic terminals where they subserve classical inhibitory autoreceptor mechanisms that suppresses excess glutamate release (Cartmell & Schoepp, 2000; Lyon et al., 2011; Schoepp, 2001). However, there is some evidence that these receptors can also be expressed on postsynaptic membranes (Tamaru, Nomura, Mizuno, & Shigemoto, 2001). mGlu receptors can also serve as heteroreceptors localized to other neuron subtypes (i.e., GABAergic), form heteromeric and homomeric dimer and trimer complexes, and have been found to be expressed on many non-neuronal cell types both within and outside of the central nervous system, such as astrocytes, microglia, oligodendrocytes, taste bud cells, and cells of the retina (D'Ascenzo et al., 2007; Mudo, Trovato-Salinaro, Caniglia, Cheng, & Condorelli, 2007; Niswender & Conn, 2010).

In the central nervous system, there is vast heterogeneity in the neuroanatomical expression patterns of mGlu receptor subtypes. The general patterns of distribution, which has primarily characterized in rodents (Niswender & Conn, 2010; Shigemoto, Nakanishi, & Mizuno, 1992; Shigemoto et al., 1993; Shigemoto et al., 1997; Spooren & Gasparini, 2004), are summarized in Table 1.

Metabotropic glutamate receptors and cognition

Table 1 Neuroanatomy of mGlu receptors.

Receptor subtype	Regions containing moderate to high expression
mGlu1	Olfactory system, cerebral cortex, thalamus, hippocampus (excluding the CA1 region), lateral septum, superior colliculus, cerebellum, dorsal striatum, hypothalamus, pallidum, ventral midbrain
mGlu2	Olfactory system, cerebral cortex, hippocampus, dorsal striatum, nucleus accumbens, amygdala, anterior thalamic nuclei, cerebellum
mGlu3	Olfactory system, cerebral cortex, hippocampus, dorsal striatum, nucleus accumbens, amygdala, anterior thalamic nuclei, cerebellum
mGlu4	Olfactory system, hippocampus (dentate gyrus), septum, thalamus, brainstem, spinal cord
mGlu5	Olfactory system, cerebral cortex, dorsal striatum, nucleus accumbens, lateral septum, hippocampal formation (CA1-CA3 regions and dentate gyrus), inferior colliculus, amygdala, spinal trigeminal nucleus
mGlu6	Retina[a]
mGlu7	Olfactory system, cerebral cortex, septum, dorsal striatum, nucleus accumbens, amygdala, thalamus, hypothalamus, cerebellum, brainstem, spinal cord
mGlu8	Olfactory system, thalamus, brainstem

[a]It has long been believed that mGlu6 receptors were exclusively expressed in the retina, primarily in neurons of the outer plexiform layer, and additional Group III mGlu receptors are also expressed in inner plexiform layer retinal neurons. Yet several recent studies have challenged the notion of retina-specific mGlu6 expression, showing moderate expression of this receptor the central nervous system of zebrafish and in brain tissue of mice carrying a fluorescent reporter protein coupled to the mGlu6 gene promoter (Huang, Haug, Gesemann, & Neuhauss, 2012; Vardi, Fina, Zhang, Dhingra, & Vardi, 2011).

3. Cognition: An executive function

The ability to execute and perform complex functions is an advanced stage of human evolution. These complex functions are classified as "executive functions," an aspect of metacognition that includes working memory, task flexibility, problem solving and conflict resolution (Frith & Dolan, 1996). Defined by any ability to plan, choose, decide, or engage in some way, executive functions allow individuals to hold and maintain information long enough to complete a set task, while incorporating ongoing environmental information (Funahashi, 2017; Funahashi & Andreau, 2013). Cognition is an executive function that refers to "the mental action or process of acquiring knowledge and understanding such knowledge through thought, experience and senses" (Dhakal & Bobrin, 2021). A variety of tasks have been designed and utilized in humans as well as preclinical models to study cognitive abilities. Of these assessments, tasks assessing working memory and task flexibility are heavily utilized in both the clinical and preclinical research setting. Working memory (WM) is essential for goal-mediated behaviors and allows an individual to hold information in mind long enough to perform a task and is considered analogous to a mental sketchpad (Baddeley, Logie, Bressi, Della Sala, & Spinnler, 1986). One prominent example is the ability to maintain a phone number in mind long enough to place it in your phone. Working memory allows for reasoning and guided decision-making and behavior, which also renders individuals capable of task flexibility. In fact, the strength or capability to perform working memory tasks often predicts an individual's cognitive flexibility (Blackwell, Cepeda, & Munakata, 2009). Behavioral and task flexibility allows individuals to adapt to situations, shifting their desired goals to encompass novel situations and ever changing environmental information. Together, these cognitive abilities rely on tightly regulated subcellular neurophysiological mechanisms across multiple brain regions, and especially within the prefrontal cortex (PFC), where alterations and abnormalities of cellular signaling are known to lead to abnormal cognition and performance deficits on working memory tasks.

4. Mediating cognition: The prefrontal cortex

The prefrontal cortex has been extensively studied for its role in mediating numerous aspects of cognition, where single cell underpinnings that mediate these functions have been examined and reported. The PFC plays a crucial role in connecting information over time by bridging gaps and behaviors, which ultimately relies on glutamatergic transmission and optimal levels of dopamine. The PFC receives inputs from multiple brain regions, with specific types of information relayed from the contralateral PFC, thalamus, hippocampus, and amygdala. For example, reciprocal connectivity between the amygdala and medial PFC (mPFC) gives rise to object associations, whereas mPFC-hippocampal connectivity plays a role in contextual associations (Lang et al., 2009). Information from these afferents is integrated in a highly organized and layer-specific fashion (Anastasiades & Carter, 2021), which relies on optimal neuromodulatory activity, such as glutamate release, GABAergic tone, as well as dopaminergic input from the ventral tegmental area (Ott & Nieder, 2019). Alterations in both glutamate and dopamine levels can give rise to deficits in neurocognitive function that is linked to altered neurophysiological properties.

On a cellular level, synaptic connections between cells of the PFC as well as its reciprocal connections are tightly regulated, and disruptions in these synaptic connections can have profound effects on cell connectivity, regional excitability as well as cellular/dendritic morphology (Anastasiades & Carter, 2021). Groundbreaking studies have demonstrated that working memory and the ability to perform cognitive tasks is mediated by prefrontal cells that maintain "persistent" firing/activity while information is maintained during a working memory task (Goldman-Rakic, 1995). Currently, this persistent activity is thought to occur due to network synaptic properties, cellular intrinsic properties and synchronized rhythmic activity between brain regions (Dembrow, Chitwood, & Johnston, 2010; Goldman-Rakic, 1995). Interestingly, disruptions in this connectivity often results in impaired cognitive abilities, a phenotype observed across multiple neuropsychiatric disorders including fragile X syndrome (FXS), schizophrenia, Alzheimer's disease (AD), Parkinson's disease (PD), and PTSD. In fact, changes in synaptic connectivity (i.e., synaptic plasticity) across numerous brain regions have been associated with each of these disorders, where the role of mGlu receptors in regulating this plasticity has been highlighted for its role in promoting long-lasting synaptic changes, causing disruptions in cognitive function.

5. Synaptic plasticity and mGlu receptors

One major aspect of neuroscience that allows us to assess changes in cellular and molecular processes underlying behavior is the study of synaptic plasticity. Altering the strength of synaptic connections, where they are strengthened or weakened over time, can indicate increases and decreases in their synaptic activity and thus, their ability to contribute to overarching behaviors. Long-lasting changes in plasticity are categorized as long-term potentiation (LTP) and long-term depression (LTD). LTP is characterized by processes involving persistent strengthening of synapses, increasing the synaptic connectivity between neurons and thus, signal transmission and excitability (Lynch, 2004). LTD is characterized by processes involving the weakening of synapses, decreasing the connectivity between neurons and thus, dampening signal transmission and excitability (Ito, 1989). The cellular mechanisms underlying these processes have been extensively studied and are predominantly driven by changes in glutamate release and actions at postsynaptic receptors, especially NMDA and AMPA receptors (Malenka & Nicoll, 1993). These bidirectional changes in plasticity are activity dependent and essential for creating and modifying memories as well as the integration of ongoing environmental information.

Briefly, both LTP and LTD rely on the activation of NMDA receptors, which ultimately leads to either insertion or removal of AMPA receptors from the postsynaptic membrane, respectively. An increase in AMPA receptors within the postsynaptic surface increases cellular excitability by enhancing ionic conductance mediated by glutamate, whereas a decrease in the number of AMPA receptors at the synapse hinders excitability and thus, interneuronal communication (Castellani, Quinlan, Cooper, & Shouval, 2001). LTP is induced via high frequency activation of NMDA receptors located on dendritic spines, which promotes calcium efflux and activates various intracellular signaling pathways. These intracellular pathways have been shown to activate and promote CaMKII activity, recruiting signaling molecules to enlarge the dendritic spine head and AMPA receptor accumulation at the synapse (Castellani et al., 2001). LTD is mediated by activation of NMDA receptors (NMDAR-LTD) at a low frequency stimulation (Nicoll, Oliet, & Malenka, 1998; Oliet, Malenka, & Nicoll, 1997), as well as signaling at mGlu and endogenous cannabinoid receptors. At low frequency stimulations, smaller magnitudes of calcium enter the cell—as compared to LTP—which in prevents effective relief of the magnesium block at NMDA receptors, and promotes LTD by removing AMPA receptors from the plasma membrane (Nicoll et al., 1998; Oliet et al., 1997). While it has been established that LTP is associated with promoting cognitive function and that LTD at synapses is associated with cognitive deficits (Matynia, Kushner, & Silva, 2002), it is important to note that LTP and LTD processes also underlie maladaptive cognitive processes (i.e., fear responses) as well as normal brain development. Here, we have highlighted the associations between LTP/LTD and cognition yet acknowledge that these processes do not always imply "positive" or "negative" effects on brain function.

Aside from NMDA-induced LTD, LTD can also be induced through activation of mGlu receptors. However, these two forms of LTD are not mutually exclusive from one another (Bashir, 2003; Cavalier et al., 2019). mGlu-induced LTD is promoted via stimulation of Group I mGlus (mGlu1 and mGlu5) which also results in endocytosis of synaptic AMPA receptors (Nicoll et al., 1998). While the end result of reduced AMPA surface expression is similar to that observed with NMDA-induced LTD, the two mechanisms are known to affect different AMPA receptor subunits, therefor resulting in the endocytosis of different AMPA receptor subtypes (Snyder et al., 2001). Furthermore, presynaptically located Group II mGlu receptors (mGlu2 and mGlu3) play a key role in reducing presynaptic neurotransmitter release when activated, which also plays a key role in stimulating LTD in glutamatergic neurons (Altinbilek & Manahan-Vaughan, 2009; Grueter & Winder, 2005; Nicholls et al., 2006). Together, these results demonstrate that Group I and Group II mGlu receptors can effectively modulate neuronal connectivity as well as long-term synaptic changes that underlie as alterations in cognition. Interestingly, the integrity of mGlu-LTD is crucial to maintain normal cognitive function, and any alteration in mGlu receptor signaling or expression could significantly alter cognitive processes (Gravius, Pietraszek, Dekundy, & Danysz, 2010; Ramiro-Cortés, Hobbiss, & Israely, 2014). Later, we highlight alterations in mGlu expression and function across multiple psychiatric and neurological disorders and diseases, which in part could explain the cognitive deficits often observed in patients suffering from these conditions.

6. mGlu receptors and cognition: Diseases and associations

Several studies have established that group I mGlu receptors play a pertinent role in executive functioning, including cognitive abilities. For example, mGlu1 receptors have been shown to be heavily involved in spatial and associative learning, highlighted by experiments in mice demonstrating that hippocampal mGlu1 receptor activation enables associative learning via activity-dependent synaptic plasticity, which relies on the induction of LTP (Gil-Sanz, Delgado-García, Fairén, & Gruart, 2008). With expression on mPFC GABAergic interneurons, mGlu1 receptors play an important role in regulating glutamatergic excitability within this region, which when unregulated has been shown to lead to psychosis and cognitive deficits in individuals with schizophrenia (Yohn et al., 2020). Further, potentiating mGlu1 function has been shown to enhance prefrontal inhibitory transmission and reverses NMDA receptor hypofunction in mouse models of cognitive impairment (Luessen et al., 2022; Maksymetz et al., 2021).

Together, these findings highlight the role of mGlu1 receptors in mediating important aspects of cognitive functioning.

mGlu5 receptors are densely expressed within corticolimbic brain areas that control higher cognitive and incentive functions including the medial prefrontal and orbitofrontal cortices, hippocampus, cingulate, striatum, and septal nuclei (Bell, Richardson, & Lee, 2002; Homayoun & Moghaddam, 2010; Romano et al., 1995; Shigemoto et al., 1993), and have been shown to be heavily involved in cognition. It has been proposed that mGlu5 activation has direct excitatory effects on excitatory postsynaptic potentials in the subthalamic nucleus (STN), hippocampus, substantia nigra pars reticulata, and septohippocampal cholinergic neurons promoting cognitive function through LTP induction and enhanced cellular excitability (Awad, Hubert, Smith, Levey, & Conn, 2000; Fitzjohn, Kingston, Lodge, & Collingridge, 1999; Homayoun & Moghaddam, 2010; Marino et al., 2001; Wu, Hajszan, Xu, Leranth, & Alreja, 2004). In addition, mGlu5 antagonists dose-dependently impair performance in several cognitive tasks including working memory, instrumental learning, and spatial learning (Balschun & Wetzel, 2002; Homayoun & Moghaddam, 2010; Homayoun, Stefani, Adams, Tamagan, & Moghaddam, 2004), and mGlu5 positive allosteric modulators have been shown to improve cognitive performance during novel object recognition, five-choice serial reaction time, and Morris water maze tasks, (Ayala et al., 2009; Balschun & Wetzel, 2002; Homayoun & Moghaddam, 2010; Liu et al., 2008). Together, these findings suggest that mGlu5 activation across several brain regions is essential to properly perform cognitive tasks.

Several studies have established that Group II metabotropic receptors, mGlu2 and mGlu3, are involved in hippocampal function and cognition (Krystal et al., 2005; Lyon et al., 2011; Moghaddam & Adams, 1998; Patil et al., 2007), where activation of mGlu2/3 receptors reversed spatial working memory deficits in rats, effects that were also observed in humans (Krystal et al., 2005; Lyon et al., 2011; Moghaddam & Adams, 1998). Additionally, hippocampal dependent spatial memory was impaired in GRM2/3 double knockout mice (Lyon, Kew, Corti, Harrison, & Burnet, 2008; Lyon et al., 2011). Thus, mGlu2 and mGlu3 play an important role in hippocampal spatial memory-related functions.

Compared to Group I and II mGlu receptors, group III mGlus are less well-studied for their roles in mediating cognition, which is partly due to an initial lack of subtype-selective pharmacological agents (O'Connor, Finger, Flor, & Cryan, 2010). Of Group III receptors, mGlu7 has been found to regulate both glutamate and GABA release, depending on the type of neuron in which the receptor is located (Niswender & Conn, 2010). Assessing the role of mGlu7 receptors in cognition has been facilitated by the availability of mGlu7 knockout mice. These animals have been shown to have hindered working memory as well as impaired fear responses (O'Connor et al., 2010); however, it is of note that developmental compensatory effects may promote these phenotypes (Cryan & Holmes, 2005). These phenotypes have been linked to impaired synaptic transmission within the amygdala and hippocampus where short-term potentiation is hindered and may affect the temporary storage of information required for long-term memory (Hölscher et al., 2004). In fact, short-term potentiation in mGlu7 knockout mice and its link to cognitive impairment has been demonstrated in multiple studies (Bushell, Sansig, Collett, van der Putten, & Collingridge, 2002; Hölscher et al., 2004). Further, mGlu7 potentiation can rescue cognitive deficits in mouse models of Rett syndrome (Gogliotti et al., 2017), major depressive disorder (Jaso et al., 2017) and PD (Amalric, 2015). mGlu4 receptors have also been shown to mediate antipsychotic-like effects in rodents (Sławińska et al., 2013), and have the ability rescue MK-801 induced cognitive impairment, an effect that requires activation of 5-HT1A receptors (Wieronska et al., 2015). mGlu4 receptors have been shown to form functional heterodimers with mGlu2 receptors selectively at inputs from the thalamus targeting the PFC where they critically regulate thalamocortical input (Xiang et al., 2021) and cognitive functioning. mGlu8 receptors have related and overlapping functions with those of mGlu4, with similar behavioral phenotypes observed in mGlu8 and mGlu4 deficient mice (Davis, Duvoisin, & Raber, 2013; Davis et al., 2013). While less is known regarding the direct effects of mGlu8 receptor activation or inhibition on cognition, these receptors have been proposed for treating anxiety, where positive allosteric modulators (PAMs) targeting mGlu8 receptors have been shown to exert anxiolytic effects (Duvoisin et al., 2010). Lastly, given their functional and anatomical overlap with mGlu4 receptors, it is feasible that these receptors play a similar role in mediating cognitive function.

Each of the studies discussed above highlight and provide foundational evidence for the role of mGlu receptors in mediating a variety of cognitive processes, which is hypothesized to be driven by their ability to regulate synaptic plasticity. Below, we will highlight studies associating alterations in the function and expression of mGlu receptors with a variety of neuropsychiatric disorders and diseases and their associated cognitive deficits, and behavioral phenotypes.

6.1 Fragile X syndrome

Fragile X syndrome (FXS) is the most commonly inherited form of mental retardation caused by a trinucleotide repeat expansion (TRE) within the X-linked Fragile X messenger ribonucleoprotein 1 (*FMR1*) gene (Bear, 2005; Bear, Huber, & Warren, 2004; O'Donnell & Warren, 2002). TRE prevents the production of fragile X mental retardation protein (FMRP), the protein encoded by *FMR1* (Bear et al., 2004; O'Donnell & Warren, 2002). The lack of FMRP causes abnormal brain development leading to the constellation of symptoms observed in patients with FXS (Bakker & Oostra, 2003; Bear et al., 2004). Central neurological symptoms include moderate to severe mental retardation, attention deficit with hyperactivity, anxiety with mood lability, obsessive-compulsive behaviors, autistic behaviors, poor motor coordination and an enhanced risk of epilepsy (Bakker & Oostra, 2003; Bear, 2005; Bear et al., 2004), whereas peripheral neurological symptoms include increased sensitivity to tactile irritation as well as loose bowel movements (Bakker & Oostra, 2003; Bear, 2005; Bear et al., 2004). Other symptoms also include facial abnormalities (i.e., elongated face and enlarged ears), obesity, hyperextensible joints, and enlarged testes in post-pubescent males (Bakker & Oostra, 2003; Bear, 2005; Bear et al., 2004).

Neurophysiological changes in excitatory synaptic transmission, such as LTD, correlate to neuronal morphological alterations including enlarged and immature dendritic spines and have been observed in patients and animal models of FXS. Such alterations in spine abnormalities are also observed with human mental retardation of unknown etiology, Down's syndrome, and Rett syndrome (Bear et al., 2004; Kaufmann & Moser, 2000; Purpura, 1974). As discussed above, changes in mGlu receptor stimulation can lead to altered synaptic plasticity across multiple brain regions. The relationship between FMRP, mGlu receptors and the phenotype associated with FXS has been shown in several studies. Specifically, FMRP is contained within ribonucleoprotein granules that transport certain mRNAs including *FMR1* mRNA to dendritic spines, where alterations in synaptic plasticity occurs (Bear, 2005). Evidence suggests that FMRP is required for the synthesis of certain dendritic proteins but that it also represses the synthesis of other dendritic proteins, and that dendritic transport of FMRP is mediated by mGlu receptors (Bear, 2005). For example, activation of Group I mGlu receptors with the selective agonist (*R,S*)-3,5-dihydroxyphenylglycine (DHPG) induces transport of FMRP to dendrites and promotes the expression of

FMRP as well as postsynaptic density protein 95 (PSD-95) in cultured hippocampal neurons and cultured cortical neurons, respectively. Interestingly, in *FMR1* knockout (KO) mice, DHPG does not enhance PSD-95 within the dendritic spine (Bear, 2005; Todd, Mack, & Malter, 2003). Additionally, polyribosome assembly in synaptosomes prepared from *FMR1* KO mouse cortex does not occur with DHPG mGlu activation, further suggesting that FMRP plays a critical role in the synthesis of dendritic proteins, and as such, synaptic plasticity (Bear, 2005; Weiler et al., 2004). FMRP also appears to facilitate the repression of other dendritic proteins, such as the dendritic microtubule associated protein 1b (MAP1b), and lack of FMRP significantly elevates MAP1b levels—which is observed in FMR1 KO mice as well as patients with FXS (Bear, 2005, Bear et al., 2004).

As discussed above, cognitive development is reliant on experience-dependent modifications of the synaptic connections between neurons within the cerebral cortex (Bear, 2005). The moderate to severe mental retardation associated with FXS is thought to be due to alterations in experience-dependent LTD, which has been associated with mGlu5 activation and the removal of AMPA receptors from dendritic spines. In wildtype animals, stable expression of mGlu receptors is required for mGlu-dependent LTD, and FMRP serves to limit LTD by inhibiting mGlu-dependent translation of synaptic mRNAs that promote LTD. However, in *FMR1* KO mice, mGlu-associated LTD is enhanced due to the inability of FMRP to limit mGlu-dependent translation of LTD inducing synaptic mRNAs (Huber, Gallagher, Warren, & Bear, 2002; Laggerbauer, Ostareck, Keidel, Ostareck-Lederer, & Fischer, 2001; Zhang et al., 2001), further supporting the notion that FMRP normally represses the protein synthesis associated with mGlu-dependent LTD (Bear, 2005; Huber et al., 2002). Thus, mGlu5 signaling, its regulation by FMRP, and its role in modifying synaptic plasticity likely plays a significant role in the cognitive phenotype associated with FXS and that these receptors may serve as future treatment targets to mitigate cognitive abnormalities observed in these patients (Huber et al., 2002).

In addition to experiencing impaired cognitive development, the majority of patients with FXS also meet the diagnostic criteria for anxiety disorders (Bear, 2005; Cordeiro, Ballinger, Hagerman, & Hessl, 2011). While the complete neurobiological mechanisms of anxiety disorders are still unknown, much of the research in this area involves the control of the hypothalamic-pituitary-adrenal axis by the amygdala, a brain region extensively involved in fear-associated behaviors (Bear, 2005). Experiments have demonstrated that when an auditory tone is paired with a painful shock, an animal will experience fear when exposed to the tone alone (Bear, 2005; Maren & Quirk, 2004). In conjunction with these findings, there is also evidence that tone-shock pairing results in LTP of the synapses relaying auditory information to the lateral amygdala and that this process requires the activation of mGlu5 (Bear, 2005; Maren & Quirk, 2004; Rodrigues, Bauer, Farb, Schafe, & LeDoux, 2002; Rodrigues, Schafe, & LeDoux, 2004). Thus, while it remains to be tested, it can be hypothesized that the mechanism underlying this LTP is dependent on the translation of pre-existing mRNA by FMRP, and that because of this, this process will be altered in *FMR1* KO mice (Bear, 2005). Despite this gap in the literature, mGlu5 antagonists have shown to be very effective anxiolytics in preclinical studies, and there is reason to believe that these medications could be effective for patients with FXS who experience anxiety (Bear, 2005; Spooren & Gasparini, 2004).

In addition to cognitive symptoms, mGlu-mediated changes in synaptic plasticity also underlie other symptoms of FXS (Bear, 2005). For example, uncoordinated voluntary movements have been associated with increased mGlu1-dependent LTD at synapses within the cerebellum between parallel fiber neurons and Purkinje neurons (Bear, 2005; Bear, Cooper, & Ebner, 1987; Ito, 1989), which is enhanced *FMR1* KO mice that also exhibit impaired motor learning similar to that of patients with FXS (Bear, 2005; Endo & Launey, 2003; Karachot, Shirai, Vigot, Yamamori, & Ito, 2001; Steinberg, Huganir, & Linden, 2004). The repetitive obsessive-compulsive-like behaviors seen in patients with FXS is correlated with decreased activation of mGlu1 and/or mGlu5 within the striatum (Bear, 2005; Graybiel, Canales, & Capper-Loup, 2000; Graybiel & Rauch, 2000; Rauch et al., 1997). Other symptoms with similar links to mGlu-mediated (especially mGlu5-mediated) changes in synaptic plasticity include hyperalgesia, irritable bowel, and obesity (Bear, 2005; Hunt & Tougas, 2002; Liu & Kirchgessner, 2000; Saper, Chou, & Elmquist, 2002; van den Pol, Romano, & Ghosh, 1995). Thus, the use of mGlu receptor antagonists as potential treatments for these symptoms may ultimately provide beneficial results.

6.2 Schizophrenia

Schizophrenia is a chronic severe mental disorder with a prevalence of about 1% (Lupták, Michaličková, Fišar, Kitzlerová, & Hroudová, 2021). It involves widespread brain dysconnectivity resulting in cognitive, behavioral, and emotional dysfunctions (Lupták et al., 2021). The clinical features of this disorder are grouped into positive symptoms, negative symptoms, and cognitive deficits (Muguruza, Meana, & Callado, 2016). Positive, also known as psychotic, symptoms include delusions, hallucinations, and thought disorder (Muguruza et al., 2016). Negative symptoms include psychomotor retardation, social withdrawal, impairments in initiative and motivation, and an impaired ability to recognize and express emotions (Muguruza et al., 2016). Cognitive deficits include disturbances in selective attention, working memory, executive control, episodic memory, language comprehension, cognitive flexibility, and social-emotional processing (Muguruza et al., 2016). Currently, pharmacological treatments primarily alleviate positive symptoms but are less effective at targeting at alleviating negative symptoms such as cognitive deficits (Miyamoto, Miyake, Jarskog, Fleischhacker, & Lieberman, 2012; Muguruza et al., 2016). Schizophrenia is a multifactorial disease that has both genetic and environmental components, yet a complete understanding of how the two produce the etiology of the disease is unknown. Classically, it has been explained via the "dopamine hypothesis" in which abnormalities in dopamine levels in certain brain regions cause the symptoms of the disorder (Carlsson, Kehr, & Lindqvist, 1976; Creese, Burt, & Snyder, 1976; Curran, Byrappa, & McBride, 2004; Davis, Kahn, Ko, & Davidson, 1991; Lieberman, Kane, & Alvir, 1987; Seeman, Lee, Chau-Wong, & Wong, 1976). However, this theory and its variations throughout the years have been unable to fully explain the pathology and clinical course of this disorder, which has led to explorations of other neurotransmitters involved in the etiology, such as glutamate (Muguruza et al., 2016).

The "glutamate hypothesis" of schizophrenia has stemmed from findings demonstrating that NMDA receptor antagonists such as phencyclidine (PCP) and ketamine induce schizophrenia-like symptomology in both humans and preclinical animal models. There is currently some evidence that schizophrenia may be associated with auto-antibodies directed against the NMDA receptor. Since the origination of this hypothesis, mGlu receptors and their role in the etiology of schizophrenia have been an important subject of exploration. Such studies have revealed that deleterious nonsynonymous single nucleotide polymorphisms (SNPs) are found in *GRM1*, the gene encoding mGlu1, in patients with schizophrenia (Muguruza et al., 2016). Additionally, higher levels of mGlu1α mRNA and protein expression within the dorsolateral prefrontal cortex has been observed in patients with schizophrenia compared to healthy controls (Gupta et al., 2005; Muguruza et al., 2016; Volk, Eggan, & Lewis, 2010). Interestingly, while most studies have shown no alterations in mGlu5 protein expression or mRNA levels in post-mortem tissue from schizophrenia patients compared to controls, regardless of brain region (Corti et al., 2011; Gupta et al., 2005; Ohnuma et al., 2000; Volk et al., 2010), others have demonstrated an increased mGlu5 protein expression without an increase in mGlu5 mRNA levels in the prefrontal cortex (Matosin, Fernandez-Enright, Lum, & Newell, 2017; Matosin et al., 2015; Muguruza et al., 2016). This suggests that while mGlu5 mRNA expression may remain constant, altered mGlu5 protein levels within specific brain regions, such as the prefrontal cortex, may give rise to symptoms experienced by patients with schizophrenia (Matosin et al., 2015; Muguruza et al., 2016). Other investigators have looked into the role of mGlu5 genetic variants in schizophrenia. For example, rs60954128 and rs3824927, two independent variants of the gene *GRM5* which encodes mGlu5, are associated with right hippocampal volume reduction and cognitive impairments in patients with schizophrenia (Matosin et al., 2018; Muguruza et al., 2016). Although further investigation is needed, some investigators have postulated the functional role of mGlu5 in schizophrenia to involve reciprocal interplay between mGlu and NMDA receptors. This hypothesis has been primarily based on findings that a reduction in mGlu5 signaling is accompanied by reductions in GluNR2 phosphorylation (Muguruza et al., 2016; Wang et al., 2020), an effect observed in the dorsolateral prefrontal cortex of patients with schizophrenia (Wang et al., 2020).

Group II mGlu receptor-related abnormalities are also implicated in schizophrenia, where several investigators have reported an association between schizophrenia and SNPs in the gene for mGlu3, *GRM3* (Chen et al., 2005; Cherlyn et al., 2010; Egan et al., 2004; Sartorius et al., 2008). However, some population-based genetic studies have been unable to replicate these findings (Martí, Cichon, Propping, & Nöthen, 2002; Tochigi et al., 2006). It has proven difficult to predict the physiological consequences of these SNPs as most of them occur in non-coding regions of the gene, yet carriers of the hCV11245618 *GRM3* intron variant have shown reduced performance on cognitive tests (Egan et al., 2004). Additionally, postmortem brain samples from patients with schizophrenia with this variant also have lower mRNA levels for the EAAT2 glutamate transporter, a main transporter responsible for the reuptake of glutamate at the synaptic cleft (Egan et al., 2004, Muguruza et al., 2016). Together, these findings suggest that the relationship between the *GRM3* genotype and schizophrenia risk is at least in part due to differences in EAAT2 expression and glutamate neurotransmission (Egan et al., 2004, Muguruza et al., 2016).

Results of studies characterizing differences in Group II mGlu receptor protein expression, as opposed to genetic variants, in schizophrenia have been less clear (Muguruza et al., 2016). Early postmortem studies using antibodies targeting both mGlu2 and mGlu3 receptor subtypes demonstrated inconsistent results, with some studies showing increased expression of these receptors in patients with schizophrenia relative to healthy controls, whereas others show no difference in expression (Crook, Akil, Law, Hyde, & Kleinman, 2002; Gupta et al., 2005; Muguruza et al., 2016). These results were also dependent on the area of the brain being assessed (Crook et al., 2002, Gupta et al., 2005, Muguruza et al., 2016). More recently, studies selectively targeting mGlu2 and mGlu3 receptors across multiple brain regions showed no difference in protein expression of mGlu2 and inconsistent results for mGlu3 expression (Corti et al., 2007a, 2007b; Dean, Duncan, & Gibbons, 2019; García-Bea, Bermudez, Harrison, & Lane, 2017; Ghose, Gleason, Potts, Lewis-Amezcua, & Tamminga, 2009; Muguruza et al., 2016). Interestingly, mGlu2 receptors have been found to couple with Gq/11–coupled serotonin 2A (5–HT2A) receptors to form a heteromeric receptor complex in both mouse and human frontal cortex (González–Maeso et al., 2008; Moreno et al., 2012). Antipsychotic drugs such as clozapine and risperidone are known to exert their therapeutic effects by targeting 5–HT2A receptors, and the heteromeric complex it forms with mGlu2 receptors suggests that their interaction may play an important role in the therapeutic effects of these compounds and the etiology of schizophrenia. Further, 5–HT2A receptors are downregulated in patients with schizophrenia, likely reducing the number of heteromeric complexes and thus, cell excitability which likely promotes cognitive dysfunction in patients. Interestingly, these same investigators had also previously shown that mGlu2 and mGlu3 signaling requires the expression of 5–HT2A and mGlu2 heteromers (Moreno et al., 2016; Muguruza et al., 2016).

Group III mGlu receptors primarily act as autoreceptors, maintaining proper glutamatergic tone. Preclinical studies have shown that agonists and PAMs of these receptors show antipsychotic-like effects in animal models of schizophrenia (Cieślik et al., 2018; Conn, Lindsley, & Jones, 2009; Poels et al., 2014; Wieronska, Zorn, Doller, & Pilc, 2016). Specifically, peripheral administration of ACPT-I, a group III mGlu receptor agonist, reduced MK801- and amphetamine-induced hyperlocomotion. Further, LSP1-2111, a specific mGlu4 specific agonist, appeared to improve negative symptoms as well as cognition in a mouse model of schizophrenia. However, of novel importance and as discussed above, mGlu4 receptors can form heterodimers with mGlu2 receptors, and their interaction has the ability to inhibit glutamatergic transmission within the PFC (Xiang et al., 2021). Targeting these heterodimers with LuAF21934 exerts an antipsychotic-like effect (Wieronska et al., 2015), and thus may serve as a new therapeutic target for treating positive symptoms associated with schizophrenia. Additional associations with Group III genetic variations and schizophrenia have also been observed, where genetic variations in a *GRM7* polymorphism, rs1396409, can be correlated to cognitive deficits of those patients with schizophrenia (Chaumette et al., 2020). The use of Group III mGlu receptor NAMs can rescue schizophrenia-like phenotypes in rodents, decreasing MK801-induced locomotor activity (Cieślik et al., 2018; Kalinichev et al., 2013), and these results are echoed by other studies demonstrating that *GRM7* knockout mice exhibit reductions in amphetamine-induced locomotor activity (Fisher et al., 2020, 2021). Lastly, while mGlu8 receptors have been somewhat understudied, a recent literature review highlights that deletion of *GRM8* fails to produce replicable endophenotypes of schizophrenia (Dogra & Conn, 2022). For an in-depth review of mGlu receptors as targets for schizophrenia treatment, we direct the readers to the recent review of Dogra and Conn (2022).

Taken together, the studies above demonstrate that abnormalities in mGlu expression and signaling is altered in patients with schizophrenia. However, the direct correlation of these findings with cognitive dysfunction remain understudied. It is reasonable to assume that alterations in mGlu expression, as observed in these patients, would lead to cognitive dysfunction as highlighted in other disorders such as FXS discussed above. Yet, this gap in knowledge requires additional preclinical and clinical studies exploring the correlation between the two as it relates to the symptomology observed with schizophrenia.

6.3 Parkinson's disease

Parkinson's disease (PD) is the second most common neurodegenerative disease and is characterized by both motor and non-motor symptoms (Azam et al., 2022). Motor symptoms include tremor, rigidity, and slowness of movements (Gelb, Oliver, & Gilman, 1999) brought about due to the degeneration of dopaminergic neurons within the substantia nigra pars compacta (SNC), which play a key role in the motor, limbic, and cognitive functions of the basal ganglia (Fearnley & Lees, 1991; Masilamoni & Smith, 2018). Non-motor symptoms include dementia and dysautonomia, which are induced by the degeneration of noradrenergic neurons in the locus coeruleus (LC) and adjoining areas (Fornai, di Poggio, Pellegrini, Ruggieri, & Paparelli, 2007; Vermeiren & De Deyn, 2017). Currently, the most effective treatment for the symptoms of PD has been dopamine (DA) replacement therapy, but long-term therapy results in undesired side effects such as motor and non-motor fluctuations, dyskinesias, and psychosis (Obeso, Olanow, & Nutt, 2000). This, combined with the fact that there is a significant time lag between the start of SNC neuron degeneration and the development of motor symptoms, highlight an opportunity for new neuro-protective treatments that halt or slow down the progression of neu-rodegeneration (Fearnley & Lees, 1991; Obeso et al., 2000). More recently, investigators have highlighted mGlu receptors as potential thera-peutic targets for neuroprotection due to their vast functions and expression across brain regions involved in the pathogenesis of PD (Conn & Pin, 1997; Johnson, Conn, & Niswender, 2009).

The relationship between the non-motor symptoms of PD and mGlu receptors has received very little attention (Amalric, 2015). Additionally, because the actions of glutamatergic and dopaminergic systems often oppose one another in the striatum and PFC, it has been hypothesized that positive modulation of mGlu receptors (especially mGlu2, mGlu3, and mGlu5) in these areas may be beneficial for mitigating the cognitive symptoms of PD (Ayala et al., 2009; De Leonibus et al., 2009; Fowler et al., 2013).

Group I mGlu receptors are expressed throughout the basal ganglia, including on DAergic terminals in the SNC (Conn, Battaglia, Marino, & Nicoletti, 2005; Martin, Blackstone, Huganir, & Price, 1992; Testa, Friberg, Weiss, & Standaert, 1998; Testa, Standaert, Young, & Penney, 1994). Experiments have highlighted changes in mGlu1 protein expression to be associated with rat models of PD (Yamasaki et al., 2016), parkinsonian monkeys (Kaneda et al., 2005), as well as in patients with PD (Kumar et al., 2014). Functionally, mGlu1 receptor activation promote an excitatory inward cationic current, resulting in cytotoxicity and neuronal degenera-tion, similar to that observed in PD (Cucchiaroni et al., 2010). Relative to mGlu5 receptors, mGlu1 receptors are relatively understudied in PD and warrant future exploration. Experiments with the mGlu5 antagonist 2-methyl-6-(phenylethynyl)pyridine (MPEP) in mice demonstrated that systemic administration of this drug protected dopaminergic terminals within the striatum from methamphetamine-induced toxicity (Battaglia et al., 2002). Additionally, mice lacking mGlu5 due to knockout of the encoding gene *GRM5* showed improved survival of nigrostriatal DA neurons following the administration of 1-methyl-4-phenyl-1,2,3,6-tetrahydropyridine (MPTP), a substance utilized to generate preclinical animal models of PD (Battaglia et al., 2004). These results were later confirmed in a study of non-human primates that underwent systemic administration of 3-[(2-methyl-1,3-thiazol-4-yl)ethynyl] pyridine (MTEP), an mGlu5 negative allosteric modulator, which revealed a neuroprotective effect against the neurodegenerative effects of MPTP (Masilamoni et al., 2011). This study also demonstrated that MTEP treatment was protective against MPTP toxicity-induced degeneration of noradrenergic and serotonergic neurons of the locus coeruleus and dorsal raphe, two brain areas whose degeneration leads to non-motor symptoms of PD (Fornai et al., 2007; Masilamoni et al., 2011). The mechanism for the neuroprotective effect of mGlu5 antagonism is not fully understood, however, some have proposed that these mechanisms likely involve a reduction in intracellular calcium levels in dopaminergic neurons caused by decreased mGlu5-mediated activation of intracellular IP_3 receptors, as well as a reduction of the potentiating effects of mGlu5 on NMDA receptors (Xia et al., 2015). Such mechanisms are thought to reduce overactive glutamatergic inputs from the subthalamic nucleus (STN) to neurons of the SNC in the PD state. Lastly, mGlu5 inhibition has been associated with reduction in the harmful effects of neuroinflammation, and/or reduction of MPTP-induced toxicity in midbrain DAergic neurons (Alagarsamy et al., 1999; Conn et al., 2005; Masilamoni et al., 2011). Together, these findings highlight that Group I mGlu receptors may promote neurodegeneration and associated symptomology, including cognitive deficits in patients with PD.

Group II mGlu receptors seem to have an effect opposite effect of mGlu5 receptors with regards to neurodegeneration, in that agonism seems to be neuroprotective (Masilamoni & Smith, 2018). Group II mGlu receptors are differentially expressed within the basal ganglia, with the striatum showing the highest concentration (Conn et al., 2005; Masilamoni & Smith, 2018; Testa et al., 1998). In the striatum, mGlu2 receptors are found on glutamatergic axons and terminals, cholinergic interneurons, and astrocytes, and they regulate glutamate release from the cortex and thalamus as well as dopamine release from the ventral midbrain (Conn et al., 2005; Johnson et al., 2009; Testa et al., 1998, 1994). Studies have shown that group II mGlu receptors reduce glutamatergic transmission in the STN, globus pallidus, and substantia nigra and because of this, group II mGlu agonists may help restore basal ganglia circuitry imbalance in PD by attenuating the output of glutamatergic corticostriatal and subthalamofugal synapses through inhibition of striatal cholinergic neurons (Conn et al., 2005; Shen & Johnson, 2003). However, the differential role of mGlu2 and mGlu3 are still under investigation (Masilamoni & Smith, 2018). The results of one study using a mGlu2/3 receptor agonist, LY379268 in mGlu2 and mGlu3 receptor deficient mice suggested that the neuroprotective properties of this agonist in mice treated with MPTP were completely mediated by the activation of astrocytic mGlu3 and that these effects were amplified in the absence of neuronal mGlu2 expression (Masilamoni & Smith, 2018). There is also evidence that mGlu3 increases the production of many neurotrophic factors including TGF-β and glial derived nerve growth factor (GDNF), which have been demonstrated to have neurotrophic and neuroprotective effects on nigral dopaminergic neurons in preclinical PD models (d'Anglemont de Tassigny, Pascual, & López-Barneo, 2015). Therefore, these results imply that group II mGlu receptors may mediate a neuroprotective effect, whereby agonists and PAMs targeting these receptors may prove beneficial therapeutic effects.

Group III mGlu receptors are expressed throughout the basal ganglia with expression of mGlu8 being less than that of mGlu4 and mGlu7 (Masilamoni & Smith, 2018). Activation of these receptors reduces the excitatory glutamatergic drive to the STN and SNC through presynaptic mechanisms (Valenti, Mannaioni, Seabrook, Conn, & Marino, 2005), and may partly explain the neuroprotective effects of mGlu4 PAMs against toxin-induced degeneration of midbrain dopaminergic neurons in PD models (Battaglia et al., 2006). Experiments have suggested that activation of these receptors protects against NMDA-induced toxicity in both cultured neurons as well as in vivo, and that this effect does not occur in mice lacking mGlu4 (Bruno et al., 2000). Several other experiments using animal models have also suggested a neuroprotective effect for group III mGlu receptors (Austin et al., 2010; Battaglia et al., 2006; Vernon et al., 2005), where mechanisms rely on direct or indirect inhibition of glutamate release in the SNC (Austin et al., 2010; Bruno et al., 2000; Conn et al., 2005; Masilamoni et al., 2011). Additionally, activation of mGlu4 at striatopallidal synapses inhibit GABA release, which would place neurons into an excitatory state by reducing inhibitory tone (Austin et al., 2010; Conn et al., 2005). This likely results in some of the antiparkinsonian effects of mGlu4 PAMs on the bradykinesia, among other motor symptoms, of PD. Additionally, since the overactivity of the indirect pathway is linked to the motor symptoms of PD and the degeneration of nigral dopaminergic neurons induced by excitotoxic effects of overactive glutamatergic projections from the STN, the downregulation of GABAergic striatopallidal transmission is potentially another mechanism by which mGlu4 PAMs elicit neuroprotection (Austin et al., 2010; Conn et al., 2005). Lastly, Group III mGlu receptors exert anti-inflammatory effects, another mechanism that may contribute to their neuroprotective role (Besong et al., 2002; Pinteaux-Jones et al., 2008; Taylor, Diemel, & Pocock, 2003). In summation, experiments with mGlu receptors in PD models have demonstrated the therapeutic potential of these targets, with the main effect of preventing glutamate-driven excitotoxicity. Despite the large amount of research on the relationships between mGlu receptors and the motor symptoms of PD, there is a lack of research regarding the non-motor/cognitive symptoms of this disease that is an important future avenue of research.

6.4 Alzheimer's disease

Alzheimer's disease (AD) is the most common neurodegenerative disease and results in progressive cognitive decline as well as behavioral and personality changes (Abd-Elrahman, Hamilton, Albaker, & Ferguson, 2020; Srivastava, Das, Yao, & Yan, 2020). It is characterized by gradual structural and functional degeneration of the cortex and hippocampus that result from excessive amyloid-β (Aβ) aggregation and deposition, neurofibrillary tangles, and progressive neuronal loss (Liu, Chang, Song, Li, & Wu, 2019; Thome et al., 2018). Neuronal loss is believed to be due, at least in part, to glutamate excitotoxicity in which excessive glutamate levels cause glutamate receptors to be overactivated, leading to large calcium influxes into neurons which lead to cell death via changes in mitochondria function, intracellular and extracellular signaling cascades, and oxidative stress (Jiang et al., 2018). Along with iGluRs, mGlu receptors have been implicated in the mechanism of this excitotoxicity, whereby Aβ oligomers interact both directly and indirectly with iGluRs and mGlu receptors and can result in synaptic deficits (Forner, Baglietto-Vargas, Martini, Trujillo-Estrada, & LaFerla, 2017). Further, altered mGlu receptor levels lead to the changes in glutamatergic synaptic functions and altered stimulation of glutamate receptors, which have been associated with the AD phenotype (Revett, Baker, Jhamandas, & Kar, 2013). Reductions in mGlu1 and mGlu5 expression within the hippocampus and cortex have been found in rodent models (Bie, Wu, Foss, & Naguib, 2019), which are region and receptor subtype-specific, with reduced mGlu1 levels being especially apparent within the CA1 region of the hippocampus. However, the magnitude of these changes may be dependent on the progression of AD (Bie et al., 2019; Liu et al., 2019). For example, one study did not find significant differences in mGlu1 expression between patients with AD and control groups, which was especially apparent in patients who were in the early stages of AD (Ishibashi, Miura, Toyohara, Ishiwata, & Ishii, 2019; Liu et al., 2019). However, these investigators did not rule out that later stages of the disease could demonstrate reduced mGlu1 levels (Ishibashi et al., 2019). Along these lines, a radioligand-binding assay study reported reduced levels of mGlu1 and mGlu5 in the cerebral cortex of patients with AD, and that the reduction of mGlu1 expression was correlated with the progression of AD, which indicate progressive neuropathological changes of the disease (Albasanz, Dalfó, Ferrer, & Martín, 2005). These investigators also found that expression levels of a group I mGlu receptor effector enzyme, phospholipase Cβ1, are also significantly decreased in patients with AD (Albasanz et al., 2005), together demonstrating that group I mGlu abnormalities may be a primary cause the cognitive impairment and dementia associated with AD. To this end, treatment with group I mGlu agonists and PAMs may prove beneficial as some studies have demonstrated that activation of group I mGlu receptors accelerates non-amyloidogenic processing of Aβ precursor protein (AβPP) by α-secretase (Jolly-Tornetta, Gao, Lee, & Wolf, 1998; Meziane et al., 1998).

Contrasting the directionality of the changes in expression of group I mGlu receptors seen in AD, the expression of mGlu2 is increased in the hippocampal CA1 and CA3 pyramidal neurons (Lee et al., 2004), and immunostaining of mGlu2 was significantly correlated with hyperphosphorylated tau-positive neurons in the brains of AD patients, implicating a role of mGlu2 in the pathogenesis of AD (Lee et al., 2004). Further, the non-specific mGlu2/3R agonist, LY379268, increases tau phosphorylation in rat primary cortical neurons (Ferraguti, Baldani-Guerra, Corsi, Nakanishi, & Corti, 1999; Lee et al., 2009), which could promote cognitive decline and AD progression. However, it should be noted that tau phosphorylation has been suggested by some to be a protective mechanism rather than a detrimental process (Morsch, Simon, & Coleman, 1999). Regardless of whether tau proteins are protective or detrimental, other investigations have shown that activation of Group II mGlu receptors (mGlu2 and mGlu3) exerts a protective effect against excitotoxicity by inhibiting the release of glutamate (Buisson & Choi, 1995; Buisson, Yu, & Choi, 1996; Kingston et al., 1999), which complicates our current understanding of their roles. These findings become more complex when examining the individual roles of mGlu2 and mGlu3, however, as it appears that AD is associated with reductions in mGlu3 but increases in mGlu2 activation. Interestingly, activation of mGlu2, but not mGlu3, inhibits GABA release and places neurons in a hyperexcitable state, promoting excitotoxic effects (Corti et al., 2007a, 2007b). Thus, targeting group II mGlu receptors, specifically inhibiting mGlu2 receptors may be considered as promising therapeutic targets for reducing accumulation of Aβ at synapses (Srivastava et al., 2020). Collectively, the findings here suggest that the pathophysiology and risk of AD may be dependent on the ratio of mGlu2 to mGlu3 expression and their activation patterns (Srivastava et al., 2020).

The role of group III mGlu receptors in AD is still under investigation. Some investigators have proposed a neuroprotective role by their ability to inhibit glutamate release and thus prevent glutamate excitotoxicity (Srivastava et al., 2020). In support of this, group III agonists have been shown to reduce the levels of pro-inflammatory cytokines derived from activated astrocytes, as well as reduce the intracellular calcium levels known to accompany Aβ-induced neurotoxicity (Zhao et al., 2009). Specifically, mGlu4 agonists have demonstrated the ability to protect degenerating neurons in the caudate nucleus of wild-type mice but not in GRM4 deficient mice, suggesting a neuroprotective role for mGlu4 activation (Ribeiro, Vieira, Pires, Olmo, & Ferguson, 2017). Studies investigating mGlu7 receptors have highlighted this receptor's promise as a therapeutic agent for AD as it has been shown to have a neuroprotective effect against NMDAR-mediated excitotoxicity in the basal forebrain (Niswender & Conn, 2010; Srivastava et al., 2020). Furthermore, GRM8 deficient mice demonstrate behavioral changes including diminished acclimatization to novel surroundings, interrupted context-based freezing response against fear, hyperactivity, and anxiety, each of which are associated with cognitive deficits (Gerlai, Adams, Fitch, Chaney, & Baez, 2002). Given that nerve terminals targeting the hippocampus contain mGlu8 receptors and are thought to tightly regulate neurotransmitter release onto GABAergic neurons, it is plausible that GRM8 deficient mice exhibit altered behavioral phenotypes due to abnormal GABAergic regulation within the hippocampus that would promote excitotoxicity (Ferraguti & Shigemoto, 2006; Ferraguti et al., 2005). Since hippocampal abnormalities are more pronounced in AD, further exploration of mGlu8 and their role in cognitive function is essential for developing pharmacotherapies for patients with AD and cognitive disorders.

In summary, experimental findings regarding the role of mGlu receptors in AD suggest that they play an important role in regulating Aβ aggregation and deposition as well as tau protein phosphorylation, and thus, may serve as effective targets to combat the symptoms accompanying AD.

6.5 Post-traumatic stress disorder (PTSD)

PTSD is an anxiety disorder that develops after a traumatic experience such as domestic violence, natural disasters, or combat (Mahan & Ressler, 2012). Re-experiencing, avoidance, and hyperarousal are the three types of symptoms that accompany the over-generalized fear that characterizes PTSD (Mahan & Ressler, 2012). Re-experiencing involves flashbacks, nightmares, and frightening thoughts about the trauma which can result in physical symptoms such as headaches (Mahan & Ressler, 2012). Avoidance includes avoiding reminders of the experience, feeling emotionally numb, losing interest in previously enjoyable activities, and learning and memory deficits (Mahan & Ressler, 2012). Hyperarousal includes being easily startled, feeling tense, difficulty sleeping, and angry outbursts (Mahan & Ressler, 2012). The prevalence and severity of PTSD is variable and not all patients who experience trauma will develop PTSD (Mahan & Ressler, 2012). Symptom severity in patients with PTSD depends on a multitude of factors that impact an individual's resistance to trauma including genetics, predisposition, social support, and early life experiences (Mahan & Ressler, 2012). Patients with PTSD exhibit significantly different responses to fear conditioning relative to trauma victims without PTSD (Mahan & Ressler, 2012). The brain areas critical to fear conditioning belong to the limbic system, which includes the amygdala, hippocampus, and PFC, each of which are known to be impacted by PTSD (Heimer & Van Hoesen, 2006). The role of mGlu receptors in regulating synaptic plasticity in these brain regions makes them essential for the consolidation of fear conditioning as well as fear extinction (Mahan & Ressler, 2012). And, although still under investigation, several studies have indicated that mGlu antagonists and genetic deletion of mGlu receptors in the limbic regions of the brain appear to impair both consolidation and extinction of fear conditioning (Mahan & Ressler, 2012).

Several studies have indicated that group I mGlu receptors play an essential role in fear conditioning, where contextual fear conditioning causes an increase in mGlu5 expression within the hippocampus, a limbic system brain region that functions to form a representation of the context in this process (Rodrigues et al., 2002). Similarly, additional experiments utilizing transgenic mice lacking mGlu5 demonstrate that these animals also have impaired contextual fear conditioning (Lu et al., 1997). In addition to the hippocampus, contextual fear conditioning also involves the amygdala, which functions to form an association between the conditioned stimulus and unconditioned stimulus, linking the hippocampal representation of the context and an aversive stimuli (i.e., shock) (LeDoux, 2000; Mahan & Ressler, 2012). More specifically, in auditory fear conditioning, a subtype of cued fear conditioning where a fearful stimuli (i.e., shock) is paired with an auditory stimulus requires integration within the lateral nucleus of the amygdala (LA) (Mahan & Ressler, 2012). Interestingly, antagonizing mGlu5 receptors utilizing the mGlu5 antagonist MPEP impairs long-term potentiation of thalamic inputs targeting the LA, and impairs fear conditioning (Mahan & Ressler, 2012). The same study also demonstrated that intra-amygdala administration of MPEP dose-dependently impairs the acquisition of, but not expression or consolidation, of auditory and contextual fear conditioning (Mahan & Ressler, 2012). These results suggest that mGlu5 plays a significant role in fear conditioning and the synaptic plasticity necessary for this process, which may underlie multiple aspects of cognition in humans (Neki et al., 1996; Schaffhauser et al., 1998). While little has been explored with regards to mGlu1 receptors and PTSD, one study has found that mice exposed to social isolation, a paradigm known to induce anxiety-like phenotypes, show reduced levels of mGlu1 protein expression within the PFC (Ieraci, Mallei, & Popoli, 2016). A reduction of mGlu1 protein expression within the hippocampus has also been reported in male mice previously exposed to prenatal chronic mild stress (Wang et al., 2015). Thus, anxiety phenotypes exhibited by individuals with PTSD may be exacerbated by a reduction in mGlu1 expression across several brain regions.

Group II mGlu receptors are also densely expressed within the limbic system, and in particular the amygdala. One study on conducted in rat brain slices indicated that activation of group II receptors induces depotentiation in the LA (Lin, Lee, Huang, Wang, & Gean, 2005). Behaviorally, intra-amygdala infusion of a group II mGlu agonist was capable of blocking the consolidation of fear memory in rats, suggesting that depotentiation within the LA may underlie consolidation impairment. Another study using intra-amygdala infusions of the group II receptor agonist (1S,2S,5R,6S)-2-aminobicyclo[3.1.0]hexane-2,6-dicarboxylic acid (LY354740) validated these findings by also showing significantly disrupted fear-potentiated startle responses in rats (Walker, Rattiner, & Davis, 2002). Investigators speculate that group II agonists inhibit amygdala function via inhibitory actions at postsynaptic receptors, based on findings by other investigators that LY354740 induced hyperpolarization of basolateral projection neurons (Walker et al., 2002). Furthermore, a study utilizing the group II mGlu2/3 antagonist, LY341495, microinjected into the lateral amygdala, demonstrated that these receptors are essential for the retention of fear extinction

(Kim et al., 2015). More recently the mGlu2/3 antagonist, BCI-838, was found to modify the anxiety response and reverse the PTSD-related behaviors of rats exposed to a series of low-level blast exposures that modeled a mild TBI in humans (Perez-Garcia et al., 2018). It is noteworthy that there was a lack of studies that differentiated the role of mGlu2 and mGlu3, and thus, it is possible that these receptor subtypes may play differential roles in phenotypic behaviors. Further investigation into the mechanism of how group II mGlu receptors play a role in the phenotype of PTSD is warranted.

Studies of group III mGlu receptors, particularly mGlu4, have indicated an important role in fear conditioning. Specifically, *GRM4* deficient mice demonstrate enhanced cued fear conditioning (Davis, Haley, Duvoisin, & Raber, 2012), yet showed no difference in contextual fear conditioning compared to wild-type mice (Davis et al., 2012). The investigators postulated that anatomical differences may explain this because contextual fear conditioning requires the hippocampus and both cued and contextual fear conditioning require the amygdala (Davis, Duvoisin, & Raber, 2013; Davis et al., 2013). Another investigation by the same investigators demonstrated that activation of mGlu4 with an agonist, LSP1-2111, inhibited acquisition of cued fear conditioning (Davis, Duvoisin, & Raber, 2013; Davis et al., 2013). Collectively, these findings support a role for mGlu4 in fear learning and memory, issues that are prominent in humans with PTSD. Findings regarding mGlu7 suggest a similar role for this receptor, where rats treated with the mGlu7 positive allosteric modulator, AMN08, show impaired acquisition and enhanced extinction of conditioned fear (Palucha et al., 2007). This effect was also observed in rats with diminished mGlu7 expression, achieved using a short interfering RNA (Fendt et al., 2008). Collectively, the studies and findings described here demonstrate the importance of mGlus in modulating fear conditioning and highlight a need for further investigation into their potential as therapeutics for PTSD.

7. Targeting mGlu receptors as a therapeutic for cognitive therapy

Above, we have highlighted the many roles that mGlu receptors play in regulating synaptic plasticity and how alterations in their expression and function as well as their downstream effects that can give rise to cognitive dysfunction associated with several psychiatric disorders including schizophrenia, FXS, AD, PD, and PTSD. Given that they play a role in mediating cognitive dysfunction, pharmacologically targeting mGlu receptors may serve as a potential therapeutic option for treating cognitive dysfunction in individuals with such disorders. In fact, the use of mGlu allosteric modulators such as PAMs and negative allosteric modulators (NAMs), as well as mGlu selective agonists and antagonists, have proven beneficial for cognitive therapy in these disorders in clinical trials (Downing et al., 2014; Kinon et al., 2011; Patil et al., 2007). PAMs and NAMs acting at various mGlu receptors have been suggested to be beneficial in the treatment of neuropsychiatric disorders, as most are low molecular weight blood–brain barrier permeable molecules, and therefor can be administered to target CNS receptors with a more favorable pharmacokinetic profile (Azam et al., 2020). Such compounds also allow for a more tolerable side-effect profile and greater overall control of drug potency and efficacy.

The recent targeting of an allosteric modulatory site on mGlu5 receptors, which potentiates receptor function, has been shown to improve cognitive performance of mice on the Morris water maze, five-choice serial reaction time test and novel object recognition test (Campbell et al., 2004; Schlumberger, Pietraszek, Gravius, & Danysz, 2010). Specifically, the mGlu5 PAM 3-cyano-N-(1,3-diphenyl-1H-pyrazon-5-yl)benzamide (CDPPB) has been found to promote bursting activity and enhance firing capabilities of neurons within the PFC (Lecourtier, Homayoun, Tamagnan, & Moghaddam, 2007). Such an effect would elevate neuronal excitation, intracellular calcium levels and subsequent gene expression, which would help induce LTP. In individuals with cognitive dysfunction, where mGlu-LTD has been observed, this effect would reverse the mGlu-directed LTD effect and increase cognitive capabilities. However, such an effect remains to be established. In addition, CDPPB has been shown to reverse the effects of NMDA receptor antagonists (i.e., ketamine and PCP), which alone induce schizophrenia-like symptoms (LaCrosse et al., 2015; Schlumberger et al., 2010). Thus, perhaps through their indirect actions at NMDA receptors, mGlu5 PAMs may also alleviate cognitive deficits observed in patients with psychotic disorders.

In addition to CDPPB, another mGlu5 PAM known as ADX47273 has shown efficacy as a cognitive-enhancing antipsychotic in preclinical animal models through its ability to elicit mGlu5 potentiation of NMDA receptor activity and calcium mobilization in the PFC, similar to that of CDPPB (Wang, Walsh, Rowan, Selkoe, & Anwyl, 2004). Despite minimal reporting on its chemical properties, preclinical studies have established ADX47273 to be effective in the treatment of cognitive dysfunction associated with schizophrenia in preclinical mouse models of the disease (Liu et al., 2008). Another study found ADX47273 to attenuate cognitive deficits in rats withdrawn from binge-like ethanol exposure, as evaluated by the Barnes maze task (Marszalek-Grabska et al., 2018). Specifically, animals treated with ADX 47273 following ethanol withdrawal were found to have improvements in reversal learning, a measure of cognitive flexibility (Marszalek-Grabska et al., 2018). Interestingly, researchers found that animals treated with ADX47273 displayed increase expression of an mGlu5 subunit, GluN2B, within the PFC and hippocampus, areas essential for normal cognitive performance. It has been proposed that this increased expression, along with ADX47273-potentiated mGlu5 signaling, ultimately results in the GluN2B phosphorylation necessary for the regulation of NMDA receptor trafficking, and synaptic plasticity underlying cognitive function.

The mGlu4 selective PAM, N-phenyl-7-(hydroxyimino)cyclopropa(b)chromen-1a-carboxamide (PHCCC), has been shown to reverse dyskinesias associated with PD (Campbell et al., 2004). This is due to the presynaptic involvement of mGlu4 at the subthalamic nucleus (STN) of the basal ganglia, such that these receptors, when stimulated, decrease downstream STN glutamatergic projections on GABAergic output nuclei. As expected, mGlu4 PAMs like that of PHCCC have thus displayed attenuation of the basal ganglia mediated overexcitation seen with PD, alongside the resulting inhibition in thalamocortical projections secondary to this overexcitation. Although PAMs to mGlu4 have not displayed a direct role in treating the cognitive impairment in such patients, the use of such allosteric modulators as an alternative treatment for PD may indirectly improve cognition by circumventing the cognitive side-effects seen with conventional dopamine-related treatments.

Studies of various mGlu receptor NAMs have also displayed a protective role in treating cognitive dysfunction associated with neuropsychiatric disorders. Evaluation of the compound N-(3-chlorophenyl)-N'-(1-methyl-4-oxo-4,5-dihydro-1H-imidazol-2-yl)urea (fenobam), an mGlu5 NAM, is one such example. In a previous phase I clinical trial involving 12 FXS patients, individuals were subjected to the Carolina Fragile X Project Continuous Performance Test (FXCPT), a continuous performance test used to evaluate cognitive processes such as attention and impulsivity. Interestingly, fenobam treatment was found to be correlated to a decrease in patient impulsivity (Berry-Kravis et al., 2009). However, overall analyses did not show improvement on the FXCPT following Fenobam treatment, which the study denotes to ceiling effects of the drug, small sample size and short-term treatment (Berry-Kravis et al., 2009). Of note, fenobam has also displayed potential use as an anxiolytic by reducing glutamatergic activity in cortical regions related to anxiety (Porter et al., 2005). A clear concern with the use of mGlu5 NAMs is the overextending cognitive impairment that may present with their long-term use via downstream NMDA receptor inhibition. This has been noted by previous studies recognizing potentiation of cognitive impairment by fenobam similar to that induced by NMDA receptor antagonists such as PCP (Gould et al., 2016; Homayoun et al., 2004; Kinney et al., 2003). It is worth mentioning that preclinical studies of partial allosteric antagonists for mGlu5, such as 2-[2-(3-methoxyphenyl)ethynyl]-5-methylpyridine (M-5MPEP), have been found to produce decreased levels of downstream NMDAR inhibition while maintaining antidepressant-like effects, as analyzed using paired-associate learning tasks in animal studies (Holter et al., 2021). With this understanding, further research into partial allosteric antagonists may bring light to additional methods for targeting mGlu5, as opposed to full allosteric modulators, for treatment of cognitive impairments with a more favorable side-effect profile.

As a receptor for prion proteins and Aβ oligomers, mGlu5 has also been a target for the treatment of cognitive dysfunction associated with AD (Hamilton et al., 2016). The compound 2-chloro-4-((2,5-dimethyl-1-(4-(trifluoromethoxy)phenyl)-1H-imidazol-4-yl)ethynyl) pyridine (CTEP), an mGlu5 NAM, is currently being explored as a cognitive therapeutic. To date, CTEP has been found to decrease Aβ plaque composites and cognitive decline in animal models of AD (Hamilton et al., 2016). This is believed to be a result of preventing the Aβ-mediated inhibition of high-frequency stimulation (HFS) induced LTP in hippocampal regions (Wang et al., 2004). Further, preclinical studies have displayed mGlu5 NAMs ability to prevent the resultant downstream inflammatory cytokine mediated neurotoxicity associated with AD. However, studies on some mGlu5 NAMs, such as that of CTEP, have suggested that chronic administration, but not acute, is necessary for improvement in memory deficits, as assessed via the Morris water maze (Hamilton et al., 2016). Similar to CTEP, 2-chloro-4-[1-(4-fluoro-phenyl)-2,5-dimethyl-1H-imidazol-4-ylethynyl]-pyridine (basimglurant) has also been utilized in phase II clinical development, and was found to be most beneficial for patients with FXS and major depressive disorder (MDD) (Youssef et al., 2018), where it alleviated some cognitive impairment, though more evaluation of this particular drug with respect to cognitive-directed paradigms is needed to determine its efficacy in treating additional disorders.

As mentioned above, orthosteric agonists/antagonists targeting mGlu receptors may also prove beneficial for cognitive improvements. Orthosteric agonists and antagonists bind directly to the active site of the receptor and either activate or inhibit receptor activity, respectively. Nonspecific agonists at Group II mGlu receptors such as LY354740, which activates both mGlu2 and mGlu3 subtypes, have previously displayed efficacy as anxiolytics and antipsychotics for disorders such as schizophrenia (Grillon, Cordova, Levine, & Morgan, 2003). Additionally, stimulation of group II receptors leads to an improvement in cognition via disinhibition at the mediodorsal thalamus (MD) in rodents (Copeland, Neale, & Salt, 2015), likely mediated by the abundance of PFC and amygdala synapses in this region. These findings highlight the need for additional studies characterizing the effects of Group II mGlu receptor stimulation on cognition.

In contrast to LY354740, the Group I specific agonist DHPG has been shown to dampen cognitive function (Liu et al., 2008; Wang et al., 2018). Specifically, DHPG was found to stimulate internalization of NMDA receptors and thus attenuation of NMDA receptor-mediated excitatory post-synaptic currents in CA1 neurons of the hippocampus, a region well known to be involved in cognition (Liu et al., 2008, Wang et al., 2018). This internalization, along with the activation of protein kinase D1 (PKD1), has been shown to precipitate significant downregulation of NMDA receptors (Wang et al., 2018) which may promote LTD. Thus, as expected, preclinical evaluation has found DHPG to impair cognition in animals, as assessed utilizing the Morris water maze as well as novel object discrimination (NOD) tests (Wang et al., 2018). The mGlu5 receptor antagonist MPEP has been shown to prevent DHPG-mediated impairments in object recognition and NMDA receptor internalization, highlighting an additional method by which it mitigates cognitive dysfunction (Wang et al., 2018). Downregulation of such signaling may prevent the associated Group I mGlu receptor-mediated LTD thought to play a role in the pathophysiology of cognitive decline in patients with FXS and AD (Wang et al., 2018). Furthermore, the mGlu5 receptor antagonist, AFQ056, has been found to attenuate the decreased synthesis of FMRP linked with FXS, preventing deficits in protein-dependent plasticity associated with FMRP loss (Sourial, Cheng, & Doering, 2013). The use of AFQ056 in phase III clinical trials has been shown to resolve hyperactivity and cognitive deficits associated with FMRP loss by increasing downstream signaling of the protein, ultimately leading to the inhibition of LTD associated with cognitive dysfunction (Sourial et al., 2013). Studies exploring the effects of DHPG, MPEP and AFQ056 have led to a better understanding of role Group I mGlu receptors, particularly mGlu5, plays in mediating cognition.

8. Conclusion

Studies evaluating the physiology of mGlu receptor subtypes, their role in cognition and compounds that target these receptors has brought light to pharmacological methods of treating cognitive impairment across a variety of neuropsychiatric disorders. The development of mGlu receptor-specific orthosteric agonists and antagonists as well as NAMs and PAMs, along with further research into the physiological roles of mGlu subtypes, opens way for novel treatment options directed toward mitigating cognitive dysfunction. However, research into receptor-specific compounds such as those described above is essential to gain a complete understanding of compound pharmacokinetics, and to explore potential side effects such including consequences of long-term exposure, drug-drug interactions, hepatic function, and teratogenicity. Overall, continued research to better understand mGlu receptor physiology, alongside receptor-specific modulators, will allow for further development of cognitive protecting and promoting pharmacotherapeutics.

Funding

This work was supported by Public Health Service grants F32AA 027962 to J.M.L.J., as well as DA043172 and AA025590 to M.F.O.

References

Abd-Elrahman, K. S., Hamilton, A., Albaker, A., & Ferguson, S. S. G. (2020). mGluR5 contribution to neuropathology in Alzheimer mice is disease stage-dependent. *ACS Pharmacology & Translational Science*, *3*(2), 334–344. https://doi.org/10.1021/acsptsci. 0c00013.

Alagarsamy, S., et al. (1999). Activation of NMDA receptors reverses desensitization of mGluR5 in native and recombinant systems. *Nature Neuroscience*, *2*(3), 234–240. https://doi.org/10.1038/6338.

Albasanz, J. L., Dalfó, E., Ferrer, I., & Martín, M. (2005). Impaired metabotropic glutamate receptor/phospholipase C signaling pathway in the cerebral cortex in Alzheimer's disease and dementia with Lewy bodies correlates with stage of Alzheimer's-disease-related changes. *Neurobiology of Disease*, *20*(3), 685–693. https://doi.org/10.1016/j.nbd.2005. 05.001.

Altinbilek, B., & Manahan-Vaughan, D. (2009). A specific role for group II metabotropic glutamate receptors in hippocampal long-term depression and spatial memory. *Neuroscience*, *158*(1), 149–158. https://doi.org/10.1016/j.neuroscience.2008.07.045.

Amalric, M. (2015). Targeting metabotropic glutamate receptors (mGluRs) in Parkinson's disease. *Current Opinion in Pharmacology*, *20*, 29–34. https://doi.org/10.1016/j.coph. 2014.11.001.

Anastasiades, P. G., & Carter, A. G. (2021). Circuit organization of the rodent medial prefrontal cortex. *Trends in Neurosciences*, *44*(7), 550–563. https://doi.org/10.1016/j.tins.2021.03.006.

Austin, P. J., et al. (2010). Symptomatic and neuroprotective effects following activation of nigral group III metabotropic glutamate receptors in rodent models of Parkinson's disease. *British Journal of Pharmacology*, *160*(7), 1741–1753. https://doi.org/10.1111/j.1476-5381.2010.00820.x.

Awad, H., Hubert, G. W., Smith, Y., Levey, A. I., & Conn, P. J. (2000). Activation of metabotropic glutamate receptor 5 has direct excitatory effects and potentiates NMDA receptor currents in neurons of the subthalamic nucleus. *The Journal of Neuroscience*, *20*(21), 7871–7879.

Ayala, J. E., et al. (2009). mGluR5 positive allosteric modulators facilitate both hippocampal LTP and LTD and enhance spatial learning. *Neuropsychopharmacology*, *34*(9), 2057–2071. https://doi.org/10.1038/npp.2009.30.

Azam, S., et al. (2020). G-protein-coupled receptors in CNS: A potential therapeutic target for intervention in neurodegenerative disorders and associated cognitive deficits. *Cell*, *9*(2). https://doi.org/10.3390/cells9020506.

Azam, S., et al. (2022). Group I mGluRs in therapy and diagnosis of Parkinson's disease: Focus on mGluR5 subtype. *Biomedicine*, *10*(4). https://doi.org/10.3390/biomedicines10040864.

Baddeley, A., Logie, R., Bressi, S., Della Sala, S., & Spinnler, H. (1986). Dementia and working memory. *The Quarterly Journal of Experimental Psychology A*, *38*(4), 603–618.

Bakker, C. E., & Oostra, B. A. (2003). Understanding fragile X syndrome: insights from animal models. *Cytogenetic and Genome Research*, *100*(1–4), 111–123. https://doi.org/10.1159/000072845.

Balschun, D., & Wetzel, W. (2002). Inhibition of mGluR5 blocks hippocampal LTP in vivo and spatial learning in rats. *Pharmacology, Biochemistry, and Behavior*, *73*(2), 375–380. https://doi.org/10.1016/s0091-3057(02)00847-x.

Bashir, Z. I. (2003). On long-term depression induced by activation of G-protein coupled receptors. *Neuroscience Research*, *45*(4), 363–367. https://doi.org/10.1016/s0168-0102(03)00002-6.

Battaglia, G., et al. (2002). Selective blockade of mGlu5 metabotropic glutamate receptors is protective against methamphetamine neurotoxicity. *The Journal of Neuroscience*, *22*(6), 2135–2141.

Battaglia, G., et al. (2004). Endogenous activation of mGlu5 metabotropic glutamate receptors contributes to the development of nigro-striatal damage induced by 1-methyl-4-phenyl-1,2,3,6-tetrahydropyridine in mice. *The Journal of Neuroscience*, *24*(4), 828–835. https://doi.org/10.1523/JNEUROSCI.3831-03.2004.

Battaglia, G., et al. (2006). Pharmacological activation of mGlu4 metabotropic glutamate receptors reduces nigrostriatal degeneration in mice treated with 1-methyl-4-phenyl-1,2,3,6-tetrahydropyridine. *The Journal of Neuroscience*, *26*(27), 7222–7229. https://doi.org/10.1523/JNEUROSCI.1595-06.2006.

Baude, A., et al. (1993). The metabotropic glutamate receptor (mGluR1 alpha) is concentrated at perisynaptic membrane of neuronal subpopulations as detected by immunogold reaction. *Neuron*, *11*(4), 771–787. https://doi.org/10.1016/0896-6273(93)90086-7.

Bear, M. F. (2005). Therapeutic implications of the mGluR theory of fragile X mental retardation. *Genes, Brain, and Behavior*, *4*(6), 393–398. https://doi.org/10.1111/j.1601-183X.2005.00135.x.

Bear, M. F., Cooper, L. N., & Ebner, F. F. (1987). A physiological basis for a theory of synapse modification. *Science*, *237*(4810), 42–48. https://doi.org/10.1126/science.3037696.

Bear, M. F., Huber, K. M., & Warren, S. T. (2004). The mGluR theory of fragile X mental retardation. *Trends in Neurosciences*, *27*(7), 370–377. https://doi.org/10.1016/j.tins.2004.04.009.

Bell, M. I., Richardson, P. J., & Lee, K. (2002). Functional and molecular characterization of metabotropic glutamate receptors expressed in rat striatal cholinergic interneurones. *Journal of Neurochemistry*, *81*(1), 142–149. https://doi.org/10.1046/j.1471-4159.2002.00815.x.

Berry-Kravis, E., et al. (2009). A pilot open label, single dose trial of fenobam in adults with fragile X syndrome. *Journal of Medical Genetics*, *46*(4), 266–271. https://doi.org/10.1136/jmg.2008.063701.

Besong, G., et al. (2002). Activation of group III metabotropic glutamate receptors inhibits the production of RANTES in glial cell cultures. *The Journal of Neuroscience*, *22*(13), 5403–5411. https://doi.org/20026585.

Bie, B., Wu, J., Foss, J. F., & Naguib, M. (2019). Activation of mGluR1 mediates C1q-dependent microglial phagocytosis of glutamatergic synapses in Alzheimer's rodent models. *Molecular Neurobiology*, *56*(8), 5568–5585. https://doi.org/10.1007/s12035-019-1467-8.

Blackwell, K. A., Cepeda, N. J., & Munakata, Y. (2009). When simple things are meaningful: Working memory strength predicts children's cognitive flexibility. *Journal of Experimental Child Psychology*, *103*(2), 241–249. https://doi.org/10.1016/j.jecp.2009.01.002.

Bruno, V., et al. (2000). Selective activation of mGlu4 metabotropic glutamate receptors is protective against excitotoxic neuronal death. *The Journal of Neuroscience*, *20*(17), 6413–6420.

Buisson, A., & Choi, D. W. (1995). The inhibitory mGluR agonist, S-4-carboxy-3-hydroxy-phenylglycine selectively attenuates NMDA neurotoxicity and oxygen-glucose deprivation-induced neuronal death. *Neuropharmacology*, *34*(8), 1081–1087. https://doi.org/10.1016/0028-3908(95)00073-f.

Buisson, A., Yu, S. P., & Choi, D. W. (1996). DCG-IV selectively attenuates rapidly triggered NMDA-induced neurotoxicity in cortical neurons. *The European Journal of Neuroscience*, *8*(1), 138–143. https://doi.org/10.1111/j.1460-9568.1996.tb01174.x.

Bushell, T. J., Sansig, G., Collett, V. J., van der Putten, H., & Collingridge, G. L. (2002). Altered short-term synaptic plasticity in mice lacking the metabotropic glutamate receptor mGlu7. *ScientificWorldJournal*, *2*, 730–737. https://doi.org/10.1100/tsw.2002.146.

Campbell, U. C., et al. (2004). The mGluR5 antagonist 2-methyl-6-(phenylethynyl)-pyridine (MPEP) potentiates PCP-induced cognitive deficits in rats. *Psychopharmacology*, *175*(3), 310–318. https://doi.org/10.1007/s00213-004-1827-5.

Carlsson, A., Kehr, W., & Lindqvist, M. (1976). The role of intraneuronal amine levels in the feedback control of dopamine, noradrenaline and 5-hydroxytryptamine synthesis in rat brain. *Journal of Neural Transmission*, *39*(1–2), 1–19. https://doi.org/10.1007/BF01248762.

Cartmell, J., & Schoepp, D. D. (2000). Regulation of neurotransmitter release by metabotropic glutamate receptors. *Journal of Neurochemistry*, *75*(3), 889–907. https://doi.org/10.1046/j.1471-4159.2000.0750889.x.

Castellani, G. C., Quinlan, E. M., Cooper, L. N., & Shouval, H. Z. (2001). A biophysical model of bidirectional synaptic plasticity: Dependence on AMPA and NMDA receptors. *Proceedings of the National Academy of Sciences of the United States of America*, *98*(22), 12772–12777. https://doi.org/10.1073/pnas.201404598.

Cavalier, M., et al. (2019). Disturbance of metabotropic glutamate receptor-mediated long-term depression (mGlu-LTD) of excitatory synaptic transmission in the rat hippocampus after prenatal immune challenge. *Neurochemical Research*, *44*(3), 609–616. https://doi.org/10.1007/s11064-018-2476-0.

Chaumette, B., et al. (2020). A polymorphism in the glutamate metabotropic receptor 7 is associated with cognitive deficits in the early phases of psychosis. *Schizophrenia Research*. https://doi.org/10.1016/j.schres.2020.06.019.

Chen, Q., et al. (2005). A case-control study of the relationship between the metabotropic glutamate receptor 3 gene and schizophrenia in the Chinese population. *Schizophrenia Research*, *73*(1), 21–26. https://doi.org/10.1016/j.schres.2004.07.002.

Cherlyn, S. Y., et al. (2010). Genetic association studies of glutamate, GABA and related genes in schizophrenia and bipolar disorder: A decade of advance. *Neuroscience and Biobehavioral Reviews*, *34*(6), 958–977. https://doi.org/10.1016/j.neubiorev.2010.01.002.

Cieślik, P., et al. (2018). Mutual activation of glutamatergic mGlu4 and muscarinic M4 receptors reverses schizophrenia-related changes in rodents. *Psychopharmacology*, *235*(10), 2897–2913. https://doi.org/10.1007/s00213-018-4980-y.

Conn, P. J., Battaglia, G., Marino, M. J., & Nicoletti, F. (2005). Metabotropic glutamate receptors in the basal ganglia motor circuit. *Nature Reviews. Neuroscience*, *6*(10), 787–798. https://doi.org/10.1038/nrn1763.

Conn, P. J., Lindsley, C. W., & Jones, C. K. (2009). Activation of metabotropic glutamate receptors as a novel approach for the treatment of schizophrenia. *Trends in Pharmacological Sciences*, *30*(1), 25–31.

Conn, P. J., & Pin, J. P. (1997). Pharmacology and functions of metabotropic glutamate receptors. *Annual Review of Pharmacology and Toxicology*, *37*, 205–237. https://doi.org/10.1146/annurev.pharmtox.37.1.205.

Copeland, C. S., Neale, S. A., & Salt, T. E. (2015). Neuronal activity patterns in the mediodorsal thalamus and related cognitive circuits are modulated by metabotropic glutamate receptors. *Neuropharmacology*, *92*, 16–24. https://doi.org/10.1016/j.neuropharm.2014.12.031.

Cordeiro, L., Ballinger, E., Hagerman, R., & Hessl, D. (2011). Clinical assessment of DSM-IV anxiety disorders in fragile X syndrome: Prevalence and characterization. *Journal of Neurodevelopmental Disorders*, *3*(1), 57–67. https://doi.org/10.1007/s11689-010-9067-y.

Corti, C., et al. (2007a). The use of knock-out mice unravels distinct roles for mGlu2 and mGlu3 metabotropic glutamate receptors in mechanisms of neurodegeneration/neuroprotection. *The Journal of Neuroscience*, *27*(31), 8297–8308. https://doi.org/10.1523/JNEUROSCI.1889-07.2007.

Corti, C., et al. (2007b). Altered dimerization of metabotropic glutamate receptor 3 in schizophrenia. *Biological Psychiatry*, *62*(7), 747–755. https://doi.org/10.1016/j.biopsych.2006.12.005.

Corti, C., et al. (2011). Altered levels of glutamatergic receptors and Na+/K+ ATPase-alpha1 in the prefrontal cortex of subjects with schizophrenia. *Schizophrenia Research*, *128*, 7–14. https://doi.org/10.1016/j.schres.2011.01.021.

Creese, I., Burt, D. R., & Snyder, S. H. (1976). Dopamine receptor binding predicts clinical and pharmacological potencies of antischizophrenic drugs. *Science*, *192*(4238), 481–483. https://doi.org/10.1126/science.3854.

Crook, J. M., Akil, M., Law, B. C., Hyde, T. M., & Kleinman, J. E. (2002). Comparative analysis of group II metabotropic glutamate receptor immunoreactivity in Brodmann's area 46 of the dorsolateral prefrontal cortex from patients with schizophrenia and normal subjects. *Molecular Psychiatry*, *7*(2), 157–164. https://doi.org/10.1038/sj.mp.4000966.

Cryan, J. F., & Holmes, A. (2005). The ascent of mouse: Advances in modelling human depression and anxiety. *Nature Reviews. Drug Discovery*, *4*(9), 775–790. https://doi.org/10.1038/nrd1825.

Cucchiaroni, M. L., et al. (2010). Metabotropic glutamate receptor 1 mediates the electrophysiological and toxic actions of the cycad derivative beta-N-methylamino-L-alanine on substantia nigra pars compacta DAergic neurons. *The Journal of Neuroscience*, *30*(15), 5176–5188. https://doi.org/10.1523/JNEUROSCI.5351-09.2010.

Curran, C., Byrappa, N., & McBride, A. (2004). Stimulant psychosis: Systematic review. *The British Journal of Psychiatry, 185,* 196–204. https://doi.org/10.1192/bjp.185.3.196.

d'Anglemont de Tassigny, X., Pascual, A., & López-Barneo, J. (2015). GDNF-based therapies, GDNF-producing interneurons, and trophic support of the dopaminergic nigrostriatal pathway. Implications for Parkinson's disease. *Frontiers in Neuroanatomy, 9,* 10. https://doi.org/10.3389/fnana.2015.00010.

D'Ascenzo, M., et al. (2007). mGluR5 stimulates gliotransmission in the nucleus accumbens. *Proceedings of the National Academy of Sciences of the United States of America, 104*(6), 1995–2000.

Davis, M. J., Duvoisin, R. M., & Raber, J. (2013). Related functions of mGlu4 and mGlu8. *Pharmacology, Biochemistry, and Behavior, 111,* 11–16. https://doi.org/10.1016/j.pbb.2013.07.022.

Davis, M. J., Haley, T., Duvoisin, R. M., & Raber, J. (2012). Measures of anxiety, sensorimotor function, and memory in male and female mGluR4$^-$/$^-$ mice. *Behavioural Brain Research, 229*(1), 21–28. https://doi.org/10.1016/j.bbr.2011.12.037.

Davis, K. L., Kahn, R. S., Ko, G., & Davidson, M. (1991). Dopamine in schizophrenia: A review and reconceptualization. *The American Journal of Psychiatry, 148*(11), 1474–1486.

Davis, M. J., et al. (2013). Role of mGluR4 in acquisition of fear learning and memory. *Neuropharmacology, 66,* 365–372. https://doi.org/10.1016/j.neuropharm.2012.07.038.

De Leonibus, E., et al. (2009). Metabotropic glutamate receptors 5 blockade reverses spatial memory deficits in a mouse model of Parkinson's disease. *Neuropsychopharmacology, 34*(3), 729–738. https://doi.org/10.1038/npp.2008.129.

Dean, B., Duncan, C., & Gibbons, A. (2019). Changes in levels of cortical metabotropic glutamate 2 receptors with gender and suicide but not psychiatric diagnoses. *Journal of Affective Disorders, 244,* 80–84. https://doi.org/10.1016/j.jad.2018.10.088.

Dembrow, N. C., Chitwood, R. A., & Johnston, D. (2010). Projection-specific neuromodulation of medial prefrontal cortex neurons. *The Journal of Neuroscience, 30*(50), 16922–16937. https://doi.org/10.1523/JNEUROSCI.3644-10.2010.

Dhakal, A., & Bobrin, B. D. (2021). *Cognitive deficits.* Treasure Island, Florida: StatPearls Publishing, LLC.

Dogra, S., & Conn, P. J. (2022). Metabotropic glutamate receptors as emerging targets for the treatment of schizophrenia. *Molecular Pharmacology, 101*(5), 275–285. https://doi.org/10.1124/molpharm.121.000460.

Downing, A. M., et al. (2014). A double-blind, placebo-controlled comparator study of LY2140023 monohydrate in patients with schizophrenia. *BMC Psychiatry, 14,* 351. https://doi.org/10.1186/s12888-014-0351-3.

Duvoisin, R. M., et al. (2010). Acute pharmacological modulation of mGluR8 reduces measures of anxiety. *Behavioural Brain Research, 212*(2), 168–173. https://doi.org/10.1016/j.bbr.2010.04.006.

Egan, M. F., et al. (2004). Variation in GRM3 affects cognition, prefrontal glutamate, and risk for schizophrenia. *Proceedings of the National Academy of Sciences of the United States of America, 101*(34), 12604–12609. https://doi.org/10.1073/pnas.0405077101.

Endo, S., & Launey, T. (2003). ERKs regulate PKC-dependent synaptic depression and declustering of glutamate receptors in cerebellar Purkinje cells. *Neuropharmacology, 45*(6), 863–872. https://doi.org/10.1016/s0028-3908(03)00210-7.

Fearnley, J. M., & Lees, A. J. (1991). Ageing and Parkinson's disease: Substantia nigra regional selectivity. *Brain, 114*(Pt 5), 2283–2301. https://doi.org/10.1093/brain/114.5.2283.

Fendt, M., et al. (2008). mGluR7 facilitates extinction of aversive memories and controls amygdala plasticity. *Molecular Psychiatry, 13*(10), 970–979. https://doi.org/10.1038/sj.mp.4002073.

Ferraguti, F., Baldani-Guerra, B., Corsi, M., Nakanishi, S., & Corti, C. (1999). Activation of the extracellular signal-regulated kinase 2 by metabotropic glutamate receptors. *The European Journal of Neuroscience, 11*(6), 2073–2082.

Ferraguti, F., & Shigemoto, R. (2006). Metabotropic glutamate receptors. *Cell and Tissue Research, 326*(2), 483–504.

Ferraguti, F., et al. (2005). Metabotropic glutamate receptor 8-expressing nerve terminals target subsets of GABAergic neurons in the hippocampus. *The Journal of Neuroscience, 25*(45), 10520–10536. https://doi.org/10.1523/JNEUROSCI.2547-05.2005.

Fisher, N. M., et al. (2020). Phenotypic profiling of mGlu7 knockout mice reveals new implications for neurodevelopmental disorders. *Genes, Brain, and Behavior, 19*(7), e12654. https://doi.org/10.1111/gbb.12654.

Fisher, N. M., et al. (2021). A GRM7 mutation associated with developmental delay reduces mGlu7 expression and produces neurological phenotypes. *JCI Insight, 6*(4), e143324. https://doi.org/10.1172/jci.insight.143324.

Fitzjohn, S. M., Kingston, A. E., Lodge, D., & Collingridge, G. L. (1999). DHPG-induced LTD in area CA1 of juvenile rat hippocampus; characterisation and sensitivity to novel mGlu receptor antagonists. *Neuropharmacology, 38*(10), 1577–1583. https://doi.org/10.1016/s0028-3908(99)00123-9.

Fornai, F., di Poggio, A. B., Pellegrini, A., Ruggieri, S., & Paparelli, A. (2007). Noradrenaline in Parkinson's disease: From disease progression to current therapeutics. *Current Medicinal Chemistry, 14*(22), 2330–2334. https://doi.org/10.2174/092986707781745550.

Forner, S., Baglietto-Vargas, D., Martini, A. C., Trujillo-Estrada, L., & LaFerla, F. M. (2017). Synaptic impairment in Alzheimer's disease: A dysregulated symphony. *Trends in Neurosciences, 40*(6), 347–357. https://doi.org/10.1016/j.tins.2017.04.002.

Fowler, S. W., et al. (2013). Effects of a metabotropic glutamate receptor 5 positive allosteric modulator, CDPPB, on spatial learning task performance in rodents. *Neurobiology of Learning and Memory, 99*, 25–31. https://doi.org/10.1016/j.nlm.2012.10.010.

Frith, C., & Dolan, R. (1996). The role of the prefrontal cortex in higher cognitive functions. *Brain Research. Cognitive Brain Research, 5*(1–2), 175–181. https://doi.org/10.1016/s0926-6410(96)00054-7.

Funahashi, S. (2017). Working memory in the prefrontal cortex. *Brain Sciences, 7*(5). https://doi.org/10.3390/brainsci7050049.

Funahashi, S., & Andreau, J. M. (2013). Prefrontal cortex and neural mechanisms of executive function. *Journal of Physiology, Paris, 107*(6), 471–482. https://doi.org/10.1016/j.jphysparis.2013.05.001.

García-Bea, A., Bermudez, I., Harrison, P. J., & Lane, T. A. (2017). A group II metabotropic glutamate receptor 3 (mGlu3, GRM3) isoform implicated in schizophrenia interacts with canonical mGlu3 and reduces ligand binding. *Journal of Psychopharmacology, 31*(12), 1519–1526. https://doi.org/10.1177/0269881117715597.

Gelb, D. J., Oliver, E., & Gilman, S. (1999). Diagnostic criteria for Parkinson disease. *Archives of Neurology, 56*(1), 33–39. https://doi.org/10.1001/archneur.56.1.33.

Gerlai, R., Adams, B., Fitch, T., Chaney, S., & Baez, M. (2002). Performance deficits of mGluR8 knockout mice in learning tasks: The effects of null mutation and the background genotype. *Neuropharmacology, 43*(2), 235–249. https://doi.org/10.1016/s0028-3908(02)00078-3.

Ghose, S., Gleason, K. A., Potts, B. W., Lewis-Amezcua, K., & Tamminga, C. A. (2009). Differential expression of metabotropic glutamate receptors 2 and 3 in schizophrenia: A mechanism for antipsychotic drug action? *The American Journal of Psychiatry, 166*(7), 812–820. https://doi.org/10.1176/appi.ajp.2009.08091445.

Gil-Sanz, C., Delgado-García, J. M., Fairén, A., & Gruart, A. (2008). Involvement of the mGluR1 receptor in hippocampal synaptic plasticity and associative learning in behaving mice. *Cerebral Cortex, 18*(7), 1653–1663. https://doi.org/10.1093/cercor/bhm193.

Gogliotti, R. G., et al. (2017). mGlu7 potentiation rescues cognitive, social, and respiratory phenotypes in a mouse model of Rett syndrome. *Science Translational Medicine, 9*(403), eaai7459. https://doi.org/10.1126/scitranslmed.aai7459.

Goldman-Rakic, P. S. (1995). Cellular basis of working memory. *Neuron*, *14*(3), 477–485. https://doi.org/0896-6273(95)90304-6.

González-Maeso, J., et al. (2008). Identification of a serotonin/glutamate receptor complex implicated in psychosis. *Nature*, *452*(7183), 93–97. https://doi.org/10.1038/nature06612.

Gould, R. W., et al. (2016). Partial mGlu5 negative allosteric modulators attenuate cocaine-mediated behaviors and lack psychotomimetic-like effects. *Neuropsychopharmacology*, *41*(4), 1166–1178. https://doi.org/10.1038/npp.2015.265.

Gravius, A., Pietraszek, M., Dekundy, A., & Danysz, W. (2010). Metabotropic glutamate receptors as therapeutic targets for cognitive disorders. *Current Topics in Medicinal Chemistry*, *10*(2), 187–206.

Graybiel, A. M., Canales, J. J., & Capper-Loup, C. (2000). Levodopa-induced dyskinesias and dopamine-dependent stereotypies: A new hypothesis. *Trends in Neurosciences*, *23*(10 Suppl), S71–S77. https://doi.org/10.1016/s1471-1931(00)00027-6.

Graybiel, A. M., & Rauch, S. L. (2000). Toward a neurobiology of obsessive-compulsive disorder. *Neuron*, *28*(2), 343–347. https://doi.org/10.1016/s0896-6273(00)00113-6.

Grillon, C., Cordova, J., Levine, L. R., & Morgan, C. A., 3rd. (2003). Anxiolytic effects of a novel group II metabotropic glutamate receptor agonist (LY354740) in the fear-potentiated startle paradigm in humans. *Psychopharmacology*, *168*(4), 446–454.

Grueter, B. A., & Winder, D. G. (2005). Group II and III metabotropic glutamate receptors suppress excitatory synaptic transmission in the dorsolateral bed nucleus of the stria terminalis. *Neuropsychopharmacology*, *30*(7), 1302–1311. https://doi.org/10.1038/sj.npp.1300672.

Gupta, D. S., et al. (2005). Metabotropic glutamate receptor protein expression in the prefrontal cortex and striatum in schizophrenia. *Synapse*, *57*(3), 123–131. https://doi.org/10.1002/syn.20164.

Hamilton, A., et al. (2016). Chronic pharmacological mGluR5 inhibition prevents cognitive impairment and reduces pathogenesis in an Alzheimer disease mouse model. *Cell Reports*, *15*(9), 1859–1865. https://doi.org/10.1016/j.celrep.2016.04.077.

Heimer, L., & Van Hoesen, G. W. (2006). The limbic lobe and its output channels: Implications for emotional functions and adaptive behavior. *Neuroscience and Biobehavioral Reviews*, *30*(2), 126–147. https://doi.org/10.1016/j.neubiorev.2005.06.006.

Hermans, E., & Challiss, R. A. (2001). Structural, signalling and regulatory properties of the group I metabotropic glutamate receptors: Prototypic family C G-protein-coupled receptors. *The Biochemical Journal*, *359*(Pt 3), 465–484.

Hölscher, C., et al. (2004). Lack of the metabotropic glutamate receptor subtype 7 selectively impairs short-term working memory but not long-term memory. *Behavioural Brain Research*, *154*(2), 473–481. https://doi.org/10.1016/j.bbr.2004.03.015.

Holter, K. M., et al. (2021). Partial mGlu5 negative allosteric modulator M-5MPEP demonstrates antidepressant-like effects on sleep without affecting cognition or quantitative EEG. *Frontiers in Neuroscience*, *15*, 700822. https://doi.org/10.3389/fnins.2021.700822.

Homayoun, H., & Moghaddam, B. (2010). Group 5 metabotropic glutamate receptors: Role in modulating cortical activity and relevance to cognition. *European Journal of Pharmacology*, *639*(1–3), 33–39. https://doi.org/10.1016/j.ejphar.2009.12.042.

Homayoun, H., Stefani, M. R., Adams, B. W., Tamagan, G. D., & Moghaddam, B. (2004). Functional interaction between NMDA and mGlu5 receptors: Effects on working memory, instrumental learning, motor behaviors, and dopamine release. *Neuropsychopharmacology*, *29*(7), 1259–1269. https://doi.org/10.1038/sj.npp.1300417.

Huang, Y. Y., Haug, M. F., Gesemann, M., & Neuhauss, S. C. (2012). Novel expression patterns of metabotropic glutamate receptor 6 in the zebrafish nervous system. *PLoS One*, *7*(4), e35256. https://doi.org/10.1371/journal.pone.0035256.

Huber, K. M., Gallagher, S. M., Warren, S. T., & Bear, M. F. (2002). Altered synaptic plasticity in a mouse model of fragile X mental retardation. *Proceedings of the National Academy of Sciences of the United States of America, 99*(11), 7746–7750. https://doi.org/10.1073/pnas.122205699.

Hunt, R. H., & Tougas, G. (2002). Evolving concepts in functional gastrointestinal disorders: Promising directions for novel pharmaceutical treatments. *Best Practice & Research. Clinical Gastroenterology, 16*(6), 869–883. https://doi.org/10.1053/bega.2002.0356.

Ieraci, A., Mallei, A., & Popoli, M. (2016). Social isolation stress induces anxious-depressive-like behavior and alterations of neuroplasticity-related genes in adult male mice. *Neural Plasticity, 2016*, 6212983. https://doi.org/10.1155/2016/6212983.

Ishibashi, K., Miura, Y., Toyohara, J., Ishiwata, K., & Ishii, K. (2019). Unchanged type 1 metabotropic glutamate receptor availability in patients with Alzheimer's disease: A study using. *Neuroimage: Clinical, 22*, 101783. https://doi.org/10.1016/j.nicl.2019.101783.

Ito, M. (1989). Long-term depression. *Annual Review of Neuroscience, 12*, 85–102. https://doi.org/10.1146/annurev.ne.12.030189.000505.

Jaso, B. A., et al. (2017). Therapeutic modulation of glutamate receptors in major depressive disorder. *Current Neuropharmacology, 15*(1), 57–70. https://doi.org/10.2174/1570159x14666160321123221.

Jiang, S., et al. (2018). Mfn2 ablation causes an oxidative stress response and eventual neuronal death in the hippocampus and cortex. *Molecular Neurodegeneration, 13*(1), 5. https://doi.org/10.1186/s13024-018-0238-8.

Johnson, K. A., Conn, P. J., & Niswender, C. M. (2009). Glutamate receptors as therapeutic targets for Parkinson's disease. *CNS & Neurological Disorders Drug Targets, 8*(6), 475–491. https://doi.org/10.2174/187152709789824606.

Jolly-Tornetta, C., Gao, Z. Y., Lee, V. M., & Wolf, B. A. (1998). Regulation of amyloid precursor protein secretion by glutamate receptors in human Ntera 2 neurons. *The Journal of Biological Chemistry, 273*(22), 14015–14021. https://doi.org/10.1074/jbc.273.22.14015.

Kalinichev, M., et al. (2013). ADX71743, a potent and selective negative allosteric modulator of metabotropic glutamate receptor 7: In vitro and in vivo characterization. *The Journal of Pharmacology and Experimental Therapeutics, 344*(3), 624–636. https://doi.org/10.1124/jpet.112.200915.

Kaneda, K., et al. (2005). Down-regulation of metabotropic glutamate receptor 1alpha in globus pallidus and substantia nigra of parkinsonian monkeys. *The European Journal of Neuroscience, 22*(12), 3241–3254. https://doi.org/10.1111/j.1460-9568.2005.04488.x.

Karachot, L., Shirai, Y., Vigot, R., Yamamori, T., & Ito, M. (2001). Induction of long-term depression in cerebellar Purkinje cells requires a rapidly turned over protein. *Journal of Neurophysiology, 86*(1), 280–289. https://doi.org/10.1152/jn.2001.86.1.280.

Kaufmann, W. E., & Moser, H. W. (2000). Dendritic anomalies in disorders associated with mental retardation. *Cerebral Cortex, 10*(10), 981–991. https://doi.org/10.1093/cercor/10.10.981.

Kim, J., et al. (2015). mGluR2/3 in the lateral amygdala is required for fear extinction: Cortical input synapses onto the lateral amygdala as a target site of the mGluR2/3 action. *Neuropsychopharmacology, 40*(13), 2916–2928. https://doi.org/10.1038/npp.2015.145.

Kingston, A. E., et al. (1999). Neuroprotective actions of novel and potent ligands of group I and group II metabotropic glutamate receptors. *Annals of the New York Academy of Sciences, 890*, 438–449. https://doi.org/10.1111/j.1749-6632.1999.tb08022.x.

Kinney, G. G., et al. (2003). Metabotropic glutamate subtype 5 receptors modulate locomotor activity and sensorimotor gating in rodents. *The Journal of Pharmacology and Experimental Therapeutics, 306*(1), 116–123. https://doi.org/10.1124/jpet.103.048702.

Kinon, B. J., et al. (2011). A multicenter, inpatient, phase 2, double-blind, placebo-controlled dose-ranging study of LY2140023 monohydrate in patients with DSM-IV schizophrenia. *Journal of Clinical Psychopharmacology, 31*(3), 349–355. https://doi.org/10.1097/JCP.0b013e318218dcd5.

Krystal, J. H., et al. (2005). Preliminary evidence of attenuation of the disruptive effects of the NMDA glutamate receptor antagonist, ketamine, on working memory by pretreatment with the group II metabotropic glutamate receptor agonist, LY354740, in healthy human subjects. *Psychopharmacology, 179*(1), 303–309. https://doi.org/10.1007/s00213-004-1982-8.

Kumar, J. S. D., et al. (2014). Alterations of mGluR1 in Parkinson's disease and suicide depression: An autoradiography study with [18F]MK1312. *Journal of Nuclear Medicine, 55*, 1830.

LaCrosse, A. L., et al. (2015). mGluR5 positive allosteric modulation and its effects on MK-801 induced set-shifting impairments in a rat operant delayed matching/non-matching-to-sample task. *Psychopharmacology, 232*(1), 251–258. https://doi.org/10.1007/s00213-014-3653-8.

Laggerbauer, B., Ostareck, D., Keidel, E. M., Ostareck-Lederer, A., & Fischer, U. (2001). Evidence that fragile X mental retardation protein is a negative regulator of translation. *Human Molecular Genetics, 10*(4), 329–338. https://doi.org/10.1093/hmg/10.4.329.

Lang, S., et al. (2009). Context conditioning and extinction in humans: Differential contribution of the hippocampus, amygdala and prefrontal cortex. *The European Journal of Neuroscience, 29*(4), 823–832. https://doi.org/10.1111/j.1460-9568.2009.06624.x.

Lecourtier, L., Homayoun, H., Tamagnan, G., & Moghaddam, B. (2007). Positive allosteric modulation of metabotropic glutamate 5 (mGlu5) receptors reverses N-methyl-D-aspartate antagonist-induced alteration of neuronal firing in prefrontal cortex. *Biological Psychiatry, 62*(7), 739–746. https://doi.org/10.1016/j.biopsych.2006.12.003.

LeDoux, J. E. (2000). Emotion circuits in the brain. *Annual Review of Neuroscience, 23*, 155–184. https://doi.org/10.1146/annurev.neuro.23.1.155.

Lee, H. G., et al. (2004). The role of metabotropic glutamate receptors in Alzheimer's disease. *Acta Neurobiologiae Experimentalis (Wars), 64*(1), 89–98.

Lee, H. G., et al. (2009). The effect of mGluR2 activation on signal transduction pathways and neuronal cell survival. *Brain Research, 1249*, 244–250. https://doi.org/10.1016/j.brainres.2008.10.055.

Lieberman, J. A., Kane, J. M., & Alvir, J. (1987). Provocative tests with psychostimulant drugs in schizophrenia. *Psychopharmacology, 91*(4), 415–433. https://doi.org/10.1007/BF00216006.

Lin, C. H., Lee, C. C., Huang, Y. C., Wang, S. J., & Gean, P. W. (2005). Activation of group II metabotropic glutamate receptors induces depotentiation in amygdala slices and reduces fear-potentiated startle in rats. *Learning & Memory, 12*(2), 130–137. https://doi.org/10.1101/lm.85304.

Liu, J., Chang, L., Song, Y., Li, H., & Wu, Y. (2019). The role of NMDA receptors in Alzheimer's disease. *Frontiers in Neuroscience, 13*, 43. https://doi.org/10.3389/fnins.2019.00043.

Liu, M., & Kirchgessner, A. L. (2000). Agonist- and reflex-evoked internalization of metabotropic glutamate receptor 5 in enteric neurons. *The Journal of Neuroscience, 20*(9), 3200–3205.

Liu, F., et al. (2008). ADX47273 [S-(4-fluoro-phenyl)-{3-[3-(4-fluoro-phenyl)-[1,2,4]-oxadiazol-5-yl]-piperidin-1-yl}-methanone]: A novel metabotropic glutamate receptor 5-selective positive allosteric modulator with preclinical antipsychotic-like and procognitive activities. *The Journal of Pharmacology and Experimental Therapeutics, 327*(3), 827–839. https://doi.org/10.1124/jpet.108.136580.

Lu, Y. M., et al. (1997). Mice lacking metabotropic glutamate receptor 5 show impaired learning and reduced CA1 long-term potentiation (LTP) but normal CA3 LTP. *The Journal of Neuroscience, 17*(13), 5196–5205.

Luessen, D. J., et al. (2022). mGlu1-mediated restoration of prefrontal cortex inhibitory signaling reverses social and cognitive deficits in an NMDA hypofunction model in mice. *Neuropsychopharmacology.* https://doi.org/10.1038/s41386-022-01350-0.

Lujan, R., Nusser, Z., Roberts, J. D., Shigemoto, R., & Somogyi, P. (1996). Perisynaptic location of metabotropic glutamate receptors mGluR1 and mGluR5 on dendrites and dendritic spines in the rat hippocampus. *The European Journal of Neuroscience, 8*(7), 1488–1500.

Lupták, M., Michaličková, D., Fišar, Z., Kitzlerová, E., & Hroudová, J. (2021). Novel approaches in schizophrenia-from risk factors and hypotheses to novel drug targets. *World Journal of Psychiatry, 11*(7), 277–296. https://doi.org/10.5498/wjp.v11.i7.277.

Lynch, M. A. (2004). Long-term potentiation and memory. *Physiological Reviews, 84*(1), 87–136. https://doi.org/10.1152/physrev.00014.2003.

Lyon, L., Kew, J. N., Corti, C., Harrison, P. J., & Burnet, P. W. (2008). Altered hippocampal expression of glutamate receptors and transporters in GRM2 and GRM3 knockout mice. *Synapse, 62*(11), 842–850. https://doi.org/10.1002/syn.20553.

Lyon, L., et al. (2011). Fractionation of spatial memory in GRM2/3 (mGlu2/mGlu3) double knockout mice reveals a role for group II metabotropic glutamate receptors at the interface between arousal and cognition. *Neuropsychopharmacology, 36*(13), 2616–2628. https://doi.org/10.1038/npp.2011.145.

Mahan, A. L., & Ressler, K. J. (2012). Fear conditioning, synaptic plasticity and the amygdala: Implications for posttraumatic stress disorder. *Trends in Neurosciences, 35*(1), 24–35. https://doi.org/10.1016/j.tins.2011.06.007.

Maksymetz, J., et al. (2021). mGlu1 potentiation enhances prelimbic somatostatin interneuron activity to rescue schizophrenia-like physiological and cognitive deficits. *Cell Reports, 37*(5), 109950. https://doi.org/10.1016/j.celrep.2021.109950.

Malenka, R. C., & Nicoll, R. A. (1993). NMDA-receptor-dependent synaptic plasticity: Multiple forms and mechanisms. *Trends in Neurosciences, 16*(12), 521–527.

Maren, S., & Quirk, G. J. (2004). Neuronal signalling of fear memory. *Nature Reviews. Neuroscience, 5*(11), 844–852. https://doi.org/10.1038/nrn1535.

Marino, M. J., et al. (2001). Activation of group I metabotropic glutamate receptors produces a direct excitation and disinhibition of GABAergic projection neurons in the substantia nigra pars reticulata. *The Journal of Neuroscience, 21*(18), 7001–7012.

Marszalek-Grabska, M., et al. (2018). ADX-47273, a mGlu5 receptor positive allosteric modulator, attenuates deficits in cognitive flexibility induced by withdrawal from 'binge-like' ethanol exposure in rats. *Behavioural Brain Research, 338*, 9–16. https://doi.org/10.1016/j.bbr.2017.10.007.

Martí, S. B., Cichon, S., Propping, P., & Nöthen, M. (2002). Metabotropic glutamate receptor 3 (GRM3) gene variation is not associated with schizophrenia or bipolar affective disorder in the German population. *American Journal of Medical Genetics, 114*(1), 46–50. https://doi.org/10.1002/ajmg.1624.

Martin, L. J., Blackstone, C. D., Huganir, R. L., & Price, D. L. (1992). Cellular localization of a metabotropic glutamate receptor in rat brain. *Neuron, 9*(2), 259–270.

Masilamoni, G. J., & Smith, Y. (2018). Metabotropic glutamate receptors: Targets for neuroprotective therapies in Parkinson disease. *Current Opinion in Pharmacology, 38*, 72–80. https://doi.org/10.1016/j.coph.2018.03.004.

Masilamoni, G. J., et al. (2011). Metabotropic glutamate receptor 5 antagonist protects dopaminergic and noradrenergic neurons from degeneration in MPTP-treated monkeys. *Brain, 134*(Pt 7), 2057–2073. https://doi.org/10.1093/brain/awr137.

Matosin, N., Fernandez-Enright, F., Lum, J. S., & Newell, K. A. (2017). Shifting towards a model of mGluR5 dysregulation in schizophrenia: Consequences for future schizophrenia treatment. *Neuropharmacology, 115*, 73–91. https://doi.org/10.1016/j.neuropharm.2015.08.003.

Matosin, N., et al. (2015). Alterations of mGluR5 and its endogenous regulators Norbin, Tamalin and Preso1 in schizophrenia: Towards a model of mGluR5 dysregulation. *Acta Neuropathologica, 130*(1), 119–129. https://doi.org/10.1007/s00401-015-1411-6.

Matosin, N., et al. (2018). Effects of common GRM5 genetic variants on cognition, hippocampal volume and mGluR5 protein levels in schizophrenia. *Brain Imaging and Behavior, 12*(2), 509–517. https://doi.org/10.1007/s11682-017-9712-0.

Matynia, A., Kushner, S. A., & Silva, A. J. (2002). Genetic approaches to molecular and cellular cognition: A focus on LTP and learning and memory. *Annual Review of Genetics, 36*, 687–720. https://doi.org/10.1146/annurev.genet.36.062802.091007.

Meziane, H., et al. (1998). Memory-enhancing effects of secreted forms of the beta-amyloid precursor protein in normal and amnestic mice. *Proceedings of the National Academy of Sciences of the United States of America, 95*(21), 12683–12688. https://doi.org/10.1073/pnas.95.21.12683.

Miyamoto, S., Miyake, N., Jarskog, L. F., Fleischhacker, W. W., & Lieberman, J. A. (2012). Pharmacological treatment of schizophrenia: A critical review of the pharmacology and clinical effects of current and future therapeutic agents. *Molecular Psychiatry, 17*(12), 1206–1227. https://doi.org/10.1038/mp.2012.47.

Moghaddam, B., & Adams, B. W. (1998). Reversal of phencyclidine effects by a group II metabotropic glutamate receptor agonist in rats. *Science, 281*(5381), 1349–1352. https://doi.org/10.1126/science.281.5381.1349.

Moreno, J. L., et al. (2012). Identification of three residues essential for 5-hydroxytryptamine 2A-metabotropic glutamate 2 (5-HT2A·mGlu2) receptor heteromerization and its psychoactive behavioral function. *The Journal of Biological Chemistry, 287*(53), 44301–44319. https://doi.org/10.1074/jbc.M112.413161.

Moreno, J. L., et al. (2016). Allosteric signaling through an mGlu2 and 5-HT2A heteromeric receptor complex and its potential contribution to schizophrenia. *Science Signaling, 9*(410), ra5. https://doi.org/10.1126/scisignal.aab0467.

Morsch, R., Simon, W., & Coleman, P. D. (1999). Neurons may live for decades with neurofibrillary tangles. *Journal of Neuropathology and Experimental Neurology, 58*(2), 188–197. https://doi.org/10.1097/00005072-199902000-00008.

Mudo, G., Trovato-Salinaro, A., Caniglia, G., Cheng, Q., & Condorelli, D. F. (2007). Cellular localization of mGluR3 and mGluR5 mRNAs in normal and injured rat brain. *Brain Research, 1149*, 1–13.

Muguruza, C., Meana, J. J., & Callado, L. F. (2016). Group II metabotropic glutamate receptors as targets for novel antipsychotic drugs. *Frontiers in Pharmacology, 7*, 130. https://doi.org/10.3389/fphar.2016.00130.

Neki, A., et al. (1996). Pre- and postsynaptic localization of a metabotropic glutamate receptor, mGluR2, in the rat brain: An immunohistochemical study with a monoclonal antibody. *Neuroscience Letters, 202*(3), 197–200. https://doi.org/10.1016/0304-3940(95)12248-6.

Nicholls, R. E., et al. (2006). mGluR2 acts through inhibitory Galpha subunits to regulate transmission and long-term plasticity at hippocampal mossy fiber-CA3 synapses. *Proceedings of the National Academy of Sciences of the United States of America, 103*(16), 6380–6385. https://doi.org/10.1073/pnas.0601267103.

Nicoletti, F., et al. (2011). Metabotropic glutamate receptors: From the workbench to the bedside. *Neuropharmacology, 60*(7–8), 1017–1041. https://doi.org/10.1016/j.neuropharm.2010.10.022.

Nicoll, R. A., Oliet, S. H., & Malenka, R. C. (1998). NMDA receptor-dependent and metabotropic glutamate receptor-dependent forms of long-term depression coexist in CA1 hippocampal pyramidal cells. *Neurobiology of Learning and Memory, 70*(1–2), 62–72. https://doi.org/10.1006/nlme.1998.3838.

Niswender, C. M., & Conn, P. J. (2010). Metabotropic glutamate receptors: Physiology, pharmacology, and disease. *Annual Review of Pharmacology and Toxicology, 50*, 295–322. https://doi.org/10.1146/annurev.pharmtox.011008.145533.

Obeso, J. A., Olanow, C. W., & Nutt, J. G. (2000). Levodopa motor complications in Parkinson's disease. *Trends in Neurosciences, 23*(10 Suppl), S2–S7. https://doi.org/10.1016/s1471-1931(00)00031-8.

O'Connor, R. M., Finger, B. C., Flor, P. J., & Cryan, J. F. (2010). Metabotropic glutamate receptor 7: At the interface of cognition and emotion. *European Journal of Pharmacology, 639*(1–3), 123–131. https://doi.org/10.1016/j.ejphar.2010.02.059.

O'Donnell, W. T., & Warren, S. T. (2002). A decade of molecular studies of fragile X syndrome. *Annual Review of Neuroscience, 25*, 315–338. https://doi.org/10.1146/annurev.neuro.25.112701.142909.

Ohnuma, T., et al. (2000). Gene expression of PSD95 in prefrontal cortex and hippocampus in schizophrenia. *Neuroreport, 11*, 3133–3137. https://doi.org/10.1097/00001756-200009280-00019.

Oliet, S. H., Malenka, R. C., & Nicoll, R. A. (1997). Two distinct forms of long-term depression coexist in CA1 hippocampal pyramidal cells. *Neuron, 18*(6), 969–982. https://doi.org/10.1016/s0896-6273(00)80336-0.

Ott, T., & Nieder, A. (2019). Dopamine and cognitive control in prefrontal cortex. *Trends in Cognitive Sciences, 23*(3), 213–234. https://doi.org/10.1016/j.tics.2018.12.006.

Palucha, A., et al. (2007). Activation of the mGlu7 receptor elicits antidepressant-like effects in mice. *Psychopharmacology, 194*(4), 555–562. https://doi.org/10.1007/s00213-007-0856-2.

Patil, S. T., et al. (2007). Activation of mGlu2/3 receptors as a new approach to treat schizophrenia: A randomized phase 2 clinical trial. *Nature Medicine, 13*(9), 1102–1107.

Perez-Garcia, G., et al. (2018). PTSD-related behavioral traits in a rat model of blast-induced mTBI are reversed by the mGluR2/3 receptor antagonist BCI-838. *eNeuro, 5*(1). https://doi.org/10.1523/ENEURO.0357-17.2018.

Pinteaux-Jones, F., et al. (2008). Myelin-induced microglial neurotoxicity can be controlled by microglial metabotropic glutamate receptors. *Journal of Neurochemistry, 106*(1), 442–454. https://doi.org/10.1111/j.1471-4159.2008.05426.x.

Poels, E. M., et al. (2014). Imaging glutamate in schizophrenia: Review of findings and implications for drug discovery. *Molecular Psychiatry, 19*(1), 20–29. https://doi.org/10.1038/mp.2013.136.

Porter, R. H., et al. (2005). Fenobam: A clinically validated nonbenzodiazepine anxiolytic is a potent, selective, and noncompetitive mGlu5 receptor antagonist with inverse agonist activity. *The Journal of Pharmacology and Experimental Therapeutics, 315*(2), 711–721. https://doi.org/10.1124/jpet.105.089839.

Purpura, D. P. (1974). Dendritic spine "dysgenesis" and mental retardation. *Science, 186*(4169), 1126–1128. https://doi.org/10.1126/science.186.4169.1126.

Ramiro-Cortés, Y., Hobbiss, A. F., & Israely, I. (2014). Synaptic competition in structural plasticity and cognitive function. *Philosophical Transactions of the Royal Society of London. Series B, Biological Sciences, 369*(1633), 20130157. https://doi.org/10.1098/rstb.2013.0157.

Rauch, S. L., et al. (1997). Probing striatal function in obsessive-compulsive disorder: A PET study of implicit sequence learning. *The Journal of Neuropsychiatry and Clinical Neurosciences, 9*(4), 568–573. https://doi.org/10.1176/jnp.9.4.568.

Revett, T. J., Baker, G. B., Jhamandas, J., & Kar, S. (2013). Glutamate system, amyloid ß peptides and tau protein: Functional interrelationships and relevance to Alzheimer disease pathology. *Journal of Psychiatry & Neuroscience, 38*(1), 6–23. https://doi.org/10.1503/jpn.110190.

Ribeiro, F. M., Vieira, L. B., Pires, R. G., Olmo, R. P., & Ferguson, S. S. (2017). Metabotropic glutamate receptors and neurodegenerative diseases. *Pharmacological Research, 115*, 179–191. https://doi.org/10.1016/j.phrs.2016.11.013.

Rodrigues, S. M., Bauer, E. P., Farb, C. R., Schafe, G. E., & LeDoux, J. E. (2002). The group I metabotropic glutamate receptor mGluR5 is required for fear memory formation and long-term potentiation in the lateral amygdala. *The Journal of Neuroscience, 22*(12), 5219–5229.

Rodrigues, S. M., Schafe, G. E., & LeDoux, J. E. (2004). Molecular mechanisms underlying emotional learning and memory in the lateral amygdala. *Neuron, 44*(1), 75–91.

Romano, C., et al. (1995). Distribution of metabotropic glutamate receptor mGluR5 immunoreactivity in rat brain. *The Journal of Comparative Neurology, 355*(3), 455–469. https://doi.org/10.1002/cne.903550310.

Saper, C. B., Chou, T. C., & Elmquist, J. K. (2002). The need to feed: Homeostatic and hedonic control of eating. *Neuron, 36*(2), 199–211. https://doi.org/10.1016/s0896-6273(02)00969-8.

Sartorius, L. J., et al. (2008). Expression of a GRM3 splice variant is increased in the dorsolateral prefrontal cortex of individuals carrying a schizophrenia risk SNP. *Neuropsychopharmacology, 33*(11), 2626–2634. https://doi.org/10.1038/sj.npp.1301669.

Schaffhauser, H., et al. (1998). In vitro binding characteristics of a new selective group II metabotropic glutamate receptor radioligand, [3H]LY354740, in rat brain. *Molecular Pharmacology, 53*(2), 228–233.

Schlumberger, C., Pietraszek, M., Gravius, A., & Danysz, W. (2010). Effects of a positive allosteric modulator of mGluR5 ADX47273 on conditioned avoidance response and PCP-induced hyperlocomotion in the rat as models for schizophrenia. *Pharmacology, Biochemistry, and Behavior, 95*(1), 23–30. https://doi.org/10.1016/j.pbb.2009.12.002.

Schoepp, D. D. (2001). Unveiling the functions of presynaptic metabotropic glutamate receptors in the central nervous system. *The Journal of Pharmacology and Experimental Therapeutics, 299*(1), 12–20.

Seeman, P., Lee, T., Chau-Wong, M., & Wong, K. (1976). Antipsychotic drug doses and neuroleptic/dopamine receptors. *Nature, 261*(5562), 717–719. https://doi.org/10.1038/261717a0.

Shen, K. Z., & Johnson, S. W. (2003). Group II metabotropic glutamate receptor modulation of excitatory transmission in rat subthalamic nucleus. *The Journal of Physiology, 553*(Pt 2), 489–496. https://doi.org/10.1113/jphysiol.2003.052209.

Shigemoto, R., Nakanishi, S., & Mizuno, N. (1992). Distribution of the mRNA for a metabotropic glutamate receptor (mGluR1) in the central nervous system: An in situ hybridization study in adult and developing rat. *The Journal of Comparative Neurology, 322*(1), 121–135.

Shigemoto, R., et al. (1993). Immunohistochemical localization of a metabotropic glutamate receptor, mGluR5, in the rat brain. *Neuroscience Letters, 163*(1), 53–57.

Shigemoto, R., et al. (1997). Differential presynaptic localization of metabotropic glutamate receptor subtypes in the rat hippocampus. *The Journal of Neuroscience, 17*(19), 7503–7522.

Shigeri, Y., Seal, R. P., & Shimamoto, K. (2004). Molecular pharmacology of glutamate transporters, EAATs and VGLUTs. *Brain Research. Brain Research Reviews, 45*(3), 250–265. https://doi.org/10.1016/j.brainresrev.2004.04.004.

Sławińska, A., et al. (2013). The antipsychotic-like effects of positive allosteric modulators of metabotropic glutamate mGlu4 receptors in rodents. *British Journal of Pharmacology, 169*(8), 1824–1839. https://doi.org/10.1111/bph.12254.

Snyder, E. M., et al. (2001). Internalization of ionotropic glutamate receptors in response to mGluR activation. *Nature Neuroscience*, *4*, 1079–1085.

Sourial, M., Cheng, C., & Doering, L. C. (2013). Progress toward therapeutic potential for AFQ056 in fragile X syndrome. *Journal of Experimental Pharmacology*, *5*, 45–54. https://doi.org/10.2147/JEP.S27044.

Spooren, W., & Gasparini, F. (2004). mGlu5 receptor antagonists: A novel class of anxiolytics? *Drug News & Perspectives*, *17*(4), 251–257.

Srivastava, A., Das, B., Yao, A. Y., & Yan, R. (2020). Metabotropic glutamate receptors in Alzheimer's disease synaptic dysfunction: Therapeutic opportunities and Hope for the future. *Journal of Alzheimer's Disease*, *78*(4), 1345–1361. https://doi.org/10.3233/JAD-201146.

Steinberg, J. P., Huganir, R. L., & Linden, D. J. (2004). N-ethylmaleimide-sensitive factor is required for the synaptic incorporation and removal of AMPA receptors during cerebellar long-term depression. *Proceedings of the National Academy of Sciences of the United States of America*, *101*(52), 18212–18216. https://doi.org/10.1073/pnas.0408278102.

Tamaru, Y., Nomura, S., Mizuno, N., & Shigemoto, R. (2001). Distribution of metabotropic glutamate receptor mGluR3 in the mouse CNS: Differential location relative to pre- and postsynaptic sites. *Neuroscience*, *106*(3), 481–503.

Tanaka, J., et al. (2000). Gq protein alpha subunits Galphaq and Galpha11 are localized at postsynaptic extra-junctional membrane of cerebellar Purkinje cells and hippocampal pyramidal cells. *The European Journal of Neuroscience*, *12*(3), 781–792. https://doi.org/10.1046/j.1460-9568.2000.00959.x.

Taylor, D. L., Diemel, L. T., & Pocock, J. M. (2003). Activation of microglial group III metabotropic glutamate receptors protects neurons against microglial neurotoxicity. *The Journal of Neuroscience*, *23*(6), 2150–2160.

Testa, C. M., Friberg, I. K., Weiss, S. W., & Standaert, D. G. (1998). Immunohistochemical localization of metabotropic glutamate receptors mGluR1a and mGluR2/3 in the rat basal ganglia. *The Journal of Comparative Neurology*, *390*(1), 5–19.

Testa, C. M., Standaert, D. G., Young, A. B., & Penney, J. B. J. (1994). Metabotropic glutamate receptor mRNA expression in the basal ganglia of the rat. *The Journal of Neuroscience*, *14*(5 Pt 2), 3005–3018.

Thome, A. D., et al. (2018). Functional alterations of myeloid cells during the course of Alzheimer's disease. *Molecular Neurodegeneration*, *13*(1), 61. https://doi.org/10.1186/s13024-018-0293-1.

Tochigi, M., et al. (2006). No association between the metabotropic glutamate receptor type 3 gene (GRM3) and schizophrenia in a Japanese population. *Schizophrenia Research*, *88*(1–3), 260–264. https://doi.org/10.1016/j.schres.2006.07.008.

Todd, P. K., Mack, K. J., & Malter, J. S. (2003). The fragile X mental retardation protein is required for type-I metabotropic glutamate receptor-dependent translation of PSD-95. *Proceedings of the National Academy of Sciences of the United States of America*, *100*(24), 14374–14378. https://doi.org/10.1073/pnas.2336265100.

Valenti, O., Mannaioni, G., Seabrook, G. R., Conn, P. J., & Marino, M. J. (2005). Group III metabotropic glutamate-receptor-mediated modulation of excitatory transmission in rodent substantia nigra pars compacta dopamine neurons. *The Journal of Pharmacology and Experimental Therapeutics*, *313*(3), 1296–1304. https://doi.org/10.1124/jpet.104.080481.

van den Pol, A. N., Romano, C., & Ghosh, P. (1995). Metabotropic glutamate receptor mGluR5 subcellular distribution and developmental expression in hypothalamus. *The Journal of Comparative Neurology*, *362*(1), 134–150.

Vandenberg, R. J., & Ryan, R. M. (2013). Mechanisms of glutamate transport. *Physiological Reviews*, *93*(4), 1621–1657. https://doi.org/10.1152/physrev.00007.2013.

Vardi, T., Fina, M., Zhang, L., Dhingra, A., & Vardi, N. (2011). mGluR6 transcripts in non-neuronal tissues. *The Journal of Histochemistry and Cytochemistry, 59*(12), 1076–1086. https://doi.org/10.1369/0022155411425386.

Vermeiren, Y., & De Deyn, P. P. (2017). Targeting the norepinephrinergic system in Parkinson's disease and related disorders: The locus coeruleus story. *Neurochemistry International, 102*, 22–32. https://doi.org/10.1016/j.neuint.2016.11.009.

Vernon, A. C., et al. (2005). Neuroprotective effects of metabotropic glutamate receptor ligands in a 6-hydroxydopamine rodent model of Parkinson's disease. *The European Journal of Neuroscience, 22*(7), 1799–1806. https://doi.org/10.1111/j.1460-9568.2005.04362.x.

Volk, D. W., Eggan, S. M., & Lewis, D. A. (2010). Alterations in metabotropic glutamate receptor 1α and regulator of G protein signaling 4 in the prefrontal cortex in schizophrenia. *The American Journal of Psychiatry, 167*(12), 1489–1498. https://doi.org/10.1176/appi.ajp.2010.10030318.

Walker, D. L., Rattiner, L. M., & Davis, M. (2002). Group II metabotropic glutamate receptors within the amygdala regulate fear as assessed with potentiated startle in rats. *Behavioral Neuroscience, 116*(6), 1075–1083. https://doi.org/10.1037//0735-7044.116.6.1075.

Wang, Q., Walsh, D. M., Rowan, M. J., Selkoe, D. J., & Anwyl, R. (2004). Block of long-term potentiation by naturally secreted and synthetic amyloid beta-peptide in hippocampal slices is mediated via activation of the kinases c-Jun N-terminal kinase, cyclin-dependent kinase 5, and p38 mitogen-activated protein kinase as well as metabotropic glutamate receptor type 5. *The Journal of Neuroscience, 24*(13), 3370–3378. https://doi.org/10.1523/JNEUROSCI.1633-03.2004.

Wang, Y., et al. (2015). Sexual differences in long-term effects of prenatal chronic mild stress on anxiety-like behavior and stress-induced regional glutamate receptor expression in rat offspring. *International Journal of Developmental Neuroscience, 41*, 80–91. https://doi.org/10.1016/j.ijdevneu.2015.01.003.

Wang, W., et al. (2018). Hippocampal protein kinase D1 is necessary for DHPG-induced learning and memory impairments in rats. *PLoS One, 13*(4), e0195095. https://doi.org/10.1371/journal.pone.0195095.

Wang, H. Y., et al. (2020). mGluR5 hypofunction is integral to glutamatergic dysregulation in schizophrenia. *Molecular Psychiatry, 25*(4), 750–760. https://doi.org/10.1038/s41380-018-0234-y.

Weiler, I. J., et al. (2004). Fragile X mental retardation protein is necessary for neurotransmitter-activated protein translation at synapses. *Proceedings of the National Academy of Sciences of the United States of America, 101*(50), 17504–17509. https://doi.org/10.1073/pnas.0407533101.

Wieronska, J. M., Zorn, S. H., Doller, D., & Pilc, A. (2016). Metabotropic glutamate receptors as targets for new antipsychotic drugs: Historical perspective and critical comparative assessment. *Pharmacology & Therapeutics, 157*, 10–27. https://doi.org/10.1016/j.pharmthera.2015.10.007.

Wieronska, J. M., et al. (2015). The antipsychotic-like effects in rodents of the positive allosteric modulator Lu AF21934 involve 5-HT1A receptor signaling: Mechanistic studies. *Psychopharmacology, 232*(1), 259–273. https://doi.org/10.1007/s00213-014-3657-4.

Wu, M., Hajszan, T., Xu, C., Leranth, C., & Alreja, M. (2004). Group I metabotropic glutamate receptor activation produces a direct excitation of identified septohippocampal cholinergic neurons. *Journal of Neurophysiology, 92*(2), 1216–1225. https://doi.org/10.1152/jn.00180.2004.

Xia, N., et al. (2015). Blockade of metabotropic glutamate receptor 5 protects against DNA damage in a rotenone-induced Parkinson's disease model. *Free Radical Biology & Medicine, 89*, 567–580. https://doi.org/10.1016/j.freeradbiomed.2015.09.017.

Xiang, Z., et al. (2021). Input-specific regulation of glutamatergic synaptic transmission in the medial prefrontal cortex by mGlu. *Science Signaling*, *14*(677). https://doi.org/10.1126/scisignal.abd2319.

Yamasaki, T., et al. (2016). Dynamic changes in striatal mGluR1 but not mGluR5 during pathological progression of Parkinson's disease in human alpha-synuclein A53T transgenic rats: A multi-PET imaging study. *The Journal of Neuroscience*, *36*(2), 375–384. https://doi.org/10.1523/JNEUROSCI.2289-15.2016.

Yohn, S. E., et al. (2020). Activation of the mGlu. *Molecular Psychiatry*, *25*(11), 2786–2799. https://doi.org/10.1038/s41380-018-0206-2.

Youssef, E. A., et al. (2018). Effect of the mGluR5-NAM basimglurant on behavior in adolescents and adults with fragile X syndrome in a randomized, double-blind, placebo-controlled trial: FragXis phase 2 results. *Neuropsychopharmacology*, *43*(3), 503–512. https://doi.org/10.1038/npp.2017.177.

Zhang, Y. Q., et al. (2001). Drosophila fragile X-related gene regulates the MAP1B homolog Futsch to control synaptic structure and function. *Cell*, *107*(5), 591–603. https://doi.org/10.1016/s0092-8674(01)00589-x.

Zhao, L., et al. (2009). Activation of group III metabotropic glutamate receptor reduces intracellular calcium in beta-amyloid peptide [31-35]-treated cortical neurons. *Neurotoxicity Research*, *16*(2), 174–183. https://doi.org/10.1007/s12640-009-9068-3.